Neanderthals Revisited

Vertebrate Paleobiology and Paleoanthropology

Edited by

Eric Delson
Vertebrate Paleontology, American Museum of Natural History,
New York, NY 10024, USA

Ross D. E. MacPhee
Vertebrate Zoology, American Museum of Natural History,
New York, NY 10024, USA

Focal topics for volumes in the series will include systematic paleontology of all vertebrates (from agnathans to humans), phylogeny reconstruction, functional morphology, Paleolithic archaeology, taphonomy, geochronology, historical biogeography, and biostratigraphy. Other fields (e.g., paleoclimatology, paleoecology, ancient DNA, total organismal community structure) may be considered if the volume theme emphasizes paleobiology (or archaeology). Fields such as modeling of physical processes, genetic methodology, nonvertebrates or neontology are out of our scope.

Volumes in the series may either be monographic treatments (including unpublished but fully revised dissertations) or edited collections, especially those focusing on problem- oriented issues, with multidisciplinary coverage where possible.

Published and forthcoming titles in this series are listed at the end of this volume.

A Volume in

The Max-Planck-Institute Subseries in Human Evolution

Coordinated by

Jean-Jacques Hublin
Max-Planck-Institute for Evolutionary Anthropology,
Department of Human Evolution, Leipzig, Germany

Neanderthals Revisited: New Approaches and Perspectives

Edited by

Katerina Harvati
Max-Planck-Institute for Evolutionary Anthropology
Leipzig, Germany

Terry Harrison
Department of Anthropology, New York University
New York, USA

Springer

A C.I.P. Catalogue record for this book is available from the Library of Congress.

ISBN-13 978-1-4020-5120-3 (HB)
ISBN-13 978-1-4020-5121-0 (e-book)

Published by Springer,
P.O. Box 17, 3300 AA Dordrecht, The Netherlands.

www.springer.com

Printed on acid-free paper

Cover image by Philipp Gunz, adapted from a CT scan image
of La Ferrassie 1 by Jean-Jacques Hublin.

First edition published 2007
Reprinted with corrections 2008

This volume is dedicated to the memory of W.W. Howells (1908–2005) for his remarkable and pioneering contributions to the study of human evolution, especially his role in the greater understanding and appreciation of the Neanderthals. He was mentor and source of inspiration to generations of anthropologists, and his work continues to be a tremendous resource for research in human variation and evolution.

Contents

Contributors

J.C.M. Ahern
Department of Anthropology
University of Wyoming
Laramie, Wyoming, USA
jahern@uwyo.edu

S.E. Bailey
Center for the Study of Human Origins
Department of Anthropology
New York University, New York
New York, USA
sbailey@nyu.edu

M. Bastir
Hull York Medical School
The University of York
Heslington, York, United Kingdom
markus.bastir@hyms.ac.uk

G. Bräuer
Abteilung für Humanbiologie
Universität Hamburg
Hamburg, Germany
guenter.braeuer@uni-hamburg.de

H. Broeg
Abteilung für Humanbiologie
Universität Hamburg
Hamburg, Germany

E. Bruner
Dipartimento di Biologia Animale e dell'Uomo
Università La Sapienza
Rome, Italy
emiliano.bruner@uniroma1.it

A.M. Busby
Department of Anthropology
University of New Mexico
Albuquerque, New Mexico, USA
mandybusby@yahoo.com

S.E. Churchill
Department of Biological Anthropology and Anatomy
Duke University
Durham, North Carolina, USA
churchy@acpub.duke.edu

R.M. Cordero
Department of Anthropology
University of New Mexico
Albuquerque, New Mexico, USA
rc17767@hotmail.com

A. García-Tabernero
Departament de Paleobiología
Museo Nacional de Ciencias Naturales
Madrid, Spain
agarciatabernero@mncn.csic.es

K. Harvati
Abteilung für Humanevolution
Max-Planck-Institut für
Evolutionäre Anthropologie
Leipzig, Germany
harvati@eva.mpg.de

T. Harrison
Center for the Study of Human Origins
Department of Anthropology
New York University
New York, New York, USA
terry.harrison@nyu.edu

J. Hawks
Department of Anthropology
University of Wisconsin–Madison
Madison, Wisconsin, USA
jhawks@wisc.edu

T.W. Holliday
Department of Anthropology
Tulane University
New Orleans, Louisiana, USA
thollid@tulane.edu

J.-J. Hublin
Abteilung für-Humanevolution
Max-Planck-Institut für
Evolutionäre Anthropologie
Leipzig, Germany
hublin@eva.mpg.de

C. Lalueza-Fox
Departamento de Biologia Animal
Universitat de Barcelona
Barcelona, Spain
carles.lalueza@cexs.upf.es

G. Manzi
Dipartimento di Biologia
Animale e dell'Uomo
Università La Sapienza
Rome, Italy
giorgio.manzi@uniroma1.it

C. Martínez-Maza
Departament de Paleobiología
Museo Nacional de Ciencias Naturales
Madrid, Spain
mcncm565@mncn.csic.es

W.A. Niewoehner
Department of Anthropology,
California State University-San Bernardino
San Bernardino, California, USA
wniewoeh@csusb.edu

S. Pääbo
Abteilung für Evolutionare Genetik
Max-Planck-Institut für Evolutionäre Anthropologie
Leipzig, Germany
paabo@eva.mpg.de

O.M. Pearson
Department of Anthropology
University of New Mexico
Albuquerque, New Mexico, USA
ompear@unm.edu

M.S. Ponce de León
Anthropologisches Institut und Museum
Universität Zuerich-Irchel
Zurich, Switzerland
marcia@aim.unizh.ch

A. Rosas
Departament de Paleobiología
Museo Nacional de Ciencias Naturales
Madrid, Spain
mcnar537@mncn.csic.es

J.H. Schwartz
Departments of Anthropology and the History and Philosophy of Science
University of Pittsburgh
Pittsburgh, Pennsylvania, USA
jhs+@pitt.edu

D. Serre
McGill University and Genome Quebec Innovation Center
Montreal, Quebec, Canada
david.serre@mail.mcgill.ca

C.B. Stringer
Department of Palaeontology
Natural History Museum
London, United Kingdom
c.stringer@ nhm.ac.uk

I. Tattersall
Division of Anthropology
American Museum of Natural History
New York, New York, USA
iant@amnh.org

J.-L. Voisin
Institut de Paléontologie Humaine
Paris, France
jeanlucv@mnhn.fr

T.D. Weaver
Abteilung für Humanevolution
Max-Planck-Institut für Evolutionäre Anthropologie
Leipzig, Germany
weaver@eva.mpg.de

C.P.E. Zollikofer
Anthropologisches Institut und Museum
Universität Zuerich-Irchel
Zurich, Switzerland
zolli@aim.unizh.ch

Preface

The question of modern human origins and the role of Neanderthals in human evolution have fascinated paleoanthropologists and the public alike for more than a century. Recent years have witnessed important scientific breakthroughs in these areas of investigation as a result of ancient DNA studies, the application of new imaging and analytical tools in the study of morphology, and novel theoretical and conceptual approaches in evolutionary biology. These exciting and important developments have transformed our understanding and appreciation of the paleobiology, environmental adaptations, evolutionary relationships, and extinction of the Neanderthals. Although a number of influential edited volumes have previously dealt with these questions, there have been no recent compendia on the subject that have allowed leading international scholars specializing in Neanderthal research to present their latest ideas and to explore new methods and approaches to these long-standing problems.

With the 150th anniversary of the discovery of the first Neanderthal specimen drawing near, and while Katerina Harvati was still an Assistant Professor at the Anthropology Department at New York University (NYU), we conceived of a conference that would provide an opportunity for those scientists actively involved in all aspects of Neanderthal research to present their latest findings and to discuss the implications of these advances for understanding the evolutionary history of Neanderthals. The "*Neanderthals Revisited: New Approaches and Perspectives*" conference was held at NYU on January 27–29, 2005. An important aim was to unite researchers in the early stages of their careers with more established authorities, thereby encouraging a fresh look at some enduring problems using innovative new methods and perspectives. We invited some 35 scholars from the USA, Europe and the Near East to present their research or to act as discussants, most of who were able to attend. This edited volume is the outcome of that conference, and it presents, in more detailed fashion, the cutting-edge research that was showcased in New York. From the close of the conference to submitting the final manuscript to the Press has taken just slightly more than one year, so we are confident that the collection of papers included in this volume present up-to-date research and current ideas on Neanderthals.

The conference program included two days of presentations and discussions open to the academic community and to the public, followed by a half-day workshop restricted to conference participants. Contributions ranged from the re-evaluation of Neanderthal and modern human anatomy, inferred Neanderthal adaptations and habitual activities, developmental patterns, phylogenetic relationships, and Neanderthal extinction. Participants also applied new methods, including computer tomography, 3D geometric morphometrics, experimental growth studies, genetic and paleogenetic analyses, and presented new perspectives and approaches, including dental analysis, cladistic methodologies, bioenergetics, and broader comparative analyses. One of the most remarkable aspects of this gathering was that the venue provided an opportunity for researchers to present contrasting views on modern human origins without the rancour that has characterized much of this debate in past years. At the close of the conference there was general agreement that we were close to reaching a consensus on the critical issue of the Neanderthal-modern human relationship, while other key questions, such as the relationship of development with morphological change, the relationship between genetics and morphology, and the re-evaluation of presumed Neanderthal adaptations, were given closer critical scrutiny.

We would like to thank all the contributors to this volume for participating in the "*Neanderthals Revisited*" conference and for providing us with uniformly outstanding papers. We are also grateful to the invited discussants at the conference: Susan Antón, Jean-Jacques Hublin, Clifford Jolly, Giorgio Manzi and Milford Wolpoff, as well as Alison Brooks and Randy White. They all provided insightful comments and initiated lively discussion during the meeting. Bob Franciscus, Yoel Rak, and Mark Stoneking participated in the conference, but were unable to contribute to this volume due to other pressing professional commitments. Special thanks go to Jen LeClair, Jennie Tichenor, Myriam Haas, and Allison Cleveland for their invaluable help in organizing the conference, creating the conference poster, and formatting the manuscripts. Several NYU students graciously volunteered their time to help with the smooth running of the conference, thereby ensuring a minimum of technical and organizational hiccups. Among these, Tom Rein, Maja Seselj, Connie Fellmann, Tim Cavaretta, Joe Califf, Andres Link, Gisselle Garcia, Laura Gaydosh, Deena Emera, and Ilana Soloman deserve special mention.

Finally, we are grateful to all the colleagues, students, members of the press, and public at large who attended the conference in large numbers and showed their remarkable enthusiasm for all things Neanderthal.

Financial support for the conference was provided by the Center for the Study of Human Origins at NYU, the Department of Human Evolution of the Max-Planck-Institute for Evolutionary Anthropology, and the Dean for Social Sciences at NYU. Fred Myers, Richard Foley, Jean-Jacques Hublin and Eric Delson were especially supportive and instrumental in the realization of both the conference and the volume. We are very grateful to Springer and to the editors of the Vertebrate Paleobiology and Paleoanthropology series, Eric Delson and Ross MacPhee, for agreeing to publish this volume, and to all the colleagues who devoted their time to reviewing the manuscripts and providing helpful input. Finally, we owe the greatest gratitude to our families, and particularly to our spouses, Elias and Terri, for their support and encouragement in this and all endeavors.

Katerina Harvati
Terry Harrison

March 2006

1. Neanderthals revisited

K. HARVATI
Max-Planck-Institute for Evolutionary Anthropology
Leipzig D-04103
Germany
harvati@eva.mpg.de

T. HARRISON
Center for the Study of Human Origins,
Department of Anthropology
New York University
New York, NY 10003, USA
terry.harrison@nyu.edu

Keywords: Neanderthals, phylogeny, evolution, taxonomy, paleobiology, species

Abstract

Neanderthals are the best represented and most studied group in the fossil human record. The relatively large number of Neanderthal fossils and their good preservation offers the possibility of robust inferences about their evolution and paleobiology. Nevertheless, debate still continues on important issues, and this suggests that deeper theoretical and methodological differences lie at the root of the lack of consensus. Such disagreements are not likely to be resolved by additional fossil findings, but rather require critical re-evaluation of the evidence at hand and the application of novel techniques and perspectives. This is the premise and main goal of this volume. The major debates in Neanderthal research are re-examined with the use of innovative state-of-the art methods and exciting new theoretical and conceptual approaches. The diverse contributions presented here offer fresh insights and advances that move us closer to reaching a consensus.

As the contributions to this volume illustrate, the Neanderthals are the best represented, most comprehensively studied, and most thoroughly understood group of fossil hominins. The wealth of specimens currently available to the scientific community, including dozens of relatively complete crania and partial skeletons from across a broad geographic range, affords scholars the opportunity to develop well-informed and robust inferences about the anatomy, phylogenetic relationships, taxonomy, and paleobiology of the Neanderthals. Equally importantly, we know a great deal about their archaeology, paleoecology, paleoenvironment, and zoogeography, all of which offer key evidence for interpreting their paleobiology in a broader environmental, behavioral, and phylogenetic context. Paleoanthropologists studying earlier parts of the human fossil record are less fortunate,

K. Harvati and T. Harrison (eds.), Neanderthals Revisited: New Approaches and Perspectives, 1–8.

having to work with taxa that are much more poorly represented, and in some cases known only by a few fragmentary specimens. It is certainly an enviable position to be in, one in which most vertebrate paleontologists, who universally lament the shortcomings of the fossil record as an impediment to resolving key problems, would be most content to find themselves.

Nevertheless, despite the quality and weight of the evidence, there continue to be major debates (that have lasted for 150 years) about a number of contentious issues, especially whether or not Neanderthals should be included in the same species as anatomically modern humans, and what is the precise phylogenetic relationship between these two forms. Our inability to agree on these fundamental questions is a matter of serious concern for paleoanthropologists: it leads to the inevitable conclusion that if we are unable to come to a decision about the nature of the relationship between Neanderthals and modern humans, how can we have confidence in our ability to resolve relationships in the earlier, much more scanty, fossil human record. However, the lack of unanimity is unrelated to the quality of the material. It is more a consequence of deeper theoretical and conceptual issues that relate to how different researchers analyze and interpret the anatomical and genetic evidence, and to the manner in which these are ultimately situated in the broader context of how biological systems operate in the natural world. If this is the case, then it will take some time before a consensus can be reached, regardless of the amount of fossil material available for study. One way forward is to explore new methods and theoretical approaches in order to better understand the paleobiology and phylogenetic relationships of Neanderthals.

The main theme of this volume is to revisit the major debates concerning the place of Neanderthals in human evolution. How morphologically distinct are the Neanderthals from modern humans, and what do these distinctions mean in terms of their paleobiology and phylogeny? How genetically distinct are Neanderthals from modern humans, and what does this mean for interpreting the population dynamics, taxonomy and phylogenetic structure of Late Pleistocene hominins in Europe? Were Neanderthals and modern humans capable of interbreeding, and can they be considered the same or different species? What were the paleoenvironmental and paleoecological contexts of Neanderthals, and how did this impact on their paleobiology, evolution, and extinction? All of these issues are tackled head-on in this volume. By presenting new evidence, using innovative and state-of-the art techniques and methods, and exploring exciting new theoretical and conceptual approaches, the contributors gain fresh insights into these issues, and ultimately succeed in edging the debate closer to a consensus. However, we leave it up to the reader to decide just how far we still have to go in order to attain a satisfactory solution to some of these long-term problems.

As editors of this volume, our aim was to assemble a collection of papers written by leading international researchers who have tackled many of these important questions using a variety of novel approaches. Equally importantly, as can be discerned from the chapter titles and the content of this volume, we have also tried to accommodate a diversity of opinions and perspectives that reflect the plurality of viewpoints among contemporary scholars. The range of topics covered include phylogeny, taxonomy, speciation, development, lifeways and adaptation, population genetics, extinction, paleoecology and archaeology, while the methods adopted include morphological analyses (i.e., traditional comparative morphology, dental anthropology, developmental biology, unilinear measurements, and three-dimensional geometric morphometrics), genetics (i.e., mtDNA, microsatellite data), experimental modeling, and computer imaging.

This volume is not organized in formal sections, but rather it follows the logic of the general themes addressed by the contributors. It starts with the Middle-Late Pleistocene human fossil record and the evolution of the Neanderthal morphotype; continues with an examination of Neanderthal and modern human ontogeny, bioenergetics, and paleobiology, and their implications for inferring behavior; followed by genetic perspectives on Neanderthals and the utility of mtDNA and cranial morphological data in reconstructing phylogeny, the possibility of Neanderthal-modern human interbreeding and its taxonomic implications; and concludes with a review of the factors that may have contributed to the extinction of the Neanderthals.

In the opening chapter, Tattersall and Schwartz (Chapter 2) review the abundant morphological and genetic evidence supporting the distinctiveness of Neanderthals from modern humans, but also from earlier Middle Pleistocene hominins. To them, this evidence clearly confirms the status of Neanderthals as a separate clade. Does this mean that they are a different species? This is a difficult, perhaps even impossible, question to settle, because as Tattersall and Schwartz highlight, nature does not come neatly packaged and there are no absolute criteria by which to recognize species, especially in the fossil record. Nevertheless, Tattersall and Schwartz make the crucial observation (echoed by other authors in the volume) that Neanderthals as a group constitute a clear-cut morphologically and historically individuated entity, evidently equivalent to those commonly recognized today as species. They further point out that the morphological variability in the European Middle Pleistocene is little understood, and provocatively propose that several hominin clades might have been contemporaries in Europe during this period. This chapter sets the stage for the subsequent discussion of both the taxonomic position of Neanderthals and the tempo and mode of their evolution.

The next two chapters address the appearance of the Neanderthal morphotype and its evolution. Bruner and Manzi (Chapter 3) reassess the Saccopastore 1 cranium, which is correlated with oxygen isotope stage (OIS) 5, and is commonly considered to be an "early Neanderthal". Even though the specimen has been known since 1929, Bruner and Manzi are able to gain new insights into the endocranial morphology of the specimen using computer tomography. Their observations on cranial capacity, degree of pneumatization, and inner ear morphology support previous conclusions that this specimen exhibits a Neanderthal-like morphology despite its small size, thereby pinpointing the appearance of this morphotype to at least 130–100 ka. The authors suggest that the demographic impact of OIS 6 was probably catalytic in the evolution of full-blown Neanderthal features through genetic drift. Rosas, Bastir, Martínez-Maza, García-Tabernero and Lalueza-Fox (Chapter 4) propose the "organismic model" for Neanderthal evolution as an alternative hypothesis to the widely accepted "accretion model," drawing insight from their work on the extensive Spanish Middle and Late Pleistocene material. The authors postulate a two-phase evolutionary process in the European Middle-Late Pleistocene fossil record. The first phase is proposed to involve an increase in body size, greater postcranial robusticity, and increased midfacial prognathism. The second phase in turn would represent a true speciation event at about 300–250 ka, corresponding with a major re-organization of cranial architecture in Neanderthals relative to their Middle Pleistocene precursors. Their hypotheses point to a promising direction of research in the study of human evolution in Europe in the Middle Pleistocene.

Chapters 5 and 6 compare Neanderthal and modern human ontogeny from several different viewpoints. Ponce de León and Zollikofer (Chapter 5) obtain three-dimensional data from computer tomography scans and analyze them using geometric morphometric methods, in order to compare the ontogenetic trajectories of two sets of sister taxa: Neanderthals and modern humans, as opposed to chimpanzees and bonobos. Their analysis indicates that the two human taxa share a common ontogenetic trajectory, but have different perinatal morphologies resulting from differences in prenatal growth. The two species of *Pan*, although overall more similar in shape to each other than the human groups, differ not only in the length of their ontogeny, but also in the direction. As Ponce de León and Zollikofer observe in their concluding comments, one of the most important findings of their study is that "spatial and temporal differences in growth and development not only generate distinct adult morphologies, but also give rise to taxon-specific life histories". This will surely be a very fruitful avenue of future research that will dramatically improve our understanding of the phylogenetic relationships and paleobiology of fossil hominins. In the following chapter, Zollikofer and Ponce de León (Chapter 6) use computer modeling of their 3-D data to simulate cranial growth under diverse conditions. This approach allows them to explore the ways in which a simple developmental system can be modified to produce different outcomes. Their results demonstrate the complexity of developmental processes, with intricate patterns potentially arising from simple changes and vice versa. The differences in the developmental pattern between Neanderthals and modern humans suggest that a change in the initial conditions may result in subsequent differences in their developmental trajectories.

Various aspects of Neanderthal anatomy, and their implications for understanding Neanderthal behavior, are explored in the next four chapters. Churchill (Chapter 7) applies a bioenergetics approach, coupled with experimental modeling of the Neanderthal body form. His innovative analysis indicates that the capacious Neanderthal ribcage may have been related to heat production, rather than to heat retention, as is commonly postulated under Bergmann's rule. Churchill's results also suggest a very high-caloric diet, with important implications for Neanderthal hunting abilities, ranging behavior, and demographics. Even though Neanderthals appear to have had bodies better adapted to generate and conserve heat than early modern Europeans and modern-day cold-adapted populations, the finding that Neanderthals occupied sites with warmer winter temperatures than early modern humans, suggests that they were less able to tolerate extreme glacial conditions. This may reflect a greater capability by early modern humans to capture sufficient calories for sustaining adequate heat generation or the use of clothing or shelters with higher insulative values. Pearson, Cordero and Busby (Chapter 8) re-assess Neanderthal habitual activities, commonly thought (based on anatomical differences) to differ markedly from those of modern humans in their extreme activity levels and foraging inefficiency. They compare Neanderthal upper and lower limb robusticity to those from several recent human foraging groups, and conclude that Neanderthals do not appear unique, but instead are quite similar to modern foraging peoples that exploit limited territories. The authors conclude that these results, far from indicating foraging inefficiency, may instead imply a more intensive form of foraging.

Niewoehner (Chapter 9) uses three-dimensional geometric morphometric methods to evaluate Neanderthal hand morphology compared to that of Early and Late Upper Paleolithic modern humans. He relates the observed shape differences to differences in inferred habitual grip positions, possibly suggesting differences in hafting technology and preference for wood as a raw material. In contrast, Upper Paleolithic human hand morphology is consistent with the archaeologically observed expansion of the technological repertoire that would have required increased emphasis on precision handling and shifts in manipulative postures. His results suggest a gradual transition in hand morphology from the Middle to the Late Paleolithic. Bailey and Hublin (Chapter 10) re-examine the isolated dental remains associated with the

Châtelperronian levels of the Grotte du Renne (Arcy-sur-Cure) site in France. These were previously considered taxonomically unidentifiable, but using a new dental scoring method developed by Bailey, the authors are able to establish the Neanderthal identity of the dental assemblage from this site. Their findings substantiate earlier inferences that Neanderthals are associated with the Châtelperronian industry in Western Europe, and that they were most likely the makers of these archaeological assemblages. Equally importantly, the recognition that isolated teeth from Late Pleistocene sites can be identified taxonomically opens up the possibility of investigating the mode and tempo of human evolution in Europe with much better sampling and a finer-grained temporal resolution than was previously possible.

Chapters eleven and twelve consider aspects of Neanderthal and modern human genetics. Serre and Pääbo (Chapter 11) present a new method for ancient DNA recovery. This resolves the problem of contamination by modern human DNA, which leads to the inability to detect modern-human like genetic material from fossil humans. Among the specimens examined under this protocol, all Neanderthals yielded Neanderthal-like mtDNA sequences, while all early modern Europeans yielded only modern human like mtDNA. The authors interpret their findings as indicating a minimal degree of possible Neanderthal contribution to the modern human gene pool. They also show that major demographic changes occurred in Late Pleistocene mammal species that coincide temporally with the extinction of Neanderthals. Such analyses highlight the importance of the study of population history for understanding Neanderthal evolution, and for providing important clues as to the timing and causes of their extinction. In the next chapter, Hawks (Chapter 12) critically re-examines the conclusions derived from mtDNA evidence about the phylogenetic relationships between Neanderthals and modern humans. In particular, he questions whether previous models predicated on the assumption of selective neutrality are valid, and proposes an alternative hypothesis that human mtDNA may have recently undergone a "selective sweep," possibly related to climate adaptation. Hawks suggests that it was positive selection rather than population replacement that explains the disappearance of archaic mtDNA variants. In this case, the observed differences between the mtDNA in Neanderthals and modern humans would be rendered phylogenetically uninformative.

The relationship between genetics and morphology in modern humans is explored further in chapter thirteen. Harvati and Weaver (Chapter 13) evaluate the usefulness of different cranial regions (i.e., face, vault, and temporal bone) in reconstructing the phylogenetic placement of Neanderthals. They assess the degree to which morphological differences (represented by three-dimensional geometric morphometric data) among recent human populations correspond to known neutral genetic differences (as represented by microsatellite data) and/or to climatic differences. Although facial morphology alone shows a relationship with climate, both vault and temporal bone morphology track neutral genetics, with the temporal bone tracking older events more successfully. The authors conclude that temporal bone morphology may be most appropriate for reconstructing the phylogeny of Neanderthals and early modern humans. Their analysis does not support a unique phylogenetic link between Neanderthals and early modern Europeans.

The issue of Neanderthal-modern human relationships, and the possibility of interbreeding between these populations, is taken up in greater detail in chapters fourteen and fifteen, with conflicting opinions expressed. Ahern (Chapter 14) addresses the question of whether Neanderthals and Upper Paleolithic Europeans differ in a significantly greater number of distinct morphological traits than do two modern human populations: a "replacing" (European Americans) and "replaced"

(Native Americans) group. The author is unable to reject the hypothesis that Neanderthals and modern humans were conspecific for most of the features used, although he does add a note of caution by acknowledging that additional traits or combinations of features might eventually falsify the single morphospecies hypothesis. Bräuer, Broeg and Stringer (Chapter 15) address the same question by re-examining the most complete crania from Mladeč in the Czech Republic. These are among the earliest modern European specimens known and they have often been suggested to exhibit Neanderthal-like features. However, the univariate and multivariate statistical analyses of frontal bone metric data presented by Bräuer and his colleagues offer no support for the claim that the Mladeč individuals might represent hybrids.

Further discussion of the concepts of speciation and interbreeding is explored by Holliday and Voisin in the next two chapters. These contributions draw on studies of other vertebrates to reframe the species question in a broader comparative perspective. Holliday (Chapter 16) reviews the literature on hybridization among mammal species, and applies it to the Neanderthal-early modern human case. He demonstrates that among interbreeding mammal species, those that have diverged as recently as these two human taxa are still able to produce fertile hybrids. He concludes from this evidence that Neanderthal-modern human hybridization was possible in all likelihood, even though there is no evidence from the genetic data or the fossil record to confirm that it actually took place. Building on ideas and concepts developed in previous work by Clifford Jolly, Holliday argues that Neanderthals and modern humans are perhaps best considered "allotaxa," good morphological species that may still have been able to interbreed. He invokes the concept of syngameon – that closely related interbreeding species can be grouped into larger taxa – as a useful model for interpreting Late Pleistocene European hominins. As a flip side to this study, it would be interesting to examine how many non-reproducing pairs of large mammal species have diverged in the past 700 ka. We suspect that it would be very few; a finding that would serve to underscore just how rapidly Neanderthals and modern humans diverged. Neanderthals are evidently autapomorphic, but it needs to be recalled that much of the distinction between Neanderthals and modern humans stem from the highly autapomorphic nature of modern humans.

Voisin (Chapter 17) extends this approach by adding a broader comparative dimension using recent models of speciation in birds. In particular, he focuses on the ring species *Phylloscopus trochiloides* (greenish warbler) from central Asia as an analogy for European hominin migration and evolution. His model of speciation by distance and temporal overlap is offered to explain the East-West morphological cline observed in Neanderthals, as well as the possible intermediate morphology of some Central European Upper Paleolithic modern humans. Voisin suggests that modern humans arriving in Eurasia during the Late Pleistocene were able to interbreed to some degree with the less derived Neanderthals in Western Asia and Central Europe, but hybridization was not possible or was extremely limited with the more highly derived Neanderthal populations at the furthest extreme of the morphological cline in Western Europe. In other words, Neanderthals and modern humans behaved as two distinct biological species in Western Europe, but not elsewhere.

Finally, Stringer (Chapter 18) presents new data on the chronology of the appearance of modern humans in Europe and the extinction of the Neanderthals. Current paleoclimatic evidence indicates a much harsher and increasingly unstable climate during the period of Neanderthal extinction (OIS 3). Stringer shows that the greatest climatic stress would have been experienced around 30 ka, and he argues that this stress probably played a key role in the demise of the Neanderthals. Stringer's contribution highlights the challenge of integrating the diverse kinds of data that are becoming increasingly available to fully test hypotheses of Neanderthal and early modern human population histories.

The contributors to this volume provide important new insights that help us to better appreciate and understand the evolution and paleobiology of the Neanderthals. These advances have been brought about not through the discovery of startling new fossil finds, but through the application of exciting new methods and technologies, and a critical questioning of established theoretical and conceptual paradigms. This is certainly the way forward, and we echo Chris Stringer's concluding remarks that we undoubtedly have some exciting years ahead of us in Neanderthal and modern human origins research. Finally, although this volume is devoted entirely to the study of Neanderthals, it should not be overlooked that it is through comparisons with them that we are able to recognize and reflect on the uniqueness and remarkable peculiarities of our own evolutionary history.

2. The distinctiveness and systematic context of *Homo neanderthalensis*

I. TATTERSALL
Division of Anthropology
American Museum of Natural History
New York, NY 10024, USA
iant@amnh.org

J.H. SCHWARTZ
Departments of Anthropology and the History and Philosophy of Science
University of Pittsburgh
Pittsburgh, PA 15260, USA
jhs+@pitt.edu

Keywords: *Homo neanderthalensis*, Neanderthals, *Homo heidelbergensis*, Mauer, Arago, Atapuerca, Sima de los Huesos, reconstructed skeleton

Abstract

The "packaging" of the diverse living world is untidy, with the result that there are no absolute criteria for recognizing in all contexts the bounded historical entities we call species. However, there is no doubt whatsoever that *Homo neanderthalensis* is as clear-cut a morphological entity as any in the hominid fossil record: one that is characterized by a whole host of cranial apomorphies. Further, a recent full-skeleton reconstruction further emphasizes just how different Neanderthal body structure was from that of *Homo sapiens*, not simply in numerous anatomical details, but in the proportions of the thorax and its relation to the pelvic region. These bodily proportions would have given these extinct hominids a very distinctive appearance on the landscape, and enhance the likelihood that we are dealing here with a reproductively differentiated entity. Still, *Homo neanderthalensis* is not unique in all those features that distinguish it from *Homo sapiens*. Many "Neanderthal" cranial features are shared with various middle Pleistocene European hominids, notably the Steinheim specimen and, to a lesser extent, the Sima de los Huesos hominids from Atapuerca. Indeed, it appears that, far from being an isolated phenomenon, *Homo neanderthalensis* formed part of a larger endemic European hominid clade. This clade seems to have existed contemporaneously in Europe with at least one other hominid lineage or clade, exemplified by the *Homo heidelbergensis* fossils from Mauer, Arago and Petralona.

Introduction

Over the century and a half since the first description of a Neanderthal fossil, an impressive record of these extinct hominids has accumulated. Indeed, not only has the morphological distinctiveness of *Homo neanderthalensis* for long been vastly better documented than that of any other of our fossil relatives (Tattersall, 1986), but we are also in a better position than in the case of any other extinct hominid to appreciate the morphological

K. Harvati and T. Harrison (eds.), Neanderthals Revisited: New Approaches and Perspectives, 9–22.

variations (around a very distinctive mean) that are shown by the various populations of this form over time and space (Tattersall and Schwartz, 2000). Yet many paleoanthropologists continue to equivocate over the question of whether the Neanderthals actually constitute a bounded historical entity (Ghiselin, 1974) of the kind that warrants recognition as a species. Since this problem appears to be related, at least partly, to more general difficulties of species definition and recognition, it seems appropriate to begin our discussion of the status of the Neanderthals with a brief consideration of the nature of the boundaries that exist in the living world.

Species as Bounded Historical Entities

It must be very clear to anyone concerned with the luxuriant variety of living organisms that at some level Nature is "packaged." The biosphere is composed of a mass of discrete (but nested) units. At higher taxonomic levels there is no problem distinguishing these units: all horses are distinct from all whales by any definition. But as we approach finer degrees of distinction, particularly at intrageneric levels, difficulties proliferate. These difficulties are reflected in the extraordinary plethora of definitions of the species, by practice and by common consent the basic systematic unit, that is currently on offer. As Hey (2001) observed, literally dozens of new such definitions have been proposed in recent decades. This is not the place to trawl yet again through this lengthening list, but perhaps it is appropriate to point out that it is vanishingly unlikely that any single definition of the term "species" will ever fit all cases. This is not only because any universal definition would have to fit both living and extinct species, which offer us different information sets; it is also because speciation, the process by which individuated, nonreticulating units come about, is not a unitary mechanism. It is not, for example, simply an inevitable, passive, consequence of the morphological differentiation of populations over time (though this routine if poorly understood process certainly furnishes the basis for the morphological differences by which species may often be distinguished). Instead, speciation is a *result* (individuation, expressed most essentially among living populations as reproductive independence, but always seen *a posteriori*: Tattersall, 1994), which may eventuate from shifts in developmental regulation at many different levels (Schwartz, 2005).

Human beings are instinctively reductionist creatures; but for all these reasons, and more, we may be unrealistic in expecting Nature to be *neatly* packaged. Ultimately, the boundaries defining historically (evolutionarily) independent units must lie in their (effective or absolute) reproductive isolation. But even reproductive behaviors may not provide us with a golden bullet. The studies of Clifford Jolly and his colleagues (e.g. Jolly, 2001) have shown that evidence of quite extensive hybridization between adjacent populations of baboons that are sometimes well differentiated to the eye may often be readily observed; yet evidence is still lacking that such behaviors are necessarily associated with the progressive integration of what evidently continue to be distinctive gene pools. Jolly (2001: 17) has, indeed, penetratingly observed that baboon allotaxa may at one and the same time be " 'phylogenetic' species, but 'biological' subspecies." And if this is truly the case, from a historical (evolutionary) perspective morphological differentiation becomes much the most significant factor to consider. Still, this hardly simplifies matters much. For, from the phenotypic standpoint, remarkable amounts of geographically (or artificially) maintained morphological variety may accumulate within a species without the disruption of reproductive continuity – although, at the same time, the latter can occur in the absence of readily detectable phenotypic change. Indeed, Schwartz (1999) has emphasized that there are no biological reasons whatever for expecting that morphological change, deriving ultimately from changes in communication at the molecular level, should be associated with reproductive incompatibility unless it specifically affects such factors as protein recognition between sperm and ova, gametic or zygotic viability, or reproductive organ morphology. Nonetheless, in the fossil record morphological characteristics are almost invariably all we have to go on.

Given these awkward realities, we have to accept the unfortunate fact that even among sympatric or parapatric living forms, whose natural behaviors we can directly observe, it will sometimes never be certain whether two close relatives are "biologically" specifically distinct or not. And the difficulties only multiply when fossils are involved, for no specifiable or quantifiable degree of morphological differentiation can be associated with speciation; and in almost all cases fossils can offer us only morphology as a basis for making species judgments. It might be argued that the Neanderthals are actually a partial exception to this, given that fragments of their (distinctive) mitochondrial genome are now available (Krings et al., 1997, 2000; Ovchinnikov et al., 2000; Schmitz et al., 2002); but while the large mtDNA differences between available Neanderthal samples and all living *Homo sapiens* populations tested are indeed strongly suggestive, it nonetheless remains true that DNA differences suffer from analogous limitations to morphology in the context of assessing "species" status.

How, then, do we determine whether two clearly related forms represent independent (individual) entities? It has recently become fashionable to argue that species are most usefully or practically equated with "basal (smallest) clusters of diagnosably distinct populations" (Cracraft, 2002: 130). But while this approach may have its possibly Siren attractions for the working systematist, it still leaves paleontologists in a quandary. Because while systematists studying living populations will often already know from direct evidence what the geographical boundaries (and the morphological variation occurring within those boundaries) of their populations happen to be, the paleontologist has to work backwards, inferentially, from morphology alone. Moreover, distinctive morphologies frequently characterize geographic variants that are not reproductively/historically individuated. Given this constraint, it would appear appropriate for paleontologists to exercise caution, and to err on the side of inclusivity when using morphological differentiation as a basis for inferring the existence of reproductively and historically independent entities. Where, then, do the Neanderthals fit within this perspective?

Homo neanderthalensis as a Morphologically and Historically Individuated Entity

There has been remarkably little debate over which hominid fossils are Neanderthal or not Neanderthal (as witnessed, for example, by the remarkable unanimity on this question, at least, among the very heterogeneous assortment of contributors to the classic volume edited by Smith and Spencer [1984]). Only at the fringes has there been any significant discussion, but this has usually focused upon whether or not a particular fossil was somehow related to the Neanderthals, as a precursor or otherwise, rather than on the matter of its inclusion within the group itself (see historical account in Tattersall, 1995). This is largely because, among all extinct hominid entities, the Neanderthal group appears by far the most clearly delimited morphologically. Instructively, of all other fossil hominids only *Homo erectus* is comparably well represented in the record, yet no end is in sight to the debate over which specimens should or should not be attributed to this species (compare, e.g., Wolpoff et al. [1994] to Schwartz and Tattersall [2000] and Antón [2003]).

Traditional cranial characters that are commonly cited as typical of Neanderthals (see e.g., Hublin, 1978, 1998; Santa Luca, 1978; Vandermeersch, 1981; Stringer et al., 1984; Schwartz and Tattersall, 1996a, b, 2005; Rak and Hylander, 2003) are numerous and include: double-arched supraorbital ridges whose surfaces roll smoothly upward from the orbital roofs and onto the frontal squama; orbits that are obliquely truncated inferomedially; a narrow lower face and a sharply retreating midface; medial projections emerging above a spinoturbinal crest that delineates a prenasal fossa lying just within the very large nasal aperture; very long and typically thin zygomatic arches; a "puffy" midface that reflects the presence internally of expanded maxillary sinuses that swell out the infraorbital and medial orbital regions; an angulation along the anterior squamosal suture, delineating distinct anterior and posterior temporal fossae; a smoothly rounded ("en bombe") cranial profile in rear view; a pitted suprainiac fossa that lies above a superior nuchal line that is undercut by the nuchal plane but poorly delineated above; a long and more or less straight parietomastoid suture that flows directly behind into an anterior lambdoid suture; widespread pneumatization within the petrosal; incomplete ossification of the ectotympanic tube laterally; and a long, narrow, ovoid foramen magnum. A useful table published by Hublin (1998, Table 1) lists several others in addition. A particularly interesting new source of cranial information has been provided by CT studies (e.g., Hublin et al., 1996; Spoor et al., 2003) showing that the bony labyrinth of the inner ear in Neanderthals is distinguished in numerous derived features from that of *Homo sapiens* (and of *Homo erectus*). In the most comprehensive such study to date, Spoor et al. (2003: 141) suggested that such differences reflected a distinctive pattern of head movements possibly related to "aspects of locomotor behaviour and the kinematic properties of their head and neck."

In the Neanderthal mandible are seen retromolar spaces; sigmoid notches that are deepest posteriorly, in front of low-set condyles, and sigmoid notch crests that terminate medial to the lateral extremities of the condyles; obliquely truncated gonial angles; symphyseal bone that, when viewed from below, is thinner from side to side than the bone distal to it. In the dentition the molars have relatively complex occlusal surfaces, with centroconids and centrocones present on the lower and upper molars respectively, and distinct talonid and trigonid basins in the relatively long and narrow lower molars. The molar occlusal surfaces are well defined peripherally by blunt crests, and are constricted in area by their inwardly sloping sides. Additional apomorphies of the Neanderthal upper and lower cheek dentitions have recently been cited by Bailey (2002, 2004) and Bailey and Lynch (2005).

Unsurprisingly, not all these and other craniodental characters typical of Neanderthals are equally strongly expressed in all unarguably Neanderthal specimens that preserve them; and indeed it is evident that, in any widely distributed group that is as close-knit phylogenetically as the genus *Homo*, there is bound to be some overlap among differentiated populations in the frequency and extent of expression, as well as in the presence/absence of particular traits. Thus, among the large assemblage of unquestionably Neanderthal fossils individuals vary, sometimes substantially, in features such as the degree of bunning of the occiput, the size of the suprainiac fossa, the length of the retromolar space, the presence or absence of a distinct postorbital plane behind the brow ridges, the prominence of the medial projections in the nasal fossa, the extent of cheektooth taurodontism, the depth of the zygomatic arches, and so forth. But it is abundantly clear – and almost universally acknowledged – that, despite their manifest individual and geographic variation, the Neanderthals represent an unusually coherent and readily recognizable group. Certain authors (e.g., Wolpoff, 1980; Frayer, 1984) have suggested that the relatively restrained expression of some of these characters in certain Neanderthals of late date indicates a degree of intermediacy with *Homo sapiens*, but there are numerous reasons for rejecting this notion (Tattersall and Schwartz, 2000).

Still, as impressive as the list of craniodental apomorphies of the Neanderthals undoubtedly is, the extent to which these hominids are distinguished morphologically from *Homo sapiens* is made particularly evident by the full-skeleton reconstruction (Figures 1, 2) recently reported by Sawyer and Maley (2005). Reconstituted from the remains of a half-dozen partial skeletons from four countries, this postcranial reconstruction draws attention to the classically Neanderthal characteristics of the postcranial skeleton that have already been exhaustively documented by numerous authors (e.g., Boule, 1911–1913; McCown and Keith, 1939; Straus and Cave, 1957; Trinkaus, 1983).

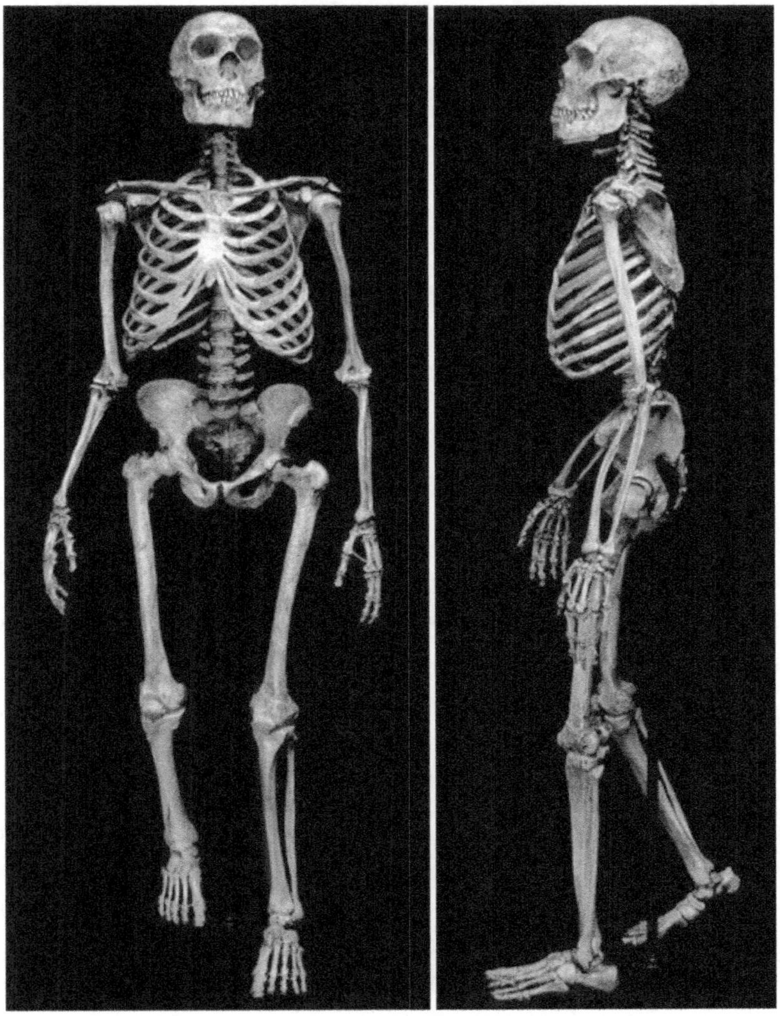

Figure 1. Front and lateral views of a complete Neanderthal skeleton as reconstructed, using elements from five partial skeletons (principally La Ferrassie 1 and Kebara 2), by G. J. Sawyer and B. Maley. Photo courtesy of Ken Mowbray.

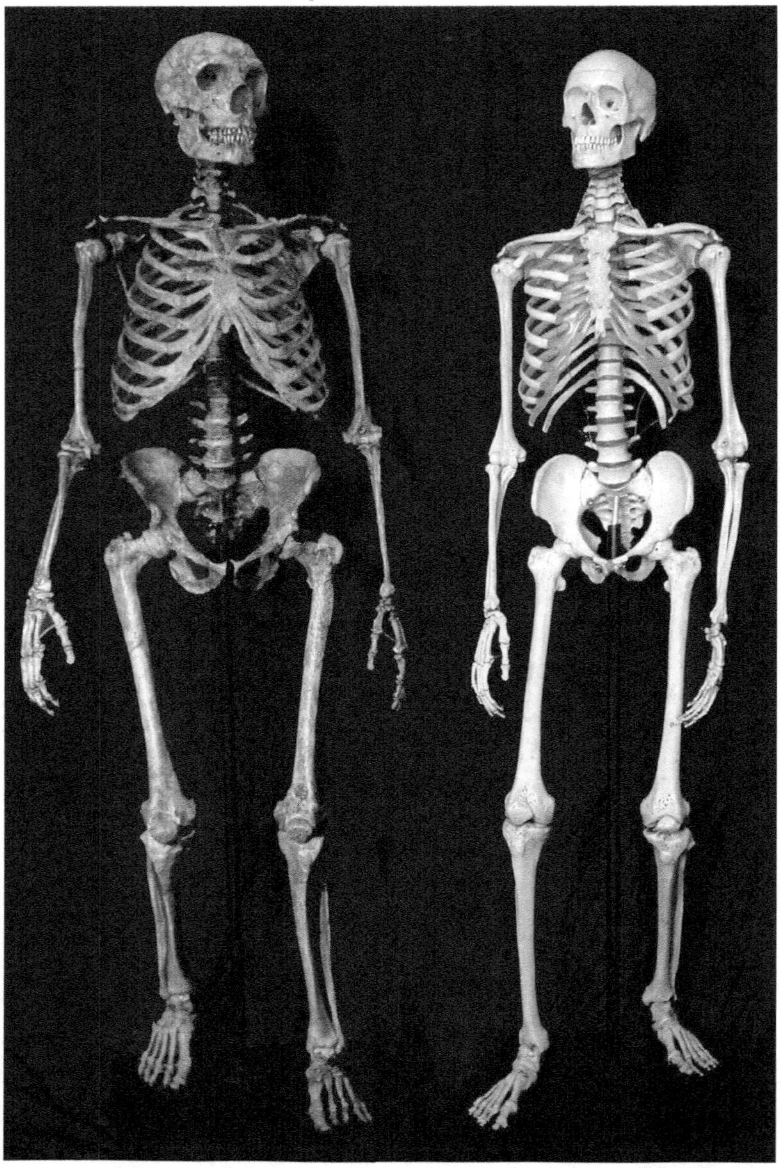

Figure 2. Front view of reconstructed Neanderthal skeleton compared with a skeleton of a modern *Homo sapiens* male of similar stature. Photo courtesy of Ken Mowbray.

Among the long-noted apomorphies that are also especially apparent from individual skeletal elements are, of course, the large limb articular surfaces, the flaring iliac blades and long, thin pubic rami of the pelvis; but the reconstruction also dramatically underlines the proportional differences in the thorax and its relationship to adjacent regions that emerge when Neanderthals and modern humans are compared (Figure 2). In the Neanderthal the rib cage is constricted above and flares dramatically out and down to match the broad pelvic bowl below. This conformation contrasts dramatically with the typical "barrel-shaped" *Homo sapiens* condition in which the thorax tapers up and also down to match a narrower pelvis. Further, as Sawyer and Maley also remark, the base of the Neanderthal vertebral column "sit[s] much lower in the pelvic basin than is the case among modern humans"

(Sawyer and Maley, 2005: 30), contributing to an extreme shortness of the waist in the former that, in limiting thoracic torsion, would have had significant consequences for gait, as well as for appearance. Altogether, this reconstruction makes it abundantly evident that in life the Neanderthals would have cut a very different figure on the landscape from *Homo sapiens*.

This major *Gestalt* difference adds, of course, to the existing morphological basis for surmising that the two kinds of hominid were significantly differentiated reproductively as well as morphologically. And, while not demonstrating this conclusively, from the perspective of "inclusive," rather than "exclusive," species concepts it also adds weight to the inference that specific mate recognition systems would have differed significantly between the two. This does not, of course, eliminate the possibility that occasional instances of coupling might have occurred when the two kinds of hominids came into contact; but it does very strongly buttress the already substantial reasons furnished by the fossil record for inferring that no significant integration of the two populations ever took place. Overwhemingly, then, the probability must be that the two kinds of hominid, Neanderthal and modern, were/are independent historical as well as morphological entities. In which case, we are fully justified, under virtually any set of criteria, in regarding the former as constituting the species *Homo neanderthalensis* (see also Harvati et al., 2004).

Neanderthals in Wider Systematic Context

The morphological distinctiveness of *Homo neanderthalensis* is hardly surprising when one considers that both molecular (Krings et al., 1997, 2000) and paleontological (e.g., Stringer and Gamble, 1993; Hublin, 1998; Tattersall and Schwartz, 2000) studies suggest that this species last shared a common ancestor with *Homo sapiens* at least a half-million years ago. For close to a century now paleoanthropologists have sought the roots of the Neanderthals far back in time, and in one guise or another many have discerned evidence in Europe for what Hublin (1998: 297) has recently termed a "Neandertalization Process," linking forms from the early Middle Pleistocene (Mauer, Tautavel, etc.) to the latest (stages 5–3) "classical" Neanderthals, via "Holstein-Hoxnian" (Bilzingsleben, Sima de los Huesos, etc.) and "Saalian" (Biache-Saint-Vaast, Ehringsdorf, etc.) intermediates. Still, if *Homo neanderthalensis* really was the end product of a steady course of phyletic modification, as this notion of a process implies, what is perhaps most surprising is how clearly the morphological boundaries of the species *Homo neanderthalensis* actually appear to be drawn.

Hublin himself finds it difficult as a matter of logic to exclude any of the "specimens involved in the Neandertalization process" from *Homo neanderthalensis*, "even if they display only a few derived Neandertal features" (Hublin 1998: 302). But from our own examination of the fossils concerned, what seems most remarkable in the longer-established record is that in strictly morphological terms there is really only one potentially questionable case of attribution to *Homo neanderthalensis*, apart from such apparently permanently inscrutable specimens as the Fontéchevade fragments (see Vallois, 1949; Trinkaus, 1973). This is the rather fragmentary assemblage from Ehringsdorf which, despite its limited size, shows unusual variation in morphology (Vlček, 1993; Schwartz and Tattersall, 2005). Thus, one occipital (Ehr H9 1032/69) lacks a suprainiac fossa, and shows instead twinned depressions both below and above the occipital "torus," while the Ehr H3 1026/69 parietal is also atypical for a Neanderthal in showing a "tent-shaped" coronal profile in rear view. And the temporal Ehr H3 1026/69 shows the straight and long parietomastoid suture of a Neanderthal, but uncharacteristically for this group the posterior root of its zygomatic arch diverges strongly from the cranial wall. Still, if we provisionally regard this material as early Neanderthal, it serves to emphasize the relative homogeneity of the rest of the Neanderthal hypodigm.

None of this is to say, however, that all "Neanderthal" characteristics of the kind we listed in the last section are confined to the large group of fossils that we may comfortably regard as belonging to *Homo neanderthalensis*. This is because in the middle Pleistocene of Europe we can indeed, like Hublin and most others, identify a variety of hominid fossils displaying some, but not all, of the features that typify Neanderthals (Hublin, 1998; Schwartz and Tattersall, 2005). Perhaps most notable among these is the Steinheim cranium (probably stage 7, around 225 ka), which possesses a fairly standard Neanderthal-like morphology of the upper face, with separately arching and smoothly rolled supraciliary ridges over orbits that have truncated inferomedial margins. The large nasal fossa and the presence of a prenasal fossa between well defined lateral and spino-turbinal crests are characteristics shared with Neanderthals, as are the angulation apparent along the anterior squamous suture, the long, straightish parietomastoid and anterior lambdoid sutures, the (rather faint) suprainiac depression, the horizontal occipital "torus" that is only fully defined below, and the rather rounded posterior profile of the braincase. There are even the rudiments of a vertically oriented medial projection faintly evident within the nasal cavity. On the other hand, the puffy midface of the Neanderthals is absent from the Steinheim specimen, as are the sharply retreating zygomas, with their laterally rising anterior roots, that give Neanderthal faces their highly characteristic allure. The poorly inflated Steinheim braincase departs from the Neanderthal condition in having fairly vertical side walls in coronal section and in showing a smoothly rounded lateral profile at the rear. As a result, despite the many Neanderthal resemblances of this specimen, it has been more or less universally regarded (see reviews by Day, 1986; Schwartz and Tattersall, 2002) as an example of "archaic *Homo sapiens*" rather than as of Neanderthal affinity. Few have ever called Steinheim a Neanderthal, and it would certainly be inaccurate to do so. However, based on the constellation of characters it exhibits, it would seem entirely reasonable to consider the Steinheim specimen as representative of the sister taxon to *Homo neanderthalensis*. Conceivably this taxon was directly ancestral to the Neanderthals; but to make that claim would involve a variety of assumptions that we would prefer to avoid here (see discussion by Tattersall and Eldredge, 1977).

Another German specimen with many of the characteristic Neanderthal cranial traits is the frustratingly incomplete partial calvaria from Reilingen, which is possibly penecontemporaneous with that from Steinheim (Ziegler and Dean, 1998). Early analyses resulted in attributions to *Homo erectus* (Czarnetzki, 1989) or to "archaic *Homo sapiens*" (Adam, 1989, Schott, 1990), but more recent contributions have focused on the Neanderthal affinities of this specimen (e.g., Condemi, 1996; Dean et al., 1998). Like Neanderthals, this fossil possesses expanded petrosal pneumatization; an occipital "torus" that is fully delineated only below, with a suprainiac depression above; an *en bombe* coronal profile of the expansive braincase; and incomplete lateral ossification of the ectotympanic tube. As in Steinheim, though, the sagittal profile of the occiput is quite rounded, and the weakly undercut occiput is fairly narrow, though the cranial vault itself appears considerably flatter and more Neanderthal-like than that of Steinheim. This specimen exemplifies the difficulty of categorizing members of this wider group strictly on characters of the cranial rear, though it is abundantly clear that Reilingen is either *Homo neanderthalensis* or a close relative.

The impressive assemblage of hominid fossils from the Sima de los Huesos at Atapuerca in Spain, now thought to be around 400 ka or possibly more (Bischoff et al., 2003), appears to be relatively homogeneous. All of this material has been referred by its describers (e.g., Arsuaga et al., 1997) to the species *Homo heidelbergensis*. This species is, however, based on the Mauer jaw, a specimen to which none of the mandibles known from the Sima appears to bear notable similarities. At the same time, Arsuaga and colleagues have noted that various features of the Sima fossils are "transitional" to Neanderthal morphology, and have concluded that these hominids are early members of the Neanderthal lineage, as well as simultaneously linked to other European middle Pleistocene fossils (Arsuaga et al., 1997). And certainly, while the Sima hominids do show fewer components of the Neanderthal character constellation than Steinheim does, the number of apparent Neanderthal synapomorphies that they possess is nonetheless quite extensive. Cranially, such resemblances to *Homo neanderthalensis* include: bilaterally arced supraorbital tori with tall and evenly rounded anterior surfaces; orbits with obliquely truncated inferomedial corners; a large nasal aperture showing a distinct prenasal fossa with a continuous internal margin; some projection of the frontal processes around the nasal aperture; an angulation along the anterior squamous suture; a long, straight parietomastoid suture; incompletely laterally ossified ectotympanic tubes; and a pitted suprainiac depression. Like those of Neanderthals, the Sima mandibles display medial pterygoid tubercles on the inner surface of the ramus, and have sigmoid notch crests that terminate just lateral to the midline of the condyle.

At the same time, however, the Sima hominids are cranially less derived and quite distinct from *Homo neanderthalensis*. In the structure of the face, differences from the latter include: an uninflated infraorbital region; horizontal conchal crests just within the aperture of the nasal cavity in lieu of vertical medial projections; no sharply retreating and inferiorly tapering midface; and anterior zygomatic roots that angle out more sharply laterally. Farther posteriorly, the Sima hominid shows deep zygomatic arches, a very short anterior lambdoid suture, and there is no clearly undercut occipital "torus." In sagittal profile the braincase is smoothly rounded, and in coronal profile it has parallel sides and a sagittal peak. Postcranially the robust Sima pelvis shows greatly flaring iliac blades, long pubic rami and a capacious pelvic canal. These pelvic features recall the Neanderthals; but they are likely to represent the primitive condition for the genus *Homo*, or at least for the subclade that contains both the Neanderthals and the Sima hominids.

Based purely on the material discussed up to this point, it might (just) be possible to argue that this "Neanderthal clade," fairly well defined in terms of synapomorphies, included a chronological succession of taxa of which *Homo neanderthalensis* was the terminal outcome. But from the evidence presented below, it is evident that the larger story of middle Pleistocene hominid evolution in Europe was more complex than the notion of a single evolving lineage can accommodate.

For the conclusion that hominid evolution in Europe prior to the abrupt arrival of *Homo sapiens* was *not* an essentially unilineal affair is dramatically reinforced by a survey of the entire variety of hominid fossils known from Europe in the period centering on 400 ka. Such fossils include the specimens from Swanscombe, in England (about 400 ka: Stringer and Hublin, 1999), the Arago hominids from southern France (perhaps 450 ka: Iacumin et al., 1996); the German Mauer jaw (around 500 ka: Cook et al., 1982); the Bilzingsleben hominids, also from Germany (300–400 ka: Schwarcz et al., 1988), and possibly the Vérteszöllös occipital from Hungary which might be as old as 350–250 ka (Cherdyntsev, 1971) though Schwarcz and Latham (1990) consider it younger.

Morphologically there is strong justification for associating the Greek Petralona cranium with the Arago hominids (Schwartz and Tattersall, 2005), although Grün (1996) concluded that this exceptionally poorly dated fossil most likely derives from significantly later in time, around 250–150 ka.

The committee that was originally convened to evaluate the two first-found elements of the Swanscombe cranial rear emphasized metrical comparisons to both *Homo sapiens* and Steinheim (Morant, 1938), and influential later contributions (e.g., Howell, 1960; Stringer et al., 1984) continued an essentially "presapiens" assignment. However, following the lead of Santa Luca (1978), most workers have moved toward comparisons with *Homo neanderthalensis* (see Stringer and Gamble, 1993; Stringer and Hublin, 1999). And, indeed, Swanscombe clearly does sort into the Neanderthal clade, although it does not seem to represent a typical Neanderthal, and in some ways it more closely resembles its counterparts from the Sima de los Huesos, for instance in having a relatively narrow and weakly undercut occipital "torus" and a poorly defined suprainiac fossa. Still, while having more vertical cranial walls than is usual for Neanderthals, it does also show a more rounded coronal cranial contour than is seen in the Spanish material and it does possess a fairly large, ovoid foramen magnum.

The Vérteszöllös occipital was initially announced as a representative of *Homo erectus* (Vértes, 1965) – in hindsight a virtually meaningless attribution in the European context – and was subsequently moved to "archaic *Homo sapiens*" by Stringer et al. (1979) at a time when the Petralona cranium was similarly classified. The two hominids share a long and very horizontal occipital "torus," but otherwise the Vérteszöllös specimen remains fairly enigmatic. Also rather inscrutable are the Bilzingsleben hominids, which have been ascribed to *Homo erectus* by their finders (e.g., Mania, 1983; Vlček and Mania, 1987). Stringer et al. (1984) found these fossils to be the "most *erectus*-like and the least Neanderthal or modern-like" of the "archaic *Homo sapiens*" group, a conclusion later sustained by Stringer (1989) through comparisons with the Saldanha calotte. The two Bilzingsleben crania are both fragmentary, but awkwardly they appear to display differences from one another (Schwartz and Tattersall, 2005) that are substantial enough at least to raise the question of whether they belonged to a single population. The affinities of neither are clearly evident, but there is little reason to associate all of the material with the Neanderthal clade and the net effect is to add to the growing impression of hominid diversity in the European middle Pleistocene.

The Arago and Petralona crania are nowadays widely accepted as classic European exemplars of the species *Homo heidelbergensis*, originally founded solely on the basis of the Mauer mandible. It is fortunate that the existence of both mandibles and crania in the Arago collection allows this attribution to be substantiated (Schwartz and Tattersall, 2005). There is a substantial *Gestalt* difference between the Mauer jaw and the two better preserved Arago mandibles because the latter are more gracile than the former, and lack its remarkable ramal length. However, both the Mauer and Arago 13 mandibles show a common configuration of the anteroinferior margins of the jaw. They also share excavated and rounded gonial regions, anteroposteriorly long coronoid bases, posteriorly decreasing corporal height, and tall but shallow infracondylar sulci that lie along the posterior margins of the rami. More striking, though, are the dental similarities. Notably, the anterior teeth in both mandibles were large, and the molars long and ovoid; the P_1s are obliquely truncated mesiodistally along their lingual surfaces, and are hence more mesiodistally tapering and elongate than the short, buccolingually wide and more ovoid P_2s; the protoconids on both the P_1s and P_2s are centrally placed; on P_1 the low lingual swelling lies opposite the protoconid, while on P_2 the metaconid is mesially situated relative to the protoconid, and on both premolars the lingual swelling or cusp is bounded by a very small fovea mesially and a much deeper fovea distally. In the molars M_2 is larger than both M_1 and M_3; the protoconids and metaconids are situated very mesially on the crowns, and in the same mutual relationship; M_1 shows evidence of a tiny trigonid basin, while this basin is more pronounced on M_{2-3}. On all molars the hypoconulid lies just buccal to the crown midline, and the talonid basin is or was quite long mesiodistally and truncated buccolingually, with some evidence of enamel wrinkling.

If we can allocate the Mauer and Arago lower jaws to the same species, then it appears permissible to regard the well-preserved Arago 21 fossil as the classic exemplar of the face of *Homo heidelbergensis*. In turn, we can associate with this species the more complete and potentially much more recent Petralona cranium, which shares with Arago 21 a massive and broad lower face that lies below hugely developed and superoinferiorly tall supraorbital margins that attain their maximum thickness around mid-orbit. In both fossils, the superior margin of the orbit bears a blunt edge that demarcates the anterosuperiorly twisting front surface of the torus from the shallow posttoral sulcus behind. The same features unite these European specimens with others from Africa (Bodo, Kabwe, Saldanha) and Asia (Dali, Jinniushan). Interestingly, while there is also a fair amount of variation in other features of the cranium among this cosmopolitan group (see discussion by Schwartz and Tattersall, 2005), that variation does not appear to be geographically organized in any clear-cut manner. In the present context, however, it is important to note that the upper facial features of *Homo heidelbergensis* noted above distinguish this species absolutely from *Homo neanderthalensis* as well as from broadly contemporaneous members of the latter's clade such as the Sima hominids.

Conclusion

Homo neanderthalensis is an unambiguously demarcated morphospecies, recognizable on a host of cranial and postcranial characters, that was also, as definitely as such things can be known in the fossil record, a fully individuated historical entity. Its identity did not shade into that of any other known hominid, and certainly not into that of the species *Homo sapiens* that entered its European and western Asian zone of distribution some 40 kyr ago and entirely displaced it within about a dozen millennia. Further, it clearly emerges from the evidence just adumbrated that the larger Neanderthal clade, if not the species *Homo neanderthalensis* itself, was already in existence in the period centering on 400 ka.

Inadequate specimens or poor dating make it hard to be sure whether or not multiple kinds of hominid belonging to this "Neanderthal clade" coexisted at this juncture, although the probability may lie in that direction. On the other hand, the new datings make it evident that at around this same time point there also existed in Europe at least one other hominid lineage, represented by *Homo heidelbergensis* (as defined by the Mauer/Arago fossils). This species is clearly demarcated from all members of the equally derived Neanderthal/Steinheim/ Sima clade in which this cranial region is known by the apparently derived structure of its upper face (among other features). It is beyond the scope of our discussion here to broach the many problems of apparently geography-independent morphological variation in the larger group to which the species name *Homo heidelbergensis* has been applied in recent years (e.g., Rightmire, 1990; Stringer and McKie, 1996; Schwartz and Tattersall, 2005). But it is already evident that this species, as broadly defined, somehow shared the European habitat with the Neanderthals and/or their relatives, potentially over a period of several hundred thousand years if the late dating of Petralona turns out to be accurate.

It remains uncertain how far back in time the roots of the endemic European clade or lineage that ultimately gave rise to *Homo neanderthalensis* can be traced. Bermúdez de Castro et al. (1997) have argued that the small sample of 780 kyr-old hominids from the ATD-6 levels of the Gran Dolina at Atapuerca is dentally primitive and represents the common ancestor of both the Neanderthal and modern human lineages. This proposition is, however, frustratingly difficult to demonstrate on the basis of the existing small and fragmentary sample of Gran Dolina hominids (see discussion by Schwartz and Tattersall, 2005); and certainly it is hard to read specifically Neanderthal affinities into these fossils. Neither is it possible to detect any Neanderthal apomorphies in the quasi-contemporaneous (with ATD-6) Ceprano hominid from Italy, recently made the holotype of *Homo cepranensis* by Mallegni et al. (2003). Indeed, this latter seems to compare most closely with Asfaw et al.'s (2002) million-year-old Daka cranium from Ethiopia (Schwartz and Tattersall, 2005). On current evidence, then, the earliest successful (i.e., long-term) hominid occupation of Europe was preceded by a series of early incursions that occurred around or subsequent to about one million years ago. None of the fossil forms witnessing such early occupations can be demonstrated to have definitively established itself in Europe much prior to the arrival of the first Neanderthal precursors, some time before (perhaps well before) the Sima hominids lived. When (and where) exactly the larger clade containing the Neanderthals originated remains to be determined.

Acknowledgments

We thank Katerina Harvati and Terry Harrison for their kind invitation to participate in the symposium *Neanderthals Revisited: New Approaches and Perspectives*, and all of the Symposium participants for stimulating discussion. And thanks to Ken Mowbray and Gary Sawyer for the illustrations.

References

Adam, K.D., 1989. Alte und neuer Urmenschenfunde in sudwest-Deutschland – Eine kritische Wardigung. *Quartär* 39/40, 177–190.

Antón, S.C., 2003. Natural History of *Homo erectus*. *Yrbk. Phys. Anthropol.* 46, 126–169.

Arsuaga, J.L., Bermudez de Castro, J.M., Carbonell, E., 1997. Special issue on the Sima de los Huesos hominids and site. *J. Hum. Evol.* 33 (2/3), 105–421.

Asfaw, B., Gilbert, W.H., Bayene, Y., Hart, W.K., Renne P.R., WoldeGabriel G., Vrba, E.S., White, T.D., 2002. Remains of *Homo erectus* from Bouri, Middle Awash, Ethiopia. *Nature* 416, 317–320.

Bailey, S.E., 2002. A close look at Neanderthal postcanine dental morphology: The mandibular dentition. *Anat. Rec. (New Anat.)* 269, 148–156.

Bailey, S.E., 2005. A morphometric analysis of maxillary crowns of Middle-Late Pleistocene hominins. *J. Hum. Evol.* 47, 183–198.

Bailey, S.E., Lynch, J.M., 2004. Diagnostic differences in mandibular P4 shape between Neandertals and anatomically modern humans. *Am. J. Phys. Anthropol.* 126, 268–277.

Bermúdez de Castro, J.M., Arsuaga, J.-L., Carbonell, E., Rosas, A., Martínez, I., Mosquera, M., 1997. A hominid from the lower Pleistocene of Atapuerca: Possible ancestors to Neandertals and modern humans. *Science* 276, 1392–1395.

Bischoff, J.L., Shamp, D.D., Aramburu, A., Arsuaga, J.L., Carbonell, E., Bermúdez de Castro, J.M., 2003. The Sima de los Huesos hominids date to beyond U/Th Equilibrium (>350 kyr) and perhaps to 400–500 kyr: New radiometric dates. *J. Archaeol. Sci.* 30, 275–280.

Boule, M. 1911–13. L'homme fossile de la Chapelle-aux-Saints. *Ann. Paléontol.* 6, 1–64; 7, 65–208; 8, 209–279.

Cherdyntsev, V., 1971. Uranium-234. Israel Program for Scientific Translations, Tel-Aviv.

Condemi, S., 1996. Does the human fossil specimen from Reilingen (Germany) belong to the *Homo erectus* or the Neanderthal lineage? *Anthropologie* 34, 69–77.

Cook, J., Stringer, C.B., Currant, A.P., Schwarcz, A.H. P., Wintle, A. G., 1982. A review of the chronology of the European Middle Pleistocene hominid record. *Yrbk. Phys.* Anthropol. 25, 19–65.

Cracraft, J., 2002. The seven great questions of systematic biology: An essential foundation for conservation and the sustainable use of biodiversity. *Ann. Missouri Bot. Gard.* 89, 127–144.

Czarnetzki, A., 1989. Ein archäischer Hominidencalvarium aus einer Kiesgrübe in Reilingen, Rhein-Neckar-Kreis. *Quartär* 39/40, 191–201.

Day, M.E., 1986. *A Guide to Fossil Man*, 4th Edition. University of Chicago Press, Chicago.

Dean, D., Hublin, J.-J., Holloway, R., Ziegler, R., 1998. On the phylogenetic position of the pre-Neandertal specimen from Reilingen, Germany. *J. Hum. Evol.* 34, 485–508.

Frayer, D.W., 1984. Biological and cultural change in the European late Pleistocene and early Holocene. In: Smith, F.H., Spencer, F. (Eds.), *The Origins of Modern Humans: A World Survey of the Fossil Evidence*. Alan R. Liss, New York, pp. 211–250.

Ghiselin, M.T., 1974. A radical solution to the species problem. *Syst. Zool.* 23, 536–544.

Grün, R., 1996. A re-analysis of electron spin resonance dating results associated with the Petralona hominid. *J. Hum.* Evol. 30, 227–241.

Harvati, K., Frost, S.R., McNulty, K.P., 2004. Neanderthal taxonomy reconsidered: Implications of 3D primate models of intra- and interspecific differences. *Proc. Natl. Acad. Sci. U.S.A.* 101, 1147–1152.

Hey, J., 2001. *Genes, Categories and Species*. Oxford University Press, Oxford.

Howell, F.C., 1960. European and northwest African middle Pleistocene hominids. *Curr. Anthropol.* 1, 195–232.

Hublin, J.-J. 1978. Le torus occipital transverse et les structures associées. Thesis, Université de Paris.

Hublin, J.J., Spoor, F., Braun, M., Zonnefeld, F., 1996. A late Neanderthal associated with Upper Paleolithic artifacts. *Nature* 381, 224–226.

Hublin, J.J., 1998. Climatic changes, paleogeography, and the evolution of Neandertals. In: Akazawa, T., Aoki, K., Bar-Yosef, O. (Eds.), *Neandertals and Modern Humans in Western Asia*. Plenum Press, New York, pp. 295–310.

Iacumin, P., Cominotto, D., Longinelli, A., 1996. A stable isotope study of mammal skeletal remains of mid-Pleistocene age, Arago Cave, eastern Pyrenees, France. Evidence of taphonomic and diagenetic effects. *Palaeogeogr. Palaeoclimatol. Palaeoecol.* 126, 151–160.

Jolly, C.J., 2001. A proper study for mankind: Analogies from papionin monkeys and their implications for human evolution. *Yrbk. Phys. Anthropol.* 44, 177–204.

Krings, M., Stone, A., Schmitz, R.W., Krainitzki, H., Pääbo, S., 1997. Neandertal DNA sequences and the origin of modern humans. *Cell* 90, 19–30.

Krings, M., Capelli, C., Tschentscher, F., Geisert, H., Meyer, S., von Haeseler, A., Grossschmidt, K., Possnert, G., Paunovic, M., Pääbo, S., 2000. A view of Neandertal genetic diversity. *Nat. Genet.* 26, 144–146.

Mallegni, F., Carnieri, E., Bisconti, M., Tartarelli, G., Ricci, S., Biddittu, I., Segre, A., 2003. *Homo cepranensis* sp. nov. and the evolution of African-European Middle Pleistocene hominids. *C. R. Palévol.* 2: 153–159.

Mania, D., 1983. *Homo erectus* von Bilzingsleben – Seine Kultur und Umwelt. *EAZ Ethnogr. Archäol. Z.* 34, 478–510.

McCown, T.D., Keith, A., 1939. *The Stone Age of Mount Carmel: The Fossil Remains from the Levalloiso-Mousterian*, Vol. 2. Clarendon Press, Oxford.

Morant, G.M., 1938. The form of the Swanscombe skull. *J. R. Anthropol. Soc.* 68, 67–97.

Ovchinnikov, I.V., Gotherstrom, A., Romanova, G.P., Kharitonov, V.M., Liden, K., Goodwin, W., 2000. Molecular analysis of Neanderthal DNA from the northern Caucasus. *Nature* 404, 490–493.

Rak, Y., Hylander, W.L., 2003. Neandertal facial morphology and increased jaw gape. *Am. J. Phys. Anthropol.* 120, (Suppl. 36,), 174.

Rightmire, G.P., 1990. *The Evolution of Homo erectus*. Cambridge University Press, Cambridge.

Santa Luca, A.P., 1978. A re-examination of presumed Neandertal-like fossils. *J. Hum. Evol.* 7, 619–636.

Sawyer, G.J., Maley, B., 2005. Neanderthal reconstructed. *Anat Rec.* (New Anat.) 283B, 23–31.

Schmitz, R.W., Serre, D., Bonani, G., Feine, S., Hillgruber, F., Krainitzki, H., Pääbo, S., Smith, F.H., 2002. The Neandertal type site revisited: Interdisciplinary Investigations of skeletal remains from the Neander Valley, Germany. *Proc. Natl. Acad. Sci. U.S.A.* 99, 13342–13347.

Schott, L., 1990. "*Homo erectus reilingensis*" – Anspruch und Wirklichkeit eines Schädelfundes. *Biol. Rundsch.* 28, 231–235.

Schwarcz, H.P., Latham, A.G., 1990. Absolute age determination of travertines from Vérteszöllös. In: Kretzoi, M., Dobosi, V. (Eds.), *Vérteszöllös: Site, Man and Culture*. Akadémiai Kiado, Budapest, pp.549–555.

Schwarcz, H.P., Latham, A.G., Mania, D., Brunnacker, K., 1988. The Bilzingsleben archaeological site: New dating evidence. Archaeometry 30, 5–17.

Schwartz, J.H., 1999. *Sudden Origins: Fossils, Genes, and the Emergence of Species*. Wiley, New York.

Schwartz, J.H., 2005. *The Red Ape: Orangutans and Human Origins*. Revised Edition. Westview Press, Boulder, CO.

Schwartz, J.H., Tattersall, I., 1996a. Toward distinguishing *Homo neanderthalensis* from *Homo sapiens*, and vice versa. *Anthropologie* (Brno) 34, 79–88.

Schwartz, J.H., Tattersall, I., 1996b. Significance of some previously unrecognized apomorphies in the nasal region of *Homo neanderthalensis. Proc. Natl Acad. Sci., U.S.A.* 93, 10852–10854.

Schwartz, J.H., Tattersall, I., 2000. What constitutes *Homo erectus*? *Acta Anthropol. Sinica* 19, Suppl., 18–22.

Schwartz, J.H., Tattersall, I., 2002. *The Human Fossil Record, Vol. 1: Terminology and Cranial Morphology of Genus Homo (Europe).* Wiley/Liss, New York.

Schwartz, J.H., Tattersall, I., 2005. *The Human Fossil Record, vol. 4: Craniodental Morphology of Early Hominids* (Genera *Australopithecus, Paranthropus, Orrorin*) *and Overview*. Wiley/Liss, New York.

Smith, F.H., Spencer, F. (Eds.), 1984. *The Origins of Modern Humans: A World Survey of the Fossil Evidence*. Alan R. Liss, New York.

Spoor, F., Hublin, J.J., Braun, M., Zonnefeld, F., 2003. The bony labyrinth of Neanderthals. *J. Hum. Evol.* 44, 141–165.

Straus, W.L., Cave, A.J.E., 1957. Pathology and the posture of Neanderthal man. *Quart. Rev. Biol.* 32, 348–363.

Stringer, C.B., 1989. A neglected Middle Pleistocene comparison for the Bilzingsleben hominid material. *EAZ Ethnogr.-Archäol. Z.* 30, 492–496.

Stringer, C. B., Gamble, C., 1993. *In Search of the Neanderthals.* Thames and Hudson, London.

Stringer, C.B., Hublin, J.-J., 1999. New age estimates for the Swanscombe hominid, and their significance for human evolution. *J. Hum. Evol.* 37, 873–877.

Stringer, C.B., McKie, R., 1996. *African Exodus: The Origins of Modern Humanity*. Henry Holt, New York.

Stringer, C.B., Howell, F.C., Melentis, J., 1979. The significance of the fossil hominid skull from Petralona, Greece. *J. Archaeol. Sci.* 6, 235–253.

Stringer, C.B., Hublin, J.-J., Vandermeersch, B., 1984. The origin of anatomically modern humans in western Europe. In: Smith, F., Spencer, F. (Eds.), *The Origins of Modern Humans: A World Survey of the Fossil Evidence.* Alan R. Liss, New York, pp. 51–135.

Tattersall, I., 1986. Species recognition in human paleontology. *J. Hum. Evol.* 15, 165–175.

Tattersall, I., 1994. How does evolution work? *Evol. Anthrop.* 3, 2–3.

Tattersall, I., 1995. *The Fossil Trail: How We Know What We Think We Know About Human Evolution.* Oxford University Press, New York.

Tattersall, I., Eldredge, N., 1977. Fact, theory and fantasy in human paleontology. *Am. Sci.* 65, 204–211.

Tattersall, I., Schwartz, J.H., 2000. *Extinct Humans.* Westview Press, Boulder, CO.

Trinkaus, E., 1973. A reconsideration of the Fontéchevade fossils. *Am. J. Phys. Anthropol.* 39, 25–36.

Trinkaus, E., 1983. *The Shanidar Neandertals.* Academic Press, New York.

Vandermeersch, B., 1981. *Les Hommes Fossiles de Qafzeh (Israel).* CNRS, Paris.

Vallois, H., 1949. L'origine de *l'Homo sapiens. C. R. Acad. Sci. Paris* 228, 949–951.

Vértes, L., 1965. Discovery of *Homo erectus* in Hungary. *Antiquity* 39, 303.

Vlček, E., 1993. *Fossile Menschenfunde von Weimar-Ehringsdorf.* Konrad Theiss Verlag, Stuttgart.

Vlček, E., Mania D., 1987. *Homo erectus* from Bilzingsleben (GDR) – His culture and environment. *Anthropologie* 25, 1–45.

Wolpoff, M., 1980. *Paleoanthropology*. Knopf, New York.

Wolpoff, M., Thorne, A. G., Jelinek, J.,Yinyun, Z., 1994. The case for sinking *Homo erectus*: 100 years of *Pithecanthropus* is enough! *Cour. Forschinst. Senckenberg* 171, 341–361.

Ziegler, B., Dean, D., 1998. Mammalian fauna and biostratigraphy of the pre-Neanderthal site of Reilingen, Germany. *J. Hum. Evol.* 34, 469–484.

3. Saccopastore 1: the earliest Neanderthal? A new look at an old cranium

E. BRUNER
Università "La Sapienza"
Dipartimento di Biologia Animale e dell'Uomo
P.le Aldo Moro 5, 00185 Roma, Italy
emiliano.bruner@uniroma1.it

G. MANZI
Università "La Sapienza"
Dipartimento di Biologia Animale e dell'Uomo
P.le Aldo Moro 5, 00185 Roma, Italy
giorgio.manzi@uniroma1.it

Keywords: Neanderthal, Accretion model, Cranial morphology, OIS 5, Saccopastore, Italy

Abstract

The fossil cranium known as Saccopastore 1 was recovered in 1929 near Rome (Italy) in a gravel quarry that has been later replaced by building areas within the city. Its Neanderthal morphology was established early on, and detailed successive works described a combination of features in which traits that are recurrent among Würmian Neanderthals blend with those shared by Middle Pleistocene hominines. Recent computed analyses allowed the study of the endocranial structures through tomographic and digital approaches, and improved the ectocranial comparisons by using landmark-based multivariate techniques. This paper is aimed at synthesizing and describing the current information available about the Saccopastore 1 cranial morphology, through an integration of past and present data. This specimen represents the best-preserved and most complete cranium in Europe among those dated to OIS 5. Its recognized Neanderthal identity suggests that the impact of the preceding cold stage (OIS 6, around 200–130 ka) was probably decisive in the definition of the Neanderthal phenotype, modifying the extent of genetic variation of previous European populations toward a more homogeneous gene pool.

Saccopastore, Ever Since 1929

A gravel quarry was active during the late 1920s in a locality known as Saccopastore, which was just outside Rome at the time, since the city's great expansion took place only in the last fifty years. The quarry was delimited by a meander of the river Aniene (tributary of the Tevere), about 3.5 km from the northeastern border of the city. It was removing gravels and sands pertaining to the lower and more recent terrace of a fluvial valley.

In April 1929, the workmen of the quarry found a fossilized human cranium (SCP1; Figure 1A) that was immediately removed from the deposit. The specimen was then

K. Harvati and T. Harrison (eds.), Neanderthals Revisited: New Approaches and Perspectives, 23–36.

delivered to the anthropologist Sergio Sergi (Figure 1B), who recognized its Neanderthal features (Sergi, 1929) and started a long-term series of studies (e.g., Sergi, 1934, 1944, 1947, 1948a, b, 1962). Another, less complete cranium of Neanderthal morphology (SCP2) was later discovered in the same locality by A.C. Blanc and H. Breuil, while they were visiting the abandoned quarry in the summer of 1935 (Breuil & Blanc, 1936; Blanc, 1948; Sergi, 1948c). During the following year (1936), a brief excavation campaign was carried out by the Istituto Italiano di Paleontologia Umana. Faunal and plant fossil remains were added to the material previously collected by the workmen, and a few flake instruments of Mousterian typology were also discovered (Blanc, 1948). Moreover, the stratigraphy of the area was studied in further detail (e.g., Blanc, 1948, 1957; Segre 1948), enriching the initial observations made by Koppel (1934) and Sergi (1935).

According to the stratigraphic profile furnished by A.G. Segre (1983) and reported in Figure 1C, the two human specimens appear embedded in the same stratigraphic span of gravels, which belong to the last interglacial cycle. An alluvial lens of mud, including terrestrial mollusks, separates the two crania, but it simply represents the occurrence of a rapid fluvial event interposed in the sequence of gravels. Consequently, the two crania should date to the Eemian and, more precisely, to one of the warmest phases of oxygen isotope stage 5 (OIS 5), most probably between the sub-chrons 5e and 5c (Caloi et al., 1998). This corresponds to a chronology of about 100–130 ka.

Despite some visible damage, SCP1 is a remarkably well-preserved fossil specimen. The cranium lacks the mandible and both the zygomatic processes. Many maxillary teeth and alveolar structures are broken, but both molar series have been preserved, with the exception of the crown of the right M1. The

Figure 1. Saccopastore: a) fossil cranium found in 1929; b) the site visited by S. Sergi (first on the left) after the discovery of the specimen; c) stratigraphic profile based on that elaborated by A.G. Segre (1983), where the levels referred to the OIS 5 are evidenced. These include (from above): level 3a, cemented gravels with large mammal remains; level 3b, slime lens with terrestrial mollusks; level 2a, fine gravels with faunal assemblage similar to that in 3a (except for the presence of *Palaeoloxodon antiquus*). The arrows indicate the position of the Neanderthal crania, labeled SCP1 and SCP2, discovered in 1929 and in 1935 respectively.

supraorbital region appears severely damaged, and two holes (each of about 6 cm^2) were made by the discoverers in different positions along the coronal suture. The endocranial cavity and other internal volumes are still partially filled by a concrete-like geological matrix. The cranium is markedly platycephalic, gently curved along its midsagittal outline and with an elliptical (or *en bombe*) profile in posterior view. The face is rather large and orthognathic, with a pronounced alveolar height and a combination of features related to a clear, although not particularly pronounced, midfacial prognathism. The pyriform aperture is wide and the orbits are large and circular. The high, broad, and rectangular nasal bones show a regular, but deep concavity in transverse section. The palate is narrow and extremely high, with a small palato-dental area and relatively small teeth. The cusps are severely worn according to the helicoidal pattern of attrition, and the underlying dentine is exposed.

This paper includes a brief review of the features displayed by SCP1. We will consider either features detectable by visual inspection or those revealed through the application of computer-assisted imaging techniques. Our aim is to evaluate whether or not SCP1 may be considered one of the most appropriate test cases for the "accretion model" of human evolution in Europe, viewed as a bioecological explanation for the establishment of the Neanderthal morphology (e.g., Dean et al., 1998; Hublin, 2000). Because of its ectocranial anatomy, the phenetic position of SCP1 is well established in the literature, with respect to the European fossil record at the boundary between Middle and Late Pleistocene. It shows a combination of features in which traits that are recurrent among Würmian Neanderthals blend with those shared by Middle Pleistocene hominines (e.g., Sergi, 1944; Condemi, 1992). Recent studies on endocranial morphology (Bruner et al., 2003; Bruner, 2004a) stress this conclusion further, and call for a re-evaluation of the Italian fossil specimen based upon both traditional and computerized data, with emphasis on the latter.

The Tomographic Approach

SCP1 was CT-scanned in 2001 using a Phillips Tomoscan AUEP machine (Manzi et al., 2001). Sequential and contiguous 1 mm scans were taken through transverse planes, aligned according to the Frankfurt horizontal. The scanner was set at 75 mA and 140 kV, with a FOV of 250 mm and a pixel size of 0.49 mm. Some streak artifacts occurred because of the high level of fossilization and the inclusion of stone matrices. It was, therefore, necessary to use a filter to clean the signal. Data were analyzed using Mimics 7.0.

The attenuation spectrum (Figure 2) shows three main components, with two peaks or phases. The low-density peak represents the fossilized bone matrix, mostly the neurocranial structures. The high-density phase represents the geological hard infiltrations in the endocranial cavity and frontal diploe, as well as a large part of the maxillary structures. These two components are well-separated densitometrically, although a certain amount of overlap hampers a complete distinction between the bone and the geological sediments. A very low-density component represents the tail of the spectrum, including the partial volume effect (see Spoor et al., 2000) and light sediment permeating many scattered and deep volumes, particularly localized within the right maxillary sinus. The basal structures and the petrous pyramids are the more permeated volumes, and their boundaries are hardly recognizable. Nevertheless, most of the internal structures (e.g., inner ears, dental roots, paranasal sinuses, endocranial morphology) can be reconstructed almost entirely by using thresholding, semi-automatic segmentation, and digital replicas. Some lines of fractures, visible in part on the ectocranial

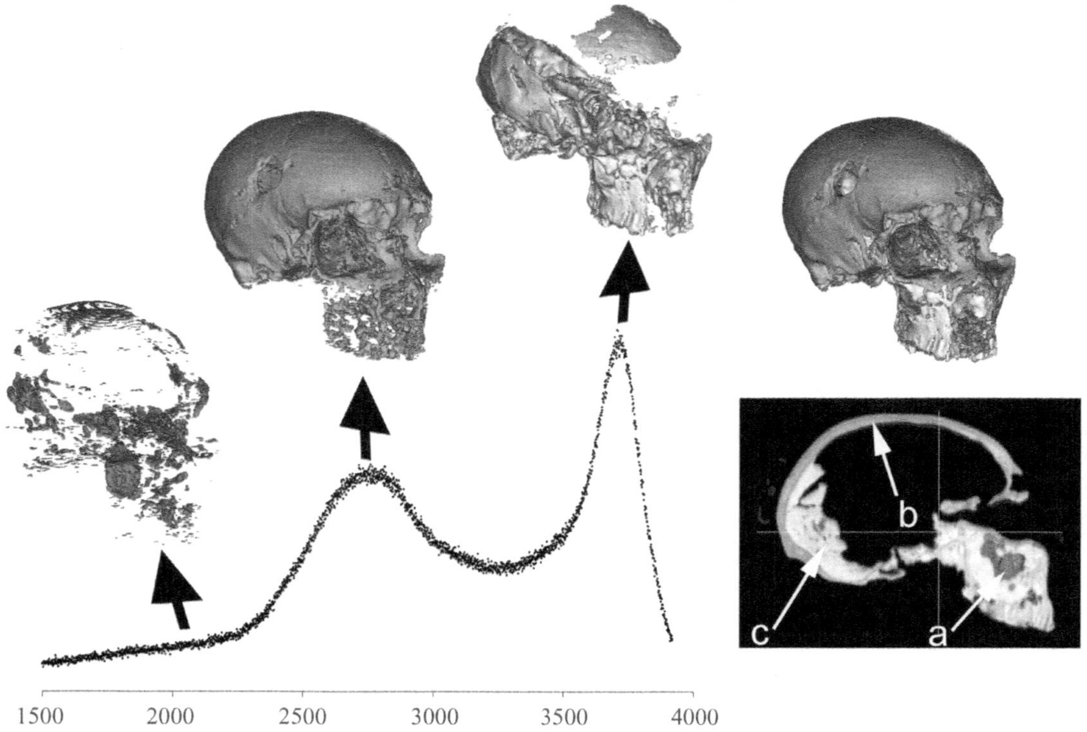

Figure 2. Attenuation spectrum of SCP1 showing two main phases. A very low-density component (a) is associated with light geological infiltration, clearly recognized within the right maxillary sinus. The low-density phase (b) represents mainly the fossil bone, mostly the neurocranial volume. The high-density phase includes highly mineralized tissues (mostly the maxillary volumes) and hard geological matrix (partially filling the endocranial cavity and the frontal diploic space).

surface, are observable within the fossil volume.

A Glimpse at the (ecto)Cranium

From an ectocranial perspective, SCP1 displays a clear Neanderthal appearance, including most of the derived traits which characterize this human group, despite the occurrence of a number of plesiomorphic features and the weak expression of some derived ones. About 25 years ago, SCP1 was restudied in detail after the discovery of new fossil evidence in Europe and in the Levant (Condemi, 1985, 1988, 1992; for other recent overviews see Stringer et al., 1984; Ascenzi et al., 1986; Manzi and Passarello, 1991; Schwartz and Tattersall, 2002; Manzi, 2004). Similar considerations are also pertinent to the second specimen from Saccopastore. The morphology of the two crania from Rome – together with those of other penecontemporaneous European fossil hominids – is the result of a combination of features encountered among the so-called "anteneanderthals" (or *Homo heidelbergensis*) with those belonging to the European Würmian Neanderthals (or *Homo neanderthalensis*).

Thus, the Saccopastore crania may represent a significant phenetic connection between two chronospecies along the same evolutionary continuum. This "intermediate" phenotype – already described by Sergi (e.g., Sergi 1962) – appears consistent with the chronology of the Italian specimens, and fits an anagenetic perspective of human evolution in Europe, that is commonly referred to as the "accretion model" (e.g., Dean et al., 1998; Hublin, 2000).

Partly following Condemi (1988, 1992), typical Neanderthal features in SCP1 are the shape of the parietals, the *en bombe* profile of the braincase in rear view, and the "double-arched" occipital torus (i.e., depressed in correspondence of the midsagittal plane). Some features, instead, are less derived: the flat and irregular suprainiac fossa, the absence of a clear *chignon* replaced by a rounded parieto-occipital midsagittal contour (but see Manzi, 1991a, b, for an alternative explanation which takes into account the occurrence of several supernumerary ossicles at lambda), the moderate midfacial projection with partially developed Neanderthal shape of the zygomatic bones, and the presence of weak concavities on the infraorbital surface of the maxilla. By contrast, definitely archaic features in SCP1 are the small dimensions, the rather developed mastoid processes (projecting more inferiorly than the occipitomastoid crests), some traits of the temporal bone (including the shape of the glenoid cavity), a *crista occipitalis externa* well-marked throughout its length, and the "*erectus*-like" appearance of the greater wings of the sphenoid.

No evidence of *crista turbinalis, crista spinalis, crista lateralis*, or nasal medial projection (see Franciscus, 1999, and references therein) can be detected in the nasal aperture, neither by visual inspection nor in the CT data. Nonetheless, it must be considered that the nasal volumes in SCP1 are largely damaged and that the geological infiltrations have thus far hampered a useful densitometric resolution between fossil and geological matrices in this anatomical region.

As noted by some authors (e.g., Condemi 1988; Rak, 1993; but see Trinkaus, 1991), this combination of features also suggests affinities between the interglacial specimens from Saccopastore and the Neanderthals from South-Western Asia – like Tabun C, Amud 1, or those from Shanidar – pointing out the key role of the paleo-biogeographic relationships between Europe and the Levant.

Landmark-based approaches using coordinates superimposition, multivariate statistics, and thin-plate spline interpolation grids (e.g., Rohlf and Marcus, 1993; Adams et al., 2004) allowed further considerations of the SCP1 cranial morphology. As far as the shape of the midsagittal vault profile (from *nasion* to *inion*) is considered, SCP1 fits the Neanderthal variation, characterized by flat parietals and enlargement of the supraorbital structures (Manzi et al., 2000). Nevertheless, this geometric morphometrics analysis indicates a peculiar platycephalic condition of SCP1 when compared to other Neanderthals, and a generalized morphological affinity with Tabun 1.

A similar shape analysis was performed along the entire midsagittal profile from prosthion to inion, using a sample of nineteen specimens from different taxa of the genus *Homo* (Bruner et al., 2004). In this analysis, SCP1 clusters with the configuration of Guattari 1 (GTT), because of the similar degree of vault flattening, reduced alveolar prognathism, and increased midfacial projection. Such a phenotypic affinity supports the interpretation of SCP1 as a small-brained Neanderthal. Taking into account the main shape variation and the values of cranial capacity, the same analysis shows that the position of "cranium 5" from Sima de los Huesos (SH5) is also rather comparable with the configuration in SCP1, suggesting a certain similarity with the European Middle Pleistocene populations as well.

In this light, the midsagittal profiles of SCP1, SH5, and Guattari, are compared directly in Figure 3 using a baseline from nasion to inion, in order to separate the neurocranium from the facial component of the profile. Superimposition and distortion grids were carried out using Morpheus et al. (Slice, 2000). Compared to the Middle Pleistocene SH5, SCP1 shows reduction of the alveolar prognathism, a similar frontal profile, and the

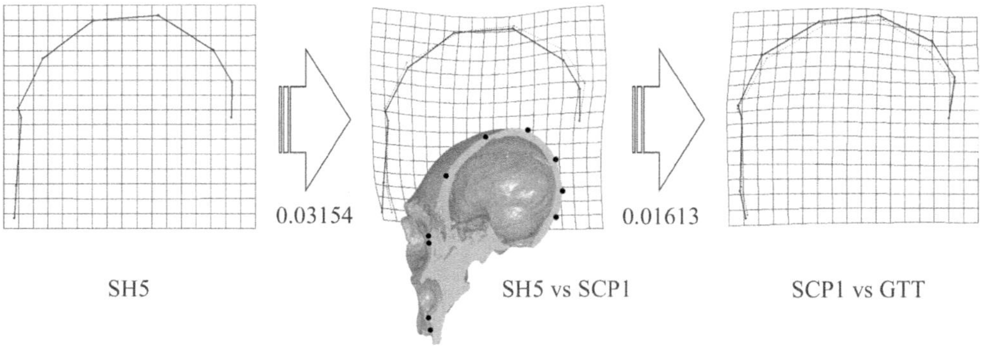

Figure 3. The midsagittal cranial profile of Sima de los Huesos cranium 5 (SH5) is superimposed on the SCP1 configuration through a nasion-inion baseline, which in turn is superimposed on the Guattari profile (GTT). Pairwise differences are visualized using thin-plate spline and interpolation grids (solid links: reference shape: dotted links: target shape), showing the respective bending energy for each comparison.

development of the parieto-occipital areas. By contrast, SCP1 and GTT display very similar configurations, with minor differences related to an even lower vault in the Würmian specimen, associated with lambdoid flattening and slight facial reduction. The bending energy – computed after Procrustes superimposition by tpsSpline 1.15 (Rohlf, 1997) – quantifies the magnitude of the distortion, revealing a phenetic distance between SH5 and SCP1 that is twice the value of that separating the two Italian Late Pleistocene specimens.

Endocranial Morphology

ENDOCAST

Because of the marked differences in the densitometric properties between the neurocranial surfaces and the geological infiltrations, the digital endocast was well reproducible by segmentation (Bruner et al., 2002). The basal surfaces and the temporal pyramids are the only poorly resolved structures, in which the boundaries between fossil bone and geological inclusions are difficult to discern. Figures 4–6 show the digital replicas of the endocast and some informative tomographic sections.

The frontal circumvolutions are clearly outlined and visible, with the *pars opercularis* and *triangularis* more defined on the left side. The right frontal lobe is slightly wider than the left one. The frontal lobes lie almost entirely on the orbital roof, without the posterior displacement described for some robust Early-to-Middle Pleistocene specimens, such as Ceprano, Kabwe, and Petralona (Seidler et al., 1997; Bruner and Manzi, 2005). The *crista galli* is pneumatized through the ethmoidal cells and tilted backward because of a hypothesized backward tension of the *falx cerebri*, associated with platycephaly and the encephalisation process *(Unpublished data)*. The cribriform plate is rather open onto the anterior fossa, namely wide and not constricted between the orbital margins, with a gradual slope between the lateral walls of the pit and the orbital roof (*type 2* morphology *sensu* Moss, 1963). The moderate right frontal petalia is associated with a marked left occipital petalia. The supramarginal area is better shaped on the right side, while the parietal volume is larger on the opposite side. The maximum endocranial width lies at the parieto-temporal boundary, rather lower on the left hemisphere. The parietal asymmetry observed on the endocast matches the ectocranial asymmetry (Sergi, 1944). The upper

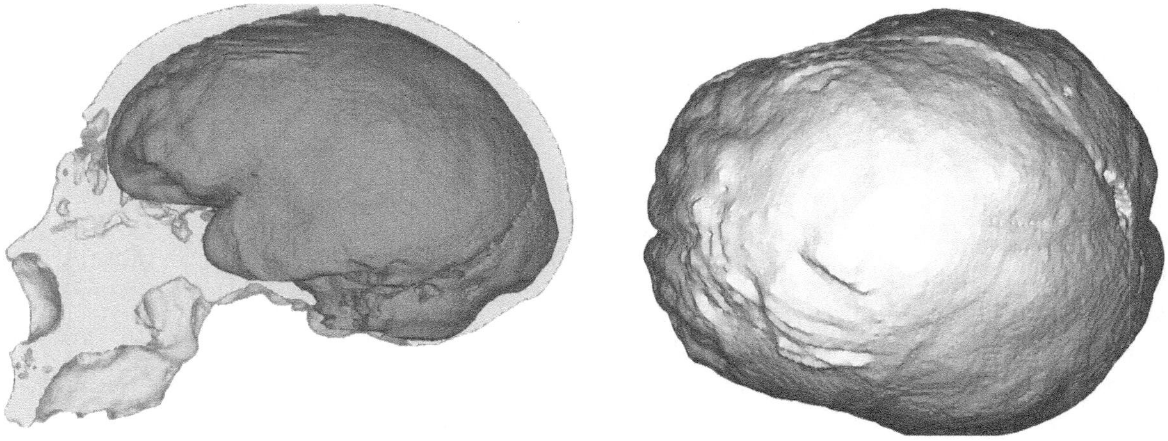

Figure 4. The digital replica of the SCP1 endocast. The sectioned cranium (left) shows the relationship between the endocast and the ectocranial structures. The dorsal view of the endocast (right) shows the well-shaped pre-frontal circumvolutions, a minor right-frontal petalia (the right lobe is slightly wider), a marked posterior asymmetry (the occipital, temporal, and parietal areas are more enlarged on the left side), and the absence of a marked narrowing of the frontal lobes when compared with endocranial maximum width.

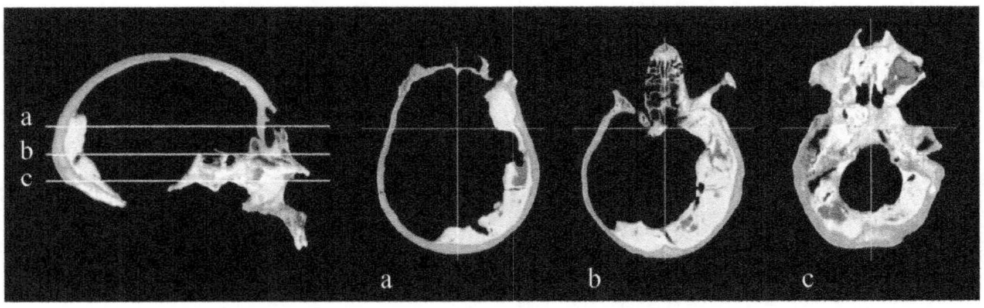

Figure 5. The midsagittal reformatted CT slice from SCP1 shows the localization of the three transverse sections at the level of the supraorbital structures (a), orbits (b), and maxillary sinuses (c). The densitometric differences between the fossil matrix (light gray) and the geological inclusions (white) are clearly visible in the endocranial posterior-upper regions, particularly on the right side, but are less defined toward the cranial base (including the temporal pyramids, the sphenoid, and the lower facial volumes); a low-density infiltration (dark gray) can also be recognized within the right maxillary sinus.

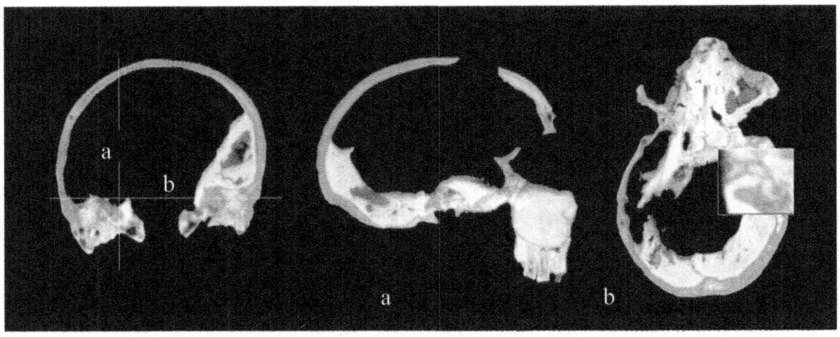

Figure 6. Even though a large percentage of the lower volumes are deeply melted with the geological infiltrations, several fossil surfaces are distinguishable, like the left maxillary sinus and the dental roots (a), or the right vestibular structures (b, magnified).

parietal circumvolutions are enlarged (Bruner et al., 2003), involving the *en bombe* profile of the Neanderthals from an endocranial perspective in posterior view. The occipital lobes are not projecting backward as in many Asian *Homo erectus* specimens. The cerebellar lobes are rather globular, almost in contact behind the brainstem. The superior sagittal sinus runs into the right transverse-sigmoid system. There are no traces of the occipito-marginal vessels. The occurrence of a spheno-parietal sinus (showing a high prevalence in the Neanderthal endocasts; Grimaud-Hervé, 1997) can be hypothesized on the left hemisphere, but its trace is not clearly defined. Imprints of the middle meningeal vessels are visible on the right side, showing a very simple network, and an obelic branch derived by the posterior ramus. A similar pattern is more frequently described in less encephalized taxa of the genus *Homo* (Saban, 1995; Grimaud-Hervé, 1997).

The main endocranial diameters are comparable to other small-brained Middle Pleistocene samples, but the general proportions show a phenetic affinity with the Neanderthals (Bruner et al., 2003; Bruner, 2004a). In particular, despite the small cranial capacity and the very low upper vault, the frontal lobes are rather lengthened antero-posteriorly, while the parietal area is short and flattened. The frontal lobes do not show a marked lateral narrowing when compared with the endocranial maximum width. Considering the general brain shape through landmarks configuration and superimposition, the similarity with the Neanderthal pattern is evident once again, despite the generalized endocranial flattening.

The cranial capacity of SCP1 was accurately estimated by Sergi at 1174 cc (Sergi, 1944). This figure is rather low considering the "classic" Neanderthal values, averaging around 1450 ml. It might be considered an individual trait, thus including sexual and idiosyncratic variants. However, when chronological and allometric factors are also taken into account the possibility of clearly identifying the Neanderthal cranial structure in such a small-brained skull is worth considering.

PNEUMATISATION

SCP1 shows large pneumatic volumes developed within the inner bony layers (e.g., Passarello and Diotallevi, 1982 and references therein). The maxillary sinuses are large, permeated by geological matrices, and easily recognizable in the tomographic sections (Figures 5C–6A). The development of the maxillary sinuses contributes to the ectocranial oncognathic morphology, although it is less evident than in the Würmian Neanderthals (Sergi, 1948b). The frontal sinuses in SCP1 can only be estimated by examining their preserved posterior boundaries, which suggest wide structures. In contrast, the frontal sinuses in the other specimen from Saccopastore (SCP2) are distinct but smaller (Manzi et al., 2001). In both cases, these structures do not develop within the frontal squama, as described for some Middle Pleistocene specimens, like Petralona or Kabwe (Seidler et al., 1997). The pneumatization of the mastoids and temporal pyramids is difficult to assess, because of the admixture with the geological matrices. The ethmoidal cells are large and well developed (Figure 5B).

INNER EAR

Although the present tomographic analysis has a limited resolution, the vestibular structures of the inner ear can be localized and recognized. The left labyrinth is less visible than the right one, because of the extreme permeation and blending of fossil and geological matrices. The boundaries are rather unclear, mostly at the anterior canal. In contrast, the right canals are more complete and better preserved (Figure 6B). A tentative description of this structure is attempted

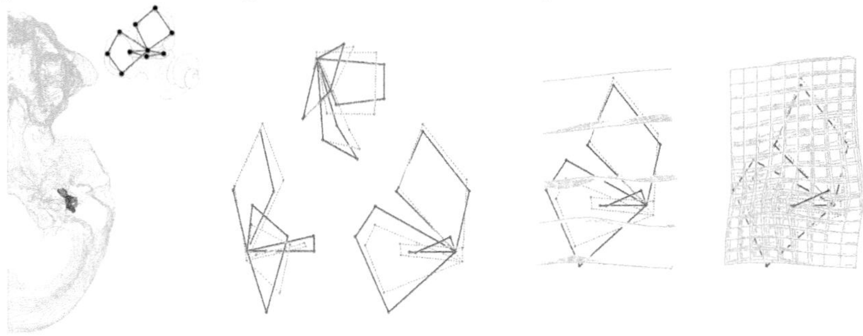

Figure 7. a) Localization of the right inner ear within the skull model using polylines; b) shape comparison between the vestibular structures in SCP1 (thinner line) and in the early Neolithic skull from Fonterossi (Abruzzo, Italy) after Procrustes superimposition, in superior (top) frontal (left) and lateral (right) views; c) grid deformations of the same comparison in lateral view using transverse and sagittal planes.

through a landmark-based direct geometrical comparison with the Early Neolithic anatomically modern specimen from Fonterossi (Bruner and Manzi, 2003). Using three-dimensional coordinates from the center of the lumen at the maximum width and maximum height of each canal (Spoor and Zonneveld, 1995, 1998), the configurations displayed by these two specimens were compared through Procrustes superimposition (Bookstein, 1991). SCP1 has a smaller posterior canal (mostly due to reduction of the upper part) and a wider transverse one. The posterior canal is rather low relatively to the plane of the lateral canal, which in turn undergoes a lateral stretching (Figure 7). Although very preliminary and exploratory because of the resolution of the tomographic data, these results are in agreement with the suggested autapomorphic morphology of the inner ear among Neanderthals (e.g., Spoor et al., 2003).

Epilogue: A Small-brained and Low-vaulted Neanderthal

The importance of SCP1 is emphasized by the fact it is the best-preserved and most complete single specimen in Europe among those referred to OIS 5. The penecontemporaneous fossil evidence is rather fragmentary, including: the two portions of cranial vault from Fontéchevade (France); the calvaria and other cranial portions from La Chaise Bourgeois-Delaunay (France); the fragmentary braincase found at Ehringsdorf (Germany); and the large but fragmentary sample from Krapina (Croatia). At the same time, no dating is available for the Apidima specimens (Greece) and the chronology of the cranium unearthed at Forbes' Quarry, Gibraltar, in 1848 – usually considered in the past of comparable age to Saccopastore (i.e., beginning of the Würm glaciation) – remains unclear.

The morphological continuum observed among European populations of the Middle and Late Pleistocene – except for the arrival of anatomically modern immigrants (e.g., Stringer, 2002) – is presently interpreted as a gradual shifting of characters within a single evolving lineage, or paleospecies, from more plesiomorphic to more derived (i.e., Neanderthal) morphologies (e.g., Hublin, 1998). Two chronospecies of this same lineage – namely *Homo heidelbergensis* (Schoetensack, 1908) and *Homo neanderthalensis* (King, 1864) – are thus considered as linked by a progressive and gradual evolution (e.g., Arsuaga et al., 1997). The "accretion model" in particular, introduced a four-step scheme to explain the Neanderthal evolution. (Dean et al., 1998; Hublin, 1998, 2000; Stringer and Hublin, 1999; but compare Hawks & Wolpoff, 2001). These steps, or progressive stages, are (according to Dean et al., 1998): "early pre-Neanderthals" (before OIS 12, i.e. before 450 Ka), "pre-Neanderthals" (OIS 11–9, roughly between 400 and 300 ka), "early Neanderthals" (OIS 7–5, between 250 and 80 ka), and "classic Neanderthals" (OIS 4–3, from 80 to 30 ka).

As noted by various authors, the prolonged anastadial-catastadial oscillation cycles, whose environmental effects were particularly dramatic in Europe during the last half million years (Figure 8), may have played a crucial role in this process. Particularly, it seems reasonable that the recurrent climatic crises (i.e., severely cold stages) may have been capable of marginalizing human populations by forcing them toward the south in Mediterranean environments, thus producing contractions in the population size over the continent, as well as repeated reductions of their morphological intra-group variability and inter-group diversity. As Hublin (1998: 307) pointed out, "such events would have produced genetic drift episodes resulting in the fixation of derived features [that] could have developed although they do not have a clear adaptive significance". In this perspective, it is probable that the impact of OIS 6 (around 200–150 ka) contributed to the definition of the phenotype that we refer to as that of the Neanderthals definitively (e.g., Stringer et al., 1984), thus modifying the extent of genetic variation toward a more homogeneous gene pool. European populations of the subsequent OIS 5 may be the earliest result of such a particularly strong bottleneck along the phenetic and phyletic lineage *Homo heidelbergensis-neanderthalensis*.

Some additional insight on this issue may be furnished by the retrieval and analysis of mitochondrial DNA (mtDNA) sequences from Neanderthal specimens (Krings et al., 1997, 2000; Ovchinnikov et al., 2000; Schmitz et al., 2002; Knight, 2003). Given that "the extent of Neandertal mtDNA diversity was similar to that of current humans" (Serre et al., 2004), it is reasonable to infer that the coalescence of the Neanderthal gene pool falls in a period close to 200 ka (e.g., Krings et al., 2000), and therefore in correspondence to the transition between OIS7 and OIS6. Hence, the genetic data are not in contrast with the above hypothesis, suggesting that the populations that were able to pass through the bottleneck caused by the climatic crisis of the OIS6 were those from which *Homo neanderthalensis* originated.

Viewed in this light, SCP1 – complemented by the other, less complete cranium from Saccopastore – has both the chronology and the appropriate ecto-endocranial morphology, as well as a number of discrete features, appropriate to be considered the best available specimen representing the earliest record of the full Neanderthal identity. Since the seminal work by Simpson (1944), species should be considered as a group of populations with a shared evolutionary history; in this light, a possible corollary may be that an evolutionary history in common is but the consequence of a shared biological model (for further discussion see Bruner, 2004b). In this framework, the Neanderthal cranial morphology must be viewed, not in terms of single traits or localized anatomical variants, but as integrated structural systems, characterized by functional relationships and biomechanical constraints (see Rosas et al., this volume). In this context, we observe that the cranial morphology of SCP1 fits the Neanderthal cranial model in terms of structural organization.

As already noted in the past, the shape of SCP1 shows the clearest phenetic affinity with the so-called "classic" Neanderthals, which is greater than that shown by other European specimens of the Middle Pleistocene. This is

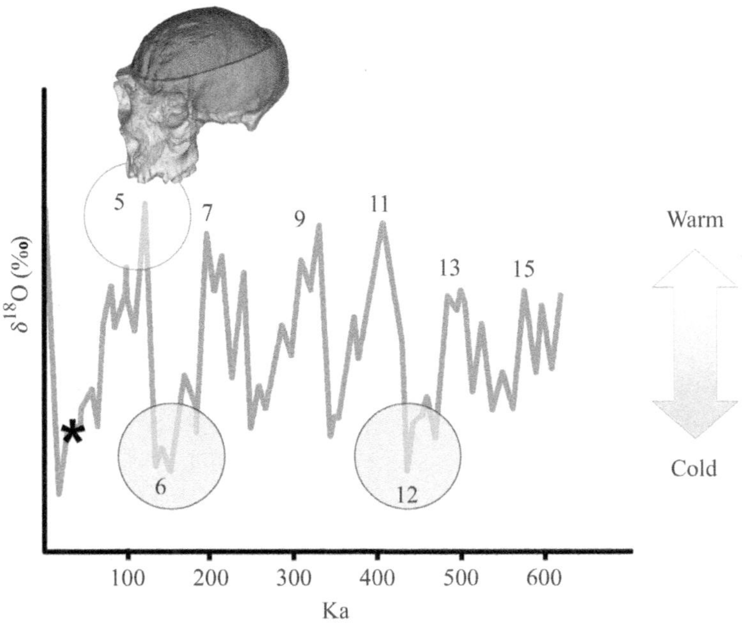

Figure 8. Schematic reproduction of the δ oxygen-18 record, based on deep-sea sediments (after Shackleton, 1967), describing the sequence of oxygen-isotope stages (OIS) in the last 600 ka. Clearly, the even-numbered OIS 12 and 6 were the most severe cold stages before the appearance of the European Würmian Neanderthals. The star marks the time period when *H. neanderthalensis* became extinct.

owing to some ectocranial features (e.g., Sergi, 1944; Condemi, 1992), despite the small size (including the small cranial capacity) and the platycephalic appearance of the cranial vault. In addition, although the endocranial variations might not match the ectocranial changes (Bookstein et al., 1999; Schäfer et al., 2001), what is visible ectocranially in SCP1 is even more clearly perceived by looking at the endocast, thus furnishing a new and more comprehensive perspective of analysis. In fact, the application of computer-assisted imaging techniques, in combination with geometric morphometric analyses, helps in considering the shape of the cranium as a complex and integrated system, far better than just looking at features that are detectable from a visual inspection. This brings to mind the words of Sergi (1934: 107): "La posizione delle ossa del cranio è il risultato di azioni esercitate su di esse durante la evoluzione e lo sviluppo. Queste azioni sono rappresentate da forze di intensità e direzione diverse, che conducono all'equilibrio definitivo degli elementi ossei. La differente posizione, pertanto, di ogni osso nei diversi tipi architetturali si può concepire come la posizione di equilibrio definitiva raggiunta, in modo diverso, per ogni tipo." [The position of the cranial bones is the result of forces acting upon them during evolution and development. These forces have different intensity and direction, which guide to the ultimate equilibrium of the bone elements. Thus, the different position of each bone in the various architectural types may be viewed as a final stage of equilibrium, reached in distinct manners by the diverse types].

Acknowledgements

This paper is dedicated to Pietro Passarello, on the occasion of his retirement from active teaching. We are grateful to Terry Harrison and Katerina Harvati – organizers, chairs, and hosts in New York – for their kind invitation to the memorable meeting of January 2005: *Neanderthals Revisited: New Approaches and Perspectives*. We are also indebted to a number of people for assistance and/or discussions, especially Gianfranco Gualdi and his staff at the Hospital Pio XI in Rome. Constructive criticisms and suggestions by Katerina Harvati and two anonymous reviewers certainly contributed to the improvement of the paper.

References

Adams, D. C., Rohlf, F. J., Slice, D. E., 2004. Geometric morphometrics: Ten years of progress following the "Revolution". *It. J. Zool.* 71, 5–16.

Arsuaga, J.L., Martínez, I., Gracia, A., Lorenzo, C., 1997. The Sima de los Huesos crania Sierra de Atapuerca, Spain. A comparative study. *J. Hum. Evol.* 33, 219–281.

Ascenzi, A., Marchetti, A.M., Micheli, M., 1986. Comparaison entre l'homme de Tautavel, les Anténéandertaliens d'Italie et l'homme de Saccopastore. *L'Anthropologie* 90, 515–537.

Blanc, A.C., 1948. Notizie sui ritrovamenti e sul giacimento di Saccopastore e sulla sua posizione nel Pleistocene laziale. *Paleontogr. Itali.* 42, 3–23.

Blanc, A.C., 1957. On the Pleistocene sequence of Rome: Palaeoecology and archaeologic correlations. *Quaternaria* 4, 95–109.

Bookstein, F.L., 1991. *Morphometric Tools for Landmark Data*. Cambridge University Press, Cambridge.

Bookstein, F., Schafer, K., Prossinger, H., Seidler, H., Fieder, M., Stringer, C., Weber, G.W., Arsuaga, J.L., Slice, D.E., Rohlf, F.J., Recheis, W., Mariam, A.J., Marcus, L.F., 1999. Comparing frontal cranial profiles in archaic and modern *Homo* by morphometric analysis. *Anat. Rec.* 257, 217–224.

Breuil, H., Blanc, A.C., 1936. Le nouveau crane de Saccopastore, Rome. *L'Anthropologie* 46, 1–16.

Bruner, E., 2004a. Geometric morphometrics and paleoneurology: Brain shape evolution in the genus *Homo*. *J. Hum. Evol.* 47, 279–303.

Bruner, E., 2004b. Evolution, actuality and species concept: A need for a paleontological tool. *Hum. Evol.* 19, 93–112.

Bruner, E., Manzi, G., 2003. Towards a re-appraisal of the Early Neolithic skeleton from Lama dei Peligni Abruzzo, Italy. Computed tomography and 3D reconstruction of the cranium. *Riv. Antropol.* 81, 69–78.

Bruner, E., Manzi, G., 2005. CT-Based description and phyletic evaluation of the archaic human calvarium from Ceprano, Italy. *Anat. Rec.* 285A; 643–658.

Bruner, E., Manzi, G., Passarello, P., 2002. The "virtual" endocast of Saccopastore 1. General morphology and preliminary comparisons by geometric morphometrics. In: Mafart, B., Delingette, H. (Eds.), *Three-Dimensional Imaging in Paleoanthropology and Prehistoric Archaeology*. British Archeol. Records IS 1049, Archeopress, Oxford, pp. 17–24.

Bruner, E., Manzi, G., Arsuaga, J.L., 2003. Encephalization and allometric trajectories in the genus *Homo*: evidence from the Neandertal and modern lineages. *Proc. Natl. Acad. Sci. U.S.A.* 100, 15335–15340.

Bruner, E., Saracino, B., Passarello, P., Ricci, F., Tafuri, M., Manzi, G., 2004. Midsagittal cranial shape variation in the genus *Homo* by geometric morphometrics. *Coll. Antropol.* 28, 99–112.

Caloi, L., Manzi, G., Palombo, M.R., 1998. Saccopastore, a stage-5-site within the city of Rome. In: The Eemian-Local Sequences, Global Perspectives. Symposium INQUA-SEQS'98, Kerkrade (abstract).

Condemi, S., 1985. Les Hommes Fossiles de Saccopastore (Italie) et leurs Relation Phylogénétiques. Ph.D. Dissertation, Université de Bordeaux I, Bordeaux.

Condemi, S., 1988. A review and analysis of the Riss-Würm Saccopastore skulls. Can they provide evidence in regard to the origin of Near Eastern Neanderthals? In: Trinkaus, E. (Ed.), *L'Homme de Neandertal*, vol. 3: L'Anatomie. Etudes Rech. Archéol. Univ. Liège 30, Liège, pp. 39–48.

Condemi, S., 1992. *Les Hommes Fossiles de Saccopastore et leur Relations Phylogénétiques*. CNRS, Paris.

Dean, D., Hublin, J.J., Holloway, R., Ziegler, R., 1998. On the phylogenetic position of the pre-Neandertal specimen from Reilingen, Germany. *J. Hum. Evol.* 34, 485–508.

Franciscus, R.G., 1999. Neandertal nasal structures and upper respiratory tract "specialization". *Proc. Natl. Acad. Sci. U.S.A.* 96, 1805–1809.

Grimaud-Hervè, D., 1997. L'évolution de l'enchéphale chez *Homo erectus* et *Homo sapiens*. CNRS, Paris.

Hawks, J.D., Wolpoff, M.H., 2001. The accretion model of Neandertal evolution. *Evolution* 55, 1474–1485.

Hublin, J.J., 1998. Climatic changes, paleogeography, and the evolution of the Neandertals. In: Akazawa, T., Aoki, K., Bar-Yosef, O. (Eds.), *Neanderthals and Modern Humans in Western Asia*. Plenum Press, New York, pp. 295–310.

Hublin, J.J., 2000. Modern-nonmodern hominid interactions: A Mediterranean perspective. In: Bar-Yosef, O., Pilbeam, D. (Eds.), *The Geography of Neanderthals and Modern Humans in Europe and the Great Mediterranean*. Peabody Museum Bulletin 8, Harvard University, Cambridge, MA, pp. 157–182.

King, W., 1864. The reputed fossil man of the Neanderthal. *Quart. J. Sci.* 1, 88–97.

Koppel, S.J., 1934. Stratigrafia e analisi della cava di Saccopastore e della regione circostante in riguardo alla posizione del cranio neandertaliano scoperto nel maggio 1929. *Riv. Antropol.* 30, 475–476.

Krings, M., Capelli, C., Tschentscher, F., Geister, H., Meyer, S., von Haeseler, A., Grossschmidt, K., Possnert, G., Paunovic, M. & Paabo, S. 2000. A view of Neandertal genetic diversity. *Nature Genetics* 26, 144–146.

Manzi, G., 1991a. Platycephaly and encephalization in the Grotta Guattari cranium: parieto-occipital morphology of a typical Neandertal. In: Piperno, M., Scichilone, G. (Eds.), *The Circeo 1 Neandertal Skull: Studies and Documentation*. Istituto Poligrafico e Zecca dello Stato, Roma, pp. 253–272.

Manzi, G., 1991b. Braincase architectural morphology of the Neandertal cranium from Grotta Guattari and the nature of the archaic/modern transition. *Quaternaria Nova* 1, 113–124.

Manzi, G., 2004. Italian prehistoric promenades: the human fossil sample. From the "earliest European" to the latest Neadertals. In: Baquedano, E., Rubio Jara, S. (Eds.), *Miscelánea* en Homenaje a Emiliano Aguirre, vol. III – Paleoantropología. Museo Archeológico Regional, Alcalá de Henares, pp. 220–230.

Manzi, G., Passarello, P., 1991. Antènèandertaliens et Nèandertaliens du Latium Italie Centrale. *L'Anthropologie* 95, 501–522.

Manzi, G., Saracino, B., Bruner, E., Passarello, P., 2000. Geometric morphometric analysis of midsagittal cranial profiles in Neandertals, modern humans, and their ancestors. *Riv. Antropol.* 78, 193–204.

Manzi, G., Bruner E., Caprasecca, S., Gualdi, G., Passarello, P., 2001. CT-scanning and virtual reproduction of the Saccopastore Neandertal crania. *Riv. Antropol.* 79, 61–72.

Moss, M.L., 1963. Morphological variations of the crista galli and medial orbital margin. *Am. J. Phys. Anthropol.* 21,159–64.

Passarello, P., Diotallevi, R., 1982. Paranasal sinus of Saccopastore I and II. *Anthropos* 21, 229–235.

Rak, Y., 1993. Morphological variation in *Homo neanderthalensis* and *Homo sapiens* in the Levant: a biogeographic model. In: Kimbel, W.H., Martin, L.B. (Eds.) *Species, Species Concepts, and Primate Evolution*. Plenum Press, New York, pp. 523–536.

Rohlf, F. J., 1997. tpsSpline 1.15. Ecology and Evolution, SUNY at Stony Brook, New York.

Rohlf, F.J., Marcus, L. F., 1993. A revolution in morphometrics. *Trends Ecol. Evol.* 8, 129–132.

Rosas, A., Bastir, M., Martínez-Maza, C., García Tabernero, A., Lalueza Fox., C., 2004. Neandertal cranio-dental evolution and development. A review in the light of new methods and fossils. In: Harvati, K., Harrison, T. (Eds.), *Neanderthals Revisited: New Approaches and Perspectives*. Springer, New York, pp. 37–70.

Saban, R., 1995. Image of the human fossil brain: endocranial casts and meningeal vessels in young and adult subjects. In: Changeaux, P., Chavaillon, J. (Eds.), *Origins of the Human Brain*. Clarendon Press, Oxford, pp. 11–38.

Schäfer, K., Seidler, H., Bookstein, F.L., Prossinger, H., Falk, D., Conroy, G., 2001. Exo- and endocranial morphometrics in mid-Pleistocene and modern humans. In: Falk, D., Gibson, K.R. (Eds.), *Evolutionary Anatomy of the Primate Cerebral Cortex*. Cambridge University Press, Cambridge, pp. 290–304.

Schoetensack, O., 1908. Der Unterkiefer des *Homo heidelbergensis* aus den Sanden von Maure bei Heidelberg. Leipzig, pp. 1–6.

Schwartz, J.H., Tattersall, I., 2002. *The Human Fossil Record – Volume One. Terminology and Craniodental Morphology of Genus Homo Europe*. Wiley-Liss, New York.

Seidler, H., Falk, D., Stringer, C., Wilfing, H., Muller, G.B., zur Nedden, D., Weber, G.W., Reicheis, W., Arsuaga, J.L., 1997. A comparative study of stereolithographically modelled skulls of Petralona and Broken Hill: implications for future studies of Middle Pleistocene hominid evolution. *J. Hum. Evol.* 33, 691–703.

Segre, A.G., 1948. Sulla stratigrafia dell'antica cava di Saccopastore presso Roma. *Rendiconti dell'Accademia Nazionale dei Lincei* 345, 743–751.

Segre, A.G., 1983. Geologia quaternaria e Paleolitico nella bassa valle dell'Aniene. *Riv. Antropol.* 62S: 87–98.

Sergi, S., 1929. La scoperta di un cranio del tipo di Neanderthal presso Roma. *Riv. Antropol.* 28, 457–462.

Sergi, S., 1934. Ossicini fontanellari della regione del lambda nel cranio di Saccopastore e nei crani neandertaliani. *Riv. Antropol.* 30, 101–112.

Sergi, S., 1935. Sulla stratigrafia di Saccopastore. *Riv. Antropol.* 30, 477–478.

Sergi, S., 1944. Craniometria e craniografia del primo paleantropo di Saccopastore. *Ricerche di Morfologia* 20/21, 733–791.

Sergi, S., 1947. Sulla morfologia della facies anterior corporis maxillae nei paleantropi di Saccopastore e del Monte Circeo. *Riv. Antropol.* 35, 401–408.

Sergi, S., 1948a. I piegamenti della base nei paleantropi di Saccopastore e del Monte Circeo. *Rendiconti dell'Accademia Nazionale dei Lincei*, Classe Sc.FMN 5 ser. 8, 306–313.

Sergi, S., 1948b. The palaeanthropi in Italy: The fossil men of Saccopastore and Circeo. *Man* 48, 61–64 (part I), 76–79 (part II).

Sergi, S., 1948c. Il cranio del secondo Paleantropo di Saccopastore. *Palaeontogr. Itali.* 42 1942–'46, 25–164.

Sergi, S., 1962. The Neanderthal Palaeanthropi in Italy. In: Howells, W.W. (Ed.), *Ideas on Human Evolution, Selected Essays 1949–1961.* Harvard University Press, Cambridge, MA, pp. 500–506.

Serre, D., Langaney, A., Chech, M., Teschler-Nicola, M., Paunovic, M., Mennecier, P., Hofreiter, M., Possnert, G., Pääbo, S., 2004. No evidence of Neandertal mtDNA contribution to early modern humans. *PLoS Biology* 2, 313–317.

Shackleton, N.J., 1967. Oxygen isotope analyses and Pleistocene temperature re-assessed. *Nature* 215, 259–265.

Simpson, G.G., 1944. *Tempo and Mode in Evolution.* Hafner, New York

Slice, D., 2000. *Morpheus et al. Ecology and Evolution*, SUNY at Stony Brook, New York.

Spoor, F., Zonneveld, F., 1995. Morphometry of the primate bony labyrinth: a new method based on high-resolution computed tomography. *J. Anat.* 186, 271–286.

Spoor, F., Zonneveld, F., 1998. Comparative review of the human bony labyrinth. *Yrbk. Phys. Anthropol.* 41, 211–251.

Spoor, F., Jeffery, N., Zonneveld, F., 2000. Using diagnostic radiology in human evolutionary studies. *J. Anat.*197, 61–76.

Spoor, F., Hublin, J.J., Braun, M., Zonneveld, F., 2003. The bony labyrinth of Neanderthals. *J. Hum. Evol.* 44, 141–165.

Stringer, C.B., 2002. Modern human origins: progress prospects. *Phil. Trans. R. Soc. London* B 357, 563–579.

Stringer, C.B., Hublin, J.J., 1999. New age estimates for the Swanscombe hominid and their significance for human evolution. *J. Hum. Evol.* 37, 873–877.

Stringer, C.B., Hublin, J.J., Vandermeersch, B., 1984. The origin of anatomically modern humans in Western Europe. In: Smith, F., Spencer, F. (Eds.), *The Origins of Modern Humans: A World Survey of the Fossil Evidence.* Alan R. Liss, New York, pp. 51–135.

Trinkaus, E., 1991. Les hommes fossiles de la grotte de Shanidar, Irak: évolution et continuité parmi les hommes archaïques tardifs du Proche-Orient. *L'Anthropologie* 9, 535–572.

4. Inquiries into Neanderthal craniofacial development and evolution: "accretion" versus "organismic" models

A. ROSAS
Departament de Paleobiología
Museo Nacional de Ciencias Naturales, CSIC.
C/ José Gutierrez Abascal 2
28006 Madrid, Spain
mcnar537@mncn.csic.es

M. BASTIR
Hull York Medical School
The University of York, Heslington,
York YO10 5DD, United Kingdom
and
Departament de Paleobiología
Museo Nacional de Ciencias Naturales, CSIC
C/ José Gutierrez Abascal 2
28006 Madrid, Spain
markus.bastir@hyms.ac.uk

C. MARTÍNEZ-MAZA
Departament de Paleobiología
Museo Nacional de Ciencias Naturales, CSIC
C/ José Gutierrez Abascal 2
28006 Madrid, Spain
mcncm565@mncn.csic.es

A. GARCÍA-TABERNERO
Departament de Paleobiología
Museo Nacional de Ciencias Naturales, CSIC
C/ José Gutierrez Abascal 2
28006 Madrid, Spain
agarciatabernero@mncn.csic.es

C. LALUEZA-FOX
Unitat d'Antropologia
Departament de Biologia Animal
Facultat de Biologia
Universitat de Barcelona, Spain
carles.lalueza@cexs.upf.es

Keywords: Neanderthal, craniofacial, organismic model, accretion model, geometric morphometrics, bone remodeling

K. Harvati and T. Harrison (eds.), Neanderthals Revisited: New Approaches and Perspectives, 37–70.

Abstract

The origin and evolutionary significance of the Neanderthals is still unresolved. Several basic questions about the biological processes determining the apomorphic morphology of this human group are still pending resolution. In this paper we present a general outline of the lines of research we are currently following in the exploration of Neanderthal anatomy. We focus our approach on three different aspects: (1) a methodological and theoretical component based on system theory; (2) specific morphological problems, such as the effects of body size and brain development on craniofacial shape in Neanderthals; and (3) the analysis of these aspects by geometric morphometric and paleohistological methods.

Presently, Neanderthals are considered the end point of an evolutionary lineage rooted in the European Middle Pleistocene fossil record. The "tempo and mode" in the evolution of the Neanderthal lineage are here evaluated by comparison of two different perspectives: the "accretion model" (Hublin, 1998a) and the "organismic model". In the former, stochastic processes are invoked to account for the "gradual tempo and a mosaic mode" for the emergence of the Neanderthal anatomy. By contrast, the "organismic model" presented here, sets out the hypothesis that the Neanderthal skull form may result from two interconnected effects at different levels of organization: (1) body-face interactions via organismic-level effects (e.g., allometry, physiological demands), and (2) brain-face interactions via integration principles.

We propose a conceptual framework in which the evolutionary process experienced by the Neanderthal lineage was primarily dominated by two distinct and successive phases. The first to appear in time was that producing midfacial prognathism, which may be related to mid-sagittal growth dynamics, in which body size and cranial base orientation are primarily involved. This evolutionary phase is accompanied by significant variation, which is still poorly understood. Here it is associated with increased body size and sexual dimorphism variation, and its concomitant craniofacial compensatory mechanisms. A second evo-devo process, largely related to the relative size increase of the occipital and temporal neural areas, affects the organization of the skull, via cascading effects resulting from integration mechanisms (at times assimilated to pleiotropic effects). In this two-phase evolutionary process, facial changes may be connected with processes affecting the variation of the sagittal cranial base, whereas temporal bone modification may be related to processes affecting bilateral, middle cranial fossa variation. Occipital modifications remain open to interpretation, although a complex interplay between increased encephalization and head position (perhaps as an indirect result of body shape) might be considered.

Taxonomically, the "accretion model" considers the Neanderthal lineage as a continuum of evolving populations, which are temporally associated with two different *nomina*: *H. heidelbergensis* and *H. neanderthalensis*, on the basis of a single chronospecies concept. In an alternative view, we favor the hypothesis that the species *H. neanderthalensis* emerges as a distinct biological entity (morph) after a speciation event (or change in morphological quality). The main point of this paper is to evaluate the "accretion model" versus a two-phase hypothesis of Neanderthal speciation in the Middle Paleolithic.

Introduction

The natural history of Neanderthals is among the oldest and most exciting topics in human evolution. Understanding the role of Neanderthals in our own ancestry has aroused many conceivable ways of approaching its study, and it has strongly challenged the intellectual capacity of scientists. Beyond attempts to clarify the genealogical position of Neanderthals in our own ancestry, from the viewpoint of morphology as a science, we face one of the most elusive and challenging case studies in human evolution.

While, for most of their scientific existence, the Neanderthals were those enigmatic people living in Europe during the Würm glaciation (Late Pleistocene, OIS 4–3), today the term Neanderthal is used in a twofold sense. On the one hand, we refer to Neanderthals as the human populations inhabiting the western end of the Eurasian continent, from approximately 80,000–30,000 years ago, also called Würm Neanderthals or "Classic" Neanderthals (further geographic subdivision may consider the Levant Neanderthals). Nowadays, a majority of scholars consider classic Neanderthals as a distinct species, *Homo neanderthalensis* (e.g.,

Tattersall, 1992; Rak et al., 1994; Schwartz and Tattersall, 1996; Arsuaga et al., 1997a; Rosas, 2000; Harvati et al., 2004; Ramírez Rozzi and Bermúdez de Castro, 2004), morphologically distinct from all other *Homo* species in their combination of: (1) their massive bodies and body proportions, (2) their projecting faces (midfacial prognathism), and (3) a considerably large cranial capacity, among other features. There is also growing evidence of ontogenetic differences between *Homo neanderthalensis* and other *Homo* representatives (e.g., *H. sapiens* and *H. heidelbergensis*) (Ponce de León and Zollikofer, 2001; Ramírez Rozzi and Bermúdez de Castro, 2004). Additionally, increasing information on mtDNA extracted from Neanderthal remains strongly suggests a species-level distinction (Krings et al., 1997; Lalueza-Fox et al., 2005).

On the other hand, the term "Neanderthal" is now sometimes applied to Middle Pleistocene populations of less clear-cut morphology. The use of the terms ante-Neanderthals (de Lumley, 1973), proto-Neanderthals (Condemi, 2003; Manzi, 2003), pre-Neanderthals (Hublin, 1998a; Dean et al., 1998, Hawks and Wolpoff, 2001), as well as primitive Neanderthals (Stringer and Hublin, 1999) or early Neanderthals (Dean et al., 1998), can be seen in the literature. The temporal range delimitation of these heuristic categories varies from one author to another, and depends, in many cases, on the statistical design of the study and availability of data. All in all, Neanderthals, rather than a single hominin species, are considered in this perspective to be an evolutionary lineage rooted in the European Middle Pleistocene fossil record (Howell, 1960; Bermúdez de Castro, 1986; Rosas and Bermúdez de Castro, 1998; Hublin, 1998a; Trinkaus, 1988; Arsuaga et al., 1997a). The date of divergence between the mtDNA of Neanderthals and contemporary humans is estimated to 465,000 years before the present, with confidence limits of 317,000 and 741,000 years (Krings et al., 1997). Such a lineage would include no less than two paleontological species: *Homo heidelbergensis* and *H. neanderthalensis*.

Within this framework, the "tempo and mode" of the evolution of the Neanderthal lineage may be conceived from two different perspectives, which we will refer to as the "accretion model" (Hublin, 1998a) and the "organismic model" (Figure 1). Under these headings we want to evaluate the hypotheses that may account for the biological processes directing the evolution of the Neanderthal craniofacial complex and its taxonomic interpretation. Our central hypothesis states that the evolution of the Neanderthal lineage is primarily dominated by a combination of two sequential evolutionary processes, the first affecting the body as a whole, and the second being the consequence of mosaic encephalization. More specifically, the first alteration of the lineage produced midfacial prognathism, which may be related to mid-sagittal growth dynamics, in which body size and cranial base are primarily involved. This stage is accompanied by high degrees of morphological variation, which is still poorly understood, despite multiple attempts at explanation (Rosas, 1997; Arsuaga et al., 1997b; Rosas and Bastir, 2004). A second transformation, largely related to the enlargement of the parietal, occipital, and temporal neural areas, affects the organization of the whole skull (and possibly the timing of development), via cascade effects resulting from integration mechanisms, at times assimilated to pleiotropic effects. On the whole, facial changes experienced by hominins of the Neanderthal lineage may be directed by and/or connected with sagittal cranial base processes. By contrast, temporal bone modification may be related to parasagittal processes (middle cranial fossae). Occipital modifications remain open to interpretation, although a complex interplay between brain enlargement and head position (perhaps as an indirect result of body shape) might be considered.

<table>
<tr><th colspan="9">Evolution of the Neanderthal lineage
Mosaic distribution of cranio-facial features</th></tr>
<tr><th colspan="5">Accretion model</th><th rowspan="3">Fossil evidence

Example specimens</th><th colspan="3">Organismic model</th></tr>
<tr><th colspan="5">Features more or less independent</th><th colspan="3">Developmental modularity
Growth integration</th></tr>
<tr><th>Neanderthal stages</th><th>Isotope stages</th><th>Derived European anatomical features</th><th colspan="2">Process</th><th>Derived European anatomical features</th><th colspan="2">Process</th></tr>
<tr><td>Stage 1
Early-pre-Neanderthals</td><td>Pre-stage 12
stage 12</td><td>Facial, mandibular and anterior dentition features</td><td rowspan="2">Elevated variability
shift in frequency</td><td rowspan="4">Chrono-species H. heidelbergensis/H. neanderthalensis</td><td>Boxgrove,
Mauer,
Arago,
Petralona
Bilzingsleben</td><td rowspan="2">Body size and form

Brain size

Pre sellar elevation</td><td rowspan="2">Elevated variability
Allometry
Compensatory mechanisms</td><td rowspan="2">Species 1. H. heidelbergensis</td></tr>
<tr><td>Stage 2
Pre-Neanderthals</td><td>Stage 11–9</td><td>Occipital area

Glabella

Sagittally oriented face</td><td>Vértesszöllös
Atapuerca SH
Swanscombe
Steinheim,
Reilingen</td></tr>
<tr><td>Stage 3
Early Neanderthals</td><td>Stage 7–5</td><td>Temporal area

Full en bombe

Increased dolichocephaly</td><td rowspan="2">Reduced variability
selection</td><td>Ehringsdorf,
Biache 1,
La Chaise S.
Lazaret,
La Chaise B-D
Saccopastore
Krapina</td><td rowspan="2">Mosaic Encephalization

Bi-lateral cranial floor (temporal fossae)</td><td rowspan="2">Reduced variability
Integration</td><td rowspan="2">Species 2. H. neanderthalensis</td></tr>
<tr><td>Stage 4
Classic Neanderthals</td><td>Stage 4–3</td><td>Temporal area

High mid-facial prognathism</td><td>Neanderthal,
Spy, Monte Circeo,
Gibraltar
La Chapelle
La Quina,
La Ferrassie</td></tr>
</table>

Figure 1. Basis of the two models discussed in this paper: the "accretion model" and the "organismic model". The accretion model (Hublin, 1998a) is based on the mosaic nature of the evolution of Neanderthals and the gradual emergence of their specialized features. Features seem to change in their frequency along the evolutionary lineage, and four steps are distinguished. *Homo neanderthalensis* is part of a chronospecies evolving during the Middle Pleistocene in Europe, and Neanderthals are the end point of a continuum of populations. In the right hand column, we propose a complementary model that we call the "Modularity Model," which is based on the hypothesis that features are organized in packages or modules. The major implication of this model is that within the European evolutionary lineage there is a speciation event that gave rise to a new species, *H. neanderthalensis*. There are grounds for considering the origination of this species as a relatively recent discrete event, in contrast to the concept of a chronospecies evolving in Europe during the Middle Pleistocene.

Testing these hypotheses may be achieved through the study of different levels of biological organization. These include the analytical hypothesis driven exploration of variation of macroscopic structures, both diachronic and intrapopulational, as well as the study of bone remodeling at a microscopic level. To support this perspective we present findings on patterns of variation (size and allometry) within mid-Pleistocene fossil populations, as well as the study of morphological covariation of craniofacial structures and their levels of integration with body size in modern humans. Also, a complete pattern of bone remodeling in mandibles of *H. heidelbergensis* is presented. Results derived from these studies are discussed in a broader anatomical context. With these aims in mind, we first present a review of the most important aspects of Neanderthal anatomy.

Evolutionary Anatomy of the Neanderthal Craniofacial System

Neanderthals are clearly distinctive in terms of their overall morphology (total morphological pattern) (Smith et al., 2005), and their characteristic morphology is seen throughout the cranial and postcranial skeleton (Wood and Richmond, 2000).

The Neanderthal craniofacial system is dominated by two major features. One is a large, low, and elongated braincase, containing a brain as large as that of living *H. sapiens* [Neanderthal endocranial capacity ranges around 1518.9 ± 169.5 ($n = 9$) (Trinkaus and Tompkins, 1990), and it was greater than that of their ancestors (Arsuaga et al., 1999a)]. The second major feature is a special configuration of the facial skeleton, usually referred to as midfacial prognathism (Rak, 1986). This craniofacial pattern is present within the context of short and stocky individuals, with voluminous chests (Franciscus and Churchill, 2002), robust limb bones (Trinkaus, 1983; Weaver, 2003), with their distal segments proportionally shorter (Trinkaus, 1981; Holliday, 1997; Churchill, 1998; Steegmann et al., 2002). However, recent analyses of the postcranial skeleton indicate that much of the *H. neanderthalensis* morphological pattern was present in their European ancestors long before true Neanderthals appear (Arsuaga et al., 1999; Carretero et al., 1997, 1999; Pearson, 2000). Also, the elevated robusticity of the Boxgrove tibia parallels that of the Neanderthals (Stringer et al., 1998).

The Neanderthal braincase is defined by rounded, laterally projecting parietal bones producing a bomb-like contour (Hublin, 1998b); a posteriorly projecting occipital bone with a suprainiac fossa, double-arched brow ridges and projecting glabella. In addition to these features, the temporal bone of Neanderthals experiences a reorganization giving rise to a set of apomorphies: the articular eminence becomes relatively flat (Martínez and Arsuaga, 1997), and the temporo-mandibular joint experiences spatial reorganization, becoming posteriorly (Arsuaga et al., 1997a) and inferiorly located (Harvati, 2003). As a consequence, the external auditory meatus obtains an elevated position relative to the zygomatic process (Stringer et al., 1984; Condemi, 1988; Harvati, 2003). The mastoid region of the temporal bone is transformed in *H. neanderthalensis*, with a small mastoid process and an anteriorly obliterated digastric groove (Martínez and Arsuaga, 1997). Furthermore, there are a large number of morphological details considered Neanderthal specializations that are distributed all over the skull (Stringer et al., 1984; Hublin, 1998b; O'Connor et al., 2005).

Based on linear distances and proportions, Arsuaga et al. (1997a) found that Neanderthal braincase variation followed a trend parallel, but different, to that of *H. erectus*/*H. ergaster*, so that Neanderthals present narrower braincases, more rounded frontal bones, less angled occipital bones, and less projecting inia.

Afro/European Middle Pleistocene specimens lie between *H. erectus* and Neanderthals. Interestingly, Neanderthals share with archaic *Homo* (*H. erectus s.l.*) a primitive endocranial anatomical pattern, falling within a single allometric trend (Bruner et al., 2003). Likewise, Neanderthals share the primitive pattern of elongated anterior cranial base with Middle Pleistocene hominins, both African and European (Lieberman, 1998; Spoor et al., 1999).

The Neanderthal facial skeleton is characterized by a large and primitive total facial prognathism (prosthion relative to mid-basicranial or temporal region) (Rak, 1986; Trinkaus, 1987, 2003; Arsuaga et al., 1997a), with a unique constellation of traits: an elevated pneumatization in the frontal, nasal and maxillary sinuses (Heim, 1974; 1976), and a characteristic midfacial projection – the so called midfacial prognathism (Stringer et al., 1984; Rak, 1986; Trinkaus, 1987). Thus, the Neanderthal facial skeleton is characterized by a derived separation of the midface and dentition in relation to the posterolateral parts of the face. The latter affect the zygomatico-ramal region and associated musculature (e.g., lateral facial retreat, Trinkaus, 1988), the posterior position of the glenoid fossa (Arsuaga et al., 1997a), and middle cranial fossa (Ponce de León and Zollikofer, 2001). Interestingly, most of the features forming what is called midfacial prognathism, such as the parasagittal orientation of the infraorbital plate, posterior position of the zygomatic root, and a flexed maxillary, among others, were already present in European Middle Pleistocene populations.

The mandible clearly reflects this pattern and presents a large number of apomorphic traits (Stringer et al., 1984; Rak et al., 1994, 2002; Rosas, 2001). Derived features related to the anterior-posterior location of mandibular structures (e.g., mental foramen position at the level of M_1, retromolar space) appear during the European mid-Pleistocene. The first evidence of Neanderthal morphology seems to emerge at around 500,000 years ago with the Mauer mandible (Rosas and Bermúdez de Castro, 1998). The buccolingual dimensions of the incisors and the morphology of the premolars, as well as metric variables related to the position of the mental foramen and retromolar space, definitely indicate that Mauer should be classified as a member of the evolutionary lineage that evolved into classic Neanderthals. On the other hand, derived features located on the internal aspect of the mandible (e.g., mylohyoid line inclination) become expressed only in classic Neanderthals (Rosas, 2001).

Neanderthals have a large anterior dentition, with strong shovelling and prominent lingual tubercles, which are probably primitive retentions (Bermúdez de Castro, 1986). The posterior dentition is relatively small, sharing a general tendency for molar reduction with modern humans and other *Homo* species. A high incidence of postcanine taurodontism (i.e., enlarged pulp chambers) is also observed in Neanderthals, together with a number of claimed uniquely derived features; among them a high frequency of the mid-trigonid crest in lower molars (Bailey, 2002) and the maxillary molar cusp placement relative to each other (Bailey, 2004).

All in all, Neanderthal anatomy represents a unique mix of primitive and derived features. Among the latter, features shared with other later *Homo* species (e.g., *H. sapiens*) may belong to general trends towards, for instance, postcanine dental reduction and general brain enlargement. A set of apomorphies are unique to European Middle Pleistocene populations, while a number of characters are recognized by most authors, though not all (Trinkaus, 1993; Franciscus and Trinkaus, 1995; Franciscus, 2003; 1999), as autapomorphies of *H. neanderthalensis* (see Table 1). Uniquely derived Neanderthal features are recognized in the morphology and growth patterns of the dentition, the mandible, internal nose, temporal bone, inner ear, brain endocast,

Table 1. Neanderthal features that are not present in their European middle pleistocene forerunners

Diagonal inclination of the mylohyoid line	(Rosas, 2001)
Inferior position of the posterior semicircular canal	(Spoor et al., 2003)
Rapid rate of dental growth; faster pace of Somatic development	(Ramírez-Rozzi and Bermúdez de Castro, 2004)
Mastoid process weakly projecting	(Martínez and Arsuaga, 1997)
Digastric groove anteriorly closed	(Martínez and Arsuaga, 1997)
Development of the upper parietal gyrus	(Bruner et al., 2003; Bruner 2004).
Posterior profile en bombe	(Stringer et al., 1984)
Unusually high frequency of bilevel (depressed) internal nasal floor	(Franciscus, 2003)
"Internal nasal margin" and a distinct swelling of the lateral nasal wall	(Schwartz and Tattersall, 1996; but see Franciscus, 1999)
Other changes in the European evolutionary lineage include: slight decrease in cranial and postcranial robusticity, and decrease in level of morphological variation.	

etc. Defining the uniqueness of their craniofacial features, and isolating species-specific differences from other hominins, has lasted for decades and the use of cladistic philosophy is a significant landmark in the attempt to identify exclusive Neanderthal features (Santa Luca, 1978; Martínez and Arsuaga, 1997; Condemi, 1988). Nowadays, it is accepted that, on the whole, Neanderthal craniofacial morphology is highly derived (Arsuaga et al., 1997a; Hublin, 2000). However, what are the processes that govern the evolution of European Pleistocene hominins that result in the emergence of such a unique morphology?

Interpretations of the Neanderthal craniofacial morphology

Despite intensive analysis lasting for more than 150 years, there is still no generally accepted evolutionary explanation for the craniofacial morphology of Neanderthals. Nearly every known evolutionary mechanism has been invoked to explain all aspects of the unique Neanderthal anatomy (see revisions by Antón, 1994; Churchill, 1998; Franciscus, 2003). The factors thought to underly the evolution of Neanderthals include biomechanical, climatic, respiratory, developmental and stochastic explanations. Historically, the diverse interpretations of the Neanderthal morphology have closely matched the prevailing evolutionary paradigms. Since the establishment of the neo-Darwinian paradigm, attempts to account for the evolutionary significance of Neanderthals can be organized into three successive schools of thought, with extensive overlap among them. These three stages may be informally defined as: (1) functional/adaptationist; (2) population-based stochastic processes; and (3) ontogenetic-based (evo-devo) approach.

The first stage was mostly dominated by the proposal of functional/adaptational hypotheses. These adaptational hypotheses are further organised into the interpretation of two craniofacial functions: respiratory and masticatory. In other words, physiological functions performed by the nasal cavity (e.g., cold adaptation, high activity levels and associated increase in body heat) and

Table 2. Summary of interpretations of Neanderthal craniofacial morphology

Class of phenomena	*General causes*	*More specific hypotheses*	*Authors: Pro*	*Authors: Con*
Mechanical adaptaion	Anterior dental loading	High bites forces	Brace, 1964; Smith, 1983; Rak, 1986; Demes, 1987; Smith and Paquette, 1989; Spencer and Demes, 1993	Antón, 1994; O'Connor et al., 2005; Couture, 1993
		Repetitive anterior dental loading		
Physiological adaptation	Cold adaptation	Nasal radiator	Howell, 1952, 1957; Coon, 1962; Ruff et al., 1997	Mann and Trinkaus, 1974; Hylander, 1977
		Respiratory moisture retention	Sergi, 1962; Franciscus and Trinkaus, 1988a, b; Trinkaus, 1989	
		Moisture retention and heat dissipator	Franciscus and Trinkaus, 1988; Trinkaus, 1989	
		Nasal respirators	Laitman et al., 1992	
Stochastic events	Genetic drift		Hublin, 1990, 1998a, 2000; Antón, 1994	Rosas et al. (this volume)
Evolutionary changes as developmental phenomena	Endocrine models	Earlier onset or greater intensity of hormones regulating adolescent growth spurt	Keith, 1925; Brothwell, 1975	
	Differential growth rates	Acceleration of *in utero* growth rates in basicranial cartilages	Green and Smith 1990; Smith and Green 1991; Smith 1991	
			Ponce de León and Zollikofer, 2001	
Structural models	Enlargement of the facial sinuses		Heim, 1976	
	Brain increment	Changes in cranial base orientation	Rosas, 1992, 1998	
		Hypostosis	Manzi et al., 2000; Bruner, 2004	
	Combined effects	Body size and brain increment	Bastir, 2004; Rosas et al. (this volume)	

biomechanical hypotheses regarding the masticatory system (Table 2; see Antón, 1994; O'Connor et al., 2005).

Cold adaptation is accepted as the best explanatory hypothesis for the postcranial skeleton and limb proportions of Neanderthals (Trinkaus, 1981; Weaver, 2003), with distal limb segments proportionally shorter (Trinkaus, 1981; Holliday, 1997; Churchill, 1998; Steegmann et al., 2002). The role of the Neanderthal face as a thermoregulatory adaptation was proposed by Howell (1952) as a response to cold, dry climates in Pleistocene Europe. The "nasal radiator hypothesis" states that their projecting nasal cavity evolved to warm inspired air potentially affecting a temperature sensitive brain (Coon, 1962); that is, distancing the brain from cold air (see Schwartz et al., 1999; Franciscus, 2003 for a revision of the subject). Under the consideration of the nasal cavity as a temperature and moisture exchanger, the original hypothesis developed into diverse variants considering cooling under conditions of heightened activity, minimizing respiratory moisture loss in cold and arid condition (Franciscus and Trinkaus, 1988a, b).

Some features of the postcranial anatomy, such as thickened cortical bone, pronounced areas of muscle attachment and bowed proximal limb segments (Sorensen and Leonard, 2001) have been interpreted as the result of elevated activity levels, implying a series of behavioral differences from modern humans (Trinkaus, 1997; Ruff et al., 1993, 1994; Churchill, 1998 and references therein). In this sense, midfacial anatomy has been related to specializations in respiratory physiology among individuals with high activity levels and associated increase in body heat (Schwartz and Tattersall, 1996; but see Franciscus, 1999), in which anatomical specializations of nasal cavity would be involved in mechanisms for shedding heat under conditions of heavy exercise in cool and moderate thermal conditions (Trinkaus, 1987). However, more recently, on the basis of an extensive study of the internal nasal floor configuration, Franciscus (2003) rejects and physiological link between nasal floor anatomy and cold adaptation. Blumenfeld (2002), Friess et al. (2002) also maintain that the facial morphology in Neanderthals is driven by factors that are mainly unrelated to climatic conditions. Therefore, while most scholars agree that the postcranial skeleton of Neanderthals can reasonably be interpreted in terms of cold adaptation, craniofacial characteristics elude a simple climatic reading of their morphology.

Biomechanical hypotheses seek to explain the evolution of Neanderthal craniofacial morphology by reference to masticatory and paramasticatory activities (see Antón, 1994; O'Connor et al., 2005). These hypotheses are based on the fact that Neanderthals present a large and heavily worn anterior dentition, together with a high incidence of degenerative remodeling of the temporomandibular joint (and shallow glenoid fossa). According to this framework, biomechanical explanations support the idea that Neanderthal facial anatomy (e.g., midfacial prognathism and topography of the infraorbital area) was an adaptive response to the heavy anterior dental loads that resulted from paramasticatory activities (Rak, 1986; Demes, 1987; Spencer and Demes, 1993). Thus, diagnostic facial features are interpreted as adaptations for effectively dissipating these loads. Presently, adaptive models have been subjected to rigorous examination by comparative, quantitative and, in some cases, experimental methods (Couture, 1993; Antón, 1994; O'Connor et al., 2005), and, on the whole, no satisfactory corroboration has been found. O'Connor et al. (2005) specifically conclude that "masticatory biomechanical adaptation is not the primary factor underlying the evolution of the Neanderthal facial architecture".

The second step in the evolutionary interpretation of Neanderthals was developed under the influence of population genetics, and invoked stochastic processes for the origin of Neanderthal anatomy (Hublin, 1990, 1998a; Antón, 1994; Bermúdez de Castro and Nicolás, 1995; Lebel et al., 2001; Franciscus, 2003). The most important line of interpretation has been the so-called "accretion model" (Hublin, 1998b) (this will be expanded on later). Under this model the paleobiogeographic scenario first proposed by Howell (1952, 1957) was updated with an improved chronological framework and a clearer understanding of the taxonomic significance of the anatomy. Endemism and genetic drift become the primary agents of Neanderthals uniqueness, everything according with the foundations of an allopatric mode of speciation (Hublin, 2000; but see Hawks and Wolpoff, 2001). According to the "accretion" model, many of the Neanderthal features have no adaptive significance.

The third step in the evolutionary explanation of Neanderthals was inaugurated after the strong influence of S.J. Gould, P. Alberch, D. Wake, B. Goodwin among others, with the rediscovery of heterochrony and the relevance of developmental constraints in the course of evolution [phylogenetic burden, (Riedl, 1975)]. Since then, the search for the ontogenetic bases for evolutionary changes in morphology has become a common research objective for scholars. The questioning of the previously omnipresent role of natural selection gave rise to the exploration of alternative causal explanations, namely those rooted in epigenetic and developmental processes. The functional/adaptational paradigm tended to give functional independent explanations for every system, or even individual traits, largely ignoring higher levels of functional, developmental and/or structural integration (Riedl, 1975). The principle of single ontogenetic shifts that have multiple morphological consequences has transformed the way of thinking in evolutionary morphology. With this relatively new perspective the ontogeny of Neanderthals has come under close scrutiny, and heterochronic modifications of a common pattern of development have become the central working hypothesis. As a result, many Neanderthal features are now thought to be the correlated result of changes in a few regulatory genes (Churchill, 1998; Franciscus, 2003).

An initial stage in this line of research has been the assessment of the relative ontogenetic (temporal) appearances of adult Neanderthal morphological characteristics. Several taxon-specific differences in cranio-mandibular shape emerge early in ontogeny (Tillier, 1988; Stringer et al., 1990; Rak et al., 1994; Schwartz et al., 1999; Krovitz, 2003; Ponce de León and Zollikofer, this volume), possibly at fetal stages (Ponce de León and Zollikofer, 2001; Franciscus, 2003, Zollikofer and Ponce de León, this volume). These include features from the temporal bone, nasal cavity, and mandible. Yet, other craniofacial features emerge late in ontogeny, especially those related to midfacial prognathism and/or antero-posterior growth of the mandible (Tillier, 1988; Arsuaga et al., 1997a; Rosas, 1998). A certain relationship between ontogenetic development of Neanderthal traits and the phylogenetic appearance of those characters has been proposed (Tillier, 1988 and references therein). Ontogeny and phylogeny seem to follow an inverse order in the development of Neanderthal apomorphic traits. The facial skeleton is the first to display these features in phylogeny, but is the latest in terms of ontogeny, whereas cranial vault derived features (including the temporal bone) appear early in ontogeny and late in phylogeny. Does this pattern have any biological meaning?

There are indications that Neanderthals had a growth pattern (and metabolism) distinct from that of modern humans, with advanced dental development, early rapid brain growth (Dean et al., 1986), and/or early cranial vault ossification (Heim, 1982; Trinkaus and LeMay, 1982). Accelerated growth has been deduced from the study of dental development, which is thought to be tightly integrated with somatic development (Smith, 1991). The study of the relative rates of growth in different organ systems, e.g., the dentition and postcranial skeleton (Thompson and Nelson, 2000), and growth of brain tissues and bone ossification (Manzi et al., 1996), have also led to similar conclusions. Sergi (1934) introduced the idea of a "conflict" between brain growth and cranial ossification among Neanderthals (Manzi, 2003), reinterpreted as an heterochrony in the development of the Neanderthal cranium; namely, a faster ossification of the cranial vault relative to brain growth rates (Manzi et al., 1996, 2000; Manzi 2003).

In the search for general processes of heterochrony that may explain the form of Neanderthals, de Beer (1958) suggested that Neanderthals were peramorphic. Nearly half a century later, Ponce de León and Zollikofer (2001) suggest a rate-hypermorphosis process to explain Neanderthal postnatal development as compared with modern humans (but see Williams et al., 2002, Zollikofer and Ponce de León, this volume). Alternatively, Williams and Krovitz (2004) pose the question of whether modern humans were paedomorphic with respect to Neanderthals. In spite of the different attempts to find an evolutionary explanation, a global heterochronic mechanism cannot be used to describe Neanderthal growth relative to modern humans (Krovitz, 1999; Williams et al., 2002). Many of these studies are based on the comparison of Neanderthals to modern humans, inferring evolutionary explanations for the differences between species, implicitly assuming an ancestor-descendant relationship. The lack of immature specimens from the Middle Pleistocene precludes a better knowledge of the expected ontogenetic shift experienced by Neanderthals. Therefore, although appealing, at present the heterochrony conceptual framework is plagued by practical and theoretical problems when dealing with highly complex organisms.

A number of studies propose that the Neanderthal singularity may well result from the allometric extension of primitive morphotypes. For instance, Bruner et al. (2003) state that the large Neanderthal brain and its morphology are consistent with the extension of an evolutionary trajectory, without specific novelties or specializations (see also Bruner, 2004). These authors identify two distinct allometric trajectories for the evolution of the structural relation between cerebral shape and encephalization in *Homo*. Neanderthals, in this instance, would be the product of the extension of a shared primitive developmental pathway (i.e., hypermorphosis). A significant morphological consequence, however, is the development of a dorsal bending of the Neanderthal encephalon (Bruner, 2004), which will be considered later. By contrast, according to a number of scholars, Neanderthal craniofacial and postcranial form represents a specialized departure from the general *Homo* pattern (Howells, 1975; Rak et al., 1994, 2002; Thompson and Nelson, 2000). From an ontogenetic perspective, Ramirez-Rozzi and Bermúdez de Castro (2004) assert that Neanderthals have undergone a major developmental shift in relation to other *Homo* species.

An application of the principle of "single ontogenetic shift-multiple morphological consequences" was developed by Smith and Green (1991) who hypothesized that the overall cranial form of Neanderthals, including an occipital bun, midfacial prognathism, and a strongly projecting supraorbital torus, results from accelerated growth of the components of the basicranium. Later, several hypotheses have been proposed to account for the facial projection seen in Neanderthals: anterior reorientation and repositioning of the anterior cranial fossa (Rosas, 1992), greater growth at mid-sphenoidal synchondrosis during fetal development (Lieberman, 1998), less basicranial flexion during early postnatal or possibly prenatal ontogeny (Lieberman et al., 2000), and scaling and non-scaling effects on craniofacial architecture (Bastir, 2004). However, although no clear picture has emerged from this research, it seems to be the most promising approach. Below, further developments within this framework will be presented.

Following the integrative evolutionary paradigm, masterfully exemplified by Wake and Larson (1987), Churchill (1996, 1998) presented a helpful discussion of the theoretical perspectives by which researchers have traditionally approached the study of Neanderthal morphology and evolution. Two distinct intellectual approaches are distinguished by Churchill that are clearly rooted in general evolutionary theory. On the one hand, structuralist theory (e.g., organism-focused) sees features as intercorrelated responses to larger-scale changes in body form (usually as the result of ontogenetic shifts). This approach stresses the phenomenon of integration; that is, change in many traits can be seen as functionally, developmentally (knock-on effects), or pleiotropically correlated to other changes (Churchill, 1996). On the other hand, the adaptationist approach, discussed above, sees morphological features as isolated entities more or less independently selected for, and, in the case of Neanderthals, biomechanical or thermoregulatory adaptations.

Based on his own results from the study of integration of upper limb, in which integration models fit the data better (but still explained no more than half of the variation), Churchill (1996) urged the need to consider both approaches in a more synthetic mode. However, the consideration of general (e.g., organismic or systemic level) versus particular factors (e.g., selection for isolated functional apparatus) responsible for evolutionary changes needs a better formalization in order to develop more accurate methods for testing hypothesis. In this context, the active consideration of integration becomes a significant element in the analyses. Antón (1994) stresses the need for the consideration of the structural interdependencies of the cranial system, and poses the question: to what degree is Neanderthal facial form explainable by and dependent upon the morphology of other cranial regions?

Models for the Evolution of Neanderthals

THE ACCRETION MODEL

Hublin (1998a) (see also Dean et al., 1998; Stringer and Hublin, 1999; Franciscus, 2003) has proposed the so called "accretion model" for explaining the evolutionary emergence of the Neanderthal morphology. This model identifies a certain correspondence between geological time and the successive appearance of the derived Neanderthal features (i.e., apomorphies), which follows a particular order in the anatomical location of the novelties. Trinkaus (1998) also recognized this pattern, and proposed "a gradual tempo and a mosaic mode" for the evolution of the Neanderthal anatomy, with a gradual increase in frequency of derived Neandertal features (Lebel et al., 2001). The sequence of the Neanderthal craniofacial evolutionary development would have started early in the mid Middle Pleistocene, with changes in the facial skeleton (frontal bone, and midfacial and mandibular elements), followed by modifications in the occipital area, and, subsequently, in the temporal bone. According to this model, full Neanderthal morphology became consolidated during OIS 7 and 5 (0.28–0.15 Ma), with the acquisition of the derived features of the temporal cranial region. "We see evidence for a slow accretion of Neanderthal characteristics beginning in early Middle Pleistocene European hominin populations, continuing through the middle Late Pleistocene, and finally resulting in the Würm classic Neanderthals" (Dean et al., 1998).

Such an evolutionary process would have developed in unstable paleogeographic conditions, which could result in isolation in Europe of Middle Pleistocene populations, precluding, or largely limiting, genetic exchange with other contemporaneous demes [the accretion model has been criticized on the basis that gene flow cannot be rejected (Hawks and Wolpoff, 2001)]. Such a situation would allow intensive dynamics of populations at the genetic level, ultimately determined by climatic instability during the European Pleistocene, which was on average colder than present, and periodically exposed to glacial periods. In agreement with that hypothesis, Neanderthals seem cold adapted (e.g., hyper-arctic), particularly in their post-cranial anatomy, as discussed above. By contrast, the mosaic distribution of craniofacial features, with different combination of derived features in individuals of the same or close populations (Rosas, 1992; Arsuaga et al., 1997a; Hublin, 1998a), does not find a consistent explanation. As an example, Arsuaga et al. (1997a) point out how Petralona shows a Neanderthal face and a primitive occipital bone, whereas Steinheim shows derived features in the occiput and a primitive face. Also, mandibles II and XIII from Arago show a fairly different distribution of apomorphies (Rosas et al., 1991; Rosas, 2000). Stochastic processes are then invoked to account for this pattern. In this perspective, the evolutionary emergence of Neanderthal features is considered mostly as the sum of small independent changes affecting specific anatomical systems (Arsuaga et al., 1997a). However, it has also been claimed that morphological variation within the *H. heidelbergensis* populations follows a systematic pattern, especially well-illustrated in the Atapuerca-SH mandibular sample (Rosas, 1995, 1997; Rosas and Bastir, 2004). In brief, while the smaller specimens can be considered morphologically primitive, the larger mandibular specimens display morphological traits typical of classic Neanderthals. Based on this size-related distribution of apomorphic features, it has been claimed that such a systematic pattern of variability contains a hitherto hidden phylogenetic signal (Rosas, 1992). Additionally, the anatomical distribution of derived features, as well as their coordinated evolutionary emergence, denotes an underlying phenotypic organization whose unraveling might be approached from an organismic point of view (Rosas, 2001).

In contrast to the "accretional" view of evolutionary change, a number of authors have put forward the concept of phenotypic integration to deal with the necessary functional and developmental interdependence among parts in functional organisms (Olson and Miller, 1958; Riedl, 1975; Lieberman, 1995; Wagner, 1996). Indeed, one of the current main goals of paleobiology is to understand how highly-integrated developmental programs are modified, producing phylogenetic change (Gould, 1977; Alberch et al., 1979; Ross and Henneberg, 1995). The study of the origin and variation of Neanderthal apomorphies serves as an example for delineating how individual combinations of derived features occur, and to look for possible causes.

THE ORGANISMIC MODEL

Here we present a so-called "organismic model", a systems-based approach to the study of evolutionary change of the Neanderthal lineage, in which the interrelations between factors controlling local anatomical and functional systems and those affecting the body as a whole are considered (Bastir, 2004). The conceptual framework presented here is based on a systems-approach to morphological analysis, and has been developed in previous epistemological and evolutionary works (Bertalanffy, 1950; Riedl, 1975; Wagner, 1996), modern craniofacial biology (Moss, 1997a, b; Enlow, 1990), and classical comparative anatomy and primatology (Dabelow, 1929, 1931; Biegert, 1957; Hofer, 1965). This theoretical framework assumes different scales of complexity in biological organisms and a hierarchical organization of the phenotype, and recognizes the necessity to analyze potential morphogenetic effects in the light of this hierarchy. The skull is comprehended as a "craniofacial system" characterized by its components (Enlow et al., 1969; Enlow et al., 1971) or modules that are structurally and functionally interwoven with each other and with the properties of the complete organism (Bastir, 2004) via morphological integration (Olson and Miller, 1958; Chernoff and Magwene, 1999). As a result, variation in the skull is exposed to and shaped by different kinds of morphogenetic influences. Two different "directions" can be identified in this respect which have been termed here "top-down effects" and "bottom-up effects" that may help one to understand and order possible relationships between the skull, its components, and the body (Figure 2).

The concept of modularity and different functional cranial components is based on the functional matrix hypothesis (Moss, 1962, 1997a, b) and Enlow's counterpart analysis (Enlow et al., 1969, 1973; Enlow and Hans, 1996). Both frameworks rest on the idea of morphological integration (Olson and Miller, 1958; Chernoff and Magwene, 1999) to explain interdependent variation patterns among spatially related structures.

Within the principles of the counterpart analysis (Enlow et al., 1969; Enlow and McNamara, 1973; Enlow and Hans, 1996), which provides a spatio-temporal model of craniofacial morphogenesis, it has been suggested that the hominin skull is developmentally patterned into different sets of craniofacial structures. The counterpart analysis organizes the face and base into two vertical columns, separated by the posterior maxillary plane (PM-plane) (Enlow, 1968; McCarthy and Lieberman, 2001). Growth counterparts anterior to the PM-plane have been termed anterior face and consist of the anterior cranial fossae (housing the frontal lobes of the brain),

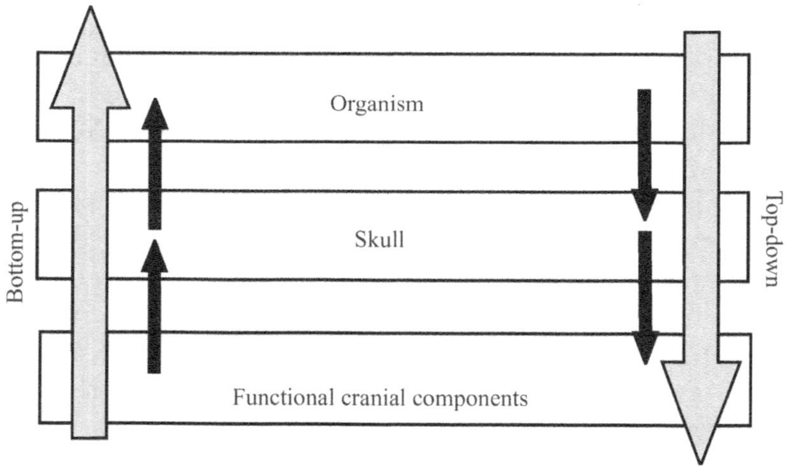

Figure 2. "Complexity scales." Our approach to the variety of features characterizing Neanderthals is based on two tightly interrelated theoretical concepts (explaining integrated packages of features); a hypothesis of hierarchical systems that attempts to order morphogenetic effects within different levels of organism complexity and a hypothesis of cranial modularity, which permits these different morphogenetic effects. Some aspects of skull morphology relate to body size (large spatial scale of organism, higher complexity), here referred to as "top-down effects." Other aspects are related to properties (i.e., size and shape) of functional cranial components, (*sensu* Moss; lower spatial scale of organism, low complexity), and are considered to have "bottom-up effects" on skull morphology.

the nasomaxillary complex, and the mandibular corpus. The posterior face, behind the PM-plane, is built by counterparts comprising the middle cranial fossae (housing the temporal lobes of the brain), the pharyngeal space, and the mandibular ramus. Therefore, the spatial aspect of this hypothesis consists of an anatomical division of facial components. In a further implementation of this perspective, two sets of structures can be coronally appreciated: those located at the sagittal plane of the skull (e.g., cribriform plate, sphenoidal plane, and spheno-occipital clivus), and those structures located bilaterally to the midline, i.e., the anterior and middle cranial fossae (frontal and temporal lobes) (Bastir et al., 2004; Bastir and Rosas, 2005). The study of the interactions between changes related to mid-sagittal dynamics and processes occurring at the lateral parts of the cranium may result in a clearer picture of the craniofacial evolution of the Neanderthal lineage.

The temporal aspect of the counterpart model is provided by an ontogenetic sequence of morphological maturation, starting with the brain, followed by the cranial base and finally the facial skeleton. Subsequent studies have supported this hypothesis of sequential maturation (Buschang, 1983; Bastir et al., 2006). In short, Enlow and co-workers (Enlow, 1968; Enlow et al., 1969) propose that the human spatio-temporal pattern of development is ultimately determined by corresponding parts of the brain, suggesting that the earlier maturing brain defines spatial conditions of the cranial base, on which the later growing facial structures attach. Therefore, limitations to the spatial extension of the face are set to some degree by the morphological properties of the brain and its different compartments and growth counterparts.

It is thus reasonable to work on the hypothesis that brain enlargement occurring in the evolution of the Neanderthals might have produced a significant spatial reorganization of the cranial base; an hypothesis previously explored or invoked by several authors (Rosas, 1992; Lieberman et al., 2000; Spoor et al., 2003; Hawks and Wolpoff, 2001; Ponce de León and Zollikofer, 2001). Brain increment – essentially mosaic in mode – should differentially affect the ontogenetic development of the integrated regions of the craniofacial complex. On the other hand, comparative and ontogenetic analyses have shown tight scaling relations between body size and the face, suggesting a somatic developmental pattern of the face rather than a neural one (Biegert, 1957; Baume et al., 1983). Thus, it seems that facial size and shape is also affected by body size increase via growth integration. Therefore, according to the organismic model, the Neanderthal skull form may result from two interconnected effects at different levels of organization: (1) body-face interactions via organismic-level effects (e.g., allometry, physiological demands); and (2) brain-face interactions via counterpart principles. With respect to the former it is suggested that eco-physiological adaptations involved in the resistance to cold stress, resulting in larger body size and possibly increased metabolic rates, might be part of a large-scale evolutionary process affecting size and shape of the face. With respect to the latter it is suggested that encephalization in Neanderthals may have modified specific proportions of the brain with consequences for lateral basicranial architecture and the attached facial and mandibular structures.

Results

BODY-FACE INTERACTIONS

Scaling and the Interaction of the Body with Craniofacial Morphology

Irrespective of early and mainly theoretical propositions about potential relationships between the size (and possibly shape) of the organism and the morphology of the skull (Dabelow, 1929; Biegert, 1957), little work has been done to advance the quantitative analysis of this problem (Bastir, 2004). However, two basic ideas may be proposed. The first, originally based on the work of Dabelow (1929), suggests a positive correlation between the size of the facial skeleton and the size of the body. In a broad comparative anatomical framework, Dabelow (1929) deduced that the size of the body and its physiological necessities would be directly reflected in the size of the masticatory system. The second idea was proposed by Biegert (1957, 1963) in the context of a primatological framework. He analyzed phylogenetic and ontogenetic parallelisms and found that the positive allometry of the face, with respect to the size of the body, has also potential consequences for the position of the facial skeleton and the overall morphology of the skull. In particular, he suggested that large body size (and the correlated increase of facial size) might produce basicranial extension and upward rotation of the anterior cranial floor and the face. This argument is grounded on the analysis of both ontogenetic and phylogenetic considerations of increasing body size, where parallel modifications in facial orientation within ontogenetic series of chimpanzees and gorillas, as well as between adult chimpanzees (smaller) and gorillas (larger), can be seen.

Therefore, from both sources the hypothesis can be proposed that increase in body size has consequences for the orientation and proportions of the facial skeleton, and thus its structural relation within the skull. These observations may be related to the increase of body size in mid-Pleistocene hominins at about 500,000 years ago (Ruff, 2002), and may serve as a basis for the analysis of such an increment in skull morphology.

In this latter regard, Enlow and Azuma (1975) suggested that the orientation of the upper and midfacial skeleton maintains a constant spatial relationship, which was recently further supported in a broader quantitative context by Lieberman and colleagues (Lieberman et al., 2000; McCarthy and Lieberman, 2001). Another hypothesis of Enlow and colleagues (Enlow, 1968; Enlow and McNamara, 1973; Enlow and Hans, 1996) suggested developmental and functional integration between the nasomaxillary skeleton and the mandible. Again, quantitative reanalysis (Bastir, 2004; Bastir and Rosas, 2004; Bastir et al., 2005) has produced further support for this idea. Nevertheless, all this recent work needs to be linked with earlier primatological studies addressing relationships between body size and facial size in order to produce a congruent picture of skull variation as part of organismic variation.

In order to explore these hypotheses, we present a preliminary analysis of a longitudinal series of modern humans for which both cranial and postcranial data are available. We used partial least squares (PLS) analysis (Rohlf and Corti, 2000; Klingenberg, 2002; Bookstein et al., 2003; Bastir et al., 2005), geometric morphometrics, and the Denver Growth Study (Maresh and Washburn, 1938; Lieberman and McCarthy, 1999; Bastir et al. 2006) to address the hypothesis of potential covariation between craniofacial data (facial proportions and orientation) and stature. The results are shown in Figure 3, where a high and significant correlation ($r = 0.89$, $P = 0.001$) between body size and facial shape is displayed. The associated regression model of shape on log transformed stature accounts for ca. 25% of variation ($F = 56.0736$, df $_{1,2} = 50, 8850$; $P = 0.0000$). Morphologically, stature-facial shape covariation is characterized by a vertical expansion and a posterior-upwards reorientation of the anterior cranial floor and the orbits (with respect to the PM plane).

These results support an hypothesis of allometric anterior cranial fossa variation and its effects on facial orientation and position.

A potential link between these findings and the fossil evidence consists of the specific difference between the anterior cranial floor in Middle Pleistocene humans and that of living people (Seidler et al., 1997). These authors have shown that the angle between orbital orientation and the Frankfurt plane is about 25° in Petralona and Kabwe, while in modern humans the angle is only 12°. Relationships to encephalization or pneumatization have been proposed, but our data, the previously outlined evidence (Biegert, 1957), and recent studies on Middle Pleistocene body size variation (Ruff, 2002), do not exclude an effect of body and facial size on this basicranial pattern of variation.

In the next sections we explore these effects specifically in the European hominin lineage through the study of mandibular variation and its evolutionary change. Particularly, the sample of mandibles from the Atapuerca-SH site will be used as a reference. The size of the mandible is a good indicator of body mass (Wood and Aiello, 1998 and references therein), and a direct relationship between mandibular size and individual body mass can be confidently assumed in the *H. heidelbergensis* and *H. neanderthalensis* samples. Furthermore, human mandibles present allometric growth coordinated with that of the face, and also with the body (Hunter, 1966; Baume et al., 1983; Ranly, 1988; see also Figure 3). In more general terms, hominid facial enlargement may follow the growth-allometry of the body (Biegert, 1957), associated functionally with the masticatory and respiratory systems (Emerson and Bramble, 1993; Enlow and Hans, 1996). Developmentally, these correlated ontogenetic trajectories might be based on hormonal control (growth hormones and other endocrinological factors [IGF] on facial and nasal cartilages [Silbernagl and Despopoulos, 1991; Lieberman 1996; O'Higgins et al., 2006]). Additionally, the volumetric expansion of the oro-naso-pharyngeal functional spaces has been suggested to drive facial growth and is conceived of as a neurotrophically-regulated

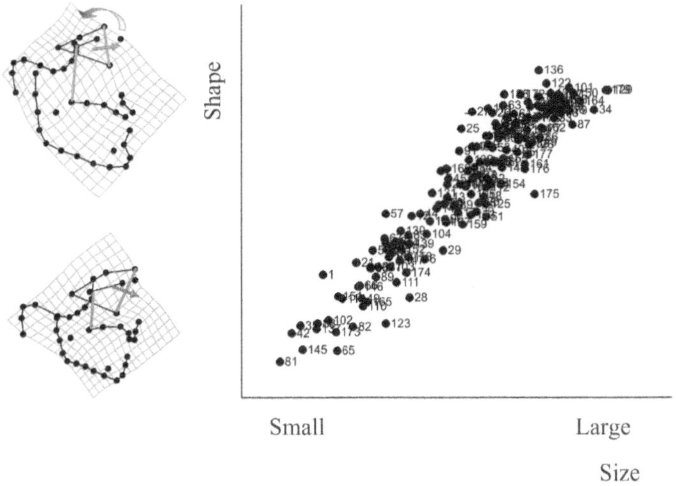

Figure 3. Analysis of the Denver Growth Study dataset, combining craniofacial and postcranial data. 180 individuals were analyzed by geometric morphometrics and Partial Least Squares analysis addressing the potential covariation between body size (stature) and facial size and shape. This analysis reflects a strong developmental integration relation between body size and facial size and position. The figure shows how vertical facial proportions and upper facial position is influenced by the ontogenetic increase of stature. Increasing body size is tightly correlated with an allometric elevation of the anterior cranial floor, indicated by the reorientation of the arrows at the orbits.

epithelial growth process (Moss, 1962; Enlow and Hans, 1996).

H. heidelbergensis Mandibular Variation

Morphological variation within *H. heidelbergensis* populations follows a systematic pattern, largely related to size, which is especially well-illustrated in the mandibles from the Atapuerca-SH site (Rosas, 1995, 1997, 2001; Rosas and Bastir, 2004) (Figure 4). Interestingly, with increasing size, there is morphological progression in the development of a retromolar space and the backward displacement of the mental foramen, a set of features usually related to midfacial prognathism (Coon, 1962; Wolpoff, 1980; Stringer et al., 1984; Rak, 1986; Trinkaus, 1993; see also Franciscus and Trinkaus, 1995). Additionally, there is a morphological progression in the symphysial region from a low and slanting symphysis in small specimens, to larger mandibles with higher symphysis featuring accentuated morphological traits on the external side, i.e., alveolar depression (Figure 4). The allometric pattern detected in the mandibles of *H. heidelbergensis* may result from local functional causes (e.g., masticatory function) or from more general processes at the craniofacial and/or organismic level. Functional masticatory explanations, nonetheless, have failed to account for the derived Neanderthal facial morphology (Antón, 1994; O'Connor et al., 2005). Accordingly, it seems biologically sound to consider that variation of the mandibles may express morphologically processes proper at the organismic level. If this is the case, why should body size/mass variation produce differences in morphological features with phylogenetic valence, e.g., retromolar space?

According to Franciscus and Trinkaus (1995) it is "a combination of relatively shorter dental arcade lengths and smaller ramus breadths in the context of maintained (or only slightly reduced) mandibular lengths which best accounts for the presence of retromolar spaces." The pattern of mandibular variation in the European mid-Pleistocene hominins seems to support this conclusion. Thus, Figure 4 shows how size-related

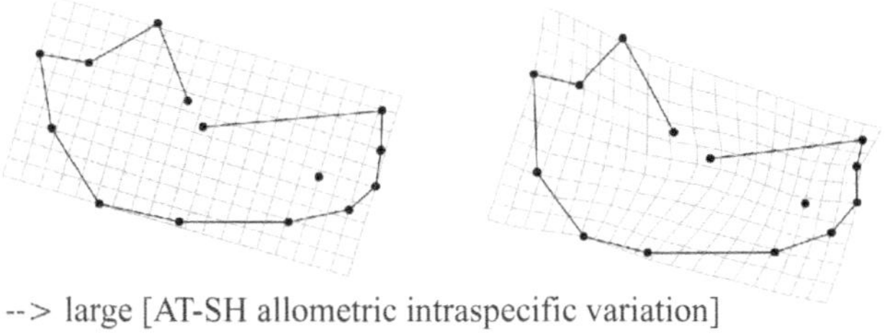

Small --> large [AT-SH allometric intraspecific variation]

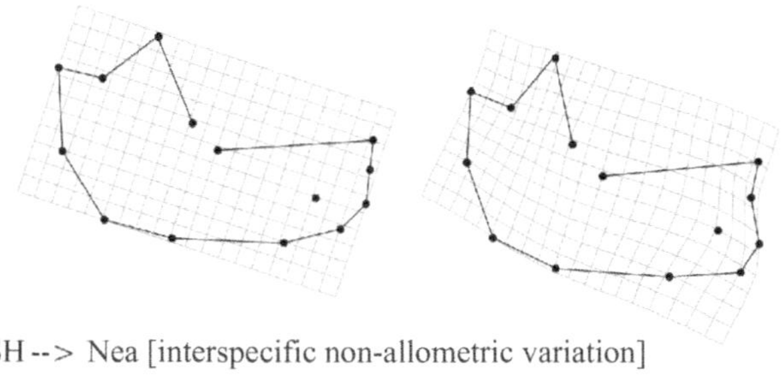

AT-SH --> Nea [interspecific non-allometric variation]

Figure 4. Allometric and non-allometric variation in mandibles. Figure A shows allometric intraspecific variability of AT-SH hominins. A small individual (upper left TPS grid) becomes transformed into a large one (upper right TPS-grid) (Rosas and Bastir, 2004). Figure B show a TPS transformation of an AT-SH hominin into a Neanderthal. The data have been standardized to a common size value by multivariate regression, so allometric variation has been removed (Bastir, 2004).

variation in the SH sample is predominantly established along an anterior-posterior direction in these hominins (Rosas and Bastir, 2004). In addition, morphological variation is mostly expressed as a forward movement of the mandibular corpus (anterior face), while the ramus (posterior face) is more stable. When related to maximum length of the mandible, the increment of the ramus breadth (i.e., posterior face) is proportionally smaller than the increment of the corpus length (i.e., anterior face). Unlike modern humans, in which facial growth follows a largely vertical increment, it can be proposed that craniofacial growth in European mid-Pleistocene hominins is predominantly oriented in a horizontal direction. Next, we approach this hypothesis with the study of bone remodeling in the mandibles from the Atapuerca-SH site.

Bone Remodeling Pattern of the Atapuerca-SH Mandibles

Following the works of Boyde (1972), Boyde and Jones (1972), Boyde and Hobdell, (1969), and Bromage (1984, 1989), we have identified different patterns of microanatomy: (1) the areas in which collagen bundles can be recognized – we have interpreted this as deposition; (2) areas in which organized depressions can be recognized – we have interpreted this as areas of osteoclast activity representing the Howship's lacunae; that is area of resorption activity; (3) areas in which a pattern of small foramina is recognized – we interpreted this as Sharpey fiber insertions; (4) areas in which a polished surface is recognized – these areas seem to correspond to areas of prolonged resting, upon which ground matrix has been mineralized (Boyde, 1972).

Figure 5 shows a reconstruction of the bone remodeling pattern in the Sima de los Huesos mandibular specimens (Rosas and Martínez-Maza, 2006). One of the most significant features detected is the variation in the pattern of remodeling at the anterior symphysis (Martínez-Maza and Rosas, 2002). The individual variation in the remodeling of the symphysis seems to follow a systematic pattern parallel to that of the morphology. The larger specimens show a remodeling distribution similar to that found in *H. sapiens* (e.g., resorption at the subalveolar depression and deposition at the base), whereas the smaller specimens display a mosaic of fields not previously documented in hominins, defined by a field of deposition at the alveolar part and an area of resorption at the basal part of the symphysis. Such remodeling field distribution gives support to the mainly horizontal growth direction of the facial skeleton in *H. heidelbergensis* as deduced from the morphology.

The corpus presents, unlike the symphysis, a constant distribution of the growth-remodeling in all the specimens. Among the most interesting features, the internal corpus presents a large field of resorption on the subalveolar plane and fossa. This resorption field suggests an intensive lateral expansion of the corpus (Figure 5), whose interpretation may be supported when considered in combination with a number of morphological features. Thus, mandibles of *H. heidelbergensis*, as well as those of classic Neanderthals, present a wide anterior mandible (Trinkaus, 1982; Stringer et al., 1984; Rosas, 1995), and a field of resorption at the corpus that may provide a mechanism for the lateral widening of the symphysis. The large mesiodistal diameters of the anterior dentition in the European hominin lineage (Bermúdez de Castro, 1986) may also be inscribed in this morphological context. If the proposed growth mechanism were the case, then it may also explain the outstanding mesial drift of the molar teeth observed in the Atapuerca-SH mandibles and maxillae, as deduced from the strong distal inclination of the roots. A lateral extension of the mandibular arcade would produce wider symphyses, which, in turn, would result in an increase of the interdental spaces. A well-known mechanism in craniofacial growth for readjusting dental alignment and occlusion is the mesial drift of the dentition (Enlow and Hans, 1996). An intensive mesial drift of the molar teeth seems to be also the case in the SH sample, as judged by the inclination of the roots mentioned above.

Moreover, several other consequences may be derived from these observations. If the complex growth process we are discussing is correct, the anterior location of the mental foramen (i.e., below M_1) observed in the mandibles of the Neanderthal lineage would also be the result of a forward migration of the dentition, especially the molars, rather than just a posterior displacement of the foramen itself, in agreement with Trinkaus (1993). Likewise, the relatively shorter dental arcade lengths, considered by Franciscus and Trinkaus (1995) to be one of the factors involved in the appearance of the retromolar space, may be due proximately to anterior broadening and concomitant molar mesial drift. In a broader morphological context, a wide anterior part of the mandible may be in causal correspondence with other features associated with the midfacial prognathism, such as a wide nose, broad glabella area, parasagittal orientation of the infraorbital plane. Along with this explanation, the aforementioned lateral expansion of the corpus may be the local expression of general processes affecting complete anatomical and/or functional systems.

Respiratory Physiology in the European Lineage

The general statement that Neanderthals had wide noses (although Middle Pleistocene hominins also have very broad noses) and spacious nasal cavities has led to the consideration of the respiratory physiology as a possible evolutionary determinant. However, the potential evolutionary meaning of respiratory physiology has been restricted to life performance under cold climatic conditions, either in dry or humid circumstances. As reviewed above, there have been several attempts to relate facial morphology of Neanderthals to climatic adaptations, arguing for mechanisms

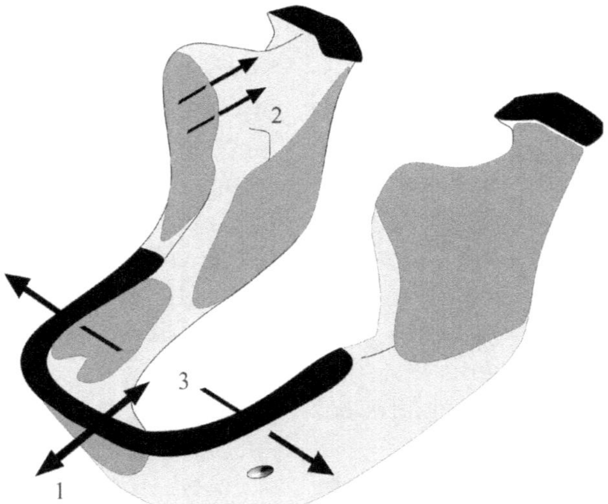

Figure 5. A reconstruction of the bone-remodeling pattern in the Atapuerca-SH mandibles. Dark gray color indicates resorption field activity. Major direction of growth and displacement are indicated by numbers. 1. A resorption field at the anterior symphysis producing alveolar depression; 2. A resorption field at the anterior border of the ramus, giving rise to ramus relocation; 3. A resorption field at the lateral body that enables a lateral expansion of the mandibular arcade.

for warming and moistening inspired air. The discovery of arguably new autapomorphies in the internal nasal cavity has further led to the proposition that Neanderthals had a highly derived respiratory anatomy (Schwartz and Tattersall, 1996; Laitman et al., 1996; Schwartz et al., 1999).

However, the simple fact that large nasal cavities in Neanderthals and their forerunners may respond to respiratory demands, referring to lung volume and energy production, has not been considered. As stressed by Laitman et al. (1996), "the acquisition and processing of oxygen and its byproducts is the primary mission of any air breathing vertebrate. Chewing, walking, reproducing, thinking are all fine but first it has to breathe". Nasal shape can be primary affected by those factors that facilitate metabolism, and growth mechanisms controlling the topology of the nasal cavity respond to demands of the respiratory system. In spite of these facts, little work has been done to examine how nose size and morphology relate via physiological factors to general body size or to differences in male-female metabolic requirements (Hall, 2005; see, nonetheless, Rosas and Bastir, 2002).

Using sexual dimorphism and metabolism as a starting point for exploring this aspect, analysis of human body composition shows a different pattern in the distribution of body fat, muscle mass, and bony tissue between males and females (Malina, 1978, 1996). Increased male body mass (mainly all muscle mass) and increased basal metabolic rates require an increased energy supply (Henry and Rees, 1991), which must be associated with an elevated oxygen input. Sexual physiological differences are well-known. For the same size and age, the pulmonary capacity is about 15% larger in males than in females (Silbernagl and Despopoulos, 1991). Owing to augmented energy requirements, male airway dimensions have to be enlarged (Enlow and Hans, 1996) in order to maintain integration between airway, relative body, and lung size. It seems that these physiological requirements can also be detected by looking at the hard tissues. Rosas and Bastir (2002) show that, for the same size, the spatial configuration and shape of the whole naso-pharyngeal region in males provides a larger volume of air intake than it does in females. In particular, males present a relatively larger naso-pharyngeal space than females, with larger piriform aperture, achieved by an increase in the angulation of the nasal bones and a downward shift of the anterior nasal floor. Male pharynx expansion was also reflected in larger choanae and a more posteriorly inclined basilar part of the occipital clivus (Rosas and Bastir, 2002). Body size and elevated sexual dimorphism (Wolpoff, 1980; Bermúdez de Castro et al., 2001; Rosas et al., 2002; but see Arsuaga et al., 1997b; Lorenzo et al., 1998) largely contribute to craniofacial variation in Middle Pleistocene populations. Morphological interaction between body size, allometry, and physiological demands linked to male/female different basal metabolism (Rosas and Bastir, 2002 and references therein) may account for a good deal of morphological diversity.

Coming back to Neanderthals, Ruff et al. (1997) estimate a mean height of 167 cm for males ($n = 7$), and mean weight of 80.8 kg; with considerable muscle mass per unit height and muscles substantially greater in mass than those in robust modern humans (Trinkaus, 1997; Churchill, 1998; Steegmann et al., 2002). Neanderthals had voluminous chests and a large lung volume, as there is a direct relationship between lung volume and body mass (Churchill and Franciscus, 2002). The large bodied Neanderthals would have needed large quantities of oxygen in order to produce the level of energy required (Jelinek, 1994). Furthermore, it is generally acknowledged that the elevated postcranial robusticity and pronounced muscle markings on the Neanderthal bones reflect heavy mechanical loading resulting from a lifestyle of intense physical work (Ruff et al., 1993, 1994; Trinkaus et al., 1994;

Abbott et al., 1996; Trinkaus, 1997). When this picture is further projected onto a climatic scenario it is inferred that under glacial conditions Neanderthals had to maintain higher metabolic rates, which would, in turn, have required a greater intake of oxygen and more frequent consumption of high calorie food (see Schwartz et al., 1999, Churchill, this volume). How these body mass/shape, as well as metabolic rate, effects relate to the anatomy of the respiratory apparatus, and how growth processes become organized in order to accommodate these physiological demands is a matter of further research. However, for the time being there is sufficient evidence to conclude that the European hominin lineage developed specific physiological demands manifested in changes of the nasal apparatus, both in size and shape. Functional and structural integration of these novelties within the craniofacial complex contribute to the development of the specific morphological pattern.

To complete the picture, Ramirez-Rozzi and Bermúdez de Castro (2004) see Neanderthals as a species of *Homo* adapted to particular environmental conditions in which a high calorie diet and a high metabolic rate were able to fuel fast somatic growth, as well as to grow and sustain a large brain. And, taking our inquiry further, what could have been the relevance of brain size and shape on Neanderthal craniofacial evolution?

BRAIN-FACE INTERACTIONS

Brain Evolution and Reorganization of the Cranial Base in H. neanderthalensis

Absolute brain size increases in Middle Pleistocene specimens, although there is also increased encephalization; that is, the change in cranial capacity is not attributable solely to an increase in body size (Rightmire, 2004). Additionally, cranial capacity also increases significantly within the European lineage (Hawks and Wolpoff, 2001; Rightmire, 2004). Middle Pleistocene specimens show a cranial capacity of around 1206 cc, whereas classic Neanderthals may reach up to 1650 cc (Aiello and Dean, 1990; Rightmire, 2004). Did encephalization produce changes in craniofacial structure in the European hominin? Several lines of evidence allow us to answer this question positively, but what were the specific effects?

The possible effects of encephalization on the structure and development of the craniofacial complex are of two kinds. First, the control exerted by the nervous system, as the pacemaker of somatic development (Sacher, 1975; Smith, 1989) and metabolism, which results from it being an energetically expensive tissue (Aiello and Wheeler, 1995). Second, the mechanistic effects of the growing brain on basicranial and facial structures (Anthony, 1954; Enlow, 1990; Rosas, 1998; Lieberman et al., 2000; 2002; Spoor et al., 2003; Bastir et al., 2004; Bastir and Rosas, 2005). In spite of the multiple assessments invoking substantial influence of brain evolution on craniofacial shape, no consistent picture has yet emerged. An essential line of research is the direct testing of hypotheses dealing with specific aspects of brain development and those of the cranial base (Jeffery and Spoor, 2002; Jeffery 2003). Below, we present our current research addressing covariance among bone structures, to further speculate on the possible effects of differential brain regional development on morphology.

Craniofacial biologists have suggested that basicranial variation has crucial effects on variation in mandibular ramus breadth in modern humans (Bhat and Enlow, 1985; Enlow et al., 1982). Recent studies have provided further support for this hypothesis (Bastir et al., 2004; Bastir and Rosas, 2005). However, it was found that the spatial position of the petrosal temporal, rather than the flexure of the midline cranial base, is correlated with mandibular ramus breadth. In humans a lower position of the petrosal bone (and also the glenoid fossa) was significantly correlated with a narrower mandibular ramus. In addition, these relationships did not depend on allometric variation (Bastir et al., 2004). The lateral regions of the cranial base show a much larger integration with the mandible than that shown by the midline of the cranial fossa, at least in modern humans (Bastir and Rosas, 2005). These findings are relevant for the interpretation of morphological variation in relative ramus breadth among Middle Pleistocene humans.

A recent study, building on this evidence, tested a model of allometric scaling in mandibles of *Homo heidelbergensis* and *Homo neanderthalensis*, by addressing particularly the non-allometric fraction of morphological variation between these two species (Bastir, 2004). It was found that the non-scaling difference between the species is expressed in Neanderthals by a further reduction in the relative ramus breadth, a forwardly directed mandibular condyle, and a marked development of alveolar depression. Therefore, and in the light of the previous findings of integration in the posterior face, it is possible to relate ramus breadth variation to localized, non-allometric spatial modification in posterior face structures. As stated above, the temporal bone and mandible are directly involved in the evolutionary emergence of Neanderthal craniofacial morphology. In the light of these findings, temporal bone modification may be related to parasagittal processes in the skull, including the middle cranial fossae. However, no obvious relationship links temporal bone evolutionary changes to physiological or climatic adaptations. Alternatively, the large increment in brain volume in the Neanderthal lineage may be invoked as a causal factor for these transformations. Anatomically, a significant aspect to be considered, apart from those already mentioned, is the inferior position of the posterior semicircular canal in *H. neanderthalensis* and its possible functional implications in head position and neck movements (Spoor et al., 2003).

The potential effects of mosaic brain evolution, in the temporal or parietal lobes for example (Bruner et al., 2003; Bastir et al., 2005), are potentially identified through evolutionary changes in their bone counterparts. For instance, the difference in the projection of the mastoid between mid-Pleistocene European hominins and Neandethals probably reflects the posterior-inferior expansion of the cranial cavity (Hawks and Wolpoff, 2001).

Bruner et al. (2003) identify two distinct allometric trajectories for the evolution of the structural relation between cerebral shape and encephalization in *Homo*, for which a similar expansion of endocranial size has been reached through different changes in shape. European Middle Pleistocene specimens and Neanderthals correspond with *H. erectus s.l.* and other non-modern human specimens. Neanderthals are hypothesized to be derived in their allometric variability and degree of encephalization (quantity), but plesiomorphic in their structural and functional processes (quality) (Bruner, 2004). Morphologically, the brain undergoes along the archaic trajectory a generalized dorsal bending as it enlarges, as a result of relative parietal shortening (Bruner, 2004). Dorsal bending of the brain may concomitantly pull up anterior and posterior poles of the brain, forcing underlying basicranial structures, and contributing further to the platycephalic form of the cranium.

Discussion

NEANDERTHAL EVOLUTIONARY LINEAGE: A TWO-PHASE PROCESS

A number of interrelated hypotheses dealing with potential causal processes underlying the evolution of the Neanderthal morphology have been proposed. A two-phase model has been outlined and its taxonomic interpretation has also been considered.

The first phase starts in the early Middle Pleistocene with the increase of body size and robusticity of the postcranium (Lorenzo et al.,

1998; Ruff, 2002). This may have produced changes in upper and midfacial anatomy, orientation, and proportions (Biegert, 1957; Seidler et al., 1997; Rosas, 1998). However, as upper, mid and lower facial structures are tightly integrated (Lieberman, Bastir, 2004; Bastir et al., 2004; 2005), these allometric upward rotations find a direct relationship on the allometric variation patterns in the mandibles (Rosas, 1997; 2000; Rosas and Bastir, 2004) and the associated remodeling patterns (Martínez-Maza and Rosas, 2002; Rosas and Martínez-Maza, 2006). The most outstanding morphological novelty is the emergence of midfacial prognathism, which is essentially produced by a twofold effect: a relatively shorter dental arcade length and smaller ramus breadth, in the context of maintained total facial prognathism (Franciscus and Trinkaus, 1995; Trinkaus, 2003). Physiological needs imposed by large body mass and high metabolic rates impose an enlargement of the nasal cavity giving rise to a broadening of the anterior face. This is expressed in the broad noses and flattening of the anterior mandible (Stringer et al., 1984; Rosas et al., 1991). The broadening of the face produces a larger and broader alveolar space that needs to be filled in order to maintain occlusal function. Such a circumstance gives rise to a forward movement of the posterior dentition, clearly manifested by mesial drift. On the other hand, variation in modern humans, both between and within populations, establishes a strong correlation between ramus breadth and vertical position of the petrosal bone in such a way that a descent of the external end of the petrosal significantly correlates with reduction of ramus breadth. Even though no direct testing with fossils has been attempted, it is reasonable to assume that a descent of the temporal area, as manifested by the flattening of the glenoid fossa (Martínez and Arsuaga, 1997a) has occurred. This phase of evolution of the Neanderthal lineage is defined by the species *H. heidelbergensis*.

During a second phase, a particular localized evolutionary process further modified the orientation and position of the petrosal bone, lowering the glenoid fossa with respect to other basicranial traits (Harvati, 2003). Such processes could either be related to general (Rightmire, 2004) or specific mosaic encephalization (Bruner et al., 2003), or to other localized phenomena that have yet to be identified. In addition, the spatial location of the digastric and mylohyoid muscles became altered. There is an evolutionary change in the spatial positioning of the bone and soft tissue matrix, especially affecting the suprahyoid muscles, which modify both the temporal bone and internal side of the mandible (Martínez and Arsuaga, 1997; Rosas, 2001). As a result of these processes, a distinct morphology emerged, qualitatively different from its putative ancestral condition. We interpret such a qualitative change as being related to a speciation event that gave rise to *H. neanderthalensis*.

Variability follows a contrasting pattern when comparing samples of *H. heidelbergensis* and *H. neanderthalensis*. There is a decrease in variation in late Middle to Late Pleistocene European populations, once the final morphological pattern of classic Neanderthals is achieved. This reduced level of morphologic variability has been attributed to genetic isolation. Alternatively, structural causes could be proposed. Our model states that the large variability observed in the European Middle Pleistocene craniofacial complex is mostly due to combinations of growth compensatory mechanisms (*sensu* Enlow et al., 1982; Enlow, 1990). Structural and functional imbalances in the growing craniofacial complex (e.g., dental occlusion), in the context of large body size variation and sexual dimorphism, give rise to distinct combinations of features. It is in this context that the distinction between primary processes and secondary consequences (Trinkaus, 1987, 2003; Franciscus, 1999, 2003) acquires a greater biological meaning. In classic Neanderthals, the craniofacial system becomes stabilized with decreased sexual dimorphism and variation in body size, and craniofacial imbalances and compensatory growth mechanisms become largely reduced.

SYSTEMATIC AND TAXONOMIC CONSIDERATIONS

Taxonomically, the "accretion model" considers a gradual and continuous emergence of distinctive features, whose evolutionary process would have consequently occurred throughout a single temporally variable species, paleontologically recognized as a chronospecies. Accordingly, from a biological viewpoint, the complete continuum of evolving populations, whose later representatives we name Neanderthals, should receive a single taxonomic nomen, i.e., *Homo neanderthalensis* (Stringer, 1996). The total pattern of morphology is achieved sometime in the later Middle Pleistocene, especially in the anatomy of the temporal bone and mandible. This transition has been associated with an operative use of *H. heidelbergensis* and *H. neanderthalensis* taxonomic nomina, on the basis of a single chronospecies concept (Arsuaga et al., 1997a; Rosas and Bermúdez de Castro, 1998).

In an alternative view, the appearance of Neanderthals may be identified as a proper speciation event that took place at around 0.3–0.25 Ma (Rightmire, 1998, 2001; Lalueza-Fox et al., 2005), and two real species are then recognized within the Neanderthal lineage: *H. heidelbergensis* and *H. neanderthalensis*. We favor the hypothesis that the species *H. neanderthalensis* emerged as a distinct biological entity after a speciation event (or change in morphological quality). Interestingly, such a speciation event is coincident in time with the appearance of the cultural Mode 3 (Mousterian industries) in Europe. While clearly favoring the origin of the lineage in the European middle Middle Pleistocene [or even earlier, (Rosas, 2000)], we identify the emergence of Neanderthals – in agreement with Rightmire (2001) – at around 0.3–0.25 Ma. The easiest way to explain such an event would be to invoke a population crisis or bottleneck sometime during the late Middle Pleistocene (e.g., the cold period OIS 8) from which the hominin fossil record is particularly poor (Hublin, 1998a; Dean et al., 1998). However, no real evidence is available to support such a statement, and additional data are needed. What we maintain is that Neanderthals are a morphologically distinct species (see Pigliucci, 2003). We agree with Tattersall (Tattersall, 1992; Tattersall and Schwartz, 1998, this volume) in recognizing the existence of separate species when distinct morphs (i.e., recognizable morphological entities distinguished by reasonably consistent sets of characters) are present in a fossil sample. There is no doubt that Neanderthals (both classic and Near Eastern) are recognizable as a distinct morph, not only from anatomically modern humans, but also from their putative ancestors. In potentially strong support of this view, recent studies on dental growth histology have been able to discriminate between life-history variables (e.g., rate of growth) of *H. heidelbergensis* and *H. neanderthalensis*, the latter showing an autapomorphic very rapid rate of tooth growth, and possibly an ontogenetic rapid pace and short developmental period (Ramírez-Rozzi and Bermúdez de Castro, 2004).

OPEN PROBLEMS AND FUTURE RESEARCH

When most of the available information is considered it becomes clear that craniofacial evolutionary change in the European lineage is a complex process in which a number of factors come together and result in a new "morph." Physiological and developmental factors interacting in the context of structural constraints, specific to the hominin skull, may explain the emergence of the unique morphological pattern of Neanderthals. Moreover, evolutionary transformation in the craniofacial anatomy of Neanderthals is clearly a 4-D process (three dimensional morphology plus time). Therefore, we will be able to account for the observed changes only in an actual spatio-temporal framework. Most of the studies approach the problem of Neanderthal anatomy from a lateral perspective, e.g., prognathism. However, consideration of the coronal aspects of the skull, e.g., vertical relocations of the petrosal in the lateral cranial base versus midline basicranium, may potentially clarify many issues.

It is becoming increasingly accepted that specific characters appear early in the ontogeny, as early as immature fossil specimens allow recognition, while fetal stages are invoked to explain the processes (Zollikofer and Ponce de León, this volume). It is clear that an organism belongs to its species from the very moment of conception, and it is during growth that specific features become morphologically identifiable. Increasing knowledge in craniofacial biology dictates that *topology* of the different developmental units (or modules) plays a determinant role, and it is upon these topologic relationships that later growth processes become established. Neanderthals seem to accommodate this principle in the sense that an early embryological platform establishes a predominantly horizontal facial growth. Respiratory and metabolic demands are physically achieved by means of oro-naso-pharyngeal expansion, following growth paths pre-established at pattern-formation stages.

What were the causes that produced such new conditions? Were they stochastic, adaptational, or the product of phylogenetic inertia? No firm answer can be offered for these question at present. Nevertheless, the consideration of organismic interactions, although elusive in many ways, is essential to understanding evolution.

Acknowledgements

We are especially grateful to Katerina Harvati and Terry Harrison for their kind invitation to contribute to this volume and also for organizing the meeting on "Neanderthals Revisited" held in New York City, January 2005. K. Harvati, G. Manzi and an anonymous referee gave substantial inputs for reorganizing and improving earlier versions of the manuscript. We thank Dan Lieberman, Chris Dean, Chris Ruff and Roger Siervogel for providing data and for help on the Denver Growth Study. This research is included in the framework of the Project BOS 2003–01531 of the Dirección General de Investigación of the Spanish Government.

References

Abbott, S., Trinkaus, E., Burr, D.B., 1996. Dynamic bone remodeling in later Pleistocene fossil hominids. *Am. J. Phys. Anthropol.* 99, 585–601.

Aiello, L., Dean, C., 1990. *An Introduction to Human Evolutionary Anatomy*. Academic Press, London.

Aiello, L.C., Wheeler, P., 1995. The expensive-tissue hypothesis. *Curr. Anthropol.* 36, 199–221.

Alberch, P., Gould, S.J., Oster, G.F., Wake, D.B., 1979. Size and shape in ontogeny and phylogeny. *Paleobiology* 5, 296–317.

Anthony, J., 1954. L'influence des facteurs encéphaliques sur la brisure de la base du crâne chez les Primates. *Ann. Paléontol.* 58, 3–11.

Antón, S.C., 1994. Mechanical and other perspectives on Neandertal craniofacial morphology. In: Corruccini, R.S., Ciochon, R.L. (Eds.), *Integrative Paths to the Past: Paleoanthropological Advances in Honor of F. Clark Howell*. Prentice Hall, Englewood Cliffs, NJ, pp. 677–695.

Arsuaga, J.L., Martínez, I., Gracia, A., Lorenzo, C., 1997a. The Sima de los Huesos crania (Sierra de Atapuerca, Spain). A comparative study. *J. Hum. Evol.* 33, 219–281.

Arsuaga, J.L., Carretero, J.M., Lorenzo, C., Gracia A., Martínez, I., Bermúdez de Castro, J.M., Carbonell, E., 1997b. Size variation in Middle Pleistocene humans. *Science* 277, 1086–1088.

Arsuaga, J.L., Lorenzo, C., Carretero, J.M., Gracia, A., Martínez, I., García, N., Bermúdez de Castro,

J.M., Carbonell, E., 1999. A complete human pelvis from the Middle Pleistocene of Spain. *Nature* 399, 255–258.

Bailey, S.E., 2002. A closer look at Neanderthal postcanine dental morphology, I. The mandibular dentition. *Anat. Rec.* 269, 148–156.

Bailey, S.E., 2004. A morphometric analysis of maxillary molar crowns of Middle-Late Pleistocene hominins. *J. Hum. Evol.* 47, 183–198.

Bastir, M., 2004. A geometric morphometric analysis of integrative morphology and variation in human skulls with implications for the Atapuerca-SH hominids and the evolution of Neanderthals: Structural and systemic factors of morphology in the hominid craniofacial system. Ph.D. Dissertation, Autonoma University of Madrid, Madrid.

Bastir, M., Rosas, A., 2004. Facial heights: evolutionary relevance of postnatal ontogeny for facial orientation and skull morphology in humans and chimpanzees. *Hum. Evol.* 47, 359–381.

Bastir, M., Rosas, A., 2005. The hierarchical nature of morphological integration and modularity in the human posterior face. *Am. J. Phys. Anthropol.* 128, 26–34.

Bastir, M., Rosas, A., Kuroe, K., 2004. Petrosal orientation and mandibular ramus breadth: evidence of a developmental integrated petroso-mandibular unit. *Am. J. Phys. Anthropol.* 123, 340–350.

Bastir, M., Rosas, A., Sheets, D.H., 2005. The morphological integration of the hominoid skull: A Partial Least Squares and PC analysis with morphogenetic implications for European mid-Pleistocene mandibles. In: Slice, D. (Ed.), *Modern Morphometrics in Physical Anthropology*. Kluwer Academic/Plenum Publishers, New York, pp. 265–284.

Bastir, M., Rosas, A., O'Higgins, P., 2006. Craniofacial levels and the morphological maturation of the human skull. *J. Anat.* 209, 637–654.

Bastir, M., Rosas, A., O'Higgins, P., 2005. Human evolution: relationships between the basicranium and the face. *Ann. Hum. Biol.* 32, 790.

Baume, R., Buschang, P., Weinstein, S., 1983. Stature, head height, and growth of the vertical face. *Am. J. Orthodont.* 83, 477–484.

Beer, G.R., de, 1958. *Embryos and Ancestors*. Clarendon Press, Oxford.

Bermúdez de Castro, J.M., 1986. Dental remains from Atapuerca (Spain) I. Metrics. *J. Hum. Evol.* 15, 265–287.

Bermúdez de Castro, J.M., Nicolás, M.E., 1995. Posterior dental size reduction in hominids: the Atapuerca evidence. *Am. J. Phys. Anthropol.* 96, 335–356.

Bermúdez de Castro, J.M., Sarmiento, S., Cunha, E., Rosas, A., Bastir, M., 2001. Dental size variation in the Atapuerca-SH Middle Pleistocene hominids. *J. Hum. Evol.* 41, 195–209.

Bertalanffy, L.V., 1950. An outline of General System Theory. *Brit. J. Philos.Sci. 1*, 134–165.

Bhat, M., Enlow, D.H., 1985. Facial variations related to headform type. *Angle Orthodontist* 55, 269–280.

Biegert, J., 1957. Der Formwandel des Primatenschädels und seine Beziehungen zur ontogenetischen Entwicklung und den phylogenetischen Spezialisationen der Kopforgane. *Gegenbaurs Morphologisches Jahrbuch* 98, 77–199.

Biegert, J., 1963. The evaluation of characteristics of the skull, hands and feet for primate taxonomy. In: Washburn, S.L. (Ed.), *Classification and Human Evolution*. Aldine, Chicago, pp 116–145.

Blumenfeld, J., 2002. Neandertal facial morphology and cold adaptation: a comparative approach. *Am. J. Phys. Anthropol.* 119 S35, 45.

Bookstein, F.L., Gunz, P., Mitteroecker, P., Prossinger, H., Schaefer, K., Seidler, H., 2003. Cranial integration in *Homo*: singular warps analysis of the midsagittal plane in ontogeny and evolution. *J. Hum. Evol.* 44, 167–187.

Boyde, A., 1972. Scanning electron microscope studies of bone. In: Bourne, G.H. (Ed.), *The Biochemistry and Physiology of Bone*. Academic Press, New York, pp. 259–310.

Boyde, A., Hobdell, M.H., 1969. Scanning electron microscopy of lamellar bone. *Z. Zellforsch.* 93, 213–231.

Boyde, A., Jones, S.J., 1972. Scanning electron microscopic studies of the formation of mineralizes tissues. In: Slavkin, H.C., Bavetta, L.A. (Eds.), *Developmental Aspects of Oral Biology*. Academic Press, New York, pp. 243–274.

Bromage, T.G., 1984. Surface remodeling studies on fossil bone. *J. Dent. Res.* 63, 491.

Bromage, T.G., 1989. Ontogeny of the early hominid face. *J. Hum. Evol.* 18, 751–773.

Brothwell, D., 1975. Adaptive growth rate changes as a possible explanation for the distinctiveness of the Neanderthalers. *J. Archaeol. Sci.* 2, 161–163.

Bruner, E., 2004. Geometric morphometrics and paleoneurology: brain shape evolution in the genus *Homo*. *J. Hum. Evol.* 47, 279–303.

Bruner, E., Manzi, G., Arsuaga, J.L., 2003. Encephalization and allometric trajectories in the genus *Homo*. Evidence from the Neanderthal and modern lineages. *Proc. Natl. Acad. Sci. U.S.A.* 100, 15335–15340.

Buschang, P.H., Baume, R.M., Nass, G.G., 1983. A craniofacial growth maturity gradient for males and females between 4 and 16 years of age. *Am. J. Phys. Anthropol.* 61, 373–381.

Carretero, J.M., Arsuaga, J.L., Lorenzo, C., 1997. Clavicles, scapulae and humeri from the Sima de los Huesos site (Sierra de Atapuerca, Spain). *J. Hum. Evol.* 33, 357–409.

Carretero, J.M., Lorenzo, C., Arsuaga, J.L., 1999. Axial and appendicular skeleton of *Homo antecessor. J. Hum. Evol.* 37, 459–499.

Chernoff, B., Magwene, P.M., 1999. Afterword. In: Olson E.C., Miller P.L. (Eds.), *Morphological Integration*. University of Chicago Press, Chicago, pp 319–353.

Churchill, S.E., 1996. Particulate versus integrated evolution of the upper body in Late Pleistocene humans: a test of two models. *Am. J. Phys. Anthropol.* 100, 559–583.

Churchill, S.E., 1998. Cold adaptation, heterochrony, and Neanderthals. *Evol. Anthropol.* 46, 46–60.

Coon, C.S., 1962. *The Origin of Races*. Knopf, New York.

Condemi, S., 1988. Caracteres plesiomorphes et apomorphes de l'os temporal des neanderthaliens europeens würmiens. In: Trinkaus, E. (Ed.), *L'Homme de Neandertal, l'anatomie*, 3, ERAUL 30, Liège, pp. 49–52.

Condemi, S., 2003. Les Néandertaliens. In: Susanne, C., Rebato, S., Chiarelli, B. (Eds.), *Anthropologie Biologique* de Boeck, Brussels pp. 271–279.

Couture, C., 1993. Changements de position du massif facial et de l'articulation temporomandibulaire dans la lignée néandertalienne. Organisation crâneio-maxillo-facilae des néandertalienns. *C.R. Acad. Sci. Paris* II 316, 1627–1633.

Dabelow, A., 1929. Über Korrelationen in der phylogenetischen Entwicklung der Schädelform I. Die Beziehungen zwischen Rumpf und Schädelform. *Gegenbaurs Morphologisches Jahrbuch* 63, 1–49.

Dabelow, A., 1931. Über Korrelationen in der phylogenetischen Entwicklung der Schädelform II. Die Beziehungen zwischen Gehirn und Schädelbasisform bei den Mammaliern. *Gegenbaurs Morphologisches Jahrbuch* 67, 84–133.

Dean, D., Hublin, J.J., Holloway, R., Ziegler, R., 1998. On the phylogenetic position of the pre-Neanderthal specimen from Reilingen, Germany. *J. Hum. Evol.* 4, 85–508.

Dean, M.C., Stringer, C.B., Bromage, T.G., 1986. Age at death of the Neanderthal child from Devil's Tower, Gibraltar and the implications for studies of general growth and development in Neanderthals. *Am. J. Phys. Anthropol.* 70, 301–309.

Demes, B., 1987. Another look at an old face: biomechanics of the neanderthal facial skeleton reconsidered. *J. Hum. Evol.* 16, 297–303.

Emerson, S.B., Bramble, D.M., 1993. Scaling, allometry and skull design. In: Hanken, J., Hall B.K. (Eds.), *The Skull*. University of Chicago Press, Chicago, pp. 384–421.

Enlow, D.H., 1968. *The Human Face: An Account of the Postnatal Growth and Development of the Craniofacial Skeleton*. Harpers Row, New York.

Enlow, D.H., 1990. *Facial Growth*, 3rd Edition. W.B. Saunders, Philadelphia.

Enlow, D.H., Azuma, M., 1975. Functional growth boundaries in the human and mammalian face. In: Langman, J. (Ed.), *Morphogenesis and Malformation of the Face and the Brain*. The National Foundation, New York, pp. 217–230.

Enlow, D.H., Hans M.G., 1996. *Essentials of Facial Growth*. W. B. Saunders, Philadelphia.

Enlow, D.H., McNamara, J.A., 1973. The neurocranial basis for facial form and pattern. *Am. J. Orthodont.* 43, 256–270.

Enlow, D.H., Moyers, R.E., Hunter, W.S., Jr. J.A.M., 1969. A procedure for the analysis of intrinsic facial form and growth. *Am. J. Orthodont.* 56, 6–23.

Enlow, D.H., Kuroda, T., Lewis, A.B., 1971. The morphological and morphogenetical basis for craniofacial form and pattern. *Angle Orthodont.* 41, 161–188.

Enlow, D.H., Pfister, C., Richardson, E., Kuroda, T., 1982. An analysis of black and caucasian craniofacial patterns. *Angle Orthodont.* 51, 279–287.

Franciscus, R.G., 1999. Neandertal nasal structures and upper respiratory tract "specialization". *Proc. Natl. Acad. Sci. U.S.A.* 96, 1805–1809.

Franciscus, R.G., 2003. Internal nasal floor configuration in *Homo* with special reference to the evolution of Neandertal facial form. *J. Hum. Evol.* 44, 701–729.

Franciscus, R.G., Trinkaus, E., 1988a. Nasal morphology and the emergence of *Homo erectus. Am. J. Phys. Anthropol.* 75, 517–527.

Franciscus, R.G., Trinkaus, E., 1988b. The Neanderthal nose. *Am. J. Phys. Anthropol.* 75, 209–210.

Franciscus, R.G., Trinkaus, E., 1995. Determinants of retromolar space presence in Pleistocene *Homo* mandibles. *J. Hum. Evol.* 28, 577–595.

Franciscus, R.G., Churchill, S.E., 2002. The costal skeleton of Shanidar 3 and a reappraisal of

Neandertal thoracic morphology. *J. Hum. Evol.* 42, 303–356.

Friess, M., Marcus, L.F., Reddy, D.P., Delson, E., 2002. The use of 3D laser scanning techniques for the morphometric analysis of human facial shape variation. *BAR Int. Series* 1049, 31–35.

Green, M., Smith, F.H., 1990. Neandertal craniofacial growth. *Am. J. Phys. Anthropol.* 81, 232.

Gould, S.J., 1977. *Ontogeny and Phylogeny*. Harvard University Press, Cambridge, MA.

Hall, R.L., 2005. Energetics of nose and mouth breathing, body size, body composition, and nose volume in young adult males and females. *Am. J. Hum. Biol.* 17, 321–330.

Harvati, K., 2003. Quantitative analysis of Neanderthal temporal bone morphology using three dimensional geometric morphometrics. *Am. J. Phys. Anthropol.* 120, 232–338.

Harvati, K., Frost, S.R., McNulty, K.P., 2004. Neanderthal taxonomy reconsidered: Implications of 3D primate models of intra- and interspecific differences. *Proc. Natl. Acad. Sci. U.S.A.* 101, 1147–1152.

Hawks, J.D., Wolpoff, M.H., 2001. The accretion model of Neandertal evolution. *Evolution* 55, 1474–1485.

Heim, J.L., 1974. Les hommes fossiles de La Ferrassie. (Dordogne) et le problème de la définition des Neánderthaliens classiques. *L'Anthropologie* 78, 312–78.

Heim, J.L., 1976. Les hommes fossiles de La Ferrassie I: Le gisement, les squelettes adultes (crâne et squelette du troc). *Arch. Inst. Paléont. Hum.* 35, 1–131.

Henry, C.J.K., Rees, D.G., 1991. New predictive equations for the estimation of basal metabolic rate in tropical peoples. *Eur. J. Clin. Nutrit.* 45, 177–185.

Hofer, H., 1965. Die morphologische Analyse des Schädels des Menschen. In: Heberer G. (Ed.), *Menschliche Abstammungslehre, Fortschritte der Anthropogenie, 1863–1964*. Gustav Fischer Verlag, Stuttgart, pp. 145–226.

Holliday, T.W., 1997. Postcranial evidence of cold adaptation in European Neandertals. *Am. J. Phys. Anthropol.* 104, 245–258.

Howell, F.C., 1952. Pleistocene glacial ecology and the evolution of "Classic Neandertal" man. Southwest. *J. Anthropol.* 8, 377–410.

Howell, F.C., 1957. The evolutionary significance of variation and varieties of "Neandertal" man. *Quart. Rev. Biol.* 32, 330–347.

Howell, F.C., 1960. European and Northwest african Middle Pleistocene hominids. *Curr. Anthropol.* 1, 195–228.

Howells, W.W., 1975. Neanderthal man: facts and figures. In: Tuttle, R.H. (Ed.), *Paleoanthropology: Morphology and Paleoecology*. Mouton, The Hague, pp 389–407.

Hublin, J.J., 1990. Les peuplements paléolithiques de l'Europe, Memoires du Museum de Prehistoire, France 3, 29–37.

Hublin, J.J., 1998a. Climatic changes, paleogeography, and the evolution of the Nenadertals. In: Akazawa, T., Aoki, K., Bar-Yosef, O. (Eds.), *Neandertals and Modern Humans in Western Asia*. Plenum Press, New York, pp. 295–310.

Hublin, J.J., 1998b. Les plus anciens representants de la lignee preneanderthalienne. In: E. Trinkaus (Ed.), *L'Homme de Neandertal, l'anatomie*, 3, ERAUL 30, Liège, pp. 81–94.

Hublin, J.J., 2000. Modern-nonmodern hominid interactions: a Mediterranean perspective. In: Bar-Yosef, O., Pilbeam, D. (Eds), *The Geography of Neandertals and Modern Humans in Europe and the Greater Mediterranean*. Peabody Museum of Archaeology and Ethnology, Bulletin 8, Harvard University, Cambridge, MA, pp. 157–182.

Hunter, C.J., 1966. The correlation of facial growth with body height during pubescence. *Angle Orthodont.* 36, 44–69.

Hylander, W.L., 1977. The adaptive significance of Eskimo craniofacial morphology. In: Dahlberg, A.A., Graber T.M. (Eds.), *Orofacial Growth and Development*. Mouton, The Hague, pp. 129–170.

Jeffery, N., 2003. Brain expansion and comparative prenatal ontogeny of the non-hominoid cranial base. *J. Hum. Evol.* 45, 263–284.

Jeffery, N., Spoor, F., 2002. Brain size and the human cranial base. *Am. J. Phys. Anthropol.* 118, 324–340.

Jelinek, A.J., 1994. Hominids, energy, environments, and behaviour in the late Pleistocene. In: Nitecki, M.H., Nitecki, D.V. (Eds.), *Origins of Anatomically Modern Humans*. Plenum Press, New York, pp. 67–92.

Klingenberg, C.P., 2002. Morphometrics and the role of the phenotype in studies of the evolution of developmental systems. *Gene* 287, 3–10.

Krings, M., Stone, A., Schmitz, R., Krainitzki, H., Stoneking, M., Pääbo, S., 1997. Neanderthal DNA sequences and the origin of modern humans. *Cell* 90, 19–30.

Krovitz, G.E., 1999. Three-dimensional analysis of modern human and Neandertal craniofacial growth patterns. *Am. J. Phys. Anthropol.* 108 S28, 175–176.

Krovitz, G.E., 2003. Shape and growth differences between Neandertals and modern humans: grounds for a species-level distinction? In: Thompson, J., Krovitz, G., Nelson, A. (Eds.), *Patterns of Growth and Development in the Genus Homo*, Cambridge University Press, Cambridge, pp. 320–342.

Laitman, J.T., Reidenberg, J.S., Marquez, S., Gannon, P.J., 1996. What the nose knows: New understandings of Neanderthal upper respiratory tract specializations. *Proc. Natl. Acad. Sci. USA* 93, 10543–10545.

Lalueza-Fox, C., Sampietro, M.L., Caramelli, D., Puder, Y., Martina Lari, Francesc Calafell, Martínez Maza, C., Bastir, M., Fortea, J., Rasilla, M. d.l., Bertranpetit, J., Rosas, A., 2005. Neanderthal evolutionary genetics, mitochondrial DNA data from the Iberian Peninsula. *Mol. Biol. Evol.* 22, 1077–1081.

Lieberman, D.E., 1995. Testing hypotheses about recent human evolution from skulls. *Curr. Anthropol.* 36, 159–197.

Lieberman, D.E., 1996. How and why recent humans grow thin skulls: Experimental data on systemic cortical robusticity. *Am. J. Phys. Anthropol.*101:217–236.

Lieberman, D.E., 1998. Sphenoid shortening and the evolution of modern human cranial shape. *Nature* 393, 158–162.

Lieberman, D.E., McCarthy, R.C., 1999. The ontogeny of cranial base angulation in humans and chimpanzees and its implication for reconstructing pharyngeal dimensions. *J. Hum. Evol.* 36, 487–517.

Lieberman, D.E., Pearson, O.M., Mowbray, K.M., 2000. Basicranial influence on overall cranial shape. *J. Hum. Evol.* 38, 291–315.

Lieberman, D.E., McBratney, B.M., Krovitz, G., 2002. The evolution and development of cranial form in *Homo sapiens*. *Proc. Natl. Acad. Sci. U.S.A.* 99, 1134–1139.

Lebel, S., Trinkaus, E., Faure, M., Fernandez, P., Guerin, C., Richter, D., Mercier, N., Valladas, H., Wagner, G. A., 2001. Comparative morphology and paleobiology of Middle Pleistocene human remains from the Bau de l'Aubesier, Vaucluse, France. *Proc. Natl. Acad. Sci. U.S.A.* 98, 11097–11102.

Lorenzo, C., Carretero, J.M., Arsuaga, J.L., Gracia, A., Martínez, I., 1998. Intrapopulational body size variation and cranial capacity variation in Middle Pleistocene human: the Sima de los Huesos sample (Sierra de Atapuerca, Spain). *Am. J. Phys. Anthropol.* 106, 19–33.

Lumley, M.A., de, 1973. Anténéandertaliens et Néandertaliens du Bassin Méditerranéen Occidental Européen. Etudes du Quaternaire, Mém. 2.

Malina, R.M., 1978. Growth of muscle tissue and muscle mass. In: Mourilyan, F.T. (Ed.), *Human Growth*. Plenum Press, New York, pp. 273–294.

Malina, R.M., 1996. Regional body composition: age, sex, and ethnic variation. In: Roche, A.F., Heymsfield, S.B., Lohman, T.G., (Eds.), *Human Body Composition*, pp. 217–255.

Mann, A., Trinkaus, E., 1974. Neandertal and Neandertal-like fossils from the Upper Pleistocene. *Yrbk. Phys. Anthropol.* 17, 169–193.

Manzi, G., 2003. "Epigenetic" cranial traits, Neandertals and the origin of *Homo sapiens*. *Riv. Antropol.* 81, 57–68.

Manzi, G., Vienna, A., Hauser, G., 1996. Developmental stress and cranial hypostosis by epigenetic trait occurrence and distribution: an exploratory study on the Italian Neandertals. *J. Hum. Evol.* 30, 511–527.

Manzi, G., Gracia, A., Arsuaga, J.L., 2000. Cranial discrete traits in the Middle Pleistocene humans from Sima de los Huesos (Sierra de Atapuerca, Spain). Does hypostosis represent any increase in "ontogenetic stress" along the Neanderthal lineage? *J. Hum. Evol.* 38, 425–446.

Maresh, M.M., Washburn A.H., 1938. Size of the heart in healthy normal children. *Am. J. Dis. Child.* 56, 33–60.

Martínez-Maza, C., Rosas, A., 2002. Bone remodeling in the Atapuerca-SH mandibles. Implications for growth patterns in Middle Pleistocene hominids. *Am. J. Phys. Anthropol.* 115, 107–108.

Martínez, I., Arsuaga, J.L., 1997. The temporal bones from Sima de los Huesos Middle Pleistocene site (Sierra de Atapuerca, Spain). A phylogenetic approach.*J. Hum. Evol.* 33, 283–318.

McCarthy, R., Lieberman, D.E., 2001. Posterior Maxillary (PM) plane and anterior cranial architecture in primates. *Anat. Rec.* 264, 247–260.

Moss, M., 1962. The functional matrix. In: Kraus, B., Reidel, R. (Eds.), *Vistas in Orthodontics*. Lea and Febiger, Philadelphia, pp. 85–98.

Moss, M., 1997a. The functional matrix hypothesis revisited. 3. The genomic thesis. *Am. J. Orthodont.* Dentofacial Orthopedics 112, 338–342.

Moss, M., 1997b. The functional matrix hypothesis revisited. 4. The epigenetic antithesis and the resolving synthesis. *Am. J. Orthodont. Dentofacial Orthop.* 112, 410–417.

O'Connor, C.F., Franciscus, R.G., Holton, N.E., 2005. Bite force production capability and efficiency in Neandertals and modern humans. *Am. J. Phys. Anthropol.* 127,129–151.

O'Higgins, P., Bastir, M., Kupczik, K., 2006. Shaping the human face. International Congress Series 1296, 55–73.

Olson, E.C., Miller, R.L., 1958. *Morphological Integration*. The University of Chicago Press, Chicago.

Pearson, O.M., 2000. Postcranial remains and the origin of modern humans. *Evol. Anthropol.* 9, 229–247.

Pigliucci, M., 2003. Species as family resemblance concepts: The (dis-)solution of the species problem? *BioEssays* 25, 596–602.

Ponce de León, M., Zollikofer, C., 2001. Neanderthal cranial ontogeny and its implications for late hominid diversity. *Nature* 412, 534–538.

Rak, Y., 1986. The Neanderthal: a new look at an old face. *J. Hum. Evol.* 15, 151–164.

Rak, Y., Kimbel, W.H., Hovers, E., 1994. A Neandertal infant from Amud Cave, Israel. *J. Hum. Evol.* 26, 313–324.

Rak, Y., Ginsburg, A., Geffen, E., 2002. Does *Homo neanderthalensis* play a role in modern human ancestry? The mandibular evidence. *Am. J. Phys. Anthropol.* 119, 199–204.

Ramírez Rozzi, F.V., Bermúdez de Castro, J.M., 2004. Surprisingly rapid growth in Neanderthals. *Nature* 428, 936–939.

Ranly, D.M., 1988. *A Synopsis of Craniofacial Growth* 2nd Edition. Appleton and Lange, Norwalk, CT.

Riedl, R., 1975. *Die Ordnung des Lebendigen. Systembedingungen der Evolution*. Paul Parey Verlag, Hamburg

Rightmire, G.P., 1998. Human evolution in the Middle Pleistocene: the role of *Homo heidelbergensis*. *Evol. Anthropol.* 6, 218–227.

Rightmire, G.P., 2001. Patterns of hominid evolution and dispersal in the Middle Pleistocene. *Quatern. Int.* 75, 77–84.

Rightmire, G.P., 2004. Brain size and encephalization in early to Mid-Pleistocene Homo. *Am. J. Phys. Anthropol.* 124, 109–23.

Rohlf, F.J., Corti, M., 2000. The use of Two-Block Partial Least-Squares to study covariation in shape. *Syst. Zool.* 49, 740–753.

Rosas, A., 1992. Ontogenia y filogenia de la mandíbula en la evolución de los homínidos. Aplicación de un modelo de morfogénesis a las mandíbulas fósiles humanas de Atapuerca. Ph.D. Dissertation, University Complutense of Madrid, Madrid.

Rosas, A., 1995. Seventeen new mandibular specimens from the Atapuerca/Ibeas Middle Pleistocene hominids sample (1985–1992). *J. Hum. Evol.* 28, 533–559.

Rosas, A., 1997. A gradient of size and shape for the Atapuerca sample and Middle Pleistocene hominid variability. *J. Hum. Evol.* 33, 319–331.

Rosas, A., 1998. Modelos de crecimiento en mandíbulas fósiles de homínidos. Atapuerca, un nuevo paradigma. In: Aguirre, E. (Ed.), *Atapuerca y la Evolución Humana*. Fundación Ramón Areces, Madrid, pp. 239–275.

Rosas, A., 2000. Human evolution in the last million years. The Atapuerca evidence. *Acta Anthropol. Sinica* 19, 8–17.

Rosas, A., 2001. Occurrence of Neanderthal features in mandibles from the Atapuerca-SH Site. *Am. J. Phys. Anthropol.* 114, 74–91.

Rosas, A., Bastir, M., 2002. Thin-plate spline analysis of allometry and sexual dimorphism in the human craniofacial complex. *Am. J. Phys. Anthropol.* 117, 236–245.

Rosas, A., Bastir, M., 2004. Geometric morphometric analysis of allometric variation in the mandibular morphology from the hominids of Atapuerca, Sima de los Huesos Site. *Anat. Rec.* 278A, 551–560.

Rosas, A., Bermúdez de Castro, J.M., 1998. The Mauer mandible and the evolutionary significance of *Homo heidelbergensis*. *Geobios* 31, 687–697.

Rosas, A., Martínez-Maza, C., in press. Bone remodelling pattern of *Homo heidelbergensis* mandibles. *J. Hum. Evol.*

Rosas, A., Bermúdez de Castro, J.M., Aguirre, E., 1991. Mandibules et dents d'Ibeas (Espagne) dans le contexte de l'evolution humaine en Europe. *L'Anthropologie* 95, 89–102.

Rosas, A., Bastir, M., Martínez-Maza, C., Bermúdez de Castro, J. M., 2002. Sexual dimorphism in the Atapuerca-SH hominids: the evidence from the mandibles. *J. Hum. Evol.* 42, 451–474.

Ross, C., Henneberg, M., 1995. Basicranial flexion, relative brain size, and facial kyphosis in *Homo sapiens* and some fossil hominids. *Am. J. Phys. Anthropol.* 98, 575–593.

Ruff, C.B., 2002. Variation in human body size and shape. *Ann. Rev. Anthropol.* 31, 211–232.

Ruff, C.B., Trinkaus, E., Walker, A., Larsen, C.S., 1993. Postcranial robusticity in *Homo*. I. Temporal trends and mechanical interpretation. *Am. J. Phys. Anthropol.* 91, 21–53.

Ruff, C.B., Walker, A., Trinkaus, E., 1994. Postcranial robusticity in *Homo*. III: ontogeny. *Am. J. Phys. Anthropol.* 93, 35–54.

Ruff, C.B., Trinkaus, E., Holliday, T.W., 1997. Body mass and the encephalization in Pleistocene *Homo*. *Nature* 387, 173–176.

Sacher, G.A., 1975. Maturation and longevity in relation to cranial capacity in hominid evolution. In: Tuttle, R.H. (Ed.), *Primate Functional Morphology and Evolution*, Mouton, The Hague, pp. 417–442.

Santa Luca, A.P., 1978. A re-examination of presumed Neandertal-like fossils. *J. Hum. Evol.* 7, 619–636.

Schwartz, J.H., Tattersall, I., 1996. Significance of some previously unrecognized apomorphies in the nasal region of *Homo neanderthalensis*. *Proc. Natl. Acad. Sci. U.S.A.* 93, 10852–10854.

Schwartz, J.H., Tattersall, I., Laitman, J.T., 1999. New thoughts on Neanderthal behavior: evidence from nasal morphology. In: Ullrich, H. (Ed.), *Hominid Evolution*. Edition Archaea, pp. 166–186.

Seidler, H., Falk, D., Stringer, C., Wilfing, H., Muller, G.B., zur Nedden, D., Weber, G.W., Reicheis, W., Arsuaga, J.L., 1997. A comparative study of stereolithographically modelled skulls of Petralona and Broken Hill: implications for future studies of Middle Pleistocene hominid evolution. *J. Hum. Evol.* 33, 691–703.

Sergi, S., 1934. Ossicini fontanellari della regione del lambda nei cranio di Saccopastore e nei crani neandertaliani. *Riv. Antropol.* 30, 101–112.

Silbernagl, S., Despopoulos, A., 1991. Taschenatlas der Physiologie. Georg Thieme Deuscher Taschenbuch-Verlag, Stuttgart.

Smith, B.H., 1989. Dental development as a measure of life history in primates. *Evolution* 43, 683–688.

Smith, F.H., 1991. The Neandertals: evolutionary dead ends or ancestors of modern people. *J. Anthropol. Res.* 47, 219–238.

Smith, F.H., Green M., 1991. Heterochrony, life history and Neandertal morphology. *Am. J. Phys. Anthropol.* S12, 164.

Smith, F.H., Jankovic, I., Karavanic, I., 2005. The assimilation model, modern human origins in Europe, and the extinction of Neandertals. *Quatern.* Int. 137, 7–19.

Sorensen, M.V., Leonard, W.R., 2001. Neandertal energetics and foraging efficiency. *J. Hum. Evol.* 40, 483–495.

Spencer, M.A., Demes, B., 1993. Biomechanical analysis of masticatory system configuration in Neandertals and Inuits. *Am. J. Phys. Anthropol.* 91, 1–20.

Spoor, F., O'Higgins, P., Dean, C., Lieberman, D., 1999. Anterior sphenoid in modern humans. *Nature* 397, 572.

Spoor, F., Hublin, J.J., Braun, M., Zonneveld, F., 2003. The bony labyrinth of Neanderthals. *J. Hum. Evol.* 44, 141–165.

Steegmann, A.T., Cerny, J.F.J., Holliday, T.W., 2002. Neandertal cold adaptation: physiological and energetic factors. *Am. J. Hum. Biol.* 14, 566–583.

Stringer, C.B., 1996. Current issues in modern human origins. In: Meikle, E., Howell, F.C., Jablonski, N. (Eds.), *Contemporary Issues in Human Evolution*. California Academy of Sciences, Memoir 21, San Francisco, CA, pp. 115–134

Stringer, C.B., Hublin, J.J., 1999. New age estimates for the Swanscombe hominid, and their significance for human evolution. *J. Hum. Evol.* 37, 873–876.

Stringer, C.B., Hublin, J.J., Vandermeersch, B., 1984. The origin of anatomically modern humans in western Europe. In: *The Origins of Modern Humans: A World Survey of the Fossil Evidence*. Alan R. Liss, New York, pp. 51–135.

Stringer, C.B., Dean, M.C., Martin, R.D., 1990. A comparative study of cranial and dental development within a recent British sample and among Neanderthals. In: De Rousseau, C.J. (Ed.), *Primate Life History and Evolution*. Wiley-Liss, New York, pp. 115–152.

Stringer, C.B., Trinkaus E., Roberts, M.B., Parfitt, S.A., Macphail, R.I., 1998. The Middle Pleistocene human tibia from Boxgrove. *J. Hum. Evol.* 34, 509–547.

Tattersall, I., 1992. Species concepts and species identification in human evolution. *J. Hum. Evol.* 22, 341–349.

Tattersall, I., Schwartz, J.H., 1998. Morphology, paleoanthropology, and Neanderthals. *Anat. Rec.* 253, 113–117.

Thompson, J.L., Nelson, A.J., 2000. The place of Neandertals in the evolution of hominid patterns of growth and development. *J. Hum. Evol.* 38, 475–495.

Tillier, A.M., 1988. A propos de séquences phylogéniques et ontogéniques chez les Néandertaliens. In: E. Trinkaus (Ed.), L'Homme de Neandertal, l'anatomie, 3, ERAUL 30, Liège, pp. 125–136.

Trinkaus, E., 1981. Neanderthal limb proportions and cold adaptation. In: Stringer, C.B. (Ed.), *Aspects of Human Evolution*. Taylor and Francis, London, pp. 187–224.

Trinkaus, E., 1982. The Shanidar 3 Neanderthal. *Am. J. Phys. Anthropol.* 57, 37–60.

Trinkaus, E., 1983. *The Shanidar Neandertals*. Academic Press, New York.

Trinkaus, E., 1987. The Neandertal face: Evolutionary and functional perspectives on a recent hominid face. *J. Hum. Evol.* 16, 429–443.

Trinkaus, E., 1988. The evolutionary origins of the Neanderthals or, why were there Neanderthals?

In: E. Trinkaus (Ed.), *L'Homme de Neandertal, l'anatomie*, 3, ERAUL 30, Liège, pp. 11–29.

Trinkaus, E., 1990. Cladistics and the hominid fossil record. *Am. J. Phys. Anthropol.* 83, 1–11.

Trinkaus, E., 1993. Variability in the position of the mandibular foramen and the identification of Neandertal apomorphies. *Riv. Anthropol.* 71, 259–274.

Trinkaus, E., 1997. Appendicular robusticity and the paleobiology of modern human emergence. *Proc. Natl. Acad. Sci. U.S.A.* 94, 13367–13373.

Trinkaus, E., 2003. Neandertal faces were not long; modern human faces are short. *Proc. Natl. Acad. Sci. U.S.A.* 100, 8142–8145.

Trinkaus, E., LeMay, M., 1982. Occipital bunning among later Pleistocene Hominids. *Am. J. Phys. Anthropol.* 57, 27–35.

Trinkaus, E., Tompkins, R.L., 1990. The Neanderthal life cycle: the possibility, probability, and perceptibility of contrasts with recent humans. In: De Rousseau, C.J. (Ed.), *Primate Life History Evolution*. Wiley-Liss, New York, pp. 153–180.

Trinkaus, E., Churchill, S.E., Ruff, C.B., 1994. Postcranial robusticity in *Homo*. II: Humeral bilateral asymmetry and bone plasticity. *Am. J. Phys. Anthropol.* 93, 1–34.

Wagner, G., 1996. Homologues, natural kinds and the evolution of modularity. *Am. Zool.* 36, 36–43.

Wake, D.B., Larson, A., 1987. Multidimensional analysis of an evolving lineage. *Science* 238, 42–48.

Weaver, T.D., 2003. The shape of the Neandertal femur is primarily the consequence of a hyperpolar body form. *Proc. Natl. Acad. Sci. U.S.A.* 100, 6926–6929.

Williams, F.L., Krovitz, G.E., 2004. Ontogenetic migration of the mental foramen in Neandertals and modern humans. *J. Hum. Evol.* 47,199–219.

Williams, F.L., Godfrey, L.R., Sutherland, M.R., 2002. Heterochrony and the evolution of Neandertal and modern human craniofacial form. In: Minugh-Purvis, N., McNamara, K.J. (Eds.), *Human Evolution through Developmental Change*. Johns Hopkins University Press, Baltimore, pp. 405–441.

Wolpoff, M.H., 1980. *Paleoanthropology*. Knopf, New York.

Wood, B., Aiello L.C., 1998. Taxonomic and functional implications of mandibular scaling in early hominids. *Am. J. Phys. Anthropol.* 105, 523–538.

Wood, B., Richmond B.G., 2000. Human evolution: taxonomy and paleobiology. *J. Anat.* 197, 19–60.

5. Neanderthals and modern humans – chimps and bonobos: similarities and differences in development and evolution

M.S. PONCE DE LEÓN
Anthropologisches Institut und Museum
Universität Zürich-Irchel
Winterthurerstrasse 190
CH-8057 Zürich, Switzerland
marcia@aim.unizh.ch

C.P.E. ZOLLIKOFER
Anthropologisches Institut und Museum
Universität Zürich-Irchel
Winterthurerstrasse 190
CH-8057 Zürich, Switzerland
zolli@aim.unizh.ch

Keywords: *Homo neanderthalensis, Homo sapiens, Pan paniscus, Pan troglodytes*, allometry, geometric morphometrics, allometry, heterochrony, heterotopy, human evolution, ontogeny, phylogeny

Abstract

Being our closest living relatives, chimps and bonobos provide the best available comparative evidence to study the evolutionary split between our sister taxon – the Neanderthals – and ourselves. Here, we analyze craniofacial development in these taxa from birth to adulthood using geometric morphometric methods. In both *Homo* and *Pan*, ontogenetic trajectories of sister taxa differ by their length, position and/or direction in shape space, as well as in the relationship between cranial size and shape. Modern human and bonobo ontogenies represent "abridged" versions of Neanderthal and chimp spatiotemporal developmental patterns, respectively, where "shortening" of trajectories is likely to represent evolutionary novelty. When examined in detail, however, the Neanderthal-human and chimp-bonobo splits do not represent equivalent forms of evolutionary developmental diversification. Rather, it appears that each bifurcation is the result of a different unique evolutionary event, during which the ancestral mode of growth and development was modified in a taxon-specific manner.

Introduction

For almost 150 years Neanderthals have been recognized as a fossil human taxon exhibiting a suite of morphological features that differentiate them both from earlier *Homo* and from contemporaneous anatomically modern humans (AMH) (Stringer and Gamble, 1993). With the advent of new analytical methods, the classical morphology-based view of Neanderthals has been modified and complemented. Among these new developments, two directions of research have proven especially promising: analysis of fossil mitochondrial

K. Harvati and T. Harrison (eds.), Neanderthals Revisited: New Approaches and Perspectives, 71–88.

DNA (mtDNA) and quantitative analysis of three-dimensional morphology.

Successful extraction and sequencing of ancient DNA from several Neanderthal individuals has provided evidence of an early evolutionary split between Neanderthals and anatomically modern humans (Krings et al., 1997, 1999; Ovchinnikov et al., 2000), essentially supporting the view that no significant interbreeding between these taxa occurred (Krings et al., 2000; Serre et al., 2004). New morphometric methods have been implemented concurrently. Most notably, the methodological framework of Geometric Morphometrics (GM) permits quantitative analysis of shape variability in complex three-dimensional forms (Dryden and Mardia, 1998). The growing popularity of GM methods has provided new insights into the distinction between Neanderthal and AMH cranial architecture and development (Ponce de León et al., 2001; Harvati et al., 2004). Studies at a smaller morphological scale have revealed previously unrecognized Neanderthal features, such as distinct tooth cusp patterns (Bailey, 2002, 2004), and a more rapid pace of tooth development that indicates divergent Neanderthal and modern human life histories (Guatelli-Steinberg et al., 2005; Ramirez Rozzi and Bermúdez de Castro, 2004). These studies consolidate the notion of two contemporaneous but independently evolving *Homo* lineages during the Middle to Late Pleistocene.

In this study we focus on the evolutionary bifurcation of a hypothetical common ancestor into Neanderthal and anatomically modern human populations. Evolutionary lineage splits are attributed traditionally to divergent adaptation often arising from habitat shifts. In the case of Neanderthals, the functionally adaptive value of several distinguishing features such as the size and shape of paranasal sinuses, limb proportions, and pelvic form remains highly controversial (Rak, 1986; Trinkaus, 1987, 2003; Churchill, 1998; Franciscus, 1999, 2003; Franciscus and Churchill, 2002; O'Connor et al., 2005). An evolutionary developmental perspective offers an alternative approach to investigating Neanderthal morphology. Rather than studying the potential adaptive significance of a novel trait, attention is directed toward how processes of growth and development are modified to bring about evolutionary novelty. From this perspective, Neanderthal morphology is considered as an integrated whole rather than an array of more or less independent features. This approach frees us from the shackles of adaptationist reasoning (Gould and Lewontin, 1979) since it puts more weight on the role of developmental constraints during the evolution of new morphologies. As we demonstrated in an earlier study (Ponce de León et al., 2001), Neanderthals and AMH exhibit clear morphological differences at an early stage of ontogeny, but follow a shared ancestral pattern of postnatal development. Henceforth, contrary to the widespread notion that only adult specimens exhibit the requisite autapomorphic features required to define a species, juvenile specimens are equally relevant in performing this task.

Studying development in an evolutionary context by comparing fossils encounters substantially more limitations compared to developmental studies that use extant taxa. Most notably, developmental processes and their modifications cannot be followed *in vivo* using the fossil record but must be inferred from patterns of morphological change between fossil specimens. It is essential, therefore, to carry out comparative studies addressing similar questions in living species from which more complete data can be derived. Fortunately, hominoid evolution has provided an ideal test case. Our closest relatives, the chimpanzees (*Pan troglodytes*) and bonobos (*Pan paniscus*), represent sister species with distinct morphologies and life histories (Barriel, 1997; D'Aout et al., 2004; Doran, 1993; Gagneux et al., 1999; Parish and

De Waal, 2000; Poti, 2005; Sept, 1998; Susman, 1984; Won and Hey, 2005; Yu et al., 2003; Zihlman and Cramer, 1978). The evolutionary dichotomy between these two species represents an appropriate analogy with which to study the evolutionary split between Neanderthals and modern humans.

HYPOTHESES

Considering these four taxa in conjunction, we may therefore ask the following questions: What are the commonalities and differences between the Neanderthal-human and the chimp-bonobo phyletic bifurcations? Is there a common pattern of ontogenetic modification that leads to the evolutionary split at the genus and at the species level? What can be inferred about the processes underlying the observed patterns of divergence? These questions are addressed with a comparative geometric morphometric study of cranial growth and development. The geometric morphometric approach provides an efficient tool kit for formulating testable hypotheses about differences and commonalities between taxon-specific patterns of ontogenetic shape change. These patterns are revealed in multidimensional morphospace by measuring the relative position, orientation and length of the resulting trajectories. As outlined elsewhere (Ponce de León and Zollikofer, 2001; Zollikofer and Ponce de León, 2004), differences in the position of the trajectory indicate different cranial forms, differences in trajectory orientation indicate different physical patterns of shape change, and differences in trajectory length indicate different amounts of shape change during development. Furthermore, the terminology of heterochronic and heterotopic analysis can be applied to data in morphospace. For example, divergent trajectories in morphospace indicate non-homologous (i.e., heterotopic) patterns of shape change, while collinear or parallel trajectories exhibiting different lengths indicate heterochronic modification.

Accordingly, the following set of null hypotheses is tested for each pair of sister taxa *X* and *Y*:

- H0-*XY*1: population means of taxa *X* and *Y* are identical.
- H0-*XY*2: trajectory vectors of taxa *X* and *Y* are collinear.
  - H0-*XY*2a: average trajectories of taxa *X* and *Y* have the same length (this hypothesis can only be tested if H0-*XY*2 cannot be falsified)

These hypotheses are applied to comparisons between *Homo* and *Pan*, between *H. sapiens* and *H. neanderthalensis*, and between *P. paniscus* and *P. troglodytes*.

Materials and Methods

SAMPLE

The sample includes 155 cranial specimens: Neanderthals ($N = 14$), anatomically modern humans ($N = 41$), chimps ($N = 64$) and bonobos ($N = 36$). Developmental status of all specimens was classified according to dental eruption patterns. The following four classes were defined: Pd4 (deciduous second molar 50% erupted), M1, M2, M3 (full eruption of permanent molars M1, M2, M3, respectively). Ontogenetic trajectories through shape space exhibit marked departures from linearity (e.g., distinct early and late phases of cranial development). Furthermore, no pre-Pd4 Neanderthal specimens are currently available. For these reasons, dental developmental stages prior to Pd4 were not included in these analyses.

The Neanderthal subsample comprises specimens representing the widest possible range of developmental stages currently available, where the lower end of the ontogenetic scale is defined by the two specimens from Dederiyeh (dental stage Pd4: second milk molar at least partially erupted) (Akazawa et al., 1995; Ishida et al.,

2000; Kondo and Dodo, 2000; Zollikofer et al., 2002) (Table 1). It is important to remember that the Neanderthal sample represents wide variability in geographical location and geologic time. To perform biologically reasonable comparisons, a *Homo sapiens* sample was chosen that incorporates temporal depth and geographic width, i.e., pooled-population variation in space and time (Table 1). It comprises recent specimens from various geographical locations as well as fossil specimens (Qafzeh 9 and 11, Lapedo 1). The *Pan paniscus* subsample is composed of wild-shot specimens from several localities. Most specimens come from the collections of the Royal Africa Museum (Tervuren, Belgium). *Pan troglodytes* is represented by a pooled sample comprising individuals from several populations and several subspecies, *P. t. troglodytes*, *P. t. schweinfurthii*, and *P. t. verus* (ontogenetic differences between subspecies currently are being investigated).

Table 1. Specimens

Specimen	*Status*[1]
Neanderthals	
Dederiyeh 1	r
Dederiyeh 2	r
Pech de l'Azé	c
Subalyuk 2	r
Roc de Marsal	r
Devil's Tower	r
Engis 2	r
La Quina18	c
Teshik Tash	r
Le Moustier 1	r
Amud 1	r
La Ferrassie 1	c
Forbes' Quarry	r
Tabun 1	r
fossil *H. sapiens*	
Lapedo 1	r
Qafzeh 9	r
Qafzeh 11	r
extant *H. sapiens*	
locality	*N*
Africa	6
Asia	6
Australia	2
America	4
Greenland	3
Europe	17

[1] r: reconstructed original specimen; c: cast.

DATA ACQUISITION AND MORPHOMETRIC ANALYSIS

All specimens were CT-scanned in order to obtain 3-dimensional virtual representations. Fossil specimens were restored and reconstructed with established computer-assisted methods in order to correct taphonomic distortion and to replace missing parts (Zollikofer and Ponce de León, 2005). For subsequent morphometric analyses, 3-dimensional coordinates of craniofacial landmarks were collected ($N = 41$). Landmarks (11 midsagittal, 15 bilateral pairs; see Table 2) were chosen that represent biological homology and cover all regions of the craniofacial surface in approximately equal proportions. Craniofacial shape variability was analyzed using GM methods. In GM, the form of each specimen is given by the geometry of its landmark configuration. Form is decomposed into size and shape such that size is measured as Centroid Size (S), which is defined as the square root of the sum of squared distances from the center of mass of the landmark configuration to each landmark (Bookstein, 1991), and shape is measured as the multidimensional deviation (landmark coordinate by landmark coordinate) of a specimen's landmark configuration from the sample mean configuration (consensus). The consensus configuration used here is the average configuration resulting from Generalized Least Squares (GLS) fitting of all specimens (Rohlf and Slice, 1990).

Shape variability was explored via Principal Components Analysis (PCA) (Jolliffe, 1986) of data in linearized Procustes shape space (Dryden and Mardia, 1998). PCA is mainly used as a dimension reduction technique. Typically a large proportion of the total sample variation in shape is contained within the first few PCs, which can be used for data

Table 2. Landmarks

#	*Name*	*Remarks*
1	nasion	
2	glabella	in juvenile specimens: midsagittal point at the height of the upper orbital margin
3	bregma	
4	lambda	
5	inion	
6	opisthion	posteriormost midpoint of foramen magnum
7	basion	anteriormost midpoint of foramen magnum
8	sphenobasion	midpoint of sphenobasilar suture
9	staphylion	posteriormost midpoint of palate
10	prosthion	midpoint between I1L and I1R, on alveolar margin
11	nasospinale	
12,13	maxillofrontale	anterior edge of maxillofrontal suture
14,15	foramen supraorbitale	point at the supraorbital notch (or a point at the orbital rim below the s-o. notch or foramen)
16,17	orbitale	zygomatico-maxillary suture at the orbital rim
18,19	frontomalare orbitale	fronto-zygomatic suture at the orbital rim
20,21	zygomaxillare	zygo-maxillary suture at the tuberosity
22,23	jugale	location of largest curvature of the zygomatic rim in the jugal region
24,25	foramen maxillare	midpoint at the level of the surface
26,27	m2 (PM2) buccal	buccal midpoint of the crown of deciduous m2 or permanent PM2
28,29	stylomastoid foramen	midpoint of external margin of foramen
30,31	carotid foramen	midpoint of external margin of foramen
32,33	oval foramen	midpoint of external margin of foramen
34,35	stephanion	intersection between coronal suture and temporal line
36,37	asterion	
38,39	porion	
40,41	pterion	meeting point of coronal and sphenoid suture

visualization. It is important to recall that PCs do not represent biological entities. Rather, they offer a convenient means of exploring complex patterns of shape variability. A primary advantage of PCA of shape is that it is always possible to move back and forth between the representation of shape variability in abstract shape space and in real physical space.

In the shape analyses performed here, a strong signal of ontogenetic allometry is expected, i.e., changes in craniofacial shape are typically closely correlated with changes in size. For each species, multivariate regression of the shape PCs versus log centroid size (log S) yields an allometric shape vector, which captures the portion of shape variability that can be explained statistically as a function of size (Figure 1). The *common allometric shape vector* of the entire sample is calculated as the mean of species-specific allometric shape vectors. It must be underlined that, like PCs, the common allometric vector does not represent a biological entity. Rather, it defines an axis of reference in shape space that can be used as a visualization aid.

Resampling statistics is used to test hypotheses outlined in the previous section. Essentially, statistical tests ask for significant differences in distance, orientation and length between taxon-specific ontogenetic trajectories through shape space. While resampling methods can be applied readily to data in linearized Procrustes shape space, it is important to bear in mind that they yield biologically

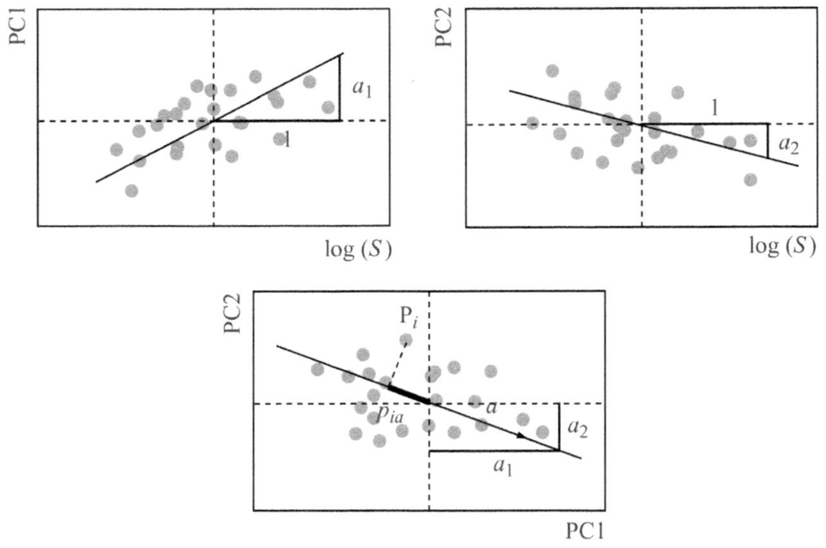

Figure 1. Calculation of the allometric shape vector. Distribution of data in multidimensional shape space is shown here for the first two principal components, PC1 and PC2. To evaluate the allometric shape vector $\boldsymbol{a}$, each PC_i is regressed against log centroid size (S), yielding slopes a_i, which together constitute the components of vector $\boldsymbol{a}$. The allometric shape score p_{ia} (bold line) of specimen P_i is calculated as its projection onto $\boldsymbol{a}$. This procedure is repeated for each species. The mean of the species-specific allometric shape vectors is the common allometric shape vector.

reasonable insights only under specific preconditions. We adhere to the following test sequence:

(1) Test for differences between taxon-specific mean values of shape in morphospace.
(2) If differences are significant, correct data to achieve mean difference = 0 between taxa.
(3) Test for divergence between taxon-specific trajectory vectors.
(4) If and only if trajectory vectors are collinear, test for differences in trajectory lengths. Otherwise, trajectory lengths cannot be compared because divergent directions in morphospace indicate non-homologous patterns of shape change in physical space.

Results

The results of PCA of shape are graphed in Figures 2–4. Figure 2 shows differences between all taxa. To visualize differences between sister species in greater detail, we performed separate PCA on the *Homo* and *Pan* subsamples (Figure 3). Statistical tests of the hypotheses stated above are reported in Tables 3 (PC variance proportions) and 4 (significance tests). Positional differences between trajectories are measured as Procrustes distance between taxon-specific mean shapes (Table 4, first row), directional differences between trajectories through shape space are measured by the divergence angle between their principal directions (Table 4, second row), and trajectory lengths are measured by the Procrustes distance between mean cranial shapes at adulthood and at dental stage Pd4 (Table 4, fourth row). Furthermore, differences in ontogenetic allometric trajectories are measured by the divergence between taxon-specific allometric shape vectors (Table 4, third row), and by the ratio between cranial sizes at adulthood and stage Pd4 (Table 4, last row).

DIFFERENCES AND COMMONALITIES BETWEEN SISTER SPECIES

Homo sapiens and Homo neanderthalensis

Postnatal trajectories of Neanderthals and modern humans exhibit the same direction through shape space, but differ in their relative lengths and positions (Figure 3A). This signifies that taxon-specific differences are established early and maintained throughout growth, such that these sister species follow essentially the same postnatal pattern of development. The Neanderthal postnatal trajectory is slightly more extended than the AMH trajectory.

It should be noted here that tracing human development further back to fetal phases indicates that the ontogenetic trajectory is curved around birth. Recalling that different directions through shape space indicate different patterns of cranial shape change, these observations suggest distinct prenatal and postnatal modes of development. It is likely that the Neanderthal trajectory exhibits a similar two-phase pattern. Work in progress on this issue, notably on the neonate Neanderthal specimen from Mezmaiskaya Cave (Golovanova et al., 1999), will help to clarify this issue.

Pan Paniscus and Pan Troglodytes

Ontogenetic trajectories of *Pan* species are graphed in Figure 3B. Postnatal trajectories differ from each other not only in their position, but also in their direction Table 4). Moreover, the trajectory of *P. paniscus* is shorter than the trajectory of *P. troglodytes*. These differences suggest species-specific postnatal modes of development. As an effect of path divergence, differences between chimpanzee and bonobo cranial morphologies increase over developmental time. This is contrary to Neanderthal and AMH trajectories, where differences remain fairly constant from birth to adulthood.

Ontogenetic Allometry

Up to this point, we compared taxon-specific ontogenetic *shape trajectories* that were measured by taxon-specific principal directions in shape space. Size-related change in shape is investigated using ontogenetic *allometric trajectories*. Taxon-specific allometric trajectories (see Figure 1) capture the portion of cranial shape change that is correlated with size, and differences between taxa can be measured as divergence angles between these trajectories.

Divergence between allometric trajectories of sister taxa is almost identical to the divergence between shape trajectories (see Table 4, rows 2 and 3). This indicates that a large portion of ontogenetic shape change is related to changes in cranial size. Expressing cranial size at dental stage Pd4 as a percentage of cranial size at adulthood further shows that *H. sapiens* exhibit slightly larger relative juvenile cranial sizes than *H. neanderthalensis*, and that *P. paniscus* show considerably larger relative juvenile cranial sizes than *P. troglodytes*.

COMPARISON BETWEEN *HOMO* AND *PAN* PATTERNS OF ONTOGENY

It can be readily recognized from Figure 2 and Table 4 that differences between ontogenetic trajectories at the genus level are more pronounced than at the species level. *Homo* and *Pan* postnatal trajectories differ considerably in their principal direction and length. Paths through shape space are shorter in Neanderthals and AMH than in chimps and bonobos, indicating less shape change from birth to adulthood.

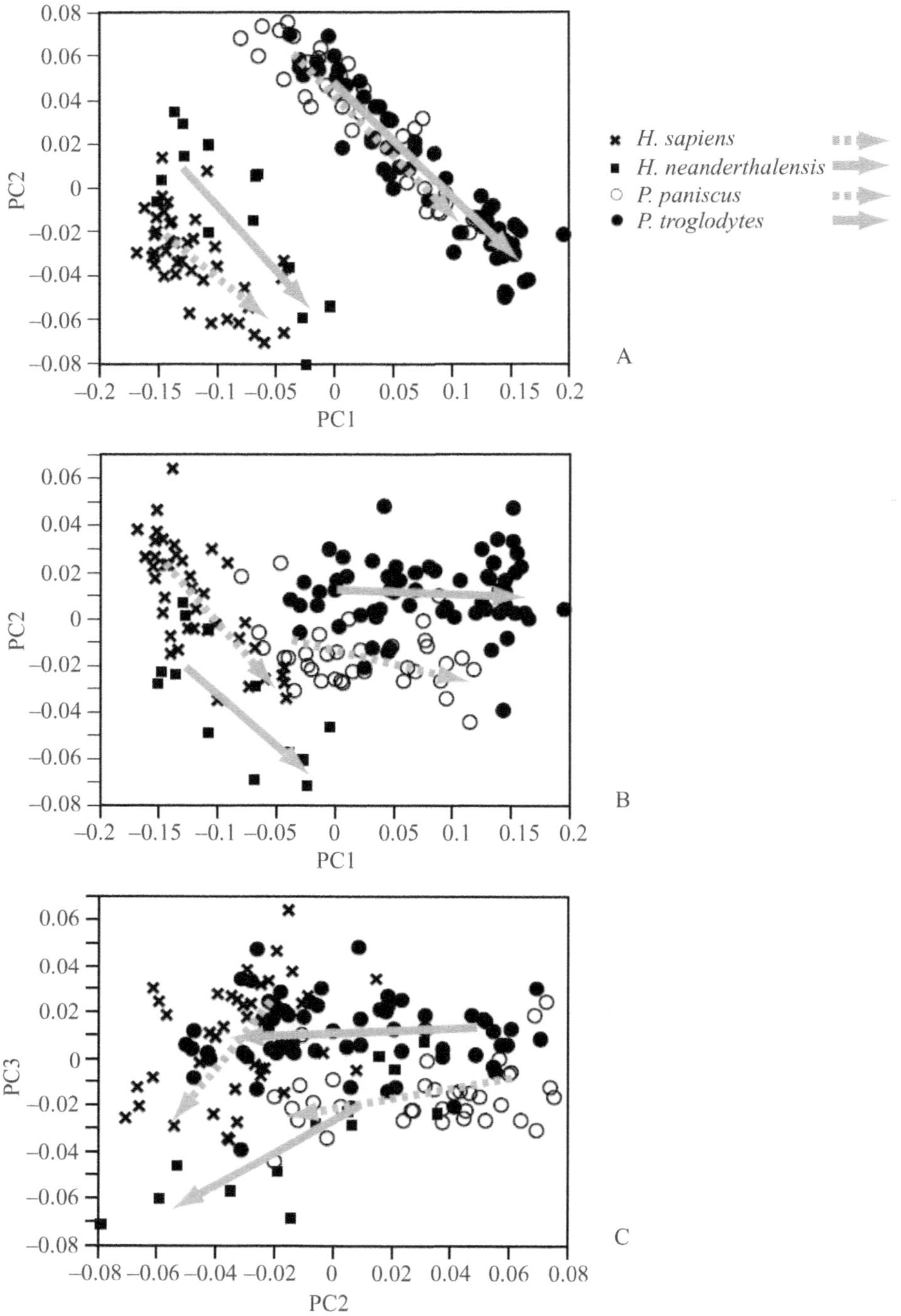

Figure 2. Ontogenetic trajectories through shape space. A: Plot of PC2 versus PC1 shows major differences between *Homo* and *Pan* trajectories. B, C: Plots of PC3 versus PC1 and PC2, respectively, exhibit differences between sister taxa. For each species, arrows point from mean cranial shape at dental stage Pd4 to mean cranial shape at adulthood.

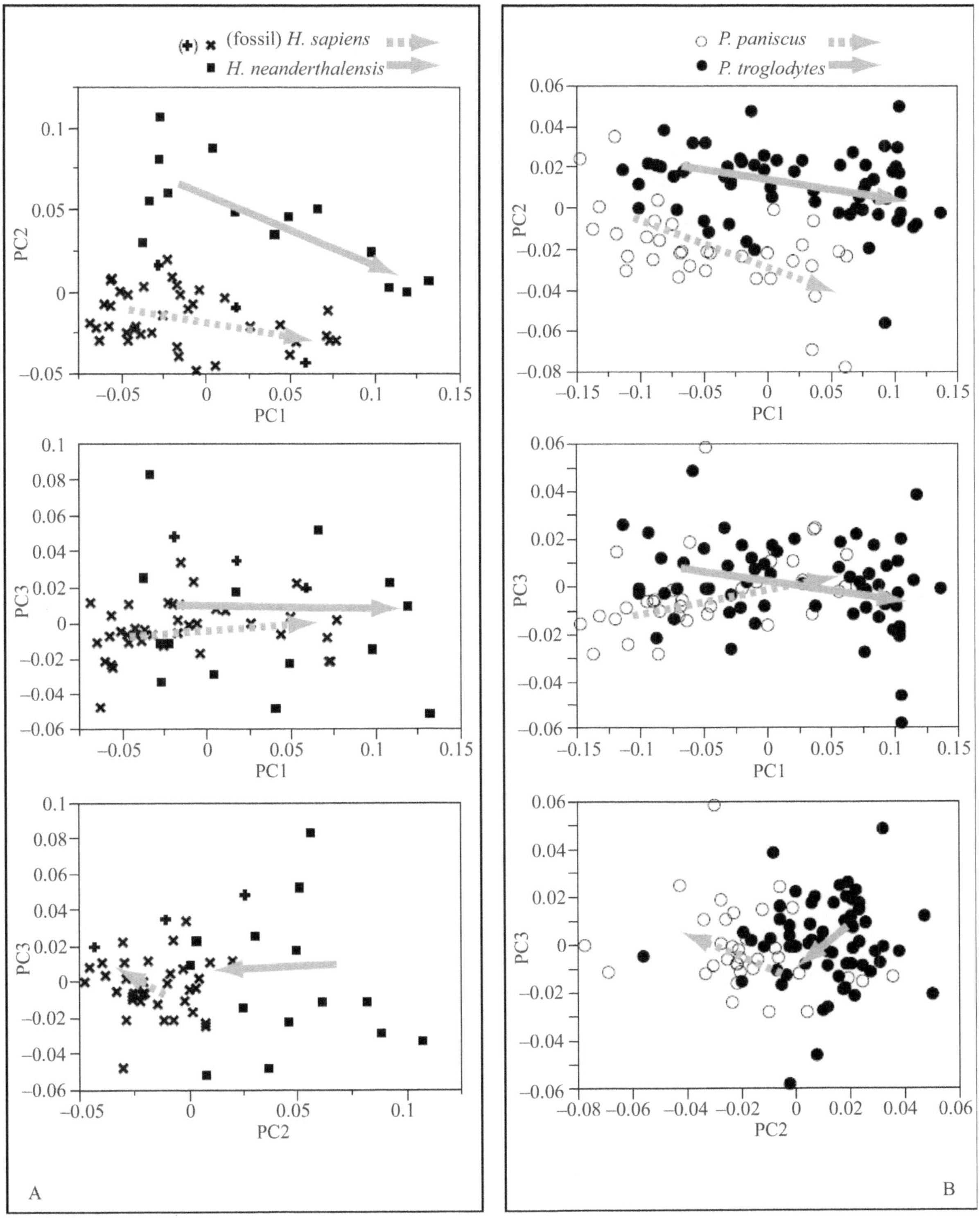

Figure 3. Ontogenetic trajectories through shape space in sister species. A: Bivariate plots of PCs 1–3 for *Homo* (+ symbols: fossil *Homo sapiens* specimens). B: Bivariate plots of PCs 1–3 for *Pan*. For each species, arrows point from mean cranial shape at dental stage Pd4 to mean cranial shape at adulthood.

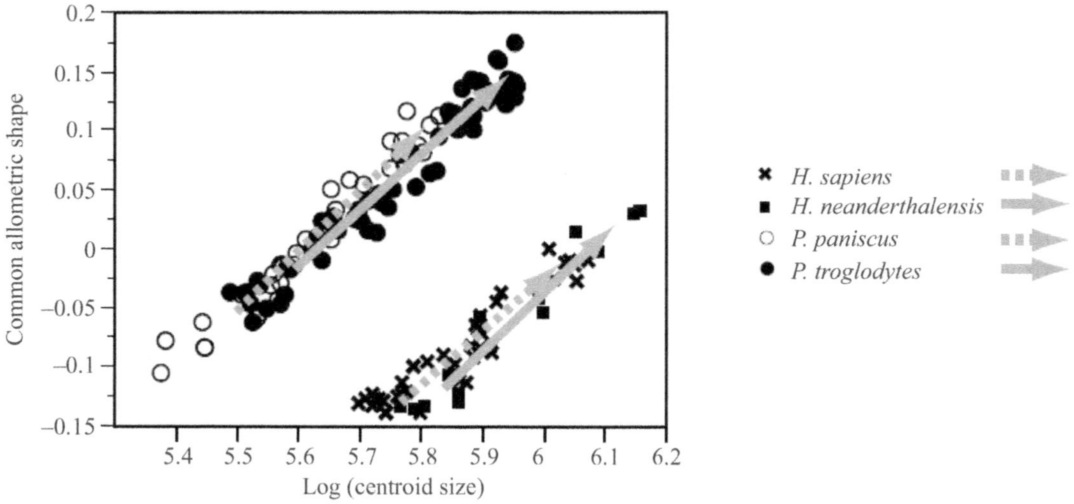

Figure 4. Graph of common allometric shape versus log centroid size. For each species, arrows point from mean cranial size/shape at dental stage Pd4 to mean cranial size/shape at adulthood.

Table 3. Variance proportions of principal components analyses

Principal component	*Homo+Pan*	*Homo*	*Pan*
PC1	0.632	0.302	0.616
PC2	0.087	0.133	0.060
PC3	0.035	0.065	0.033
PC4	0.029	0.061	0.032
PC5	0.020	0.050	0.025
PC6	0.019	0.048	0.023
PC7	0.015	0.039	0.019
PC8	0.014	0.035	0.018
PC9	0.013	0.028	0.015
PC10	0.011	0.026	0.013

There is also a trend difference in ontogenetic allometry, i.e., how change in shape during development is related to change in size. Figure 4, depicting ontogenetic allometry for all four species, shows a marked separation between *Pan* and *Homo* trajectories. A *Homo* specimen at a given allometric shape is larger than a *Pan* specimen at the same allometric shape, and a *Homo* specimen of a given size exhibits a lower allometric shape value than a similarly sized *Pan* specimen.

Finally, we may ask how patterns of divergence between *Homo* sister species differ from patterns of divergence between *Pan* sister species. Figures 2 and 3 reveal that each bifurcation follows a genus-specific pattern. In *Homo*, developmental trajectories differ primarily during prenatal ontogeny (as evinced by the distance between Neanderthal and human trajectories at age Pd4), while postnatal trajectories are statistically similar. In *Pan*, developmental trajectories differ both during prenatal *and* postnatal ontogeny. However, human-Neanderthal and chimp-bonobo bifurcations also exhibit commonalities. In each genus, sister species differ in trajectory length (Table 4). Humans and bonobos represent "shortened" Neanderthal and chimpanzee allometric

Table 4. Comparisons between trajectories through morphospace

	Homo – Pan	*H. sapiens – H. neanderthalensis*	*P. paniscus – P. troglodytes*
Procrustes distance between mean shapes	**0.175875** ($p<0.001$)	**0.078918** ($p<0.001$)	**0.074843** ($p<0.001$)
Divergence cosinus between shape trajectories	**0.775** ($p<0.001$)	0.821 ($p = 0.25$)	**0.953** ($p<0.001$)
Divergence cosinus between allometric trajectories	**0.787** ($p<0.001$)	0.817 ($p = 0.09$)	**0.950** ($p = 0.01$)

	Homo	*Pan*	*H. sapiens*	*H. neand.*	*P. paniscus*	*P. troglo.*
Trajectory length from mean Pd4 cranial shape to mean adult cranial shape	0.112	0.172	0.107	0.138	0.157	0.171
Mean Pd4 cranial size in % of mean adult cranial size	77.7	71.2	78.1	77.4	74.8	71.5

trajectories, respectively (Figure 3). Furthermore, human and bonobo trajectories are shifted towards smaller sizes and less advanced shapes relative to Neanderthal and chimpanzee trajectories, respectively (Figure 4). However, length differences between divergent trajectories must be interpreted with great care because comparisons are made between non-homologous patterns of morphological change.

Discussion

In this study, we investigate commonalities and differences between patterns of cranial ontogeny in *Homo* and *Pan* in order to gain insights into modifications of the developmental program that led to evolutionary splits between sister taxa. First, we address the question whether a possible scenario of ontogenetic modification leading to the Neanderthal/human split is comparable to that leading to the chimp/bonobo split. The results shown in Figures 2 and 3 suggest that there are relatively few commonalities between divergences within the genera. It appears that each bifurcation may have been the result of a single evolutionary event that modified a unique set of spatial and/or temporal characteristics of the ancestral human-chimp mode of ontogeny. Neanderthal and AMH postnatal trajectories are already separated at early postnatal stages and then follow parallel paths through shape space. According to the logic of geometric morphometrics, similar directions through shape space imply similar spatial patterns of shape change during development (Zollikofer and Ponce de León, 2002). Spatial similarity, in turn, hints at a shared ancestral mode of postnatal development in Neanderthals and AMH. Under these preconditions, it is possible to compare Neanderthal and AMH postnatal development in terms of heterochrony. As can be seen in Figure 3, the AMH trajectory through shape space is less extended than the Neanderthal trajectory. In other words, AMH attain "less developed" (more paedomorphic) cranial shapes at adulthood. Concomitantly, the ontogenetic allometric trajectory of AMH is less extended, such that adult AMH attain smaller cranial sizes than adult Neanderthals.

The heterochronic differences in size and shape can be interpreted in three different ways. One possibility is to see modern humans as derived paedomorphic variants, while Neanderthals conserved the ancestral speed of development. The second possibility is to see Neanderthals as derived and AMH as representing the ancestral state. Finally, the third interpretation, and in our view the most likely one, is that both AMH and Neanderthals exhibit a combination of ancestral and derived ontogenetic features. According to this hypothesis, the ancestral (cf. *Homo erectus*?) pattern of cranial shape change is preserved in both species, but AMH and Neanderthals exhibit opposite evolutionary trends with respect to velocities of development and growth. While AMH exhibit deceleration, resulting in relatively paedomorphic adult crania, Neanderthals exhibit a trend towards acceleration, resulting in more peramorphic and larger adult crania.

This interpretation is consistent with the comparative study of dental developmental velocities in mid-Pleistocene *Homo* (Ramirez Rozzi and Bermúdez de Castro, 2004). Neanderthal incisors exhibit a relatively wide spacing between adjacent perikymata compared to the incisors of potential ancestor species, while modern humans show a trend towards closer spacing. Assuming first that the timing of perikymata formation is similar in all species, and second that dental development is a good indicator of overall developmental speed, Neanderthal and AMH ontogenies appear to be accelerated and decelerated variants of the ancestral mode of development, respectively.

These findings and interpretations of Neanderthal and modern human ontogeny converge in some respects with the results of a suite of studies dedicated to the same subject, but which utilize different methods. Krovitz (2000) investigated facial ontogeny with Euclidean Distance Matrix Analysis (EDMA) (Lele and Richtsmeier, 2001), which is a geometric-morphometric method that defines form by the matrix of all inter-landmark distances in a landmark configuration and uses Principal Coordinates Analysis (PCA, a variant of multidimensional scaling) to study shape variability. These studies, as well as a recent EDMA-based analysis of facial ontogeny in a sample comprising AMH, *Australopithecus africanus*, *Pan troglodytes* and *Pan paniscus*, agree that evolution through ontogenetically early differentiation is an ancient pattern of hominoid phylogeny (Rogers Ackermann and Krovitz, 2002). In another set of studies, Neanderthal versus human craniofacial ontogeny was analyzed with classical multivariate techniques applied to sets of craniometric distance measurements (Godfrey et al., 1998; Williams, 2000; Williams et al., 2002). Interestingly, these authors converge in the conclusion that "modern humans and Neanderthals follow parallel shape changes from different points of origin" (Williams et al., 2002: 430).

Several alternative studies point to differences in postnatal patterns of ontogeny between Neanderthal and modern human lineages (Krovitz, 2003), as well as between other hominin species (Cobb et al., 2004), and between modern human populations (Vidarsdottir et al., 2002). In order to resolve the two conflicting views, it is critical to recognize the existence of differences in analytical methods, in sample composition, and in the morphological regions analyzed. Krovitz (2003) used facial measurements mostly taken from casts of Neanderthal specimens, while our study used a more complete representation of cranial morphology based on virtual reconstruction of original fossil specimens (Zollikofer and Ponce de León, 2005). Neanderthal specimens, whether original specimens or casts of originals, often exhibit considerable taphonomic deformation and reconstruction bias/inaccuracy. Failure to remove such effects not surprisingly introduces morphometric error, notably in three-dimensional measurements. Further, EDMA-based growth analyses compare mean forms of two age groups rather than entire samples representing ontogenetic trajectories. Neanderthal age groups often have sample sizes of only one specimen. Thus, differences between Neanderthal and modern human postnatal patterns of facial ontogeny reported by Krovitz (2003) conflate variation due to taxon-specific divergence with variation due to fossil distortion and individual morphology. Differences between modern human and Neanderthal postnatal facial ontogeny have been also reported by Williams et al. (2003) who used sets of linear measurements (interlandmark distances). It is problematic, however, to differentiate between size and shape when interlandmark distances are used (Bookstein, 1989). In addition, it is difficult to decide whether reported differences can be attributed to divergence of allometric trajectories or shape trajectories.

The postulated parallelism of Neanderthal and modern human shape trajectories also conflicts with studies that reveal divergence of postnatal ontogeny within human populations (Vidarsdottir et al., 2002) and hominin species (Cobb et al., 2004). The pattern of divergence between modern human populations that was revealed in the former study is complex. In particular, there is no indication that populations typically began as more similar juvenile forms and became less similar adult forms (i.e., diverge). Rather, population-specific trajectories tend to "cross" each other. Likewise, it appears that the amount of divergence actually measured strongly depends on the composition of the adult relative to the juvenile sample, which creates a caveat when interpreting between-population divergence. It is likely that many human skeletal collections exhibit a "typological Museum bias", which expresses itself as a tendency of classical field expeditions to collect the "most characteristic" specimens of a population rather than a random sample (Yoel Rak, pers. comm.). Overall, this effect tends to bias adult population-specific samples toward extreme values, thus accentuating differences between population-specific trajectories.

How can the results of these studies be compared with the results of the present study? First and foremost, our studies use pooled-population samples, which restricts direct comparisons with studies that use specific geographic populations. The Neanderthal ontogenetic series incorporates several sources of variation – residual fossil deformation, inter-population variation, and variation in evolutionary time. While modern human specimens were chosen in order to emulate this variation – it is likely that more subtle differences at the population level remain undetected within these "mixed-effect" samples. Nevertheless, it is worth noting that the divergence between Neanderthal and AMH trajectories in this study is numerically detectable (see Table 4) but not statistically significant. Whether divergence represents noise or data will be difficult to clarify unless a Neanderthal ontogenetic sample from a narrow window in geologic time and geographic space will be available.

Apart from potential postnatal differences in patterns of facial ontogeny, the key to the Neanderthal/AMH evolutionary split lies in the modification of prenatal modes of development, which most likely represent derived versions of an ancient *Homo* mode and bring about distinct cranial morphologies very early during lifetime. It is worth mentioning here that the shape distance between Neanderthal and AMH trajectories is slightly larger than that between chimp and bonobo trajectories. In other words, the *Homo* sister species are more distinct than the *Pan* sister species, a fact that has been noted previously in comparative morphometric studies of adult Neanderthals, fossil and modern humans, and great ape species (Harvati, 2003; Harvati et al., 2004).

To summarize, we postulate that the major differences between AMH and Neanderthal morphologies result from derived prenatal modes of ontogeny, while during postnatal ontogeny both taxa follow essentially the same ancestral mode of development, with only minor additional heterochronic modifications.

We may now ask whether a causal connection exists between taxon-specific perinatal cranial forms and postnatal heterochrony. Model considerations (see Zollikofer & Ponce de León, this volume) suggest that one and the same ancestral postnatal developmental processes, if "applied" to distinct perinatal morphologies, may indeed result in parallel postnatal trajectories exhibiting different lengths, *i.e.*, generate heterochronic modification similar to that observed in humans versus Neanderthals.

Possible scenarios of ontogenetic modification underlying the chimp/bonobo split, however, are more complex. Different morphologies around birth imply divergent modes of prenatal development, as postulated for *Homo*. In contrast to *Homo*, however, *Pan* species also follow divergent postnatal trajectories through shape space. This indicates derived, taxon specific modes of growth and development. Recalling that the morphologic distance between *Pan* trajectories is generally smaller than that between *Homo* trajectories, it is thus possible to consider that the complexity of evolutionary modification (i.e., shift plus divergence of trajectories) is not directly correlated with the resulting amount of morphologic modification (i.e., dissimilarity between cranial shapes).

An analogous set of questions may be weighed regarding the ontogenetic roots of the split between *Homo* and *Pan*. What is the pattern of disparity between trajectories, and what are the respective roles of prenatal and postnatal modes of ontogeny in bringing about differences at the genus level? As shown in Figure 2, *Homo* and *Pan* trajectories through shape space differ in their relative position, direction and length. Human and ape trajectories are separated by a marked distance already present at the onset of postnatal stages, and postnatally, the morphological distinction between genera increases through divergence and different lengths of trajectories. The clear morphologic divergence between genera around the time of birth is indicative of ape-specific and human-specific modes of prenatal development, while postnatal divergence between trajectories is indicative of divergent modes of development in *Homo* and *Pan*. Overall, we interpret the phyletic signal contained in these data as follows: Assuming that the *Pan* trajectory more closely reflects the ancestral condition of ontogeny than the *Homo* trajectory, it appears that the ontogeny of *Homo* is characterized by a strongly derived spatial mode of cranial shape change already at prenatal stages, and by less overall shape change during postnatal development. This notion is consistent with an earlier study of ontogenetic allometry in modern humans and chimps (Penin et al., 2002). Based on the assumption that shorter trajectories represent derived evolutionary states, it is also reasonable to postulate that AMH and bonobo ontogenies represent abridged versions of ancestral *Homo* and *Pan* trajectories, while Neanderthal and chimp ontogenies are closer to the respective ancestral conditions.

Taking into account necessary caveats (see Zollikofer & Ponce de León, this volume), it is possible to use the terminology of heterochrony and heterotopy to characterize evolutionary paths from a common ancestor of *Homo* and *Pan* towards the extant species within these genera. It is often stated that *Homo* is paedomorphic relative to *Pan* (Gould, 1977; Dean and Wood, 1984; Bjorklund, 1997; Verhulst, 1999), implying that human ontogeny represents a shortened version of the supposed ontogeny of the last common ancestor to which *Pan* is closer. Theoretical considerations demonstrate that applying the notion of heterochrony to this comparison is problematic (Zollikofer and Ponce de León, 2004). In its essence, heterochrony signifies modification of the temporal properties of an ancestral ontogeny *given that spatial properties remain the same*. As we have stated previously, the latter condition clearly is not fulfilled; *Homo* and *Pan* trajectories diverge considerably, indicating clearly distinct spatial modes of development, i.e., heterotopic divergence. Accordingly, direct comparisons between the *temporal* course of cranial development in *Homo* and *Pan* have limited biological meaning because respective *spatial* modes of development represent incommensurate entities with respect to time. This dilemma is at the core of the open debate over the question of whether humans are paedomorphic or peramorphic (Shea, 1989; Godfrey and Sutherland, 1995, 1996; Klingenberg, 1998; Gould, 2000; Smith, 2002). Arguments advanced for each view largely depend on the choice of morphologic or morphometric features, and because the spatial similarity of patterns of ontogeny is typically not verified, it is easy to provide counter-arguments based on alternative sets of measurements.

Within genera, heterochronic comparisons are less problematic. The fact that Neanderthals and AMH essentially follow parallel postnatal ontogenetic trajectories indicates shared spatial modes of development. This permits heterochronic interpretation of the subtle differences between their postnatal trajectories. From this perspective, AMH appear slightly paedomorphic compared to Neanderthals, and the reduced rate of cranial shape change is accompanied by a reduced rate of growth, i.e., less increase in size.

There is a long tradition of considering bonobos as paedomorphic variants of chimpanzees (Shea, 1983, 1984, 1988). Indeed, the shortening of the ontogenetic trajectory of *P. paniscus* relative to that of *P. troglodytes* and the smaller size of the former species relative to the latter at any given shape, indicate a heterochronic shift toward more paedomorphic cranial forms. However, the divergence of trajectories between *Pan* sister species indicates different spatial modes of cranial development, i.e., heterotopic modification of an ancestral mode of development. As a consequence, it is inadequate to consider the two *Pan* sister species as mere heterochronic variants of their last common ancestor. A study focusing on the *Pan* sister species comes to similar conclusions (Lieberman et al., 2007).

What can be learned about hominoid evolution from the patterns of ontogenetic divergence investigated here? Apparently there is more than one pathway for determining how ancestral ontogenetic processes are modified to give rise to novel modes of development and novel morphologies. It seems that the diversity of fossil and extant hominoid taxa is likely a result of variable forms of evolutionary developmental tinkering. Each bifurcation between taxa, be it at the genus or species level, has its own background of developmental modification giving rise to different ways in which descendant ontogenetic trajectories diverge from each other. One of the most fascinating implications of this idea is that spatial and temporal differences in growth and development not only generate distinct adult morphologies, but also give rise to taxon-specific life histories.

Acknowledgments

The comments of Katerina Harvati, Clive Finlayson, and two anonymous reviewers are highly appreciated. We are also grateful for the help of all curators in providing access to fossil specimens. Special thanks go to Wim van Neer (Royal Africa Museum, Tervuren), Wim Wendelen (Royal Africa Museum, Tervuren), Christoph L. Zollikofer (Kantonsspital Winterthur), Guy Marchal (University Hospital Leuven), Walter

Coudyzer (University Hospital Leuven), and Isuf Hoxha (Kantonsspital Winterthur). We thank Kris Carlson for proof-reading the manuscript. This work was supported by Swiss NSF grant 3100–067209.01 to Ch. P. E. Z.

References

Akazawa, T., Muhesen, S., Dodo, Y., Kondo, O., Mizoguchi, Y., 1995. Neanderthal infant burial. *Nature* 377, 585–586.

Bailey, S.E., 2002. A closer look at Neanderthal postcanine dental morphology: the mandibular dentition. *Anat. Rec.* 269, 148–156.

Bailey, S.E., 2004. A morphometric analysis of maxillary molar crowns of Middle-Late Pleistocene hominins. *J. Hum. Evol.* 47, 183–198.

Barriel, V., 1997. *Pan paniscus* and hominoid phylogeny: morphological data, molecular data and "total evidence." *Folia Primatol.* 68, 50–56.

Bjorklund, D.F., 1997. The role of immaturity in human development. *Psychol. Bull.* 122, 153–169.

Bookstein, F.L., 1989. "Size" and "shape": a comment on semantics. *Syst. Zool.* 38, 173–180.

Bookstein, F.L., 1991. *Morphometric Tools for Landmark Data*. Cambridge University Press, Cambridge.

Churchill, S.E., 1998. Cold adaptation, heterochrony, and Neandertals. *Evol. Anthropol.* 7, 46–61.

Cobb, S.N., O'Higgins, P., 2004. Hominins do not share a common postnatal facial ontogenetic shape trajectory. *J. Exp. Zool. B (Mol. Dev. Evol.)* 302, 302–321.

D'Aout, K., Vereecke, E., Schoonaert, K., De Clercq, D., Van Elsacker, L., Aerts, P., 2004. Locomotion in bonobos (*Pan paniscus*): differences and similarities between bipedal and quadrupedal terrestrial walking, and a comparison with other locomotor modes. *J. Anat.* 204, 353–361.

Dean, M.C., Wood, B.A., 1984. Phylogeny, neoteny and growth of the cranial base in hominoids. *Folia Primatol.* 43, 157–180.

Doran, D.M., 1993. Comparative locomotor behavior of chimpanzees and bonobos: the influence of morphology on locomotion. *Am. J. Phys. Anthropol.* 91, 83–98.

Dryden, I.L., Mardia, K., 1998. *Statistical Shape Analysis*. Wiley, New York.

Franciscus, R.G., 1999. Neandertal nasal structures and upper respiratory tract "specialization". *Proc. Natl. Acad. Sci. U.S.A.* 96, 1805–1809.

Franciscus, R.G., 2003. Internal nasal floor configuration in *Homo* with special reference to the evolution of Neandertal facial form. *J. Hum. Evol.* 44, 701–729.

Franciscus, R.G., Churchill, S.E., 2002. The costal skeleton of Shanidar 3 and a reappraisal of Neandertal thoracic morphology. *J. Hum. Evol.* 42, 303–356.

Gagneux, P., Wills, C., Gerloff, U., Tautz, D., Morin, P.A., Boesch, C., Fruth, B., Hohmann, G., Ryder, O.A., Woodruff, D.S., 1999. Mitochondrial sequences show diverse evolutionary histories of African hominoids. *Proc. Natl. Acad. Sci. U.S.A.* 96, 5077–5082.

Godfrey, L.R., Sutherland, M.R., 1995. Flawed inference: why size-based tests of heterochronic processes do not work. *J. Theor. Biol.* 172, 43–61.

Godfrey, L.R., Sutherland, M.R., 1996. Paradox of peramorphic paedomorphosis: heterochrony and human evolution. *Am. J. Phys. Anthropol.* 99, 17–42.

Godfrey, L.R., King, S.J., Sutherland, M.R., 1998. Heterochronic approaches to the study of locomotion. In: Strasser, S. (Ed.), *Primate Locomotion*. Plenum Press, New York, pp. 277–307.

Golovanova, L.V., Hoffecker, J.F., Kharitonov, V.M., Romanova, G.P., 1999. Mezmaiskaya cave: A Neanderthal occupation in the Northern Caucasus. *Curr. Anthropol.* 40, 77–86.

Gould, S.J., 1977. *Ontogeny and Phylogeny*. The Belknap Press of Harvard University Press, Cambridge, MA.

Gould, S.J., 2000. Of coiled oysters and big brains: how to rescue the terminology of heterochrony, now gone astray. *Evol. Dev.* 2, 241–248.

Gould, S.J., Lewontin, R.C., 1979. The spandrels of San Marco and the Panglossian paradigm: a critique of the adaptationist programme. *Proc. R. Soc. Lond. B* 205, 581–598.

Guatelli-Steinberg, D., Reid, D.J., Bishop, T.A., Larsen, C.S., 2005. Anterior tooth growth periods in Neandertals were comparable to those of modern humans. *Proc. Natl. Acad. Sci. U.S.A.* 102, 14197–14202.

Harvati, K., 2003. The Neanderthal taxonomic position: models of intra- and inter-specific craniofacial variation. *J. Hum. Evol.* 44, 107–132.

Harvati, K., Frost, S.R., McNulty, K.P., 2004. Neanderthal taxonomy reconsidered: implications of 3D primate models of intra- and inter-specific differences. *Proc. Natl. Acad. Sci. U.S.A.* 101, 1147–1152.

Ishida, H., Kondo, O., Muhesen, S., Akazawa, T., 2000. A new Neanderthal child recovered at Dederiyeh Cave, Syria, in 1997–1998. *Am. J. Phys. Anthropol. Suppl.* 30, 186–187.

Jolliffe, I.T., 1986. *Principal Component Analysis*. Springer, Berlin.

Klingenberg, C.P., 1998. Heterochrony and allometry: the analysis of evolutionary change in ontogeny. *Biol. Rev. Camb. Philos. Soc.* 73, 79–123.

Kondo, O., Dodo, Y., 2000. Estimation of stature from the skeletal reconstruction of an immature Neandertal from Dederiyeh Cave, Syria. *J. Hum. Evol.* 38, 457–473.

Krings, M., Stone, A., Schmitz, R.W., Krainitzki, H., Stoneking, M., Pääbo, S., 1997. Neandertal DNA sequences and the origin of modern humans. *Cell* 90, 19–30.

Krings, M., Geisert, H., Schmitz, R.W., Krainitzki, H., Pääbo, S., 1999. DNA sequence of the mitochondrial hypervariable region II from the Neandertal type specimen. *Proc. Natl. Acad. Sci. U.S.A.* 96, 5581–5585.

Krings, M., Capelli, C., Tschentscher, F., Geisert, H., Meyer, S., von Haeseler, A., Grossschmidt, K., Possnert, G., Paunovic, M., Pääbo, S., 2000. A view of Neandertal genetic diversity. *Nat. Genet.* 26, 144–146.

Krovitz, G.E., 2000. Three-dimensional comparisons of craniofacial morphology and growth patterns in Neandertals and modern humans. Ph.D. Dissertation. Johns Hopkins University, Baltimore.

Krovitz, G.E., 2003. Shape and growth differences between Neanderthals and modern humans: grounds for a species level distinction. In: Thompson, J., Krovitz, G., Nelson, A., (Eds.), *Patterns of Growth and Development in the Genus Homo*. Cambridge University Press, Cambridge, pp. 320–342.

Lele, S., Richtsmeier, J., 2001. *An Invariant Approach to the Statistical Analysis of Shapes*. Chapman and Hall, Boca Raton, FL.

Lieberman, D.E., Carlo, J.O.S., Ponce de León, M.S., Zollikofer, C.P.E., 2007. A geometric morphometric analysis of heterochrony in the cranium of chimpanzees and bonobos. *J. Juman. Evol.* 52, 647–662.

O'Connor, C.F., Franciscus, R.G., Holton, N.E., 2005. Bite force production capability and efficiency in Neandertals and modern humans. *Am. J. Phys. Anthropol.* 127, 129–51.

Ovchinnikov, I.V., Gotherstrom, A., Romanova, G.P., Kharitonov, V.M., Liden, K., Goodwin, W., 2000. Molecular analysis of Neanderthal DNA from the northern Caucasus. *Nature* 404, 490–493.

Parish, A.R., De Waal, F.B., 2000. The other "closest living relative". How bonobos (*Pan paniscus*) challenge traditional assumptions about females, dominance, intra- and intersexual interactions, and hominid evolution. *Ann. N. Y. Acad. Sci.* 907, 97–113.

Penin, X., Berge, C., Baylac, M., 2002. Ontogenetic study of the skull in modern humans and the common chimpanzees: neotenic hypothesis reconsidered with a tridimensional Procrustes analysis. *Am. J. Phys. Anthropol.* 118, 50–62.

Ponce de León, M.S., Zollikofer, C.P.E., 2001. Neanderthal cranial ontogeny and its implications for late hominid diversity. *Nature* 412, 534–538.

Poti, P., 2005. Chimpanzees' constructional praxis (*Pan paniscus, Pan troglodytes*). *Primates* 46, 103–113.

Rak, Y., 1986. The Neanderthal: a new look at an old face. *J. Hum. Evol.* 15, 151–164.

Ramirez Rozzi, F.V., Bermudez De Castro, J.M., 2004. Surprisingly rapid growth in Neanderthals. *Nature* 428, 936–939.

Rogers Ackermann, R., Krovitz, G.E., 2002. Common patterns of facial ontogeny in the hominid lineage. *Anat. Rec.* 269, 142–147.

Rohlf, F.J., Slice, D., 1990. Extensions of the Procrustes method for the optimal superimposition of landmarks. *Syst. Zool.* 39, 40–59.

Sept, J., 1998. Shadows on a changing landscape: comparing nesting patterns of hominids and chimpanzees since their last common ancestor. *Am. J. Primatol.* 46, 85–101.

Serre, D., Langaney, A., Chech, M., Teschler-Nicola, M., Paunovic, M., Mennecier, P., Hofreiter, M., Possnert, G.G., Paabo, S., 2004. No evidence of Neandertal mtDNA contribution to early modern humans. *PLoS Biol.* 2, E57.

Shea, B.T., 1983. Paedomorphosis and neoteny in the pygmy chimpanzee. *Science* 222, 521–522.

Shea, B.T., 1984. An allometric perspective on the morphological and evolutionary relationships between pygmy (*Pan paniscus*) and common (*Pan troglodytes*) chimpanzees. In: Susman, R.L. (Ed.), *The Pygmy Chimpanzee: Evolutionary Biology and Behavior*. Plenum Press, New York, pp. 89–130.

Shea, B.T., 1988. Heterochrony in primates. In: McKinney, M.L. (Ed.), *Heterochrony in Evolution: A Multidisciplinary Approach*. Plenum Press, New York, pp. 237–266.

Shea, B.T., 1989. Heterochrony in human evolution: the case for neoteny reconsidered. *Yrbk. Phys. Anthropol.* 32, 69–101.

Smith, K.K., 2002. Sequence heterochrony and the evolution of development. *J. Morphol.* 252, 82–97.

Stringer, C.B., Gamble, C., 1993. *In Search of the Neanderthals: Solving the Puzzle of Human Origins*. Thames and Hudson, London.

Susman, R.L., (Ed.) 1984. *The Pygmy Chimpanzee: Evolutionary Biology and Behavior*. Plenum Press, New York.

Trinkaus, E., 1987. The Neandertal face: evolutionary and functional perspectives on a recent hominid face. *J. Hum. Evol.* 16, 429–443.

Trinkaus, E., 2003. Neandertal faces were not long; modern human faces are short. *Proc. Natl. Acad. Sci. U.S.A.* 100, 8142–8145.

Verhulst, J., 1999. Bolkian and Bokian retardation in *Homo sapiens*. *Acta Biotheor.* 47, 7–28.

Vidarsdottir, U.S., O'Higgins, P., Stringer, C., 2002. A geometric morphometric study of regional differences in the ontogeny of the modern human facial skeleton. *J. Anat.* 201, 211–229.

Williams, F.L., 2000. Heterochrony and the human fossil record: comparing Neandertal and modern human craniofacial ontogeny. In: Stringer, C.B., Barton, R.N.E., Finlayson, J.C. (Eds.), *Neanderthals on the Edge*. Oxbow Books, Oxford, pp. 257–267.

Williams, F.L., Godfrey, L.R., Sutherland, M.R., 2002. Heterochrony and the evolution of Neandertal and modern human craniofacial form. In: Minugh-Purvis, N., McNamara, K. (Eds.), *Human Evolution through Developmental Change*. The Johns Hopkins University Press, Baltimore, pp. 405–441.

Williams, F.L., Godfrey, L.R., Sutherland, M.R., 2003. Diagnosing heterochronic perturbations in the craniofacial evolution of *Homo* (Neanderthals and modern humans) and Pan (*Pan troglodytes* and *Pan paniscus*). In: Thompson, J., Krovitz, G., Nelson, A. (Eds.), *Patterns of Growth and Development in the Genus Homo*. Cambridge University Press, Cambridge, pp. 295–319.

Won, Y.J., Hey, J., 2005. Divergence population genetics of chimpanzees. *Mol. Biol. Evol.* 22, 297–307.

Yu, N., Jensen-Seaman, M.I., Chemnick, L., Kidd, J.R., Deinard, A.S., Ryder, O., Kidd, K.K., Li, W.H., 2003. Low nucleotide diversity in chimpanzees and bonobos. *Genetics* 164, 1511–1518.

Zihlman, A.L., Cramer, D.L., 1978. Skeletal differences between pygmy (*Pan paniscus*) and common chimpanzees (*Pan troglodytes*). *Folia Primatol.* 29, 86–94.

Zollikofer, C.P.E., Ponce de León, M.S., 2002. Visualizing patterns of craniofacial shape variation in *Homo sapiens*. *Proc. R. Soc. B* 269, 801–807.

Zollikofer, C.P.E., Ponce de León, M.S., 2004. Kinematics of cranial ontogeny: Heterotopy, heterochrony, and geometric morphometric analysis of growth models. *J. Exp. Zool. (Mol. Dev. Evol.)* 302B, 322–340.

Zollikofer, C.P.E., Ponce de León, M.S., 2005. *Virtual Reconstruction: A Primer in Computer-assisted Paleontology and Biomedicine*. Wiley, New York.

Zollikofer, C.P.E., Ponce de León, M., Ishida, H., Suzuki, H., Kobayashi, Y., Tsuchiya, K., Akazawa, T., 2002. Computer-assisted reconstruction of the Dederiyeh Neanderthal infants. I: cranium and mandible. In: Ishida, H., Nakatsukasa, M., Ogiwara, N. (Eds.), *Recent Advances in Physical Anthropology and Primatology*. Kinsei-sha, Kyoto, pp. 35–40.

6. Cranial growth models: heterochrony, heterotopy, and the kinematics of ontogeny

C.P.E. ZOLLIKOFER
Anthropologisches Institut und Museum
Universität Zuerich-Irchel
Winterthurerstrasse 190
CH-8057 Zürich, Switzerland
zolli@aim.unizh.ch

M.S. PONCE DE LEÓN
Anthropologisches Institut und Museum
Universität Zuerich-Irchel
Winterthurerstrasse 190
CH-8057 Zürich, Switzerland
marcia@aim.unizh.ch

Keywords: Allometry, computer simulations, geometric morphometrics, human evolution, heterochrony, heterotopy, kinematics, morphogenetic modeling, ontogeny, phylogeny.

Abstract

In fossil hominins, phyletic diversification – the process by which ancestral species give rise to descendant ones – can only be inferred through analysis of patterns of morphological diversity displayed in the fossil record. These patterns are interpreted typically in terms of selection/adaptation and related to environmental change. From an organism-centered perspective, evolutionary modification of developmental processes is an equally important source of phyletic diversity. Here, we use model systems to simulate cranial growth and to explore how mutations in the "genes" of an "ancestral" morphogenetic system may affect "descendant" ontogenies and "adult" morphologies. Intriguingly, a model that assumes basic epigenetic interactions between developmental processes is capable of producing a wide variety of patterns of developmental modification, many of which are not foreseen in classic heterochronic theory. Also, small changes in developmental "genes" often have complex effects on patterns of ontogeny. With regard to the evolutionary split between Neanderthals and modern humans, these model considerations shall be an incentive to look at taxon-specific character complexes from the perspective of developmental as opposed to functional constraints.

Introduction

Morphological change in the cranium during human evolution is typically seen as a suite of functional adaptations to changing environmental conditions and/or specialization to an ecological niche. A paramount example can be seen in the Neanderthals, who are often thought to represent a human species adapted to the harsh climatic conditions of mid-Pleistocene Europe (Churchill, 1998; Lahr and Foley, 1998). However, establishing

K. Harvati and T. Harrison (eds.), Neanderthals Revisited: New Approaches and Perspectives, 89–112.

correlations and causal links between varying environmental factors and variation in character states faces both practical and theoretical obstacles. First, due to the incompleteness of paleoenvironmental and fossil records, reconstruction of past key events and inferences regarding their impact on the course of human evolution should remain tentative. Second, it is typically impossible to interpret changes in a particular body structure in terms of specific functional changes, i.e., adaptations to a specific task or a specific environmental condition. Rather, it appears that morphological alterations involve correlated sets of characters and reflect a network of changing functional requirements. Disentangling these networks is a challenging task because it is often difficult to discriminate between cause and effect. Third, evolutionary modification of morphology is typically channeled by developmental constraints. As a consequence, in highly integrated systems such as the hominin skull, every conceivable morphology cannot be realized, thus morphology becomes a compromise between external (environmental) constraints and requirements, and internal (developmental) constraints and requirements.

These arguments merit further attention. Using the metaphor of supply and demand, it is developmental variation that supplies evolutionary novelty and diversity, while selection reduces diversity by restricting demands to a small fraction of the supply. Accordingly, how and why the form of the hominin cranium changes over evolutionary time spans can be tackled in two complementary ways. On the one hand, we may investigate the adaptive and functional context of change, while on the other hand, we may study how change is brought about through modification of developmental programs.

The latter approach is adopted here. First, we devise a general model of cranial growth and use computer simulations to explore how changes in model parameters – which stand for "developmental genes" and epigenetic interactions – affect developmental paths and result in cranial morphological diversity. Second, we ask how our results comply with existing theoretical frameworks of evolutionary developmental modification, notably with current concepts of heterochrony, heterotopy, and ontogenetic allometry. These concepts are central to evolutionary developmental reasoning, but there is no consensus on how to define them theoretically and apply them during empirical data analysis. In the study presented here, we refrain from adopting one or the other existing theoretical framework. Rather, we use insights from simulations to reveal limitations of current concepts of heterochrony and heterotopy, and we propose a combination of geometric-morphometric analysis and kinematic analysis to quantify the widest possible diversity of patterns of evolutionary modification of developmental pathways.

Measuring and Interpreting Ontogenetic Modifications

The central role of ontogenetic modification as a promoter of phylogenetic ramification was recognized by Haeckel (1866), who coined the terms *heterochrony* and *heterotopy* to denote evolutionary modification of temporal and spatial properties of developmental programs, respectively. Analyzing ancestor-descendant relationships in terms of heterochronic and heterotopic modification of ontogeny has proven extremely fruitful in generating insights into developmental paths of phyletic diversification: Since the conceptual foundations were provided in Stephen Jay Gould's *Ontogeny and Phylogeny* (Gould, 1977), heterochronic theory has diversified into two principal schools of thought. The first, which may be named "pan-heterochrony" (McKinney and McNamara, 1991; McKinney, 1999; McNamara, 2002) proposes that most observed instances of developmental dissociation between sister taxa can be explained in terms of temporal shift and/or scaling (contraction or extension) of a common ancestral path of ontogeny. The second school of thought, which is named "spatiotemporal" here (Raff and Wray, 1989; Zelditch and Fink, 1996; Zelditch et al., 2000), proposes that developmental dissociation is normally brought about by modification of temporal *and* spatial properties of the ancestral ontogeny, the latter of which tends to be overlooked because it is typically more difficult to analyze in quantitative terms.

As we showed in an earlier study (Zollikofer and Ponce de León 2004), many aspects of the ongoing "heterochronic debate" hinge on principal conceptual ambiguities in the notions of heterochrony and heterotopy, and in the basic notions of size, shape and developmental time, which are used to measure heterochronic and heterotopic modification. Originally proposed as purely descriptive terms for *patterns* of developmental divergence, heterochrony and heterotopy are often used as explanatory terms to denote biological *processes* underlying observed patterns of divergence (Godfrey and Sutherland, 1995; Klingenberg, 1998). To avoid confusing data and inference, a clear separation is required between the description of patterns of heterochronic/heterotopic divergence and ensuing hypotheses regarding potential underlying heterochronic/heterotopic processes. In this context, it is revealing to consider Gould's last attempt to rescue the original semantics of heterochronic terminology (Gould, 2000). According to his diagnosis, heterochronic terminology was transformed from an originally purely descriptive into an explanatory category, and artificial categories were reified as biological entities. Gould went on to identify a single major source of logical confusion, namely the application of terminological categories originally designed for the description of *shape* modification to the description of *rate* modification. From a historical perspective, it appears that researchers started using Gould's shape terms, i.e., morphological features, to describe size data, i.e., rates of growth (Klingenberg, 1998). Gould's insistence on an exclusively "taxonomic" use of heterochronic terminology – to apply it to features, not rates – has its great merits as it forces us to disentangle cause and effect, to discern between pattern and process, and to separate description from explanation. However, at the same time, it downplays the role of rate change, which is central to heterochronic reasoning.

A further difficulty arises from the fact that differentiation must be made with respect to developmental modules involved in modification – in addition to the necessary conceptual differentiation between pattern and process. Depending on the level of organization and scale (organs, tissues, cells, etc.), heterochronic and/or heterotopic patterns of ontogenetic dissociation are characterized with different terms. As an example, consider Crouzon disease (Jones, 1988), a congenital malformation of the craniomaxillofacial morphology resulting from mutations in the fibroblast growth factor (FGF) receptor genes (Yu et al., 2000; Wilkie and Morriskay, 2001). These mutations affect temporal patterns of neural crest cell differentiation, and probably also their spatial migration patterns (Sarkar et al., 2001; Abzhanov et al., 2003; Santagati and Rijli, 2003). A gain-of-function mutant in an FGF receptor gene represents molecular *process heterochrony* since it generates increased rates of signal transduction, differentiation and bone deposition in neural crest cell derivatives (all peramorphic features). Growth disorders of the skull are characterized by premature neurocranial suture closure (a peramorphic feature) and retarded growth of the maxilla (a paedomorphic feature) among other symptoms (Jones, 1988; Andresen et al., 2000). Considering the maxilla alone, retarded growth represents a heterochronic shift toward paedomorphy. Considering entire cranial morphology, however, the same growth disorder

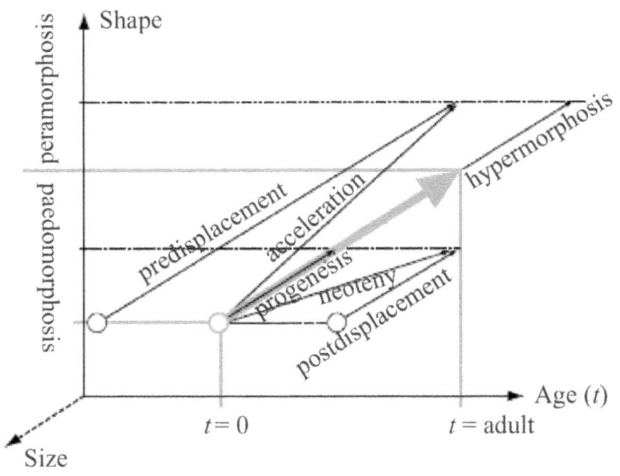

A

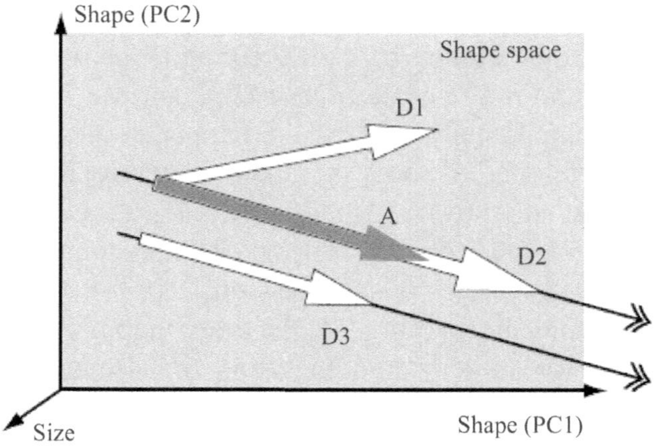

B

Figure 1. Definition of heterochrony and heterotopy. A: In classical definitions, heterochrony is a divergence between ancestral and descendant shape-time (and/or shape-size) graphs. Shape is typically thought of as a one-dimensional entity, e.g., a ratio between two linear measurements. Developmental trajectories are displayed as arrows from an early developmental phase (circles) to adulthood (arrow tips). The gray arrow denotes the ancestral trajectory. Descendant trajectories are constructed by applying basic transformations (translation: pre-/post-displacement; rotation: acceleration/neoteny; scaling: hypermorphosis, progenesis). Correspondingly, descendant adults develop shapes beyond those of ancestral adults (peramorphosis) or attain shapes, which are characteristic for juvenile ancestors (paedomorphosis). Note that various evolutionary transformations (e.g. shift along the shape axis) are not comprised in this terminology. Likewise, concomitant changes in size are not considered. B: In a geometric-morphometric definition, shape is a multidimensional measure (two dimensions are shown here, PC1 and PC2), such that ancestral (A) and descendant (D1–D3) trajectories through shape space can be compared with respect to length and direction. Heterotopic modification is exemplified by divergence between ancestral and descendant trajectories through shape space (D1). Heterochronic modification occurs as differing developmental speeds along trajectories through shape space, when segments of ancestral and descendant shape trajectories coincide (pure heterochrony, D2), or at least are collinear, i.e., point into the same direction (generalized heterochrony, D3; double-headed arrows). The size axis indicates an additional dimension of morphology – ontogenetic allometry – along which trajectories may diverge, resulting in "giantism" or "dwarfism" of descendant relative to ancestral forms.

appears as a heterotopic modification, because the spatial relationships between the maxilla and the rest of the face are disturbed. This example demonstrates that depending on the module and/or level of analysis one and the same biological phenomenon can be described with a multitude of terms, even to the point of being contradictory.

An additional difficulty arises during the application of operational definitions of heterochrony and heterotopy. Using size, shape and developmental time as fundamental variables of interest, heterochrony is typically defined as a temporal dissociation between ancestral and descendant shape-age trajectories over time. As illustrated by Stephen J. Gould's famous developmental clock (Gould, 1977), a heterochronic shift of the shape-age trajectory is typically accompanied by a shift of the size-age trajectory, i.e., shifts in ontogenetic allometric relations (this point will be discussed shortly in greater detail). An important, but often overlooked, precondition of this definition is that ancestor and descendant must follow essentially the same shape trajectories (Figure 1) (Godfrey et al., 1998). Accordingly, heterotopy is defined and measured here as the spatial dissociation between shape trajectories over time.

While there is some agreement on these basic definitions, there is less agreement on how size, shape and developmental time should be defined. We focus on size and shape because our understanding of heterochrony and heterotopy directly depends on precise quantitative definitions of these terms. Typically, a distinction is made between size measuring a biological object's extent, and shape as a measure of an object's geometry. Within this relatively loose framework, however, definitions can vary considerably depending on the biological context, the methods of measurement used, and the questions asked (Bookstein, 1989; Mosimann, 1988).

Once age, size, and shape have been defined, measuring and describing divergence between ancestral and descendant ontogenetic trajectories becomes a central issue. The classic terminology of heterochrony comprises 6 different types of modification of a *linear* ancestral shape-time and/or size-shape trajectory (Figure 1). This relatively complex terminology, however, only captures a subset of all potential modifications that may result from scaling, translation and rotation of a descendant relative to an ancestral trajectory (for example, no classic terminology exists to describe vertical shifts of the descendant relative to the ancestral trajectory, or trajectories that converge towards an adult shape). Also, it is difficult to describe modification of biologically more realistic, curvilinear ontogenetic trajectories (Rice, 1997). And finally, heterochronic terminology can only characterize the relationship between shape and age (technically termed "development"), while vocabularies describing relationships between shape and size (ontogenetic allometry) or size and age ("growth") are less complete.

Basic Questions

Considering these issues, three main questions are tackled:

(1) How are shape, size and developmental time best defined in order to detect, measure and compare modifications in ontogenetic trajectories of complex morphologies such as the hominin skull?
(2) Which terminology is most adequate to describe spatiotemporal modification of patterns and/or processes of ontogeny?
(3) What is the relationship between spatiotemporal modifications of developmental processes and the resulting spatiotemporal modification of developmental patterns?

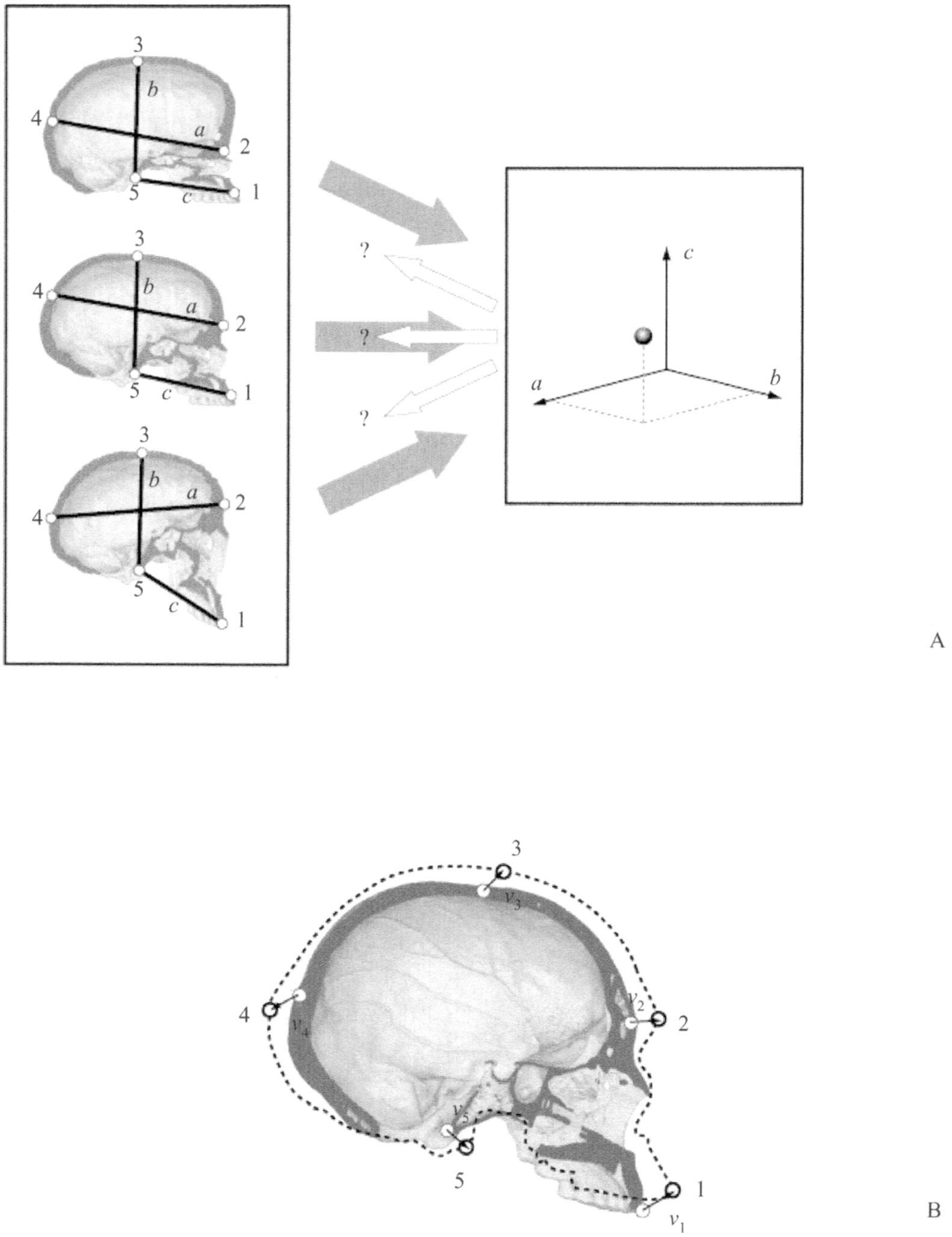

Figure 2. Measuring shape. A: Sets of linear measurements (*a–c*) taken between homologous landmarks (1–5) cannot discriminate between geometrically distinct forms. In the case illustrated here, three distinct cranial morphologies (left box) are mapped (grey arrows) onto one and the same point in feature space (right box), which is defined by the three measurements *a*, *b*, *c*. Consequently, the original object geometry cannot be recovered from such shape measurements (white arrows), unless all possible distances between landmark pairs are measured (as in Euclidean Distance Matrix Analysis; Lele et al., 2001). B: In geometric morphometrics, the shape of an object is measured as the landmark-based deviation from a reference configuration (vectors v_1–v_5). Using vector coordinates $x_1, y_1, \ldots, x_5, y_5$, the complete object geometry can be recovered from shape data.

SIZE AND SHAPE

How are size and shape best defined to measure change of form along ontogenetic trajectories? In classical morphometry, shape is typically measured as a set of ratios between linear measurements (angles can be considered to represent a special case of ratios, as they represent inverse trigonometric functions of ratios). As a proxy of size, some linear combination of original measurements is used, or alternatively the factor with the highest loading resulting from principal components analysis (PCA) on all linear measurements (Jolliffe, 1986). One major disadvantage of measuring shape as a ratio, however, is that ratios incompletely represent the object geometry from which they are derived (Figure 2A). As a consequence, changes in object geometry cannot be monitored in detail, and thus questions regarding heterochronic versus heterotopic modification are resolved only partially.

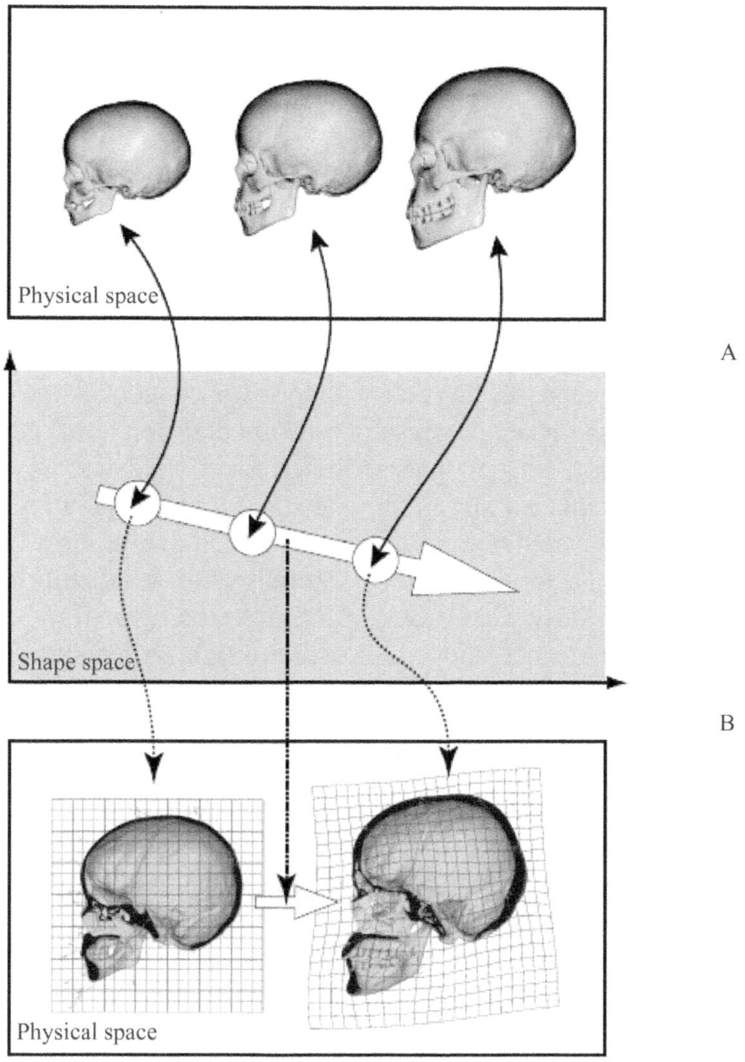

Figure 3. Shape change in the physical world and in shape space. A: Correspondence between locations in multidimensional shape space (circles, middle graph) and cranial shapes in physical space. B: Correspondence between a shape trajectory in shape space (arrow, middle graph) and a pattern of shape transformation in physical space (transformation grid from juvenile to adult).

The framework of geometric morphometrics (Bookstein, 1991; Dryden and Mardia, 1998) offers an elegant solution to these problems (Figure 2B). Shape is measured as the minimized deviation of an object's spatial configuration of anatomic landmarks from a reference configuration (e.g., the sample mean configuration), while size is measured as centroid size (the square root of the sum of squared distances between each landmark of a given configuration and its center of mass; Bookstein, 1991). The resulting shape space is multidimensional ($D = 2K$-4 for K-landmark configurations in the plane), but because each location in shape space corresponds to a well-defined landmark configuration in physical space, it is possible to switch between physical and abstract representations of shape variability in a sample (Figure 3A). This property turns out to be especially useful during developmental analyses, because it offers the possibility to characterize complex patterns of physical shape transformation as trajectories through shape space (Figure 3B). It needs to be mentioned at this point that shape space analysis has several limitations. First, shape spaces have the geometry of hyperspheres such that linearization procedures are needed before linear multivariate analyses can be applied. While such procedures tend to introduce distortions, these are typically negligible since they are within the range of variation encountered in closely related biological shapes. A second limitation, which is often overlooked, results from the definition of landmarks used to quantify morphologies. As shown in earlier studies (Zollikofer and Pance de León, 2004), alternative landmark sets lead to slightly different results, though they do not distort the overall outcome of geometric morphometric analyses.

TERMINOLOGY AND MEASUREMENT OF DEVELOPMENTAL CHANGE

As shown in Figure 1, the terminology of heterochronic and heterotopic modification is highly elaborated yet relatively poorly adapted to render the full spectrum of empirical observations on developmental modification. We therefore need a more technical set of terms that permits precise description of phenomena without interpretation.

As we proposed in an earlier study (Zollikofer and Ponce de León, 2004), it is sensible to adopt methods and terminology of kinematic analysis to investigate patterns of ontogenetic form change. Kinematic analysis provides a well-established tool kit to measure spatiotemporal change in a physical system. Similar tools can be applied to measure spatiotemporal change in organisms. In fact, close formal analogies exist between physical kinematics and "ontogenetic kinematics". While the former describes how objects move along trajectories through physical space, the latter describes how objects change their physical shape during ontogeny, and how they move along trajectories through shape space (Figure 3). The major incentive for applying kinematic concepts, however, is to clarify a fundamental issue of classic heterochronic/heterotopic analysis. A key hurdle is that developmental velocities, a concept that is used extensively, only rarely are defined explicitly. In kinematic analyses, the definition of velocity is straightforward; velocity is a *vector*, whose magnitude and direction indicate the temporal and spatial components of movement, i.e., the speed and path of a moving object (Winter, 1990). Similarly, one can define developmental velocity vectors that describe the temporal and spatial components of form change during ontogeny. As noted by several authors (Godfrey et al., 1998; Zelditch and Fink, 1996), heterochronic studies often focus on differential developmental speed, i.e., differentiating the *magnitude* of the developmental velocity vector, while fewer consider differences in vector *direction*. From a kinematic perspective, it is evident that magnitudes of ancestral and descendant velocity vectors can only be compared if they have the same direction in shape space (Figure 1B). In other words, heterochronic comparisons can be performed only if two taxa follow essentially similar *physical spatial* patterns of shape change.

It is worth noting that many of the ongoing debates about different concepts of heterochrony/heterotopy (see "pan-heterochronic" versus "spatiotemporal" approaches) have their roots in a general disagreement over how to compare developmental speeds and velocities. To appreciate the combination of kinematics and geometric-morphometrics in heterochronic/heterotopic analysis, the following metaphor is introduced. Imagine two sprinters, whose performances shall be compared. It is evident that their running speed can be compared sensibly only if they run along equally-shaped trajectories, i.e., if their velocity vectors have the same direction. This condition is met, for example, during a competition where sprinters run along parallel tracks. While physical velocity vectors have two or three spatial dimensions, morphometric shape velocity vectors are multidimensional. However, the same criteria for comparison apply: developmental speeds can be compared only if shape velocity vectors are collinear. In contrast, consider classic morphometry, where shapes usually are one-dimensional directionless quantities, e.g., ratios between linear measurements. Under these conditions, it is impossible to discern whether shape change in one species is directionally equivalent to shape change in another species. Consequently, it is impossible to verify if heterochronic comparisons are biologically meaningful at all. Returning to our sprinters, this is as if one would compare running speeds of two competitors, ignoring that one of them is running downhill, while the other is running uphill.

RELATIONS BETWEEN PROCESS AND PATTERN OF DEVELOPMENT

Having established a methodological tool kit to measure patterns of heterochronic/heterotopic modification, we may now consider the underlying processes. Here we ask how modifications in the genetic (and epigenetic) network affect the course of growth and development and, ultimately, observable spatiotemporal patterns of phenetic change. A comprehensive strategy to answer this question consists in studying the effects of developmental genes and their allelic variants on ontogenetic programs and on the resulting phenetic patterns of development. This process-oriented approach is widely used to study development in laboratory animals ranging from nematodes to vertebrates. In humans, analysis of congenital "molecular" diseases opens a similar window, albeit smaller, on the complex web of cause and effect that characterizes ontogenetic modifications (Wilkie and Morriss-Kay, 2001). However, this window remains closed during the study of human evolution because actualistic evidence is not available for fossil forms. Hence, a bottom-up approach from developmental genes to phenes is impossible. Alternatively, a top-down approach focuses on phenetic change during ontogeny, in order to construct inferences regarding potential process modifications that may underlie the observed patterns of divergence in growth and development.

As a complement to experiments in developmental genetics on the one hand, and to morphological analyses on the other, a third strategy – morphogenetic modeling – is presented here. Morphogenetic models establish *in silico* (i.e., silicon chip) genetic networks and epigenetic interactions that determine form change in model organisms. Rather than dwelling on the behavior of specific real-world developmental genes and phenes, this approach studies general properties of developmental systems and explores a wide range of consequences of modifying the genetic and epigenetic structure of the system. Growth model systems have several advantages. (1) they force the experimenter to state explicitly (in terms of algorithms) all data and hypotheses

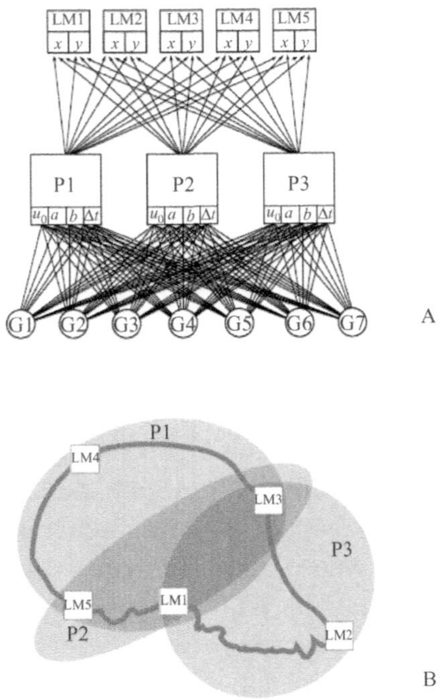

P_i:	$u_i(t) = u_{0i} \cdot e^{b_i/a_i(1-e^{-a_i(t-\Delta t_i)})}$

P1:	$u_{01}= 1$	$a_1= 1$	$b_1= 1$	$\Delta t_1= 0$
P2:	$u_{02}= 1$	$a_2= 1$	$b_2= 2$	$\Delta t_2= 0$
P3:	$u_{03}= 1$	$a_3= 1$	$b_3= 1.5$	$\Delta t_3= 0$

LM1:	$x_1(t) = 0$	$y_1(t) = 0$
LM2:	$x_2(t) = 0$	$y_2(t) = u_3(t)$
LM3:	$x_3(t) = -u_1(t)$	$y_3(t) = (u_2(t) + u_3(t))/2$
LM4:	$x_4(t) = -u_1(t)$	$y_4(t) = u_1(t)$
LM5:	$x_5(t) = u_2(t)$	$y_5(t) = 0$

C

Figure 4. A minimal model of cranial growth. A: General structure. A set of interacting genes (G1–G7, bottom) determines the key parameters (u_0, a, b, t; see Equation 4 in the text) of a set of growth processes (P1–P3; middle), which in turn interact epigenetically to determine positions (x, y) of specific anatomic landmarks in the developing morphology (LM1-LM5, top). Interactions between genes are not modeled here. B: Spatial structure of the cranial model used in computer simulations. Cranial growth in the midsagittal plane (bold outline) is determined by three processes P1, P2, P3 (shaded areas), which govern growth of the braincase, the face, and the cranial base (note overlaps between regions). Cranial shape is determined by a 5-landmark configuration (LM1–LM5). **C**: An implementation example. The top rectangle contains the logistic function used to model growth processes P1, P2 and P3, respectively (see Fig. 5). The middle rectangle contains actual parameter values for the respective process functions, $u_1(t)$, $u_2(t)$ and $u_3(t)$. The bottom rectangle contains equations describing how functions $u_1(t)$, $u_2(t)$ and $u_3(t)$ determine the xy-positions of each landmark.

concerning how an organism grows, how growth processes interact, and how growth parameters are modified during evolution; and (2) they permit control over all aspects of growth, as well as "probing" the ontogenetic system at various stages without perturbing development.

Cranial Growth Models

To simulate cranial growth *in silico*, various assumptions are required: how genes interact, how they govern morphogenetic processes, and how epigenetic interactions ultimately result in the formation and development of spatial structures. New experimental and clinical insights into cranial developmental processes demonstrate the complexity of genetic and epigenetic networks regulating bone and soft tissue formation (Wilkie and Morriss-Kay, 2001; Santagati and Rijli, 2003). Accordingly, it is difficult to directly incorporate results from experimental evidence into the design of a generalized cranial growth model.

Kauffman (1993) encountered a similar problem during the design of computer models to simulate the behavior of genetic regulatory networks. Since our empirical knowledge about the exact nature of interactions within a network of genes is usually incomplete, it is sensible to use statistical model parameters that specify states of gene activity, as well as the type and mean number of interactions between genes and their products. Model simulations based on this approach permit analysis of the general properties of genetic regulatory networks, independent of a particular real-world case (Kauffman, 2004; Kauffman et al., 2004).

In principle, Kauffman's statistical modeling approach can be adopted and expanded to devise morphogenetic models, and to use these models to analyze general properties of *morphogenetic regulatory networks*. Here, we will focus on a basic model that specifies connections between genes, epigenetic processes, and the resulting phenotype in a straightforward manner (Figure 4A). The model postulates, from top to bottom, three units of organization: a set of anatomic features defined by the position of points of reference (anatomic landmarks); a set of growth processes that determine the development of these anatomic features; and a set of genes that determine the growth processes.

To render simulations realistic, some biological assumptions are incorporated into the implementation (Figure 4B). The model assumes that the cranium consists of three basic morphogenetic units – the neurocranium, the face, and the basicranium – and that relative growth rates of associated growth processes are coupled to each other by allometric constraints. Such constraints also are incorporated into the model. Imagine two morphometric variables, u_j and u_k, that measure the outcome of growth processes j and k. Using Huxley's original definition (Huxley, 1924), allometry presupposes proportionality of specific growth rates between the two growth processes,

$$\frac{du_j}{dt}\frac{1}{u_j} = p\frac{du_k}{dt}\frac{1}{u_K}. \tag{1}$$

Integration of this equation yields the well-known static allometric equation

$$u_j = c \cdot u_K^p , \tag{2}$$

where c is a scaling factor, and p is the allometric coefficient relating u_j to u_k. Note that time vanishes upon integration, such that the allometric equation describes a time-independent correlation between u_j and u_k.

In the model system, the actual time course of u_j and u_k must be specified in detail.

Huxley's specific growth rates are assumed to exhibit an exponential decay over time, such that

$$\frac{du}{dt}\frac{1}{u} = b \cdot e^{-at} \tag{3}$$

Integration yields a logistic growth function (a so-called Gompertz function) of the following form:

$$u(t) = u_0 \cdot e^{b/a(1-e^{a(t-\Delta t)})} \tag{4}$$

where parameter u_0 is the initial value of u, a indicates the slope of the growth function at its point of inflection, b is the factor of proportionality from Equation (3), Δt describes potential temporal shifts between processes, and saturation is attained at

$$u(t \rightarrow \infty) = u_0 \cdot e^{b/a} \tag{5}$$

In computer simulations, each of the parameters u_0, a, b, and Δt is modeled to be under genetic control and thus sensitive to evolutionary modification by mutation. Figure 5 shows how changes in these parameters modify the shape of the corresponding growth curves.

We now consider a two-dimensional model in which morphology is quantified by the xy-position of $K = 5$ anatomic landmarks. According to the three postulated cranial developmental units, three growth processes of the form described in Equation 4 are

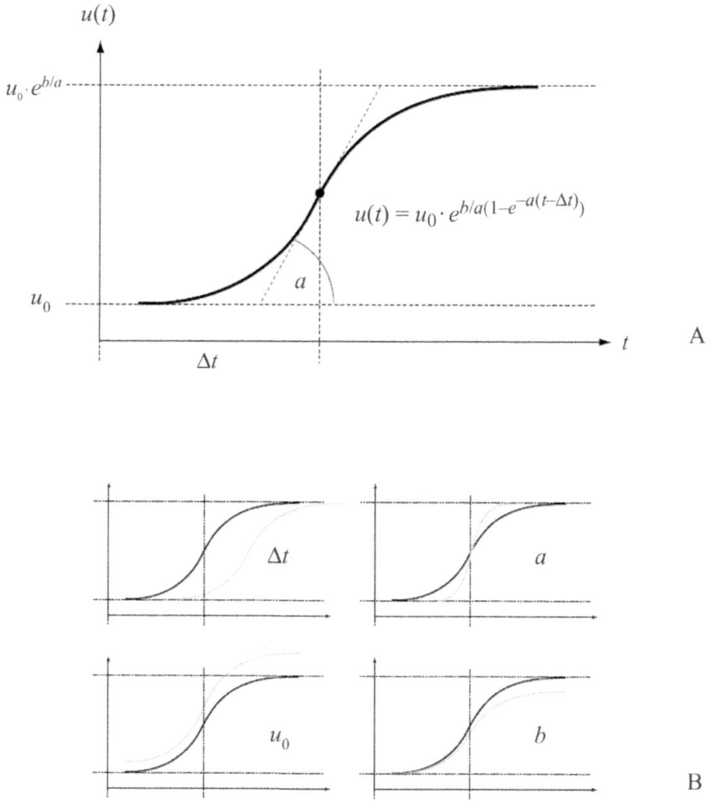

Figure 5. Logistic growth function (A) and effects of modification of its parameters (B).

introduced. As shown in Figure 4, these growth processes P1, P2, P3 govern the temporal behavior of variables u_1, u_2, and u_3, which in turn control the *xy*-positions of the 5 landmarks. Figure 4C shows an example of how landmark positions are determined by growth processes.

Evolutionary modification of developmental pathways is simulated in the following way:

(1) Define all growth and interaction parameters of an "ancestral" growth model.
(2) Simulate "ontogeny" by calculating landmark positions according to the model specifications of Equation (4) and Figure 4C.
(3) Track the resulting morphologies over developmental time and analyze patterns of form change using geometric morphometric methods.
(4) Define a "descendant" growth model by modification ("mutation") of ancestral growth parameters; modifications affect temporal and/or spatial properties of the growth processes.
(5) Simulate the descendant's "ontogeny" (as in step 2).
(6) Track the resulting patterns of form change by means of geometric morphometrics.
(7) Repeat steps (4)–(6) to explore the parameter space of the morphogenetic system.
(8) Compare "ancestral" and "descendant" ontogenetic trajectories and explore how spatial and/or temporal modifications of *process* parameters constrain the resulting ontogenetic trajectories and spatial *patterns* of shape change.

The principal objective of these studies is to simulate spatiotemporal modification at the process level and study the resulting spatiotemporal modifications at the pattern level. In other words, we explore the parameter space and the associated "developmental reaction norm" (Waddington, 1942) of the morphogenetic model system, i.e., the range of possible ontogenetic pathways and morphologies that can be produced by modification of system parameters.

To measure form change, landmark coordinates of the developing morphology were sampled at regular logarithmic time intervals. Shape variability in the resulting "longitudinal" samples of ancestral and descendant cranial landmark configurations was analyzed in linearized Procrustes shape space (Dryden and Mardia, 1998).

Results

The results of computer simulations are graphed in Figures 6–8. To interpret these figures, it is helpful to recall that a point in shape space corresponds to one specific landmark configuration in physical space, and that a direction through shape space corresponds to one specific spatial mode of shape change in the landmark configuration (a so-called deformation field, typically illustrated as a grid; see Figure 3B). The $K = 5$-landmark configurations considered here reside in a shape space with $D = 6$ dimensions, such that principal components analysis (PCA) is used as a dimension reduction technique to assist in visualizing results. In all simulations, >99% of the total shape variation in the sample is contained in the first two principal components (PC1 and PC2), thus these two components are sufficient to visualize ontogenetic trajectories. Moreover, the size of the landmark configurations is only correlated with PC1. Ontogenetic allometry, i.e., size-related change in shape, thus can be graphed as PC1 versus log centroid size. In all graphs, time (proceeding from left to right) is represented implicitly by the spacing between data points along trajectories.

In the first set of simulations, parameters were adjusted in order to define biologically reasonable boundary conditions of growth:

(1) The initial landmark configuration represents a juvenile specimen with a relatively large neurocranium and a relatively small face in comparison to the adult configuration.
(2) The face and the neurocranium exhibit positive and negative allometric growth characteristics, respectively.
(3) Linear dimensions increase by a maximum factor of 5.

Growth simulations of an "ancestral" species, according to these boundary conditions yield slightly curved ontogenetic trajectories through shape space that approach the final (adult) shape at time ($t \rightarrow \infty$). The slight curvature expresses the fact that the deformation field describing shape transformation is fairly, but not entirely, constant over time (see deformation grid in Figure 6).

To simulate descendant growth trajectories, the process parameters were modified in four different ways. First, entire sets of parameters were modified; second, single *temporal* process parameters were modified, and third, single *spatial* parameters were modified. Finally, the model was extended to simulate growth in 3-dimensional morphologies.

COORDINATED MODIFICATION OF PROCESS PARAMETERS

First, we consider ancestral and descendant trajectories under modification of entire sets of growth parameters: $\{u_{01}, u_{02}, u_{03}\}$, $\{a_1, a_2, a_3\}$, $\{b_1, b_2, b_3\}$, and $\{\Delta t_1, \Delta t_2, \Delta t_3\}$. The term *coordinated modification* indicates that all "ancestral" parameter values of a given set $\{\ldots\}$ are multiplied with the same factor to obtain "descendant" parameters. This type of modification is akin to the effect of a growth hormone influencing all growth processes similarly. The results of growth simulations are shown in Figure 6. A conspicuous commonality of all descendant trajectories is their coincidence with the ancestral trajectory. Coincidence in shape space indicates spatial identity of patterns of shape change in physical space, while contraction/extension and/or shift along the ancestral trajectory indicate temporal modifications. Associated size-shape trajectories show translation and scaling of the descendant along the ancestral trajectory for parameters a, b, and Δt. Parameters u_i deserve special consideration. Their modification results in an identical descendant trajectory through shape space, and in parallel shift of the descendant versus the ancestral size-shape trajectory, indicating simple scaling of the entire morphology without alteration of the temporal course of ontogeny.

MODIFICATION OF SINGLE TEMPORAL PROCESS PARAMETERS

In a second series of simulations we explore the effects of modifications in each of the single temporal process parameters: u_0, a, b, and Δt. Overall, modification of single growth parameters produces considerable diversity in descendant trajectories (Figure 7). In all cases, the onset of the ancestral trajectory remains fixed in shape space. Modification of u_0 results in positional shift of the trajectory, while its direction and length remain almost unaffected. Modification of b or a (data for a not shown) similarly results in divergence of the descendant from the ancestral ontogenetic trajectory, as well as alteration of trajectory length. Modification of Δt leads to seemingly opposite effects, as the end of the ancestral trajectory assumes the role of a fixed point, towards which the shortened or extended trajectories of descendant populations converge.

MODIFICATION OF SINGLE SPATIAL PROCESS PARAMETERS

In a final set of experiments, modifications of local spatial characteristics of the growth model are explored. Recall that each landmark's position results from a specific combination of the three model growth processes, i.e., from specific epigenetic interactions between them (see Figures 4B, C). Accordingly, any modification of the way in which these processes determine the landmark positions alters the spatial growth characteristics of the entire system. A typical outcome of these simulations is shown in Figure 7 (bottom graph). Ancestral and descendant trajectories tend to diverge from each other. Obviously, modifying spatial characteristics can have similar effects to modifying temporal parameters a and/or b.

INCREASING THE COMPLEXITY OF THE MODEL SYSTEM

In a last set of experiments, we ask whether increasing the complexity of the model system increases the complexity to simulation results. Complexity can be added to the model system in three ways: (1) by uncoordinated modification of an extended set of process parameters, (2) through introduction of additional growth processes, and (3) through extending landmark configurations into the third dimension. As an overall trend, more complex model systems do not lead to more complex patterns of divergence between ancestral and descendant trajectories. Figure 8 shows the results of simulating growth of a cranial model system containing 9 3-dimensional landmarks (5 in the midplane, 2 bilateral pairs) whose positions are determined by 4 growth processes (3 processes as in Figure 4B, and process 4 governing lateral growth at landmarks 6–9, and influencing landmarks 3 and 4). In this example, parameters u_{0i} of the descendant's growth processes were modified, each process with a different factor. Accordingly, the descendant's ontogenetic trajectory through shape space is translated and slightly elongated relative to the ancestor's trajectory.

TERMINOLOGICAL ISSUES

The first set of experiments (Figure 6) demonstrated the consequences of coordinated modification of an entire set of growth parameters. Temporal modification of the ancestral pattern of ontogeny was most obvious, while its spatial properties remained relatively constant. This is reflected by extension/contraction or shift of the descendant along the ancestral trajectory through shape space, and associated shifts of the size-shape trajectory. This corresponds to *pure pattern heterochrony*, as defined in Figure 1B. It is worth noting that any of the parameters describing the form and position of the growth curve can be modified in a coordinated manner to yield pattern heterochrony.

Modification of a single parameter u_0 (Figure 7, top graph), or uncorrelated modification of several parameters u_0 (Figure 8) altered the initial shape and size of the morphology under consideration, as evinced by displacement of the onset point of the descendant trajectory (which can be considered the end point of a preceding ontogenetic phase). With all other process parameters unchanged, this resulted in parallel ancestral and descendant ontogenetic trajectories. Parallelism in shape space expresses similar spatial patterns of shape transformation in real space, but applied to different initial morphologies. This corresponds to the definition of *generalized pattern heterochrony* (Figure 1B).

Changes in allometric growth characteristics (a, b) and timing (Δt) of single processes, or in the spatial organization of epigenetic interactions between processes, lead to heterotopic changes, i.e., divergence between ancestral and descendant trajectories in shape space

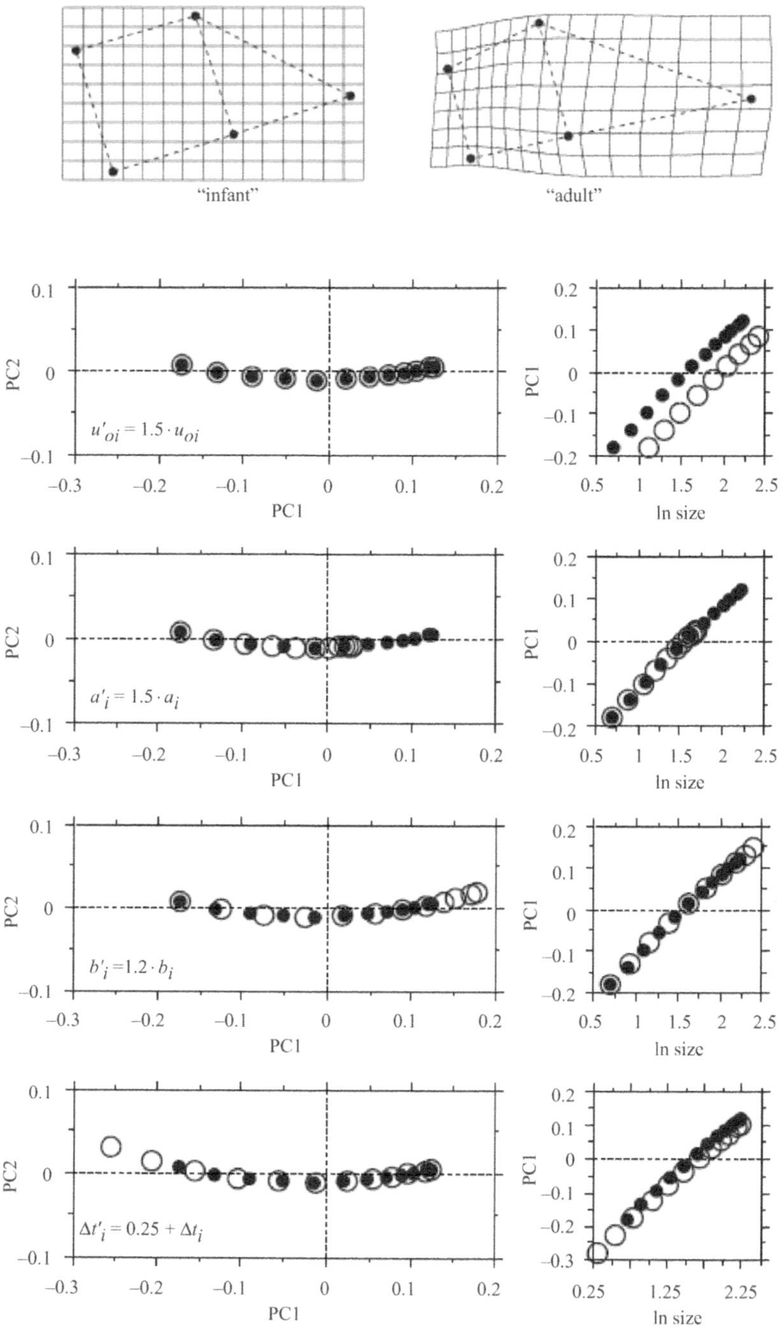

Figure 6. Simulation of cranial growth under coordinated modification of process parameters. The TPS deformation grid in the top graph visualizes shape change corresponding to advancement along the ancestral trajectory (sequence of filled circles in graphs below). Left graphs: ontogenetic trajectories through shape space (PC1 and PC2 represent the first two principal components from PCA of shape). Inset equations in each graph indicate how sets of process parameters $\{u_{01}, u_{02}, u_{03}\}$, $\{a_1, a_2, a_3\}$, $\{b_1, b_2, b_3\}$, and $\{\Delta t_1, \Delta t_2, \Delta t_3\}$ were modified. Note shift and/or scaling of the descendant trajectory (open circles) relative to the ancestral trajectory (filled circles). Right graphs: ontogenetic allometry. Note log-linear increase of size along shape axis PC1. Further explanations appear in the text.

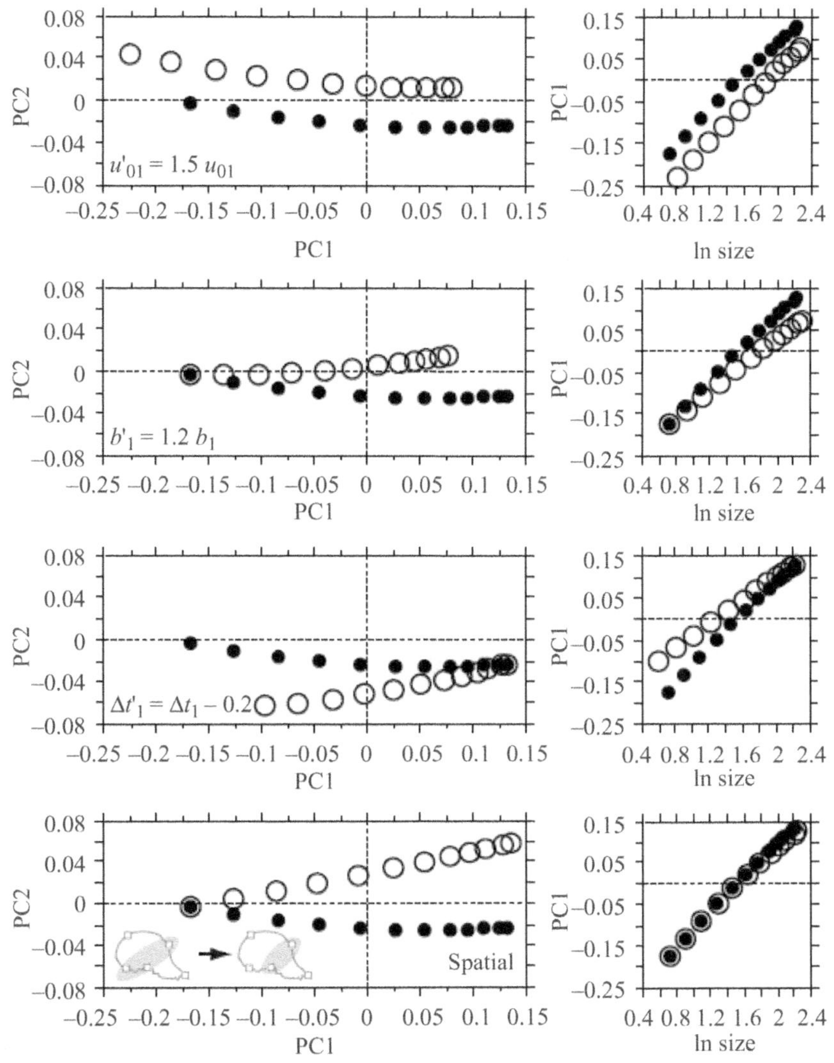

Figure 7. Simulation of cranial growth under modification of single process parameters. Inset equations indicate the type of modification of each parameter; spatial modification (bottom graph) affected the region of influence of process P2 (cf. Fig. 4B). Modifications of single local process parameters of the ancestor's ontogenetic program (filled circles) generate complex alterations in the resulting descendant trajectories (open circles) through shape space (left graphs) as well as in ontogenetic allometric trajectories (right graphs). Further explanations appear in the text.

(Figure 7). These modifications fulfill the definition of *pattern heterotopy* (Figure 1B).

Simulations also allow the relationship between process versus pattern heterochrony and heterotopy to be addressed. Modification of a single process parameter typically creates a multi-pattern effect, which leads to the divergence and scaling of descendant relative to ancestral trajectories through shape space, as well as the divergence of allometric trajectories. Conversely, simultaneous but uncoordinated modification of more than one process parameter may result in relatively simple modification of spatial patterns of ontogeny. There are no apparent straightforward connections between modifications at the process and pattern level. Simple changes on the process level may have complex effects at the pattern level,

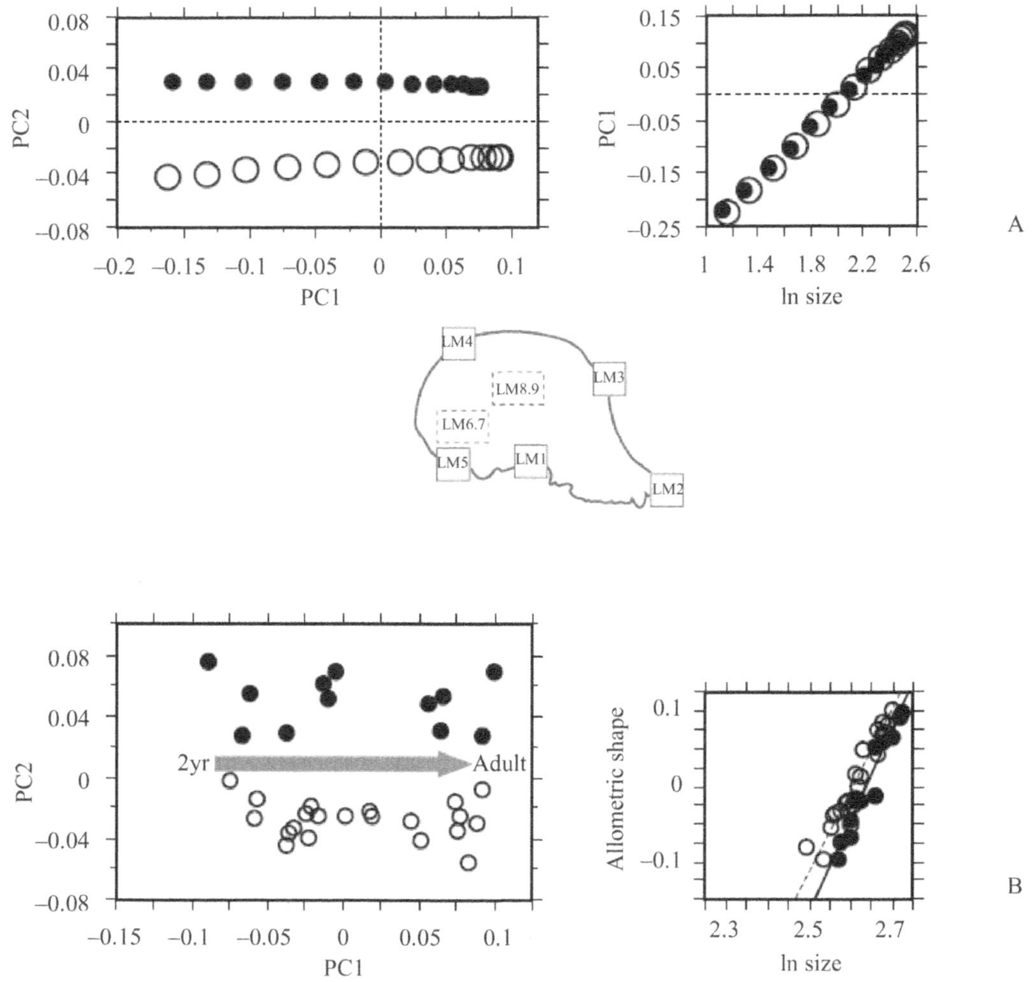

Figure 8. Growth simulation and growth data in three dimensions. A: Adding complexity to the model system. Cranial shape change (top left) and ontogenetic allometry (top right) are simulated with a 3-dimensional 9-landmark, 4-process model (inset graph; landmarks 6/7 and 8/9 are bilateral pairs). Modification of initial conditions ($u'_{01} = 0.9\, u_{01}$; $u'_{02} = 1.1\, u_{02}$; $u'_{03} = 1.0\, u_{03}$; $u'_{04} = 1.2\, u_{04}$) leads to shift of the descendant (open circles) relative to the ancestral trajectory (filled circles). B: Empirical growth data from Neanderthals (filled circles) and anatomically modern humans (AMH; open circles). Arrow indicates developmental change from individual ages of 2 years to adulthood. Further explanations appear in the text.

and vice versa, while heterochronic process modification may cause heterotopic pattern modification, and vice versa.

Discussion

WHY MODEL CRANIAL GROWTH?

What are the implications of the computer models presented here for the analysis of real biological patterns of ontogenetic modification? To answer this question, let us first consider both the potential and the limitations of the computational modeling approach. Model simulations of cranial growth assume an intermediate position between developmental genetics experimentation and morphometric analysis of the developing cranium. One major advantage of computer models is that they require precise definitions of the mechanisms, constraints and causal relationships governing growth and development. Defining and implementing a model system is thus equivalent to stating explicit hypotheses regarding the operation and minimum requirements of developmental systems. Simulations based on these premises are essentially virtual experiments that expose links between the structure of the system (i.e., its parameters) and its phenotypic outcome (i.e., patterns of development). An additional beneficial aspect of virtual experimentation is that the entire realm of cause and effect connecting process with pattern is open for analysis. Furthermore, sampling data does not interfere in the process of development, which is unlike the situation in physical experimentation.

The virtual experiments presented here serve as a practical example in which to examine two aspects of evolutionary developmental reasoning. The first concerns links between the parameters defining a developmental system and the "developmental reaction norm" of the system. The second aspect concerns the theoretical framework of heterochrony and heterotopy: Namely, we may ask whether classical terminology is suitable to describe the full range of patterns of ontogenetic modification that are produced by a general developmental model.

To tackle the first question, let us revisit the cranial growth model in Figure 4. The model is relatively straightforward, as it is composed of only a few interacting units of organization. Simulations shown in Figures 6–8 demonstrate that this basic model can generate the full variety of heterotopic and heterochronic phenomena that can be postulated on theoretical grounds (Alberch et al., 1979; Godfrey and Sutherland, 1996; Gould, 1977; Klingenberg, 1998), and that are described in empirical studies (Nehm, 2001; Roopnarine, 2001; Shea, 1988; Zelditch et al., 2003). However, in contrast with analyses that treat shape as a ratio, i.e., as a one-dimensional variable (Gould, 1977; Alberch et al., 1979), and with analyses that use non-geometric shape measures (Godfrey and Sutherland, 1996; Klingenberg, 1998), the geometric-morphometric analyses presented here reveal the kinematics of shape change at a level of detail that is necessary to decide whether heterochronic comparisons are even biologically meaningful. Model simulations of Figure 6 can be compared using the vocabulary of heterochrony because shape trajectories are collinear, thus representing equivalent patterns (and processes) of morphological shape change. In contrast, divergent ancestral and descendant trajectories (Figure 7) should not be compared in terms of heterochrony because they represent spatially dissimilar patterns of ontogenetic shape change. In this latter case, non-geometric analysis would reveal differences in trajectory lengths, but importantly, it would fail to reveal that patterns of shape change are not equivalent to each other.

If we adhere to the idea of evolutionary tinkering, the model simulations further suggest that the "developmental reaction norm" of even a comparatively simple developmental system is considerable, because of the many ways in which interactions between growth processes can be modified. Overall, it appears that developmental systems are capable of producing a host of ontogenetic paths leading to disparate adult morphologies such that modification of developmental processes and interactions provides a huge source of potential evolutionary novelty.

A related finding, which can be named the *paradox of ontogenetic modification*, concerns the relationship between process modification and resulting pattern modification. From an empirical point of view, the simulations demonstrate the principal limitations in inferring developmental processes from observed patterns of shape change. Simple effects may have complex causes, or vice versa, while heterochronic process modification can cause heterotopic effects, and vice versa. Empirical evidence for these effects was observed in Piranha development (Zelditch et al., 2003), where ontogenetic disparity is inversely proportional to the number of varying developmental parameters. As another empirical illustration, we return to Crouzon disease, where a simple genetic cause elicits a complex pattern of malformation (Neilson and Friesel, 1995; Reardon et al., 1994), and where the suite of developmental disorders characterizing this syndrome incorporates both heterochronic and heterotopic components at different levels of organization.

Heterochronic/heterotopic description offers relatively arbitrary perspectives on the kinematics of growth and development, and ultimately provides no immediate insights into the underlying mechanisms of process modification. It is therefore important to insist on clearly separating the description of a spatiotemporal pattern of development in terms of heterochrony and/or heterotopy from the search for heterochronic and heterotopic process modifications. Overall, we may state that pattern and process analyses ask questions regarding proximate and ultimate causes of ontogenetic modification, respectively, and both perspectives are biologically relevant.

Model Skulls and Real Skulls

How do these considerations impact on human evolutionary developmental studies? In a spatially complex structure such as the hominin skull, direct causal connections between growth processes themselves and observable patterns of growth and development are difficult to establish (Santagati and Rijli, 2003). Accordingly, it is difficult to establish direct connections between process modification, pattern modification, and functional implications. Model simulations, however, provide unique insights for generating hypotheses about such connections.

In an earlier comparative study on cranial growth and development in Neanderthals and anatomically modern humans (AMH) (Ponce de León et al., 2001), these sister species followed parallel ontogenetic trajectories through shape space (Figure 8B). Thus, Neanderthal and AMH cranial forms already were distinct at early developmental stages, and their postnatal development followed the same mode of development. Furthermore, it is reasonable to suggest that this mode of development represents the shared ancestral condition. Thus, development of the two species fits the definition of generalized heterochrony (see Figure 1B). The AMH trajectory is slightly shorter than the Neanderthal trajectory, implying less development of the former relative to the latter (paedomorphy). Likewise, AMHs attain a given cranial shape at smaller sizes than Neanderthals (see Figure 8B, right graph). These findings converge in many respects with studies performed on the same taxa, but using different morphometric approaches (Krovitz, 2000; Williams, 2000; Rogers Ackermann and Krovitz, 2002; Williams et al., 2002), such as Euclidean Distance Matrix Analysis (Lele and Richtsmeier, 2001), and multivariate techniques applied to sets of craniometric distance measurements (Godfrey et al., 1998). Overall, there is an overriding evolutionary trend that ontogenetically early differentiation is an ancient pattern of hominoid phylogeny. However, some studies argue that modern humans and Neanderthals follow parallel shape trajectories (Williams et al., 2002), while others suggest taxon-specific modes of postnatal development (Krovitz, 2003). In any case, there is no doubt that differences in developmental modes, rates and timing between Neanderthals and AMH may have had considerable functional and adaptive significance for the growing organism and its life history. Recent comparative studies on dental and cranial development in AMH, Neanderthals, and potential ancestral *Homo* support such a hypothesis (Coqueugniot et al., 2004; Ramirez Rozzi and Bermúdez de Castro, 2004).

Interestingly, the empirical data on AMH/Neanderthal ontogeny shown in

Figure 8B are similar to the model data shown in Figure 8A in that "sister taxa" exhibit fairly parallel ontogenetic trajectories through shape space. To address these similarities in greater detail, recall that differences between simulated trajectories arose from minor localized modifications of developmental parameters u_{0i}. Modification of these parameters changes the initial landmark configuration. It also influences the overall amount of growth (see Figure 5B), but the basic growth characteristics remain unaffected. Different values of u_{0i} generate parallel ontogenetic trajectories with slightly different lengths. Similar considerations can be applied to empirical comparative data of Neanderthal and AMH development. As a most parsimonious working hypothesis, it is sensible to postulate that observed differences between taxon-specific trajectories can be attributed to relatively small differences in initial conditions of the developmental system around birth, i.e., slightly different morphologies and growth parameters of two taxa. Furthermore, the subtle observed heterochronic shifts between species during their postnatal ontogeny may be a direct consequence of differences in perinatal cranial form, rather than the result of specific selection processes modifying the course of taxon-specific postnatal ontogenies. According to this hypothesis, shared ancestral postnatal processes of development act on slightly different perinatal cranial forms in Neanderthals and AMH. Such differences in "initial conditions" generate differences in subsequent developmental trajectories.

Another point that warrants consideration when comparing Neanderthal and AMH morphologies concerns comparisons between juvenile and adult cranial shape. Morphologists tend to emphasize adult individuals in analyses (i.e., offset points of ontogenetic trajectories), often assuming that adults display the full range of taxon-specific features. However, juxtaposing adult individuals tends to mix comparative criteria such as shift, divergence and scaling of ontogenetic trajectories. These criteria are extremely relevant when comparing cranial morphology in phyletic and cladistic analyses. Given the importance of identifying phyletically valid morphological traits that represent underlying genetic differences and similarities, it is essential to compare trajectories through shape space rather than individual points in shape space.

Using model systems to investigate developmental causes of ontogenetic diversity loosens the bindings of the restrictive "adaptationist programme" (Gould and Lewontin, 1979) and leads to a process-oriented understanding of species-specific morphologies. Theoretical models and computer-based simulation of cranial growth are finally beginning to shed light on the complex web of causation that leads from developmental genes to developmental patterns. Specifically, the empirical investigation of ontogenetic trajectories in hominins and hominoids is still in its infancy. Such investigations offer rich potential for substantiating hypotheses about the role of ontogenetic divergence as a source of evolutionary novelty. In such an endeavor, cranial growth models act like an explorer's compass during ventures into *terra incognita*.

Acknowledgments

The insightful comments of Katerina Harvati, Clive Finlayson, and of two anonymous reviewers are greatly acknowledged. This work was supported by Swiss NSF grant 3100–067209.01 to Ch. Z.

References

Abzhanov, A., Tzahor, E., Lassar, A.B., Tabin, V., 2003. Dissimilar regulation of cell differentiation in mesencephalic (cranial) and sacral (trunk) neural crest cells in vitro. *Development* 130, 4567–4579.

Alberch, P., Gould, S.J., Oster, G.F., Wake, D.B., 1979. Size and shape in ontogeny and phylogeny. *Paleobiology* 5, 296–317.

Andresen, P.R., Bookstein, F.L., Conradsen, K., Ersboll, B.K., Marsh, J.L., Kreiborg, S., 2000. Surface-bounded growth modeling applied to human mandibles. *IEEE Trans. Med. Imaging* 19, 1053–1063.

Bookstein, F.L., 1989. "Size" and "shape": a comment on semantics. *Syst. Zool.* 38, 173–180.

Bookstein, F.L., 1991. *Morphometric Tools for Landmark Data*. Cambridge University Press, Cambridge.

Churchill, S.E., 1998. Cold adaptation, heterochrony, and Neandertals. *Evol. Anthropol.* 7, 46–61.

Coqueugniot, H., Hublin, J.J., Veillon, F., Houet, F., Jacob, T., 2004. Early brain growth in *Homo erectus* and implications for cognitive ability. *Nature* 431, 299–302.

Dryden, I.L., Mardia, K., 1998. *Statistical Shape Analysis*. John Wiley, New York.

Godfrey, L.R., Sutherland, M.R., 1995. What's growth got to do with it? Process and product in the evolution of ontogeny. *J. Hum. Evol.* 29, 405–431.

Godfrey, L.R., Sutherland, M.R., 1996. Paradox of peramorphic paedomorphosis: heterochrony and human evolution. *Am. J. Phys. Anthropol.* 99, 17–42.

Godfrey, L.R., King, S.J., Sutherland, M.R., 1998. Heterochronic approaches to the study of locomotion. In: Strasser, E., Fleagle, J.G., Rosenberger, A.L., McHenry, H. (Eds.), *Primate Locomotion*. Plenum Press, New York. pp. 277–307.

Gould, S.J., 1977. *Ontogeny and Phylogeny*. The Belknap Press of Harvard University Press, Cambridge, MA.

Gould, S.J., 2000. Of coiled oysters and big brains: how to rescue the terminology of heterochrony, now gone astray. *Evol. Dev.* 2, 241–248.

Gould, S.J., Lewontin, R.C., 1979. The spandrels of San Marco and the Panglossian paradigm: a critique of the adaptationist programme. *Proc. R. Soc. Lond.* B 205, 581–598.

Haeckel, E., 1866. *Generale Morphologie der Organismen*, Berlin.

Huxley, J.S., 1924. Constant differential growth-ratios and their significance. *Nature* 114, 895–896.

Jolliffe, I.T., 1986. *Principal Component Analysis*. Springer, Berlin.

Jones, K.L., 1988. *Smith's Recognizable Patterns of Human Malformation*, 4th Edition. W.B. Saunders, Philadelphia.

Kauffman, S.A., 1993. *The Origins of Order*. Oxford University Press, Oxford.

Kauffman, S., 2004. A proposal for using the ensemble approach to understand genetic regulatory networks. *J. Theor. Biol.* 230, 581–590.

Kauffman, S., Peterson, C., Samuelsson, B., Troein, C., 2004. Genetic networks with canalyzing Boolean rules are always stable. *Proc. Natl. Acad. Sci. U.S.A.* 101, 17102–17107.

Klingenberg, C.P., 1998. Heterochrony and allometry: the analysis of evolutionary change in ontogeny. *Biol. Rev. Camb. Phils. Soc.* 73, 79–123.

Krovitz, G.E., 2000. Three-dimensional comparisons of craniofacial morphology and growth patterns in Neandertals and modern humans. Ph.D. Dissertation, Johns Hopkins University, Baltimore.

Krovitz, G.E., 2003. Shape and growth differences between Neanderthals and modern humans: grounds for a species level distinction. In: Thompson, J., Krovitz, G., Nelson, A. (Eds.), *Patterns of Growth and Development in the Genus Homo*. Cambridge University Press, Cambridge, pp. 320–342.

Lahr, M.M., Foley, R.A., 1998. Towards a theory of modern human origins: geography, demography, and diversity in recent human evolution. *Yrbk. Phys. Anthropol.* 27, 137–176.

Lele, S., Richtsmeier, J., 2001. *An Invariant Approach to the Statistical Analysis of Shapes*. Chapman and Hall, Boca Raton, FL.

McKinney, M.L., 1999. Heterochrony: beyond words. *Paleobiolgy* 23, 149–153.

McKinney, M.L., McNamara, K.J., 1991. *Heterochrony: The Evolution of Ontogeny*. Plenum Press, New York.

McNamara, K.J., 2002. Changing times, changing places: heterochrony and heterotopy. *Paleobiology* 28, 551–558.

Mosimann, J.E., 1988. Size and shape analysis. In: Kotz, L., Johnson, N.L. (Eds.), *Encyclopedia of Statistical Sciences*, Vol. 2. Holland, Dordrecht, pp. 219–239.

Nehm, R.H., 2001. The developmental basis of morphological disarmament in Prunum (Neogastropoda: Marginellidae). In: Zelditch, M.L. (Ed.), *Beyond Heterochrony: The Evolution of Development*. Wiley-Liss, New York, pp. 1–26.

Neilson, K.M., Friesel, R.E., 1995. Constitutive activation of fibroblast growth factor receptor-2 by a point mutation associated with Crouzon syndrome. *J. Biol. Chem.* 270, 26037–26040.

Ponce de León, M.S., Zollikofer, C.P.E., 2001. Neanderthal cranial ontogeny and its implications for late hominid diversity. *Nature* 412, 534–538.

Raff, R.A., Wray, G.A., 1989. Heterochrony: developmental mechanisms and evolutionary results. *J. Evol. Biol.* 2, 409–434.

Ramirez Rozzi, F.V., Bermúdez de Castro, J.M., 2004. Surprisingly rapid growth in Neanderthals. *Nature* 428, 936–939.

Reardon, W., Winter, R.M., Rutland, P., Pulleyn, L.J., Jones, B.M., Malcolm, S., 1994. Mutations in the fibroblast growth factor receptor 2 gene cause Crouzon syndrome. *Nat. Genet.* 8, 98–103.

Rice, S.H., 1997. The analysis of ontogenetic trajectories: When a change in size or shape is not heterochrony. *Proc. Natl. Acad. Sci. U.S.A.* 94, 907–912.

Rogers Ackermann, R., Krovitz, G.E., 2002. Common patterns of facial ontogeny in the hominid lineage. *Anat. Rec.* 269, 142–147.

Roopnarine, P.D., 2001. Testing the hypothesis of heterochrony in morphometric data: lessons from a bivalved mollusc. In: Zelditch, M.L. (Ed.), *Beyond Heterochrony: The Evolution of Development*. Wiley-Liss, New York, pp. 271–303.

Santagati, F., Rijli, F.M., 2003. Cranial neural crest and the building of the vertebrate head. *Nat. Rev. Genet.* 4, 806–820.

Sarkar, S., Petiot, A., Copp, A., Ferretti, P., Thorogood, P., 2001. FGF2 promotes skeletogenic differentiation of cranial neural crest cells. *Development* 128, 2143–2152.

Shea, B.T., 1988. Heterochrony in primates. In: McKinney, M.L. (Ed.), *Heterochrony in Evolution: A Multidisciplinary Approach*. Plenum Press, New York, pp. 237–266.

Waddington, C.H., 1942. Canalization of development and the inheritance of acquired characters. *Nature* 150, 563–565.

Wilkie, A.O.M., Morriss-Kay, G.M., 2001. Genetics of craniofacial development and malformation. *Nat. Rev. Genet.* 2, 458–468.

Williams, F.L., 2000. Heterochrony and the human fossil record: comparing Neandertal and modern human craniofacial ontogeny. In: Stringer, C.B., Barton, R.N.E., Finlayson, J.C. (Eds.), *Neanderthals on the Edge*. Oxbow Books, Oxford, pp. 257–267.

Williams, F.L., Godfrey, L.R., Sutherland, M.R., 2002. Heterochrony and the evolution of Neandertal and modern human craniofacial form. In: Minugh-Purvis, N., McNamara, K. (Eds.), *Human Evolution through Developmental Change*. The Johns Hopkins University Press, Baltimore, pp. 405–441.

Winter, D.A., 1990. *Biomechanics and Motor Control of Human Movement*. Wiley Interscience, New York.

Yu, K., Herr, A.B., Waksman, G., Ornitz, D.M., 2000. Loss of fibroblast growth factor receptor 2 ligand-binding specificity in Apert syndrome. *Proc. Natl. Acad. Sci. U.S.A.* 97, 14536–14541.

Zelditch, M.L., Fink, W.L., 1996. Heterochrony and heterotopy: stability and innovation in the evolution of form. *Paleobiology* 22, 241–254.

Zelditch, M.L., Sheets, H.D., Fink, W.L., 2000. Spatiotemporal reorganization of growth rates in the evolution of ontogeny. *Evolution* 54, 1363–1371.

Zelditch, M.L., Sheets, D.H., Fink, W., 2003. The ontogenetic dynamics of shape disparity. *Paleobiology* 29, 139–156.

Zollikofer, C.P.E., Ponce de León, M.S., 2004. Kinematics of cranial ontogeny: heterotopy, heterochrony, and geometric morphometric analysis of growth models. *J. Exp. Zool. (Mol. Dev. Evol.)* 302B, 322–340.

7. Bioenergetic perspectives on Neanderthal thermoregulatory and activity budgets

S.E. CHURCHILL
Department of Biological Anthropology and Anatomy,
Box 90383,
Duke University
Durham, NC 27708–0383, USA
churchy@duke.edu

Keywords: Pleistocene *Homo*, cold adaptation, human paleontology

Abstract

The study of adaptation in Neanderthals is confounded by equifinality – the existence of multiple adaptive pathways to the same morphological end state – manifest as an inability to discriminate between equally likely selective agents behind a given trait. The capacious chests of Neanderthals serve as one example, possibly representing an adaptation either to cold or to high activity levels. While single features may be adaptive in multiple contexts, their relative adaptive value may vary greatly between contexts. Without means of evaluating competing adaptive arguments, we have little hope of identifying the primary selective agents that operated on Neanderthal body form. Bioenergetics provides a basis for quantifying the costs and benefits of various adaptive solutions to a given environmental challenge – thus providing potential for resolving issues of equifinality. Evaluating claims of cold-adapted morphology in Neanderthals involves determining the energetic costs of adhering to Bergmann's and Allen's rules. Skin surface area (SA) is the major determinant of basal metabolic rate (BMR) in mammals, thus estimating Neanderthal SA allows an estimate of the caloric cost of their cold-adapted body form. Clinical equations exist for estimating SA from stature and mass, but these have never been tested on humans of extreme (i.e., "hyper-arctic") body form. A half-size reconstruction of a male European Neanderthal was used to test the utility of these formulae: results indicate that they can be used confidently to predict Neanderthal SA. Based on Neanderthals for whom mass and stature can be reasonably estimated, mean SA is greater than that of Inuit of comparable stature, and suggests higher BMRs in Neanderthals than reported in previous studies. The estimates derived here can be used to model Neanderthal daily energy budgets, and form the basis of evaluating the costs/benefits of hypothesized morphological and behavioral adaptations.

K. Harvati and T. Harrison (eds.), Neanderthals Revisited: New Approaches and Perspectives, 113–134.

Introduction

Studies of Neanderthal adaptation are increasingly adopting a bioenergetic perspective (Jelinek, 1994; Sorensen and Leonard, 2001; Steegmann et al., 2002; Aiello and Wheeler, 2003; Weaver and Steudel-Numbers, 2005), as are studies of adaptive evolution in the genus *Homo* generally (Leonard and Robertson, 1992, 1997; Aiello and Wheeler, 1995; Aiello, 1997; Aiello and Wells, 2002; Leonard and Ulijaszek, 2002). The study of the flow of energy[1] through the Neanderthal world holds the promise of a solution to some of the conundrums that exist in functional morphological or adaptive studies of these archaic humans. By evaluating adaptive mechanisms from the perspective of energy budgets – the caloric costs and benefits associated with various physical and behavioral solutions to adaptive problems – we are in a better position to identify the true selective agents at play and discard competing, less likely adaptive hypotheses.

Many attempts to explore adaptive traits in Neanderthals quickly bump up against the problem of equifinality (that there may exist many different ways to arrive at the same end state), manifest as an inability to discriminate between equally likely selective agents behind a given trait. Take, for example, the capacious chests of Neanderthals (Franciscus and Churchill, 2002; Sawyer and Maley, 2005). Were large chests an adaptation to cold (producing a somewhat more spherical body shape with a surface area/volume (SA/V) ratio more conducive to retaining heat, following Bergmann's [1847] well-known ecogeographic rule)[2] or to high activity levels (providing a greater lung capacity, thus greater ventilatory ability for sustained work output)? Were the relatively short, stout bodies of Neanderthals all about thermoregulation (again following ecogeographic rules – Bergmann's as well as the equally well-known Allen's [1877] rule),[3] or was this a body shape best suited for exerting muscular force on the environment (i.e., reflecting a musculoskeletal system with better mechanical leverage) in the accomplishment of subsistence and technological tasks (see Churchill, 1994, 1996)? Similar examples of competing adaptive hypotheses can be cited concerning Neanderthal nasal morphology, masticatory biomechanics, overall facial form, and pelvic morphology.[4]

At first blush equifinality seems less of a problem than an inconvenience. Large chests and stout bodies were probably beneficial to Neanderthals for both reasons – thermoregulation and power – and we are left only with the niggling but minor problem of being unable to determine the relative importance of the two selective agents. However, adaptive inferences can be and often are interdependent, since adaptations to one environmental problem can constrain or enhance adaptive solutions to other problems. For example, the argument that Neanderthal noses may have functioned as heat vents – that is, they were adapted to *lose heat* rather than conserve it (Trinkaus, 1987; Dean, 1988; Franciscus and Trinkaus, 1988) – depends upon the assumption of very high activity levels in Neanderthals, an assumption that in turn is based on adaptive inferences from other morphological features, like enlarged thoraces, which may in fact be adaptations to other things (such as *conserving heat*, as discussed above). Without means of evaluating such arguments, alternative adaptive hypotheses simply accrue over time with little hope of winnowing out the untenable ones.

A bioenergetic perspective provides some hope for refining our understanding of adaptive evolution in the genus *Homo*. In this paper I explore the utility of bioenergetics in research on Neanderthal adaptation by applying the approach to one aspect of their body form – their large chests. This entails first a consideration of the energetic interrelationships between ventilation, activity and thermoregulation, as well as a review of the inferences that have been made about Neanderthal energy budgets. I then report on the results of a model study conducted to refine our estimates of Neanderthal energetics. New estimates of energy expenditure are then put into context in terms of Neanderthal chest size and cold adaptation.

The Adaptive Value of Large Chests

Morphometric and anatomical comparisons of the size and shape of Neanderthal ribs relative to those of recent humans suggests that the Neanderthals had large chests for their stature (Franciscus and Churchill, 2002; Sawyer and Maley, 2005). It is less clear if Neanderthals had thoraces that were large for their mass, and which might then suggest ventilatory capacities and aerobic demands greater than that seen in modern humans. Attempts to evaluate the respiratory capacity of Neanderthals and modern humans relative to body mass are hampered by small fossil sample sizes and by a great amount of variation within modern samples. The area enclosed by the eighth ribs (calculated using the tuberculo-ventral chord [C] and subtense [S] of one of the eighth ribs [as $\pi(0.5C)S$], following Franciscus and Churchill, 2002) provides a crude measure of the respiratory area of the lower thorax (but one that regrettably cannot take into account variation in the size of the costal cartilages, or in other features that contribute to ventilatory volume, such as the size of the diaphragm and the size and shape of the other ribs: see Bellemare et al., 2001). Only two Neanderthals preserve complete eighth ribs – Shanidar 3 has a respiratory area proportional to mass (as judged by modern human comparators), while Kebara 2 has a relatively large respiratory area (Figure 1). The relative size of the eighth rib is also highly variable within samples: Aleutian Islanders – who are expected to have large chests for both thermoregulatory and activity reasons (see Churchill, 1994) – do not differ from European-Americans in this respect (Figure 1). While thoracic dimensions do in some cases reflect ventilatory capacity (as in high altitude populations: see references in Churchill, 1994), variation in the size of soft tissues may play a more important role in determining ventilatory capacities. In a radiographic study by Bellemare et al. (2001), the anteroposterior and mediolateral diameters of

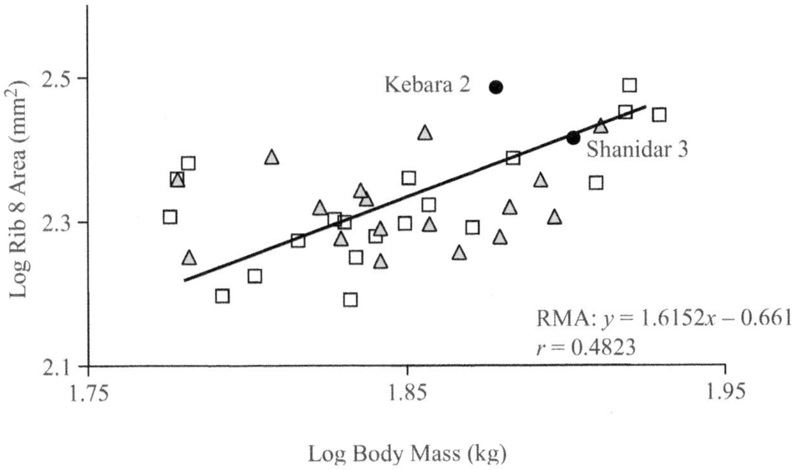

Figure 1. Log respiratory area of the eighth rib on log body mass in Neanderthal (black circles), Aleutian Islander (grey triangles) and European-American (open squares) males. Reduced major axis regression line based on the modern human individuals only.

the thorax (taken at the level of the sixth rib) only accounted for 6.9% and 26.1%, respectively, of the variation in plethysmographically-measured total lung capacity, with the height of the diaphragm accounting for the remaining 67% of the variation (in males: similar values obtained for females). At present we can only say that the overall constellation of Neanderthal rib morphology suggests they had relatively large chests, and that this may reflect ventilatory capacities that were relatively great.

To better understand variation in thoracic size, we must first consider thermogenesis and activity in endothermic animals. Figure 2 illustrates the way that energy is partitioned as it flows through an endotherm. Exchanges of energy along these pathways – either in transformations from one form of chemical potential energy to another or in the performance of work – are never completely efficient: heat is produced as a by-product of metabolic processes. In terms of performing mechanical work, metabolic heat production represents wasted energy. From a thermoregulatory perspective, however, it is a useful waste, since this heat can be used to provide much of the energy of activation necessary to initiate biochemical reactions, and thus is critical to sustaining the reactions of catabolism (i.e., breakdown of complex organic molecules into simpler molecules) and anabolism (i.e., synthesis of complex organic molecules from simpler precursors). In a hypothermic

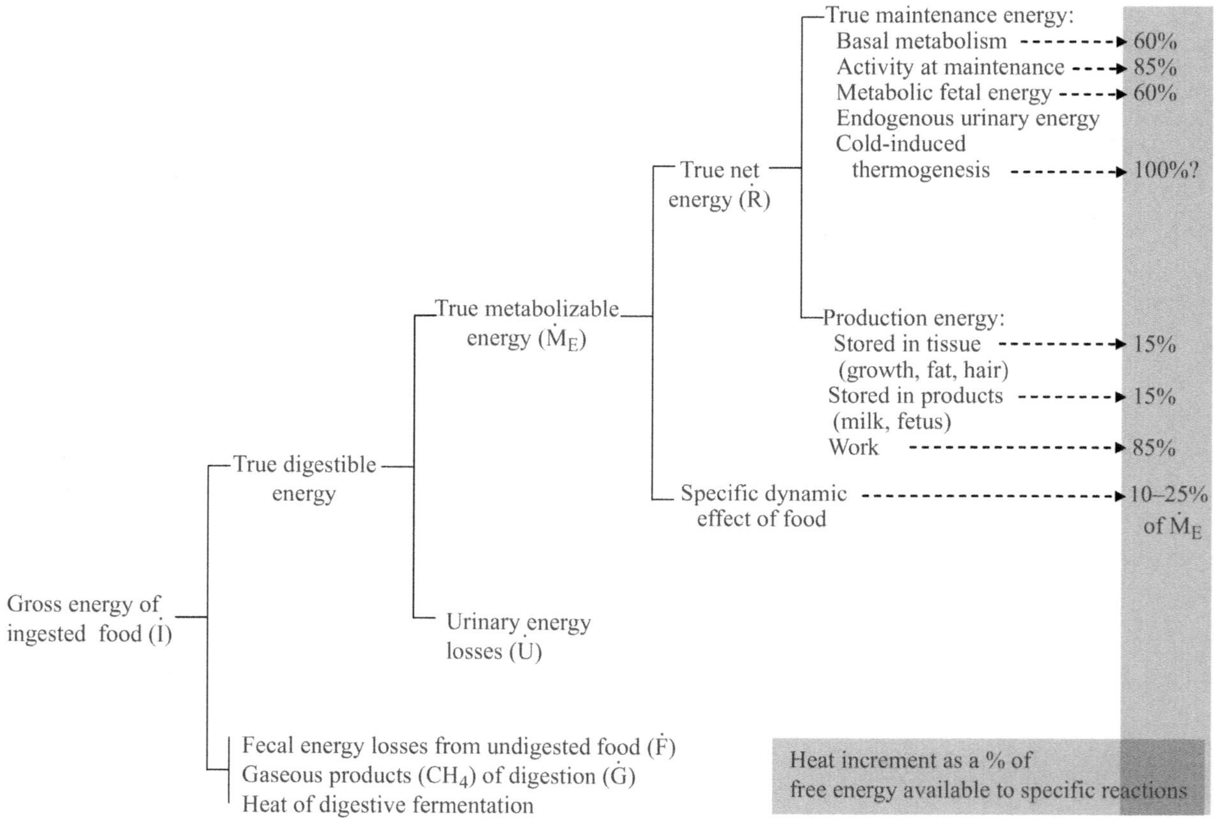

Figure 2. Energy flow through an endothermic animal. The right hand column (shaded) reflects the relative inefficiency (i.e., the amount of heat produced as by-product relative to the amount of free energy entering the reaction) or heat increment of various metabolic processes. Adapted from Moen, 1973.

environment maintenance of an internal temperature sufficient to sustain these reactions is critical to survival, and a consideration of energy flow (as in Figure 2) gives an idea of the pathways available to an endotherm for generating body heat if heat loss to the environment begins to exceed normal metabolic heat output (at what is called the lower critical temperature: see Aiello and Wheeler, 2003).

Anabolic processes, being "uphill" or endergonic (requiring input of energy for the work of production), have relatively low heat increments (Blaxter, 1989) – that is, much of the free energy entering the process is stored as free energy in the new tissues, and heat production is negligible. Basal metabolic processes, requiring the catabolic breakdown of energy-yielding substrates, are notably better at heat production. The synthesis of adenosine triphosphate (ATP), needed to drive most cellular processes, requires the exergonic (energy-yielding) oxidation of glucose, glycogen, or other substrates. The conversion of energy from one chemical form to another occurs at about 40% efficiency, thus more than half of the free energy entering the reaction is liberated as heat (Martini, 1998). Additional heat will be generated as the synthesized ATP is used to perform cellular work, but the timing and rate of this thermogenesis may not serve the organism's immediate thermoregulatory needs.

Endotherms do have more than one pathway that can be called upon to rapidly increase thermogenesis. The first is to disregard work efficiency and simply burn fuel to produce heat. Under the combined effect of the hormones thyroxine and norepinephrine, cellular respiration can be revved as much as 40% above basal levels (see references in Sorensen and Leonard, 2001: work on rats suggests that values 80% above basal metabolism are attainable by some mammals; see references in Frisancho, 1993). This response, commonly called nonshivering thermogenesis, can occur within days of exposure to hypothermic environments, and apparently is accomplished by increasing the rate of active transport of cellular substances.[5]

How much heat can be cranked out by nonshivering thermogenesis? Various methods exist for estimating basal metabolic rate (BMR) in mammals of known mass (although surface area is the better predictor of BMR: see below). Following Kleiber (1961) for placental mammals, BMR (in Watts) scales with the 3/4 power of mass (M, in kg) as:

$$BMR = 3.39M^{0.75} \tag{1}$$

Thus, a male Neanderthal of average mass (84.5 kg: Wolpoff, 1999) would be expected to have a BMR of about 94.5 W (which would burn about 1950 kcal per day).[6] Human-specific equations produce nearly identical results (e.g., BMR = 95.4 W using equation 1 in Sorensen and Leonard, 2001 and converting from kcal d^{-1} to W). Alternatively, the masses of metabolically active tissues can be estimated from body mass, and their contributions to total BMR (estimated using mass-specific organ metabolic rates) summed, as in Table 1. Assuming these tissues in combination contribute about 89% of total BMR (see Table 1 in Aiello and Wheeler, 1995), Neanderthal BMR can be estimated at about 118 W. These alternate methods thus suggest BMRs on the order of 95–118 W in Neanderthals.

Assuming heat production can be elevated 40% above BMR by nonshivering thermogenesis, hypothermic Neanderthals could likely have sustained heat production of 133–165 W (roughly equivalent to the heat put off by 150–200 W incandescent light bulbs) for days on end by metabolic processes alone. Would this be enough to keep a Neanderthal warm during a European winter? The average temperature of the coldest month during a mild interval of OIS 3 was likely about −5° C (van Andel and Tzedakis, 1996); with wind chill this may drop to −12°C (and as low as −17°C during the coldest intervals: Aiello and Wheeler, 2003). Heat loss at an ambient temperature of 10°C in naked males of

Table 1. Predicted organ mass (g) and metabolic rate ($W \cdot kg^{-1}$) in a Neanderthal of body mass (M) = 84.5 kg

Organ	*Organ mass prediction method*[1]	*Predicted mass*	*Mass-specific metabolic rate*[2]	*Total organ metabolic rate*[3]
Heart	$5.2M^{0.987}$	415	32.3	13.4
Lungs	$9.0M^{1}$	761	6.7	5.1
Gut	$17.0M^{1}$	1437	12.2	17.5
Liver	$32.2M^{0.94}$	2085	12.2	25.4
Kidney	$6.3M^{0.87}$	299	23.3	7.0
Brain[2]		1575	11.2	17.6
Skeletal muscle	$415M^{1}$	35068	0.5	17.5
Skin	$77M^{1}$	6507	0.3	2.0
Total				105.5

[1] Organ mass prediction methods and mass-specific metabolic rates from Aiello and Wheeler, 1995: Table 1 and Figure 3.
[2] Average male Neanderthal brain volume from Wolpoff, 1999: Table 102. Mass value assumes a brain tissue mass-density of 1 $g \cdot cm^{-3}$.
[3] Total organ metabolic rate in W.

European descent occurs at an average rate of $104.3 \pm 11.0\ Wm^{-2}$ (based on data presented in Budd et al., 1991). Assuming a skin surface area of about 2 m^2 (see below) and clothing insulation on the order of 0.31 m^2KW^{-1} (equivalent to the insulation of the fur of a large mammal from a temperate environment: Cena and Clark, 1978. Note that modern arctic suits provide only 0.558 m^2KW^{-1} of insulation: Brajkovic et al., 2001), at 10°C a Neanderthal would have lost heat at the rate of about 144 W. The relatively great muscle mass of Neanderthals would have provided further insulation (perhaps cutting heat loss another 5%: Glickman-Weiss, 1993), as may have a covering of body hair (a relatively sparse covering of 3.9 cm long body hair would provide about 0.156 m^2KW^{-1} of insulation: Aiello and Wheeler, 2003). Although we have no way of knowing how much body hair the Neanderthals had, its likely that some combination of insulative materials (clothing, muscle, body hair, and possibly subcutaneous fat [but see Aiello and Wheeler, 2003]) would have allowed for thermal maintenance under mild winter conditions with nonshivering thermogenesis alone. However, it is not likely that nonshivering thermogenesis would have been sufficient during the colder parts of the winter or under emergency conditions (such as accidental submersion in cold water).

Greater heat production can be won by muscular activity, which is performed at a fuel efficiency of only 15–25% (Heglund and Cavagna, 1985, 1987). Thus, muscular activity can be drawn upon – voluntarily or involuntarily – to provide a rapid source of heat. Involuntarily, heat outputs as much as three times basal metabolic rate (285–355 W) can be attained after prolonged shivering alone (Frisancho, 1993).[7] With voluntary activity heat output can be increased further still: a well-conditioned (championship-level athlete) adult male can sustain useful power outputs of 300–370 W for a couple of hours, with a corresponding heat production of 1020–1280 W (in the range of heat produced by a toaster oven) (Wilkie, 1960; Shephard, 1978). Thus, muscular activity can be called upon to contribute to homeostasis under hypothermic conditions, both with heat output an order of magnitude greater than possible by metabolic processes alone (bearing in mind that not all of this heat is available for thermoregulation, since exercise may increase convection and evaporative heat loss from the skin and losses from the respiratory passages: Blaxter, 1989), and with useful output of mechanical work on the environment. This pathway, however, is expensive – costing between 1130–1440 kcal h^{-1} – and thus periodic bouts of exercise may be an untenable strategy for keeping warm when calories are in short supply.

Here is where thoracic size comes into play, since sustained muscular activity requires the steady aerobic replenishment of muscle reserves of phosphagen (ATP and creatine phosphate), which in turn is limited by the rate of oxygen transport from the atmosphere to muscle tissue (Shephard, 1978). For this reason heat production is linearly related to maximal aerobic capacity (νO_{2max} or maximal oxygen intake: MacHattie et al., 1960). The maximum power outputs above require the absorption and transport of four to five liters of oxygen per minute (at 1 l O_2 min^{-1} = 330 W power input, corresponding to 75 and 255 W output of mechanical power and heat, respectively: Åstrand and Rodahl, 1970). Exceptional athletes can hit peak oxygen consumption rates of six liters per minute, corresponding to inputs of close to 2000 W (and heat increments of greater than 1500 W). Given that Neanderthals had muscle masses greater than average modern humans, they may have been able to sustain heat outputs comparable to or even in excess of that seen in modern elite athletes, provided they could supply their muscle tissue with adequate oxygen.

The amount of oxygen extracted in the lungs is a constant 3.1% of the respired air volume (Schmidt-Nielsen, 1984), thus to extract 4–5 liters of oxygen per minute would require alveolar ventilation (VA) of 129–161 l min^1. Well-conditioned modern human male athletes have larger-than-average vital capacities (VC: the maximum amount of air that can be expired after a maximum inhalation) by as much as 30–45% (compare the mean VC of runners [6.3 l] and swimmers [6.98 l] [Novak et al., 1968] with average males [4.8 l] [Martini, 1998]). Using the value cited above for runners, and assuming that maximum tidal volumes (VT, the amount of air moved into and out of the lungs in a single respiratory cycle) during forced breathing never surpass 50% of vital capacity (see Shephard, 1978) and that these athletes have an anatomic dead space (the volume of the trachea and other conducting passageways) of about 0.15 l (Martini, 1998), we would expect ventilatory rates of 43–54 breaths min^{-1} to attain the work and heat production described above. Higher rates of oxygen consumption (up to 6 l min^{-1}: see above) are probably obtained not by further increasing the ventilation rate, but by voluntary increases in the maximum tidal volume to greater than 50% vital capacity (at levels of respiratory discomfort intolerable to all but the best-conditioned and most dedicated athletes) (Shephard, 1978). In this respect, larger chests – and hence greater vital capacity – facilitate high levels of work and heat output.

The above analysis makes explicit the separate pathways by which climatic selection may favor something like large chests in Neanderthals – they were likely adaptive in cold environments for both heat generation (permitting higher rates of sustained heat output) and heat conservation (providing a SA/V configuration more conducive to maintaining thermal constancy) (Figure 3). The problem of equifinality emerges here in another form: the development of large chests in Neanderthals may have been genetic in origin (from selection for enhanced thermoregulatory ability) or epigenetic (from enhanced growth of the ribs and costal cartilages in response to high activity levels in childhood), or both. Enlarged thoraces have been demonstrated to be a developmental, epigenetic effect of growth in an hypoxic environment (Greksa, 1990; Frisancho, 1993). Indeed, several aspects of Neanderthal body form may be developmental effects related to highly active childhoods spent in seasonally cold environments. Leonard et al. (1994) suggested that the short stature of native Siberians is a function of

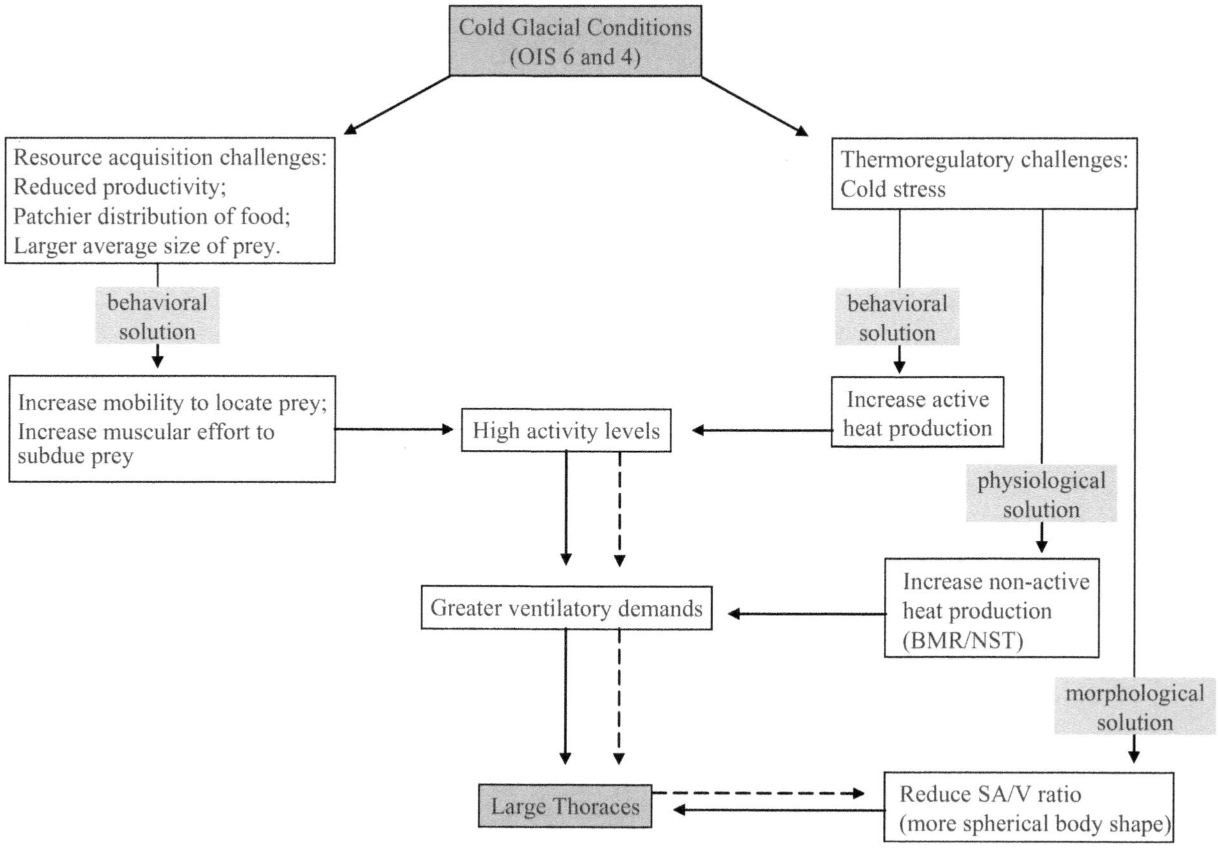

Figure 3. Large thoraces as an equifinal solution to the various adaptive challenges that stem from cold glacial conditions. Since enlarged lungs and greater respiratory volume enhance both work and heat output, and since larger chests may help maintain thermal constancy, identifying the primary adaptive advantage of Neanderthal thoracic morphology becomes difficult (Abbreviations: OIS – Oxygen Isotope Stage; BMR – basal metabolic rate; NST – nonshivering thermogenesis; SA/V – surface area/volume). The dotted lines reflect the role that developmental sequelae of high childhood oxygen demands may play in producing large adult chest size (and hence SA/V ratios) – see text for details.

their elevated active and resting metabolic rates, which in turn are aspects of their thermregulatory adaptation to cold climate. By this model, the energy demands of staying warm reduce the amount of energy that can be allocated to growth, resulting in short adult stature and foreshortened limbs (a body shape conforming to Bergmann's and Allen's rules). A similar model, but one focusing on accelerated growth of the respiratory and circulatory systems at the expense of total body growth, has been proposed for Andean highlanders (Frisancho, 1993). Teasing genetic, adaptive solutions from epigenetic, developmental effects remains a challenge, but one that bioenergetic approaches may ultimately help to resolve.

The bioenergetic perspective provides a basis for quantifying the costs and benefits of various adaptive solutions to a given environmental challenge, and draws our attention to the outstanding questions that need resolution for a more complete understanding of Neanderthal adaptation. Further consideration of thermogenesis in Neanderthals requires some estimates of their absolute and relative chest size (that is, were their lung volumes proportional to body mass, or disproportionately large?) to refine the above estimates of heat production at work and rest, and better estimates of body surface area to gauge thermal conductance under different ambient conditions. Armed with this kind of information, we can better address the various adaptive hypotheses in play. Under probable winter conditions, how does heat production compare with heat loss in a Neanderthal at work, and what levels of work would be required before purposeful loss of heat (heat venting) was necessary to prevent thermal stress? What are the costs and benefits, in terms of water loss and work of breathing, of venting excess body heat through the nose rather than the mouth under various conditions? At what temperature would a cold-stressed Neanderthal require active (muscular) heat production above BMR to maintain thermal balance? By using calories (or derived equivalents, like Watts) as a currency, a bioenergetic approach allows a much finer apprehension of the costs and benefits of adaptive modalities, and a true appreciation of the ways in which energy was budgeted in the Mousterian world.

The Cost of Being a Neanderthal

To understand the energetics and ecology of Neanderthals we need to know something about both energy capture and expenditure, beginning with estimates of foraging efficiency and the caloric costs of somatic and maturity maintenance (see Nisbet et al., 2000). Given estimates of net foraging returns and total maintenance costs, we can estimate what Neanderthals had left over for production (i.e., growth and fat storage) and reproduction (i.e., fetal growth, fetal metabolic energy, and milk production). We are a long way from knowing these parameters, but fortunately some insights about the adaptive value of certain Neanderthal features can be had with fairly rough estimates of their average total energy expenditure (TEE: kcal d^{-1}). TEE (sometimes called daily energy expenditure) is a measure of the energy needed for basic metabolic processes, metabolic thermogenesis, and productive work (of foraging, protecting oneself from the elements, avoiding predators, etc).

A number of approaches can be used to estimate TEE, the most direct being to use empirically-determined relationships between TEE and body mass (M) in primates (see Aiello and Wells, 2002). For a sample of human and nonhuman primates (Leonard and Robertson, 1997):

$$TEE = 86.0M^{0.793} \quad (2)$$

Using the mean mass, plus or minus one standard deviation, of seventeen male (77.6 ± 4.5 kg) and nine female (66.4 ± 4.8 kg) Neanderthals (from Ruff et al., 1997 supplemental data) this equation provides estimated TEE ranges of 2586–2835 kcal d^{-1} and 2258–2532 kcal d^{-1} for males and females, respectively (Figure 4). These values are low relative to reported values for extant arctic foragers (Steegmann et al., 2002), likely because nonhuman primates have lower daily energy requirements per unit body mass than do human foragers generally (Leonard and Robertson, 1997).

More reasonable estimates of TEE may be had by estimating BMR in Neanderthals and then doubling or tripling BMR (as suggested by the World Health Organization [WHO] to account for very high activity levels). Using sex-specific mean Neanderthal masses with WHO regressions for estimating BMR from mass, and adjusting for the effects of cold climate on BMR, Sorensen and Leonard (2001) estimated mean male and female BMR as 1841 and 1435 kcal d^{-1}, respectively. Doubling and tripling the BMR estimates produces TEE estimates in the range of 3682–5523 (males) and 2870–4305 kcal d^{-1} (females) (Figure 4). If TEE is estimated instead by using observed calorie consumption

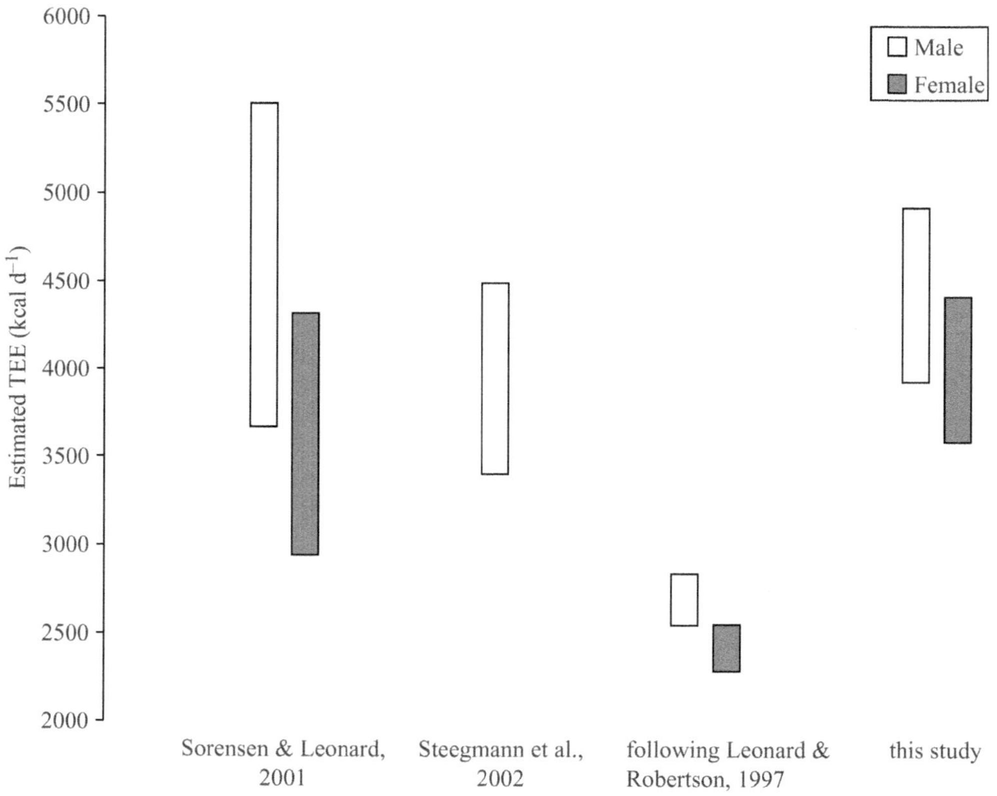

Figure 4. Estimated total energy expenditure (kcal d^{-1}) in Neanderthal males (open bars) and females (grey bars) from several recent studies.

by arctic foragers and adjusting for the larger size of Neanderthals (Steegmann et al., 2002), somewhat lower values are obtained (3360–4480 kcal d^{-1} for males; female data not reported: Steegmann et al., 2002) (Figure 4). The direct measurement of calorie consumption in Inuit that still practice a traditional foraging lifestyle indicate TEEs on the order of 2.0–2.5 times BMR (see Shephard and Rode, 1996; Steegmann et al., 2002), suggesting that the high end (3x BMR) of the estimates by Sorensen and Leonard (2001) might be too high.

Since various methods of estimating TEE produce considerably different results (note in Figure 4 that the highest estimate for males following Leonard and Robertson [1997] is below the lowest estimate for females derived by Sorensen and Leonard [2001]), I sought to refine estimates of Neanderthal BMR and to use these, along with adjustments for activity suggested by observations on extant Inuit (2.0−2.5× BMR), to refine our estimates of TEE. BMR has been shown to be proportional to skin surface area (Winslow and Herrington, 1949; Gehan and George, 1970), thus BMR estimates can be improved if accurate estimates of Neanderthal surface area can be determined.

Regression equations for estimating surface area from stature and mass are commonly employed by anesthesiologists (DuBois and DuBois, 1916; Gehan and George, 1970; Bailey and Briars, 1996), because the medicinal dosages needed for anesthetics and chemotherapy are dependent upon organ surface areas and renal (glomerular) filtration rates, which are themselves proportional to skin surface area. However, these formulae are derived from modern human samples, and it is unclear if they can accurately estimate surface area in humans that lie at the extreme ends of the range of variation in body shape (i.e., long and linear Nilotics or short and stocky Inuit). Since Neanderthal body shape was even more extreme than that of Inuits ("hyper-arctic": Holliday, 1997) , the applicability of clinical equations for estimating Neanderthal surface area remains open to question.

A half-size anatomical reconstruction of a male European Neanderthal was used to evaluate the accuracy of clinical SA estimation equations when applied to Neanderthals. Scale models have been used to good effect in other areas of paleontology (Colbert, 1962; Alexander, 1985; Paul, 1988; Farlow et al., 1995). In this case, a proportionately correct half-stature model should have one-fourth the surface area and one-eighth the volume of his full size equivalent, and thus SA can be measured directly on the model and scaled to life size. The model is based on the dimensions of the La Ferrassie 1 Neanderthal, both because this individual is fairly complete, and because his estimated stature (171 cm) is close to the mean estimated stature of male Neanderthals (165.9 ± 5.0 cm, n=16). The model was constructed of Plasticene on a rigid armature, and was built by Jeanna Novelli, Laura Gruss, Natalia Rybczynski and Michael Black under my supervision.

Construction of the Half-Size Neanderthal Model

The dimensions of the skeletal elements that contribute to the stature of La Ferrassie 1 (LF1) are given in Table 2. Note that the sum of these dimensions (162 cm) does not take into account the contribution of soft tissue – such as the intervertebral discs – to stature, thus we added nine cm (since LF1's stature estimated from femoral length = 171 cm: Ruff et al., 1997, supplemental data) to trunk and neck length (proportionally distributed between the cervical, thoracic and lumbar vertebrae) in determining the proportions of the full size Neanderthal.

Trunk length was taken as the summed body heights of T1–L5 plus the height of the anterior border of the iliac blade (following Heim, 1982, since the sacrum is not preserved) plus a 7.2 cm correction for soft tissue (79.5% of the total 9 cm correction, corresponding to the portion of the total summed vertebral body heights occupied by the lumbar and thoracic vertebrae). Anteroposterior and mediolateral trunk widths were determined in three ways. First, we estimated skeletal

Table 2. Height (cm) of elements contributing to stature in La Ferrassie 1

Element	*Height*
Cranial Height (basion-bregma)	13.7
Cervical Vertebrae (summed body heights)	9.9
Thoracic Vertebrae (summed body heights)	25.1
Lumbar Vertebrae (summed body heights)	13.0
Height of Iliac Blade (anterior border)[1]	11.5
Femur (bicondylar length)	45.8
Tibia (articular length)	35.5
Talus and Calcaneus	7.5
Stature Minimum	162.0

All measurements from Heim, 1982.
[1] Used in place of the S1 ventral body height (Heim, 1982).

bi-iliac breadth from stature using the ratio (0.1856) suggested by Ruff et al. (1997), and we also used their recommended soft tissue correction to arrive at a living bi-iliac breadth of 34.1 cm. Attempts to estimate shoulder breadth from clavicular length (following arguments in Churchill, 1994) produced values that were clearly too small (70–80% of bi-iliac breath), thus we employed a bit of artistic license to sculpt shoulders that were proportional to hip width based on reference photos of arctic modern humans (e.g., figure 13.10 in Weiss and Mann, 1985). Antero-posterior and mediolateral chest widths were estimated by taking average values from Nunivak Inuit males (Hrdlička, 1930) and correcting for the larger size of LF1 relative to average Inuit male stature (average of group means = 163 ± 3 cm: data from Holden and Mace, 1999), giving us anteroposterior and mediolateral diameters of 26.3 cm and 32.1 cm, respectively.

The model was scaled to one-half linear size. The head was sculpted on a half-sized scale model of a Neanderthal cranium (based on LF1) created by Dr. Steven Wagner. Inter- and intralimb proportions were determined from the actual long bone lengths of LF1 (Heim, 1982). The dimensions of the hands and feet were based on half-sized, scaled drawing of the manual and pedal skeletons of LF1 (Heim, 1982). To determine the appropriate muscularity, we compared the size of upper limb muscle attachment areas (scapular area, humeral epicondylar breadth and radial tuberosity area) between Neanderthals and Aleutian Islander males (data from Churchill, 1994), and determined that the Neanderthals had on average muscle attachment areas about 12.9% larger than those of the Aleuts (Aleutian Islanders were used as reference because they were the only circum-arctic group for which upper limb muscularity measures were available). Once again we used

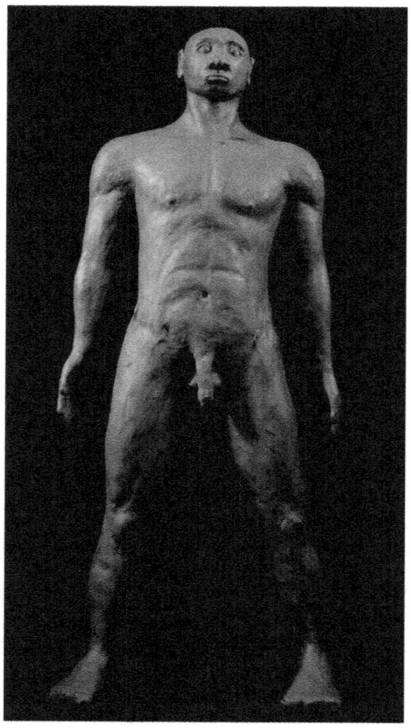

Figure 5. Half-scale model of a male Neanderthal, based on the skeleton of La Ferrassie 1.

artistic license, and created a model that was somewhat more muscular than Inuit males based on reference photos.

The completed model (Figure 5) was coated in Silastic® RTV Silicone Rubber (Type E) and cheesecloth to produce a surface area peel. The peel was cut into small sections (to lie flat on a scanner) and scanned to digital files. The area of each piece was determined with NIH Image 1.62, and the areas were summed to get a total surface area of the model. The model was also submerged in water and the amount of water displaced was used to determine its volume.

Results of the Model Study

The half-sized model displaced 10.55 l of water, corresponding to a volume for the full-sized Neanderthal of 84.4 l. Assuming an average mass density of 1 gm cm^{-3} for the human body (Katch et al., 1967), this equates with a mass of 84.4 kg – very close to the estimate of 85 kg obtained by Ruff et al. (1997) based on a combination of femoral head and bi-iliac breadth + stature methods. The skin SA of the model – 0.5308 m^2 – corresponds to a full-sized SA of 2.1232 m^2.

The surface area estimated from the model is compared with estimates derived from SA estimation equations from the clinical literature in Table 3 (using stature [171 cm] and mass [85 kg] estimates derived from La Ferrassie 1's skeletal remains). All of the clinical equations underestimate SA by 4–7% relative to the model estimate. The clinical equations used here have also been found to vary in their predictive accuracy with recent humans. Wang and colleagues (1992) evaluated the accuracy of multiple estimation methods and found those of Boyd (1935) and Gehan and George (1970) to be the most reliable. Both of these methods produced SA estimates among the closest to that derived from the model. The method of Gehan and George (1970) is a refinement of that of Boyd (Wang et al., 1992; Bailey and Briars, 1996), and as it incorporates infants and children into the estimation sample it is likely applicable to a wide range of human body shapes. For this reason it is probably the most appropriate for use with populations at the extremes of body form variation (such as Neanderthals). The relative concordance of the estimates derived from Gehan and George's method and from the model (4.4% difference) suggests that the clinical equation can be reasonably applied to Neanderthals.

Neanderthal Energy Budgets

Mass and stature estimates were used to estimate SA and BMR for 5 female and 16 male Neanderthals (Table 4). Neanderthals have absolutely greater amounts of surface area compared to modern humans – even arctic-adapted

Table 3. Comparison of clinical equations for estimating Surface Area (m^2) of La Ferrassie 1 relative to estimate derived from model

Method	*Equation*	*SA Estimate*	*Percentage Difference*[1]
DuBois and DuBois, 1916	$0.007184 \cdot St^{0.725} \cdot M^{0.425}$	1.9727	−7.1
Boyd, 1935	$0.017827 \cdot St^{0.3} \cdot M^{0.4838}$	2.0287	−4.5
Gehan and George, 1970	$0.0235 \cdot St^{0.42246} \cdot M^{0.5146}$	2.0291	−4.4
Haycock et al., 1978	$0.024265 \cdot St^{0.3964} \cdot M^{0.5378}$	2.0313	−4.3
Mosteller, 1987	$((St \cdot M)/3600)^{0.5}$	2.0094	−5.4

Based on skeletally-derived estimates of stature (St: 171 cm) and mass (M: 85 kg) following Ruff et al., 1997.
[1] Percentage difference between clinically-derived surface area and model-derived surface area estimate of 2.1232 m^2.

Table 4. Estimated mass (kg), stature (cm), surface area (m^2) and basal metabolic rate ($kcal\ d^{-1}$) in Neanderthals

	Mass	*Stature*	*Surface Area*	*BMR*
Females				
Grotte du Prince	74.8			
La Ferrassie 2	67	155	1.7222	1795
Krapina 208	68.4			
Krapina 209	63.7			
Krapina 214	62.2			
La Quina 5	71.2	163	1.8152	1891
La Quina 18		164		
Shanidar 6	59.4	148	1.5875	1654
Spy 1	67.5	161	1.7568	1831
Tabun C1	63.2	156	1.6758	1746
Female Mean	66.4	157.8	1.712	1783
sd	4.8	6.0	0.086	90
n	9	6	5	5
Males				
Amud 1	75.3	179	1.944	2025
La Chapelle 1	77.3	163	1.894	1973
La Ferrassie 1	85	172	2.034	2120
Fond de Forêt 1	83.9	161	1.965	2047
Kebara 2	75.6	166	1.887	1966
Kiik Koba 1	78.1	164	1.909	1989
Krapina 213	80.6			
Lezetziki 1	73.9	166	1.865	1943
Neanderthal 1	78.9	166	1.929	2009
Régourdou 1	72.1	164	1.832	1909
Saint Césaire 1	78.9	165	1.924	2004
Shanidar 1	80.5	172	1.978	2061
Shanidar 2	75.2	160	1.852	1930
Shanidar 3	79.9	166	1.941	2023
Shanidar 4	72	161	1.816	1892
Shanidar 5	68.5	168	1.802	1878
Spy 2	83.6	161	1.961	2044
Male Mean	77.6	165.9	1.908	1988
sd	4.5	5.0	0.064	66
n	17	16	16	16

1. All mass estimates are from Ruff et al. (1997: supplemental data).
2. Stature estimates are the average of two stature estimation techniques reported by Feldesman et al. (1990), except for: Kebara 2, Kiik Koba 1, Lezetxiki 1, Régourdou 1, Saint Césaire 1, Shanidar 2, Shanidar 3, La Quina 5, La Quina 18, Spy 1 and Tabun C1, where stature estimate was from Ruff et al. (1997: supplemental data).
3. Sex is indeterminate for Régourdou 1 and Lezetxiki 1, they are considered here as male based on their relatively high mass estimates.
4. Surface area estimated following Gehan and George (1970).
5. BMR estimate following Winslow and Herrington (1949).

modern humans – of similar stature. Inupiats are close to Neanderthals in mean stature (for males: 167.4 vs. 165.9 cm, respectively; for females: 155.9 vs. 157.8 cm, respectively: Inupiat data from Ruff et al., 2005), yet Neanderthals average about 0.1 m^2 – or about 5–6% – more skin SA regardless of sex (when sex-specific average Inupiat statures and masses are used to estimate SA). However, Neanderthals also tend to be more massive for a given stature – by 11–13% – than the Inupiat, and thus tend to have lower ratios of SA to mass (Figure 6). Of the modern human samples represented in Figure 6, only the two Inuit samples (represented by the two triangles closest to the regression line) fall close to the regression line that describes the relationship between SA and mass in Neanderthals. One sample of Evenki reindeer herders of Siberia (Galloway et al., 2000) – represented by the triangle in the lower left corner of the plot – also appears to be similar in body shape to Neandertals. Saami, ethnic Finns, Yakuts and a second sample of Evenki are all relatively cold-adapted, but fall farther away from the line (as seen in the remaining triangles). The earliest modern Europeans (Early Upper Paleolithic sample) are more similar to North Africans (North African Epipaleolithic sample) in SA/M relationships, while by the later Upper Paleolithic (Late Upper Paleolithic sample) modern Europeans had converged somewhat on the SA/M relationship characteristic of Neanderthals and Inuit. Nilotic east Africans, not surprisingly, can be seen to have high SA/M ratios.

The SA estimates in Table 4 allow the estimation of BMR following Winslow and Herrington (1949) (where BMR = 1042 kcal m^{-2} d^{-1}). This produces mean male and female BMR estimates about 8% and 19% higher, respectively, than those obtained by Sorensen and Leonard (2001). However, given the lower physical activity level multipliers used here (2.0–2.5 times BMR based on empirical data on Inuit, versus 2.0–3.0 times BMR as suggested for active peoples by

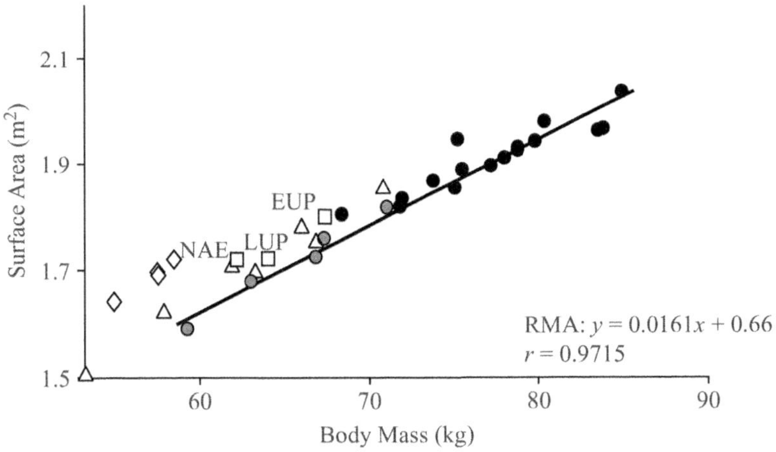

Figure 6. Skin surface area (m^2) on body mass (kg) for individual Neanderthals and means of combined-sex samples of recent and fossil modern humans. Individual Neanderthal male (black circles) and female (grey circles) data from Table 4. Surface areas for modern humans calculated from published mean stature and mass: Diamonds, warm-adapted Nilotics (Roberts and Bainbridge, 1963); Triangles, cold adapted Inuit, Evenki, Yakut, Saami and ethnic Finns (Eveleth and Tanner, 1976; Leonard et al., 1996; Galloway et al., 2000; Ruff et al., 2005; Snodgrass et al., 2005); Squares, European Early [EUP] and Late Upper Paleolithic [LUP] and North Africa Epipaleolithic [NAE] fossil modern humans (Ruff et al., 1997 supplemental data). Reduced major axis regression line based on Neanderthals only.

WHO), the estimates of TEE fall at the lower end of the range of male values and at the higher end of the range of female values reported by Sorensen and Leonard (2001) (Figure 4). Taken together, the estimates by Sorensen and Leonard (2001), Steegman et al. (2002), and this study suggest that it cost, on average, between about 3500 and 5000 kcal per day to feed an adult Neanderthal living in cold conditions.

To put the energetic needs of Neanderthals into economic perspective, an adult male Neanderthal would have probably needed to consume the equivalent of about 2 kg of caribou (in good condition, 9–10% body fat: Table 5) every day. A social group of ten Neanderthals comprised of three adult males (consuming about 4475 kcal d^{-1}), three adult females (eating about 4015 kcal d^{-1}) and four juveniles (consuming about 3000 kcal d^{-1}) would have required the equivalent of two 80 kg caribou per week. Assuming that meat was the primary item in the Neanderthal diet (Fizet et al., 1995; Bocherens et al., 1999, 2001; Richards et al., 2000), that hunting was the primary means of meat procurement, and that all the adults were hunting, a group like this would have required hunting return rates on the order of 3.8 kg $hunter^{-1}$ d^{-1}. If only the males were hunting, the group would have needed return rates of 7.6 kg $hunter^{-1}$ d^{-1} – well in excess of the 5.4–5.6 kg $hunter^{-1}$ d^{-1} recorded for wolves in northern Europe and North America (Jedrzejewski et al., 2002; Hebblewhite et al., 2003). Clearly, Neanderthal bodies were costly, and because of the energy demands of thermoregulation they were most costly at the time of year when calories were most difficult to obtain.

The Adaptive Value of Large Chests Revisited

What, from all of this, can we infer about the adaptive significance of Neanderthal body form? Recent work by Aiello and Wheeler (2003) suggests that the low SA/M bodies of Neanderthals bought them relatively little in terms of maintaining thermal constancy: they estimate the lower critical temperature (the temperature at which an unclothed human must increase internal heat production to maintain thermal constancy) for Neanderthals to be 27.3°C, while for modern humans it is 28.2°C.[8] Thus, Neanderthals appear not to have been morphologically well-suited to staying warm in winter conditions without cultural buffering (i.e., clothing, shelter and fire) and substantial increases in

Table 5. Caloric value of an 80 kg caribou in good condition: edible mass (kg), average caloric value (kcal kg^{-1}) and average available energy (kcal) per caribou of various tissues

Tissue	*Average caloric value*	*Edible Mass*	*Available calories*
Meat	1190	28.1	33,439
Viscera	1250	16.1	20,125
Adipose fat	9000	7.8	70,200
	Weighted average: 2380	Total: 52.0 (65%)	Total: 123,764

Caloric value of caribou tissues from Keene, 1985; edible mass data from Foote, 1965.
Available calories for a given tissue = average caloric value x edible mass of the tissue.
Weighted average = the sum of the average caloric value of each tissue times the percentage of the total edible mass represented by that tissue.

metabolic heat production (see above). Total energy expenditures two to two-and-a-half times BMR, gained through a combination of activity and increased metabolism of brown fat (Steegmann et al., 2002), are not unreasonable. An adult male metabolizing 4475 kcal d^{-1} (the middle of the range of estimated values) would generate 217 W throughout the day, while a male at the high end of the range (5000 kcal d^{-1}) would have maintained heat production on the order of 242 W throughout the day. Combined with insulation from clothing and shelter, this kind of heat production would likely have kept a Neanderthal warm even under relatively severe conditions.

The large chests of Neanderthals may, then, reflect a need for oxygen consumption rather than a need to adhere to Bergmann's rule. A liter of oxygen is needed for every 4.8 kcal burned, such that a Neanderthal male would have required somewhere between 830 and 1040 l O_2 d^{-1}. Given a constant O_2 capture rate of 3.1% (Schmidt-Nielsen, 1984), this equates with alveolar ventilation rates of 19–23 l of air per minute. The average adult modern human moves a half liter of air with each breath, and breathes 12 to 18 times per minute at rest, for a total alveolar ventilation of 6–9 l min^{-1}. Neanderthals would have needed to increase their average respiratory rate to 38–46 breaths min^{-1} (which is very fast for a mammal of our size, especially at rest), or they would have needed greater lung volumes (and hence greater tidal volumes). While much more work is needed on Neanderthal thoracic morphology, thermoregulation, and activity patterning, the results of this study support the idea that large thoraces would have been adaptively advantageous for Neanderthals for reasons having to do with heat production (during periods of both active and resting metabolism) more than heat retention (*a la* Bergmann's rule).

The Cost of Cold Adaptation in Neanderthals

One outcome of the Stage 3 climatic project (van Andel and Davies, 2003) was the demonstration that Neanderthals occupied sites with warmer winter temperatures, on average, than did modern humans of the Aurignacian and Gravettian (median wind chill temperatures 4.0–6.6° C warmer: Aiello and Wheeler, 2003). Thus, despite having bodies that were on average more massive (and thus better able to generate heat) and with lower SA/V ratios compared to early modern Europeans (or even Inuit, for that matter), the Neanderthals appear to have been less able to have tolerated extreme glacial conditions than were their modern human successors. It is unlikely that Neanderthals could have inhabited the areas they did without some form of insulation (Aiello and Wheeler, 2003). But it may have been that elevated metabolic heat production was a more critical component of Neanderthal cold adaptation than it was for modern humans (metabolic heat production and insulation are additive solutions to the problem of cold, and certainly the adaptive strategies of both groups relied on both components – I am merely suggesting that the relative importance of the two components differed between groups). Early modern humans may have used clothing with much higher insulative value, and hence may have had to rely less on metabolic heat production. What may have limited the environmental tolerances of the Neanderthals, then, was not their physiological ability to generate heat but rather their ability to capture calories sufficient to consistently do so. As noted by Aiello and Wheeler (2003:155) "It should be noted that physiological capability does not necessarily imply ecological viability. The costs of maintaining internal heat production at the required levels would only have been possible if Neanderthals were able to sustain a correspondingly high level of dietary energy intake."

As also noted by Aiello and Wheeler (2003), differences in energetic budgets between Neanderthals and early modern Europeans may have impacted the competitive interactions between them. In addition to limiting their climatic tolerance (and hence geographic ranges and habitat choices) and stressing their foraging systems, the higher energetic demands of Neanderthal thermoregulation would have reduced the energy available for reproduction. Even small differences in completed fertility relative to more energetically-efficient modern humans would have had a profound impact on Neanderthal demographics, and their survival as a species, when played out over several millennia.

Acknowledgements

I would like to thank Terry Harrison and Katerina Harvati for the invitation to participate in the *Neanderthal Revisited* Conference, as well as Leslie Aiello and one anonymous reviewer for providing helpful comments on an earlier draft of this paper. The work presented here has also benefited from discussions with Leslie Aiello, Alison Brooks, Bob Franciscus, Trent Holliday, Jean-Jacques Hublin, Bill Leonard, Chris Stringer, Erik Trinkaus, and many of the participants of the Conference – to all of them I am grateful. Thanks also go to Michael Black, Laura Gruss, Jeanna Novelli and Natalia Rybczynski for their skillful work on the model.

Notes

1. The term bioenergetics is used here in its broad sense, meaning the study of the flow of energy in living systems, with an almost exclusive emphasis on energy flow at the level of the organism (rather than its narrower but more widely-used sense of the study of the biochemistry and thermodynamics of cellular oxidations and syntheses).
2. Bergmann's (1847) rule holds that within a species or a closely related set of species, larger individuals will tend to be found in colder environments. The generally (but not universally) accepted explanation is that for a given shape, larger individuals will have more heat-producing mass relative to heat-losing surface area, and hence greater ability to maintain thermal constancy in cold weather.
3. Allen's (1877) rule holds that individuals living in colder environments will tend to have shorter extremities than conspecifics living in warmer climates. Since extremities (including ears and tails) have high surface area to volume ratios, shortening of extremities reduces heat loss in cold environments.
4. The problem of equifinality is by no means restricted to the study of Neanderthals – take for example the alternative explanations that have been proposed to account for the apparent increase in body size with latitude that characterizes many endothermic taxa and that forms the empirical basis for Bergmann's rule (Rosenzweig, 1968; McNab, 1971; Geist, 1987), one such alternative being a need for larger ranges – favoring larger animals – in environments with lower productivity.
5. Under normal (non-hypothermic) circumstances, active transport of sodium and potassium against their chemical gradients (the Na+/K+ pump) consumes about 12% of cellular energy (van de Graaff and Fox, 1986). Under the influence of thyroxine the pump works much faster, thus demanding an increase in ATP production (using elevated levels of circulating energy substrates liberated from the liver and adipose tissue by norepinephrine) at 40% efficiency, and also releasing heat as the ATP is used in work – moving Na+ and K+ ions "up hill" against their solution gradients. Since the cell membranes are permeable to these ions, the last bit of free energy in the process – the potential energy stored in the transported ions – is released as they re-cross the cell membrane and move down the gradient. This situation is analogous to furiously pumping water from the downstream to upstream sides of a very leaky dam – not very effective in keeping one's feet dry, but a great way to generate heat!
6. Watts (Joules sec^{-1}) are a measure of power (rate of energy expenditure) and are used here when discussing heat production and work output. The measure of food energy used here is the kilocalorie (kcal = 4186.8 J). A watt is thus equivalent to 0.2388×10^{-4} kcal sec^{-1}.
7. Aiello and Wheeler (2003) modeled Neanderthal heat loss assuming heat production equal to estimated BMR and at a heat production equal to three times BMR (based on the maximum metabolic rate that can be sustained by humans indefinitely:

Burton and Edholm, 1955). Neanderthals may have been able to sustain such an elevation in metabolic rate through the combined effects of non-shivering thermogenesis, shivering, high protein intake, and activity.

8. Aiello and Wheeler (2003) based this estimate on mean male Neanderthal BMR of 91.630 W (1891 kcal d^{-1}) and SA of 1.898 m^2. If the average values from Table 4 are used instead (BMR = 1988 kcal d^{-1}, corresponding to 96.3 W; SA = 1.908 m^2) the lower critical temperature drops only slightly (to 26.9°C).

References

Aiello, L.C., 1997. Brains and guts in human evolution: the expensive tissue hypothesis. *Brazil. J. Genet.* 20, 141–148.

Aiello, L.C., Wells, C.K., 2002. Energetics and the evolution of the genus *Homo*. *Ann. Rev. Anthropol.* 31, 323–338.

Aiello, L.C., Wheeler, P., 1995. The expensive tissue hypothesis. *Curr. Anthropol.* 36, 199–221.

Aiello, L.C., Wheeler, P., 2003. Neanderthal thermoregulation and the glacial climate. In: van Andel, T., Davies, W. (Eds.), *Neanderthals and Modern Humans in the European Landscape During the Last Glaciation.* McDonald Institute for Archaeological Research, Cambridge, pp. 147–166.

Alexander, R.M., 1985. Mechanics of posture and gait of some large dinosaurs. *Zool. J. Linn. Soc.* 83, 1–25.

Allen, J.A., 1877. The influence of physical conditions in the genesis of species. *Rad. Rev.* 1, 108–140.

Åstrand, P.O., Rodahl, K., 1970. *Textbook of Work Physiology*. McGraw-Hill, New York.

Bailey, B.J.R., Briars, G.L., 1996. Estimating the surface area of the human body. *Stat. Med.* 15, 1325–1332.

Bellemare, J.-F., Cordeau, M.-P., Leblanc, P., Bellemare, F., 2001. Thoracic dimensions at maximum lung inflation in normal subjects and in patients with obstructive and restrictive lung diseases. *Chest* 119, 376–386.

Bergmann, C., 1847. Ueber die Verhaltnisse der Warmeokonomie der thiere zu ihrer grosse. *Gottinger Studien* 3, 595–708.

Blaxter, K., 1989. *Energy Metabolism in Animals and Man.* Cambridge University Press, Cambridge.

Bocherens, H., Billiou, D., Mariotti, A., Patou-Mathis, M., Otte, M., Bonjean, D., Toussaint, M., 1999. Palaeoenvironmental and palaeodietary implications of isotopic biogeochemistry of last interglacial Neanderthal and mammal bones in Scladina Cave (Belgium). *J. Archaeol. Sci.* 26, 599–607.

Bocherens, H., Billiou, D., Mariotti, A., Toussaint, M., Patou-Mathis, M., Bonjean, D., Otte, M., 2001. New isotopic evidence for dietary habits of Neandertals from Belgium. *J. Hum. Evol.* 40, 497–505.

Boyd, E., 1935. *The Growth of the Surface Area of the Human Body*. University of Minnesota Press, Minneapolis.

Brajkovic, D., Ducharme, M.B., Frim, J., 2001. Relationship between body heat content and finger temperature during cold exposure. *J. Appl. Physiol.* 90, 2445–2452.

Budd, G.M., Brotherhood, J.R., Hendrie, A.L., Jeffery, S.E., 1991. Effects of fitness, fatness, and age on men's responses to whole body cooling in air. *J. Appl. Physiol.* 71, 2387–2393.

Burton, A.C., Edholm, O.G., 1955. *Man in a Cold Environment*. Edward Arnold, London.

Cena, K., Clark, J.A., 1978. Thermal insulation of animal coats and human clothing. *Phys. Med. Biol.* 23, 565–591.

Churchill, S.E., 1994. Human upper body evolution in the Eurasian Later Pleistocene. Ph.D. Dissertation, University of New Mexico, Albuquerque, NM.

Churchill, S.E., 1996. Particulate versus integrated evolution of the upper body in late Pleistocene humans: a test of two models. *Am. J. Phys. Anthropol.* 100, 559–583.

Colbert, E.H., 1962. The weights of dinosaurs. *Am Museum Novitates* 2076, 1–16.

Dean, M.C., 1988. Another look at the nose and the functional significance of the face and nasal mucous membrane for cooling the brain in fossil hominids. *J. Hum. Evol.* 17, 715–718.

DuBois, D., DuBois, E.F., 1916. A formula to estimate the approximate surface area if height and weight are known. *Arch. Int. Med.* 17, 863–871.

Eveleth, P.B., Tanner, J.M., 1976. *Worldwide Variation in Human Growth*. Cambridge University Press, Cambridge.

Farlow, J.O., Smith, M.B., Robinson, J.M., 1995. Body mass, bone "strength indicator," and cursorial potential of *Tyrannosaurus rex*. *J. Vert. Paleontol.* 15, 713–725.

Feldesman, M.R., Kleckner, J.G., Lundy, J.K., 1990. The femur/stature ratio and estimates of stature in mid- and late-Pleistocene fossil hominids. *Am. J. Phys. Anthropol.* 83, 359–372.

Fizet, M., Mariotti, A., Bellon, G., 1995. Effect of diet, physiology and climate on carbon and nitrogen stable isotopes of collageen in a Late Pleistocene anthropic palaeoecosystem: Marillac, Charente, France. *J. Arch. Sci.* 22, 67–79.

Foote, D.C., 1965. Exploration and resource utilization in Northwestern Arctic Alaska before 1855. Ph.D. Dissertation, McGill University, Montreal.

Franciscus, R.G., Churchill, S.E., 2002. The costal skeleton of Shanidar 3 and a reappraisal of Neandertal thoracic morphology. *J. Hum. Evol.* 42, 303–356.

Franciscus, R.G., Trinkaus, E., 1988. The Neandertal nose. *Am. J. Phys. Anthropol.* 75, 209–210.

Frisancho, A.R., 1993. *Human Adaptation and Accommodation*. University of Michigan Press, Ann Arbor.

Galloway, V.A., Leonard, W.R., Ivakine, E., 2000. Basal metabolic adaptation of the Evenki reindeer herders of central Siberia. *Am. J. Hum. Biol.* 12, 75–87.

Gehan, E.A., George, S.L., 1970. Estimation of human body surface area from height and weight. *Can. Chem. Rep.* 54, 225–235.

Geist, V., 1987. Bergmann's rule is invalid. *Can. J. Zool.* 65, 1035–1038.

Glickman-Weiss, E.L., Nelson, A.G., Hearon, C.M., Goss, F.L., Robertson, R.J., Cassinelli, D.A., 1993. Effects of body morphology and mass on thermal responses to cold water: revisited. *Eur. J. Appl. Phys.* 66, 299–330.

Greksa, L.P., 1990. Developmental responses to high-altitude hypoxia in Bolivian children of European ancestry: a test of the developmental adaptation hypothesis. *Am. J. Phys. Anthropol.* 2, 603–612.

Haycock, G.B., Schwartz, G.J., Wisotsky, D.H., 1978. Geometric method for measuring body surface area: A height-weight formula validated in infants, children, and adults. *J. Ped.* 93, 62–66.

Hebblewhite, M., Paquet, P.C., Pletscher, D.H., Lessard, R.B., Callaghan, C.J., 2003. Development and application of a ratio estimator to estimate wolf kill rates and variance in a multiple-prey system. *Wildlife Soc. Bull.* 31, 933–946.

Heglund, N.C., Cavagna, G.A., 1985. Efficiency of vertebrate locomotory muscles. *J. Exp. Biol.* 115, 283–292.

Heglund, N.C., Cavagna, G.A., 1987. Mechanical work, oxygen-consumption, and efficiency in isolated frog and rat muscle. *Am. J. Physiol.* 253, C22–C29.

Heim, J.-L., 1982. Les Hommes Fossiles de la Ferrassie: Tome II – Les Squelettes Adultes (Squelette des Membres). Archives de l'Institut de Paleontologie Humain, 38, Masson, Paris.

Holden, C., Mace, R., 1999. Sexual dimorphism in stature and women's work: a phylogenetic cross-cultural analysis. *Am. J. Phys. Anthropol.* 110, 27–45.

Holliday, T.W., 1997. Postcranial evidence of cold adaptation in European Neandertals. *Am. J. Phys. Anthropol.* 104, 245–258.

Hrdlička, A., 1930. Anthropological Survey in Alaska. 46th Annual Report of the Bureau of American Ethnology, 1928–1929.

Jedrzejewski, W., Schmidt, K., Theuerkauf, J., Jedrzejewski, B., Selva, N., Zub, K., Szymura, L., 2002. Kill rates and predation by wolves on ungulate populations in Bialowieza Primeval Forest (Poland). *Ecology* 83, 1341–1356.

Jelinek, A.J., 1994. Hominids, energy, environment, and behavior in the Late Pleistocene. In: Nitecki, M.H., Nitecki, D.V. (Eds.), *Origins of Anatomically Modern Humans*. Plenum Press, New York, pp. 67–92.

Katch, F., Michael, E.D., Horvath, S.M., 1967. Estimation of body volume by underwater weighing: description of a simple method. *J. Appl. Physiol.* 23, 811–813.

Keene, A.S., 1985. Nutrition and economy: Models for the study of prehistoric diet. In: Gilbert, R.I J., Mielke, J.H. (Eds.), *The Analysis of Prehistoric Diets*. Academic Press, Orlando, FL, pp. 155–190.

Kleiber, M., 1961. *The Fire of Life: An Introduction to Animal Energetics*. John Wiley, New York.

Leonard, W.R., Katzmarzyk, P.T., Comuzzie, A.G., Crawford, M.H., Sukernik, R.I., 1994. Growth and nutritional status of the Evenki reindeer herders of Siberia. *Am. J. Hum. Biol.* 6, 339–350.

Leonard, W.R., Katzmarzyk, P.T., Crawford, M.H., 1996. Energetics and population ecology of Siberian herders. *Am. J. Hum. Biol.* 8, 275–289.

Leonard, W.R., Robertson, M.L., 1992. Nutritional requirements and human evolution: a bioenergetics model. *Am. J. Hum. Biol.* 4, 179–195.

Leonard, W.R., Robertson, M.L., 1997. Comparative primate energetics and hominid evolution. *Am. J. Phys. Anthropol.* 102, 265–281.

Leonard, W.R., Ulijaszek, S.J., 2002. Energetics and evolution: an emerging research domain. *Am. J. Hum. Biol.* 14, 547–550.

MacHattie, L., Haab, P., Rennie, D.W., 1960. Eskimo metabolism as measured by the technique of 24-hour indirect calorimetry and graphic analysis. Arctic Aeromedical Laboratory Tech. Rep. AAL-TR-60-43.

Martini, F.H., 1998. *Fundamentals of Anatomy and Physiology*, 4th Edition. Prentice Hall, Upper Saddle River, NJ.

McNab, B.K., 1971. On the ecological significance of Bergmann's rule. *Ecology* 52, 845–854.

Moen, A.N., 1973. *Wildlife Ecology: an Analytical Approach*, W.H. Freeman and Company, San Francisco.

Mosteller, R.D., 1987. Simplified calculation of body-surface area. *New Eng. J. Med.* 317, 1098.

Nisbet, R.M., Muller, E.B., Lika, K., Kooijman, S.A.L.M., 2000. From molecules to ecosystems through dynamic energy budget models. *J. Animal Ecol.* 69, 913–926.

Novak, L.P., Hyatt, R.E., Alexander, J.F., 1968. Body composition and physiologic function of athletes. *J. Am. Med. Assoc.* 205, 764–770.

Paul, G.S., 1988. *Predatory Dinosaurs of the World*. Simon and Schuster, New York.

Richards, M.P., Pettitt, P.B., Trinkaus, E., Smith, F.H., Paunovic, M., Karavanic, I., 2000. Neanderthal diet at Vindija and Neanderthal predation: the evidence from stable isotopes. *Proc. Natl. Acad. Sci. U.S.A.* 97, 7663–7666.

Roberts, D.F., Bainbridge, D.R., 1963. Nilotic physique. *Am. J. Phys. Anthropol.* 21, 341–370.

Rosenzweig, M.L., 1968. The strategy of body size in mammalian carnivores. *Am. Midland Nat.* 80, 299–315.

Ruff, C.B., Niskanen, M., Junno, J.-A., Jamison, P., 2005. Body mass prediction from stature and bi-iliac breadth in two high latitude populations, with application to earlier higher latitude humans. *J. Hum. Evol.* 48, 381–392.

Ruff, C.B., Trinkaus, E., Holliday, T.W., 1997. Body mass and encephalization in Pleistocene *Homo*. *Nature* 387, 173–176.

Sawyer, G.J., Maley, B., 2005. Neanderthal reconstructed. *Anat. Rec.* (Part B: New Anat.) 283B, 23–31.

Schmidt-Nielsen, K., 1984. *Scaling: Why is Animal Size So Important*? Cambridge University Press, Cambridge.

Shephard, R.J., 1978. *Human Physiological Work Capacity*. Cambridge University Press, Cambridge.

Shephard, R.J., Rode, A., 1996. *The Health Consequences of "Modernization": Evidence from Circumpolar Peoples*. Cambridge University Press, Cambridge.

Snodgrass, J.J., Leonard, W.R., Tarskaia, L.A., Alekseev, V.P., Krivoshapkin, V.G., 2005. Basal metabolic rate in the Yakut (Sakha) of Siberia. *Am. J. Hum. Biol.* 17, 155–172.

Sorensen, M.V., Leonard, W.R., 2001. Neanderthal energetics and foraging efficiency. *J. Hum. Evol.* 40, 483–495.

Steegmann, A.T.J., Cerny, F.J., Holliday, T.W., 2002. Neanderthal cold adaptation: physiological and energetic factors. *Am. J. Phys. Anthropol.* 14, 566–583.

Trinkaus, E., 1987. Bodies, brawn, brains and noses: human ancestors and human predation. In: Nitecki, M.H., Nitecki, D.V. (Eds.), *The Evolution of Human Hunting*. Plenum Press, New York, pp. 107–145.

van Andel, T.H., Tzedakis, P.C., 1996. Palaeolithic landscapes of Europe and environs, 150,000–25,000 years ago: an overview. *Quat. Sci. Rev.* 15, 481–500.

van Andel, T.H., Davies, W. (Eds.), 2003. *Neanderthals and Modern Humans in the European Landscape During the Last Glaciation*. McDonald Institute for Archaeological Research, Cambridge.

van de Graaff, K.M., Fox, S.I., 1986. *Concepts of Human Anatomy and Physiology*. McGraw-Hill, Boston.

Wang, Y., Moss, J., Thisted, R., 1992. Predictors of body surface area. *J. Clin. Anesth.* 4, 4–10.

Weaver, T.D., Steudel-Numbers, K., 2005. Does climate or mobility explain the differences in body proportions between Neanderthals and their Upper Paleolithic successors? *Evol. Anthropol.* 14, 219–223.

Weiss, M.L., Mann, A.E., 1985. *Human Biology and Behavior: An Anthropological Perspective*. 4th Edition. Little Brown, Boston.

Wilkie, D.R., 1960. Man as a source of mechanical power. *Ergonomics* 3, 1–8.

Winslow, C.E.A., Herrington, L.P., 1949. *Temperature and Human Life*. Princeton University Press, Princeton, NJ.

Wolpoff, M.H., 1999. *Paleoanthropology*, 2nd Edition. McGraw-Hill, Boston.

8. How different were Neanderthals' habitual activities? A comparative analysis with diverse groups of recent humans

O.M. PEARSON
Department of Anthropology, MSC 01–1040,
University of New Mexico,
Albuquerque, NM 87131–1086, USA
ompear@unm.edu

R.M. CORDERO
Department of Anthropology, MSC 01–1040,
University of New Mexico,
Albuquerque, NM 87131–1086, USA
rc17767@hotmail.com

A.M. BUSBY
Department of Anthropology, MSC 01–1040,
University of New Mexico,
Albuquerque, NM 87131–1086, USA
mandybusby@yahoo.com

Keywords: Neanderthals, cross-sectional geometry, foraging, habitual activity

Abstract

The lifeways of Neanderthals remain poorly understood despite numerous hints from the archaeological record and from Neanderthal anatomy that their lifestyles may have differed substantially from early modern humans and from more recent populations. The distinctiveness, inefficiency, or primitiveness of Neanderthal lifestyles and patterns of activity play a central role in many adaptive scenarios that have been proposed to explain the eventual ascendancy of modern humans and the Neanderthals' demise. However, many recent studies of faunal remains accumulated by Neanderthals, stable isotope analyses of Neanderthal bones, and the energetic demands of their large body mass suggest that these archaic humans were equally efficient hunters as fully modern foragers in similar environments. This contradictory evidence inspired the present study. To gain new insights into Neanderthal behaviors we use a comparative approach, examining indices of residual bone strength and midshaft diaphyseal shapes of the humerus, femur, and tibia in several groups of Neanderthals (European and Near Eastern), early modern humans, and sixteen diverse groups of recent humans. The results indicate that Neanderthal limbs bear a close similarity to a fairly wide spectrum of more recent groups, especially those who practiced intensive forms of foraging within fairly limited territories. The Neanderthal pattern differs strikingly from those of Skhul-Qafzeh and Gravettian humans, but these early modern humans also differ substantially from the later, intensive foragers. These results are probably more indicative of the relative distribution of people to resources in the Mousterian than the relative behavioral *capacities* of modern humans and Neanderthals.

K. Harvati and T. Harrison (eds.), Neanderthals Revisited: New Approaches and Perspectives, 135–156.

Introduction

This paper presents an attempt to gain insights into the habitual activities of Neanderthals and their contemporaries, early modern (or nearly modern) humans in Israel, and anatomically modern humans from the Upper Paleolithic of Europe. This topic has been intensively investigated in the two decades since Trinkaus' (1983a, 1983b, 1984) synthesis of functional differences between Neanderthals and early modern humans, but a variety of recent studies have suggested that aspects of that synthesis may be incorrect and could profitably be revisited (Gaudzinski, 1999; Grayson and Delpech, 2002; Sorensen and Leonard, 2001; Pearson, 2000; Ruff, 2000a).

Anthropologists have long noted anatomical differences between Neanderthals and modern human postcranial skeletons. Endo and Kimura (1970; Endo, 1971) led the way by focusing attention on the functional implications of many of these differences, a theme that was elaborated upon later (e.g., Heim, 1982; Trinkaus, 1983a). Among those differences were the presence of well-developed femoral pilasters in early modern humans in contrast to nearly circular cross-sections of Neanderthal femora (McCown and Keith, 1939; Endo and Kimura, 1970; Trinkaus, 1976; Vandermeersch, 1981), and a marked "flattening" of the mid-shaft of Neanderthal humeri (Endo and Kimura, 1970; Heim, 1982; Trinkaus, 1983a). Endo and Kimura (1970) were the first to apply the principles of modeling the cross-sections of limb bones as bent beams in order to gain insights into the biomechanical strength of Neanderthal bones, and were soon emulated by others (Lovejoy and Trinkaus, 1980; Senut, 1985). By the 1990s, much of the work on functional adaptations in Neanderthals shifted to analyses of cross-sectional geometry.

The background for the present study thus begins in the mid-1980s, with Trinkaus' view of Neanderthals as extremely active and immensely strong foragers whose lifestyle – and presumably a long history of selection to allow them to efficiently follow that lifestyle – was reflected in the robusticity of their long bones, the sizes of certain muscle markings, increased leverage available to specific muscles, the thickness of the cortical bone in their limbs, and a variety of other features found throughout the skeleton. Trinkaus' (1983a, b, 1984) emphasis on anatomical and, by extension, inferred behavioral contrasts between Neanderthals and modern humans proved to be both very influential and long lasting. The view of Neanderthals as inefficient foragers who needed great physical strength and endurance to survive was heavily influenced and substantiated by Binford's (1985, 1989) conclusions that the faunal record of Neanderthals, and even contemporaneous early moderns from the Middle Stone Age of South Africa (Binford, 1984), did not show many signatures of the behavior of ethnographically documented hunter gatherers, including spatial organization within occupation sites, planning depth as indicated by the excavation pits or other features for storing food at occupation sites, the use of special purpose sites, and frequent curation of high-quality tools or raw material.

Interpretations of Neanderthal behavior have gone though numerous swings from an emphasis on their behavioral differences from modern humans to emphasis upon their similarity with modern humans (Trinkaus and Shipman, 1993). There have always been disagreements about how "modern" or "archaic" Neanderthal behavior was, and a variety of scientists disagreed with aspects of Binford's synthesis of Neanderthal behavior, as well as other contemporary scientists whose syntheses emphasized the distinctly "non-modern" character of the Neaderthal archaeological record (e.g., Mellars, 1996, 2004b; Klein, 1995, 1999, 2003). Among these dissenting voices was Chase (1986), who noted that the faunal remains from Neanderthal hunters at Combe Grenal did not differ substantially from similar faunas accumulated by later, "modern" Upper Paleolithic hunters. Similarly, both Straus and Clark argued that the archaeological record of Cantabrian Spain presented no sharp break in lithic technology or subsistence at the Middle-Upper Paleolithic boundary (e.g., Straus, 1995, 1997; Clark and Lindley, 1989; Clark, 2002).

In the 1990s and 2000s, more studies scrutinized the archaeological evidence for putative differences between Neanderthals and early modern humans. A variety of authors argued that at least some sites show evidence that Neanderthals could be highly proficient hunters, able to target and kill prime age adult reindeer or other ungulates reliably and to accumulate faunas heavily dominated by a single species (Gaudzinski, 1999; Gaudzinski and Roebroeks, 2000; Grayson and Delpech, 2002, but see Mellars, 2004a). Similarly, by refitting animal bone fragments from various species into diaphyses and counting the numbers and types of skeletal elements present, Marean demonstrated that, at the Mousterian site of Kobeh in the Zagros Mountain of Iran, the people (presumably Neanderthals) had access to high-utility elements, and thus were most likely hunters rather than scavengers (Marean and Kim, 1998, Marean, 1998). Studies of stable nitrogen preserved in Neanderthal bones have consistently shown that Neanderthals consumed a substantial quantity of meat, comparable to that eaten by contemporary wolves (Fizet et al., 1995; Richards et al., 2000; Bocherens et al., 2001; Drucker and Bocherens, 2004). Neanderthals had to have been able to procure a substantial amount of meat by hunting, scavenging, or some combination of the two.

Sorensen and Leonard's (2001) recent analysis of Neanderthal energetics provides one indication that portions of the mid-1980s synthesis of Neanderthal behavior and anatomical correlates may need revision. Using conservatively low estimates of body mass for Neanderthals (65 kg males, 55 kg females), they estimated that for a foraging time of 5.74 hours per day (the primate average), Neanderthal males would expend 3682–5523 kcal/day and females 2870–4305 kcal/day. Given a need for foraging returns of 1.35 times their daily energy expenditure, they argued that Neanderthals would have to obtain return rates comparable to modern foragers. However, a variety of authors have argued that Neanderthals and other late archaic *Homo* were substantially heavier with masses as high as 100 kg for males and 80 kg females (Kappelman, 1996; Arsuaga et al., 1999; Rosenberg et al., 1999). Given the same assumptions made by Sorensen and Leonard (2001), males that large would have burned 4860–7290 kcal/day and females 3678–5518 kcal/day, which would have required their foraging return rates to equal or exceed those of the *most* efficient modern hunter-gatherers. These correspond to very large amounts of calories, but such high consumption is not unknown from recent times. Based on a detailed diary kept by Peter Fidler in the winter of 1791–1792, Helm (1993) calculated that Chipewyan Indians in the boreal forest of Canada consumed 5140–5780 kcal (2.80–3.13 kg of meat) per person per day. To feed such an enormous appetite it is essential to be an effective hunter.

Given these multiple lines of evidence, it should be clear that there is reason to suppose that at least some aspects of the mid-1980s synthesis of Neanderthal behavior are likely to be incorrect. That realization does not help us to gain a better insight into what the habitual activities of Neanderthals were like, however, nor does it help us to understand why Neanderthals eventually lost out in competition to modern humans. In part, the synthesis from the mid-1980s was so influential because it offered answers to both questions. What can we now say with certainty about what limb bones can reveal about the habitual activities of Neanderthals?

Materials and Methods

The approach we have adopted is to compare a limited set of indices designed to capture variation in the shape and strength of three major limb bones: the humerus, femur, and tibia. It has been proposed that the cross-sectional shape and strength of limb bones may reflect different aspects of behavior (Trinkaus et al., 1991) and a number of studies have documented differences between ancient and recent populations in these or related indices (Martin and Saller, 1956; Endo and Kimura, 1970; Ruff et al., 1984; Ruff, 1987; Churchill, 1996; Churchill et al., 1996b; Trinkaus and Ruff, 1999a, b), including the distinctive aspects of Neanderthal diaphyseal shape and strength described above. We did not include measures of asymmetry in the upper limb; previous studies have clearly demonstrated that, in comparison to a broad set of comparative populations, both Neanderthals and early modern humans display a large amount of asymmetry in the upper limb that presumably reflects a markedly greater amount of functional strains applied to the right arm than to the left (Trinkaus et al., 1994; Churchill and Formicola, 1997).

Indices of limb bone residual robusticity were calculated as the ratio of (midshaft AP*ML (or maximum*minimum) diameters)/(articular breadth). AP*ML diameters were used for the femur and tibia; maximum and minimum diameters were used for the humerus. To some degree, it can be expected that the distal-most part of the deltoid tuberosity will affect the dimensions of the humeral midshaft. Therefore, the results reported here will not be completely comparable to those from studies of the cross-sectional geometry of the mid-distal humerus (e.g., Trinkaus et al., 1994; Trinkaus and Churchill, 1999; Churchill, 1994, 1996; Churchill and Formicola, 1997). The articular breadths in question comprised the vertical diameters of the femoral and humeral heads and the medio-lateral breadth of the tibial condyles. Indices of midshaft maximum to minimum (or AP vs. ML) diameters served as a measure of midshaft shape, which is often considered to be an indicator of habitual activity (e.g., Ruff et al., 1984; Ruff, 1987; Trinkaus et al., 1991; Trinkaus and Ruff, 1999a, 1999b).

We compare the values obtained for Neanderthals for each of the indices to the entire spectrum of recent humans' values, noting cases in which similarities exist. In cases in which Neanderthals closely resemble more than one recent population, and those populations share a number of cultural, behavioral, or work-related factors in common, it would tend to support (but not prove) that a similar factor could be responsible for the morphology of the Neanderthals. However, if the societies that resemble Neanderthals in terms of morphology differ greatly in most respects, it should serve as a warning that the underlying cause of the Neanderthal morphology may be difficult to establish with any degree of certainty.

We measured the humerus and lower limb bones from the right side whenever possible. Given the large amount of humeral asymmetry that exists in many samples (Churchill and Formicola, 1997), the requirement to measure the midshaft dimensions of the right humerus was strictly observed in fossil specimens; if only the left humerus was available, the specimen was excluded from the analysis. In the recent samples – which, with the exception of the Khoisan, tended to display low levels of bilateral asymmetry – this requirement was relaxed so that data from the left humerus were used in cases in which the right humerus was missing or damaged. The lower limb displays much less asymmetry than the upper limb, so in cases in which the right femur or tibia was missing or damaged, we substituted values measured from the left side. We adopted a conservative approach toward missing data, choosing not to estimate missing values. We made four exceptions, estimating the vertical head diameter of the femoral head of Qafzeh 9 and La Ferrassie 1 to be identical to their preserved horizontal diameter, and estimating the vertical diameter of the humeral head of Dolní Věstonice 14 and Chancelade from the vertical diameters of their femoral heads.

Samples

In order to evaluate patterns of Neanderthal limb bone residual strength and shape in a broad comparative context, we compared Neanderthals to 17 recent populations and six

Table 1. Samples with Geographic/Behavioral Variables

Group	*N females*	*N males*	*Total N*	*Terrain*	*Subsistence*	*Seafaring*	*Digging*	*Grinding Tools*	*Mobility*	*Climate*
Würm European Neanderthals	4	12	16	rugged	H&G	absent	absent	absent	mobile	cold
Middle Eastern Neanderthals	3	6	9	rugged	H&G	absent	absent	absent	mobile	cold
Neanderthals (pooled)	7	18	25	rugged	H&G	absent	absent	absent	mobile	cold
Skhul-Qafzeh	3	5	8	moderate	H&G	absent	absent	absent	mobile	moderate
Gravettian	3	14	17	rugged	H&G	absent	absent	absent	mobile	cold
Magdalenian	2	7	9	rugged	H&G	absent	absent	absent	mobile	cold
Epigravettian	0	7	7	rugged	H&G	absent	unknown	absent	semi-sedentary	moderate
Inuit – Riverine	10	11	21	moderate	H&G	moderate	absent	absent	semi-sedentary	cold
Inuit – Coastal	14	51	65	moderate	H&G	present	absent	absent	semi-sedentary	cold
Tierra del Fuego	9	19	28	moderate	H&G	present	some	absent	mobile	cold
Chinese	0	25	25	flat	industrial	absent	some	absent	sedentary	cold
Mesolithic 1	4	3	7	flat	H&G	some	absent	present	semi-sedentary	cold
Mesolithic 2	4	4	8	flat	H&G	some	absent	unknown	semi-sedentary	cold
Mesolithic 3	0	2	2	rugged	H&G	absent	absent	absent	mobile	cold
Sami	24	29	53	moderate	H&G-herding	absent	absent	absent	semi-sedentary	cold
Euro-Americans	25	24	49	flat	industrial	absent	some	absent	sedentary	cold
Australians	5	15	20	flat	H&G	absent	present	present	semi-sedentary	hot
Jebel Sahaba	9	12	21	flat	H&G	absent	unknown	present	semi-sedentary	hot
Khoisan	14	20	34	moderate	H&G-pastoral	absent	present	females only	mobile	hot
African Americans	30	40	70	flat	industrial	absent	some	absent	sedentary	hot
Zulu	31	31	62	moderate	horticultural	absent	present	present	sedentary	hot

"fossil" populations (Table 1). Details on other samples have been presented elsewhere (Pearson, 1997, 2000), with additional populations or exceptions as described below. In each population, males and females were treated separately. Neanderthals were divided into those from the Würm glaciation of Europe (oxygen isotope stages 4–3) and contemporaneous Neanderthals from the Near East (Israel and Iraq). Fossil specimens included in the European Neanderthal sample comprise, for males: La Chapelle-aux-Saints, La Ferrassie 1, Spy 1 and 2 (here both considered to be males), Le Régourdou 1 (measurements of casts; here considered as a male), Neanderthal 1, Kiik-Koba 1 (measurements of casts), Fond de Forêt; females: La Ferrassie 2 and La Quina 5. Fossils in the Near Eastern Neanderthal sample are as follows. Males comprise: Kebara 2, Amud 1, Shanidar 1, 3 and 4 (measurements from casts and Trinkaus, 1983a); females include: Tabun C1, Shanidar 6 and 8 (measurements from casts and Trinkaus, 1983a). Fossils in the Skhul-Qafzeh sample are, females: Skhul II, VII, and Qafzeh 9 (which, interestingly, falls closer to the males rather than females of many of the recent samples in the analyses that follow); and males: Qafzeh 8, Skhul IV, and V. Fossils in the Gravettian sample include, for males: Předmostí 1, 3, 5, 9, 14 (from Matiegka, 1938); Dolní Věstonice 13 and 14, Pavlov 1, Sungir 1 (from Debets, 1967; Khrisanfova, 1980, 1984); Grotte de Enfants 4, Barma Grande 2, 5, and 6; and for females: Předmostí 4 and 10 (from Matiegka, 1938) Dolní Věstonice 3 (from Trinkaus and Jelínek, 1997), and Grotte des Enfants 5. Specimens included in Magdalenian sample comprise males: Oberkassel 1, La Madeleine, Neuessing 2, Chancelade, and Gough's Cave 1; females: Oberkassel 2. Fossils in Epigravettian sample (all males) came from Arene Candide with one individual from Grotta Continenza.

With respect to the more recent samples of humans, we subdivided Pearson's (1997, 2000) sample of "Mesolithic Europeans" into three groups identified in the analyses that follow as "Meso 1," "Meso 2," and "Meso 3." "Meso 1" comprises a series of skeletons from the Danish Mesolithic (predominantly individuals from Ertebølle, supplemented with measurements provided by Dr. C. Meiklejohn) (Pearson, 1997). "Meso 2" includes individuals from the French Mesolithic sites of Tévic and Hoëdic from the coast of Brittany. "Meso 3" includes only two male individuals, Gramat and Rochereil (Vallois and de Félice, 1977), drawn from sites along the drainage of the Dordogne river, the same topography occupied by the earlier Neanderthals in the region.

Following Collier (1989), Inuit populations were divided between those from contexts along the northwestern coast of Alaska who practiced whaling and intensively hunted marine mammals and those from predominantly riverine contexts from southwestern Alaska (mainly along the Kuskokwim River). The two Inuit samples are as described previously (Pearson 1997, 2000), but with the addition of a large series from Point Hope (included in the "coastal" group) from the American Museum of Natural History.

The Australian sample comprises a continent-wide, pooled sample. Societies in Aboriginal Australia lived in a wide variety of habitats and varied greatly in levels of mobility, their subsistence practices, and other aspects of their lifestyle (Mulvaney and Kamminga, 1999). Aboriginal people who lived along the Murray River tended to follow intensive foraging patterns, had the smallest territories, highest population densities, and apparently, the greatest genetic differentiation among themselves and between themselves and the inhabitants of the rest of the continent (Pardoe, 1994, 1995). We wished to subdivide the Australian sample into (at least) Murray Valley and non-Murray Valley groups, but the resultant sample sizes were undesirably small, and so we pooled the Australian samples of each sex.

The Tierra del Fuego sample is also a combined sample of skeletons of Yaghan (Yamana) and Selk'nam (Ona) from Instituto de la Patagonia, Punta Arenas, Chile, Museo del Fin de Mundo, Argentina, the Museo Etnografico, Buenos Aires, Argentina, and the literature (Hyades and Deniker, 1891; Hultkrantz, 1907; Pearson and Millones, 2003, 2005).

An important aspect of the recent samples is that more is known, or can confidently be inferred, about the habitual activities of the recent groups than is the case for the fossils. We identified seven geographical and behavioral characteristics for each population (Table 1). Ruff (2000b) and Stock and Pfeiffer (2004) identify the first characteristic, terrain, as an important correlate with bone strength. We defined three categories for terrain: rugged, moderate, and flat. Trinkaus et al. (1991) argues that the second characteristic, mobility, is also an important correlate with bone strength. We characterized the mobility patterns for each group as mobile, semi-sedentary, or sedentary. Previous researchers have shown that habitual behaviors related to economic activity also show correlations with bone strength (e.g., Ruff et al., 1984; Collier, 1989; Bridges, 1989; Churchill, 1994). These activities include seafaring, use of grinding stones, and digging/horticulture. We scored each population for the presence or absence of these three characteristics. We also added the category "some" to describe groups that occasionally practice seafaring or use grinding stones.

Analysis of Relationship between these Ratios and Data from Cross-sectional Geometry

It can be argued that cross-sectional geometry provides a much better indication of biomechanical adaptations than external measurements of limb bones. However, because second moments of area – for example, J, the torsional second moment of area, which figures prominently in most analyses of cross-sectional geometry – depend upon the sum distances squared of each bit of bone to the axis of bending in a section, the external dimensions of the bone have a disproportionate influence upon the magnitude of second moments of area. Thus, there is good reason to think that the external dimensions should provide a reasonable reflection of second moments of area (Pearson, 2000). Here, we offer a series of analyses to evaluate how closely the indices used in this paper correspond to cross-sectional properties, which generally are considered to be some of the best data to use for biomechanical analyses in physical anthropology.

To evaluate the degree of correspondence between the indices based on external dimensions and cross-sectional geometry, we compared both in a pooled sex, pooled sample series of 121 skeletons of Khoisan, African American, or Zulu ancestry that has been employed in other analyses (Grine et al., 1995; Churchill et al., 1996a; Pearson and Grine, 1996a, 1996b, 1997; Pearson, 2000). The data set for these individuals comprises both external dimensions and measures of the cross-sectional geometry of the major long bones. First for each limb bone, we calculated correlations between the indices of residual robusticity described above and based on external midshaft dimensions and the adjusted torsional second moment of area (J) of the midshaft section. Adjustments were made by dividing J by the product of bone length (maximum length of each bone, but excluding the intercondylar eminence in the case of the tibia) and the same measure of epiphyseal size as employed in the ratio of residual robusticity (vertical diameter of the femoral head, and superior-inferior diameter of the humeral head). Regression analyses showed that in each case, the index based on external measurements was definitely related to adjusted *J* and the relationship was particularly close in the humerus ($R^2 = 0.66$ for the femur;

$R^2 = 0.69$ for the tibia; $R^2 = 0.81$ for the humerus).

Next, for those who are interested in estimating second moments of area from external dimensions, we calculated the estimation equations below by performing multiple regression with logged data, then transformed the resultant equations back into raw space. For J, the predicted values from the external dimensions are:

$$\text{Femur: } J = 10^{-1.0618}(F9)^{1.6852}(F10)^{2.3011} \pm 6{,}377.3 \text{ mm}^4;\ \mathrm{R}^2 = 0.857 \quad (1)$$

$$\text{Tibia: } J = 10^{-0.5184}(T6)^{2.2915}(T7)^{1.2546} \pm 5{,}021.6 \text{ mm}^4;\ \mathrm{R}^2 = 0.864 \quad (2)$$

$$\text{Humerus: } J = 10^{-0.5227}(H3)^{0.6276}(H4)^{0.1484} \pm 2{,}172.4 \text{ mm}^4;\ \mathrm{R}^2 = 0.901 \quad (3)$$

In these formulae, F9 is the AP diameter and F10 is the ML diameter of the femoral midshaft; T6 is the AP diameter and T7 is the ML diameter of the tibial midshaft; and H3 is the maximum diameter and H4 the minimum diameter of the humeral midshaft. The formulae are based on the entire pooled sex, pooled-group sample; dividing the pooled sample by sex or group (or both) would likely provide more appropriate prediction equations for samples that match various characteristics of those sub-samples.

Lastly, we explored the relationship between the midshaft shape indices based on external measurements and measures of $I_{max/min}$, an analogous variable frequently used in studies of cross-sectional geometry. As before, we calculated correlations between midshaft shape ratios and $I_{max/min}$ in the combined sample of African American, Zulu, and Khoisan limb bones. For the femur, the correlation was $r = 0.55$ ($p < 0.0001$, $n = 121$), but a problem arose because 12 individuals in the sample have a greater ML than AP breadth of the femoral midshaft. Excluding the three individuals who had the most severe departures from an AP>ML breath (i.e., an AP breadth of 80–85% that of the ML breadth) increased the correlation to $r = 0.69$. Thus, it can be concluded that the ratio of the external dimensions of the femur provides a reasonable, but certainly not perfect, approximation of $I_{max/min}$ ratios for the femoral midshaft. The same applies even more clearly to the other two limb bones in the study. For the tibia, the correlation between $I_{max/min}$ and the ratio of ML to AP midshaft breadths was $r = -0.84$ ($p < 0.0001$, $n = 118$). For the humerus, the correlation between $I_{max/min}$ and the ratio of maximum to minimum midshaft breadth was $r = 0.88$ ($p < 0.0001$, $n = 118$).

It could be argued that only analogous ratios of external breadths and ratios of second montents of area calculated around the same anatomical or geometric axes should be compared. For example, it could be argued that one should only compare ratios of AP/ML external dimensions of the femur and tibia to ratios of $I_{x/y}$ for the bones (rather than $I_{max/min}$) in which I_x corresponds to the AP plane and I_y to the ML plane of the bone. We do not agree with this argument for several reasons. First, while not precisely the same, the femoral and tibial midshaft AP and ML dimensions are close to the maximum and minimum dimensions of these bones' cross-sections. Second, comparisons in studies of cross-sectional geometry of $\mathrm{I}_{x/y}$ between the femur and tibia are not perfect. Although Ruff and Hayes (1983) presented a detailed and well-reasoned set of criteria for aligning the femur and tibia by allowing their condylar surfaces at the knee to form two of three points for their ML plane, this procedure neverthless results in the frontal planes of the two bones being only "approximately coplanar" with a frontal plane that can be expected to deviate 5°–8° from the actual anterior plane (Ruff and Hayes, 1983, p. 363). Another difficulty lies in the fact that I_x and I_y in this system correspond to axes around which a researcher assumes that the bones are bent. There is no guarantee that these intuitive assumptions are accurate and, in the case of the femoral midshaft, there is reason to suspect that the assumption of AP bending is inaccurate. Pauwels (1980) has summarized data and biomechanical models that suggest that the mode of bending in the femoral midshaft is from antero-lateral to postero-medial, approximately 45° from the AP axis. On the othe hand, $I_{max/min}$ is an inherent property of any cross-section that is invariant of the section's orientation while $I_{x/y}$ will vary with the section's orientation in all bone cross sections except those that are perfectly circular and have a concentric periosteal and endosteal perimeter. Furthermore, a fundamental tenent of research on cross-sectional geometry has been that the distribution of bone in a cross-section (i.e., its shape) reflects the loading history to which the bone has been subjected. Whether or not the tenent is true (Lieberman et al., 2004), if one accepts the logic of the proposition, it follows that a property such as $I_{max/min}$ should always reveal something about the bone's loading history while a ratio of $I_{x/y}$ could be expected to reveal similar information only if it happened to correspond to a biomechanically relevant axis, which, for all of the reasons noted above, is less certain than we might wish. Given these facts, we think it is both reasonable and sensible to compare values of $I_{max/min}$ of the long bones in this sample with their ratios of diaphyseal midshaft shape based on external measurements.

The analyses above show that there is good reason to think that the ratios of residual strength based on external measurements and those based on cross-sectional geometry will show broadly similar results, we wish to add a few words of caution, for neither approach is likely to be perfect. As illustrated in Figure 1, the product of diaphyseal breadths should follow an exponential relationship with increasing body size. Based on published formulae (e.g., those in Grine et al., 1995), body mass is likely to increase faster than articular dimensions, which serve as a proxy for mass in our analyses. As our ratios result from the combination of these two trends, we expect that the ratios of residual robusticity will make large-sized individuals (or populations) appear to be somewhat "stronger" than they actually are.

Studies of cross-sectional geometry also have flaws. First, in order to control for the product of body mass and beam (bone) length, these studies must estimate body mass. Estimates of mass based on populations that differ in physique from the target specimens will cause systematic errors in mass, and, unfortunately, the physique of most prehistoric populations must itself be estimated. In addition, with fragmentary fossils, in order to use some of the existing protocols for analyses of body size- and shape-adjusted cross-sectional geometry, researchers may be forced to use estimates of bone length and body mass that have the potential to produce large uncertainties (Pearson et al., 2001). Finally, experimental studies have shown that the way bones are bent in life may not always correspond to the way we might suppose them to have been bent on the basis of their cross-sectional geometry (Gross et al., 1992; Demes et al., 1998; 2001; Lieberman et al., 2004), although this criticism necessarily applies to both studies of cross-sectional geometry, it also applies to studies that evaluate external dimensions, as in the present case.

Results

UPPER LIMB

Inspection of Table 2 reveals that many of the aspects of Neanderthal humeral strength and cross-sectional shape that have seemed remarkable to various observers can be matched by several recent populations. With regard to residual strength, European Neanderthal males' mean is matched (or exceeded) by the males of several recent populations including Tierra del Fuego, Tévic-Hoëdic Mesolithic

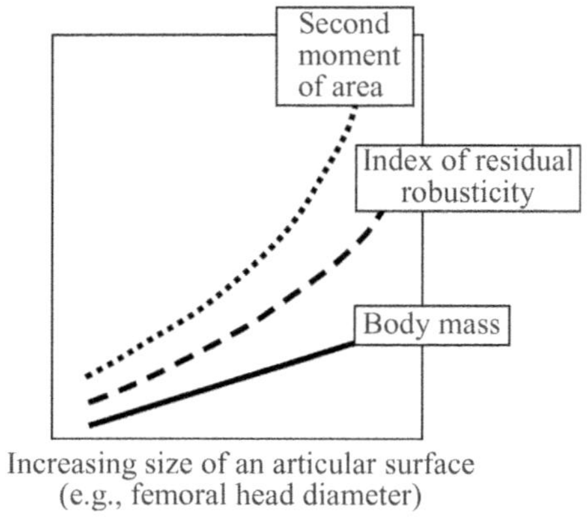

Figure 1. Relationships between increasing articular dimensions and body mass, indices of residual robusticity, and second moments of area (e.g., J).

French, Australians, Danish Mesolithic, Zulu, African Americans, and European Americans. The same holds for the Neanderthal females. Although much has been made of the fact that Neanderthals appear to have had stronger humeri relative to their body size than early modern humans from Skhul and Qafzeh, as well as, to a lesser degree, Early-Middle Upper Paleolithic modern humans from Europe (Gibbons, 1996; Trinkaus et al., 1998; Churchill and Formicola, 1997), the implication of the present finding is that Neanderthals are not particularly unusual relative to many modern groups with respect to the strength of their arms (once adjusted for body size); rather, it is the early moderns who stand out as unusual, as Trinkaus and Churchill (1999) have noted previously.

The other aspect of Neanderthal humeral morphology that can be assessed within a broader context is the "flatness" (platybrachia) of the humeral shaft. Neanderthals

Table 2. Humerus – Indices of Residual Robusticity and Midshaft Shape

	Residual Robusticity		*Midshaft shape index*	
	Males	*Females*	*Males*	*Females*
Group	Mean ± 1SD (n)	Mean ± 1SD (n)	Mean ± 1SD (n)	Mean ± 1SD (n)
Würm European Neanderthals	8.72 ± 1.04 (3)	6.93—(1)	139.1 ± 6.3 (5)	138.8 ± 7.7 (2)
Middle Eastern Neanderthals	12.34—(1)	7.90—(1)	136.6 ± 5.3 (3)	130.2—(1)
Neanderthals (pooled)	9.63 ± 2.00 (4)	7.42 ± 0.68 (2)	138.1 ± 5.7 (8)	136.0 ± 7.4 (3)
Skhul-Qafzeh	8.46 ± 0.03 (2)	10.47—(1)	117.1 ± 7.5 (3)	125.2 ± 0.9 (3)
Gravettian	8.24 ± 1.07 (10)	7.22 ± 0.16 (2)	126.0 ± 8.3 (11)	128.3 ± 10.7 (3)
Magdalenian	8.13 ± 1.41 (5)	7.97 ± 0.15 (2)	129.2 ± 10.7 (5)	125.6 ± 13.6 (2)
Epigravettian	8.57 ± 1.02 (6)	—	136.4 ± 7.4 (6)	—
Inuit – Riverine	9.23 ± 1.58 (9)	7.63 ± 0.84 (10)	140.6 ± 11.5 (6)	135.9 ± 8.7 (10)
Inuit – Coastal	9.61 ± 1.28 (30)	7.76 ± 0.86 (14)	133.2 ± 11.8 (32)	137.5 ± 8.5 (14)
Tierra del Fuego	10.28 ± 1.14 (10)	8.58 ± 2.16 (8)	134.0 ± 9.8 (16)	140.5 ± 6.9 (9)
Chinese	8.45 ± 1.26 (24)	—	134.2 ± 10.1 (25)	—
Mesolithic 1	9.83 ± 0.95 (2)	6.53—(1)	124.0 ± 10.7 (3)	138.0 ± 10.7 (3)
Mesolithic 2	10.19 ± 0.92 (3)	6.74 ± 1.02 (2)	138.4 ± 10.2 (4)	127.7 ± 6.4 (3)
Mesolithic 3	7.96 ± 0.40 (2)	—	127.3 ± 2.5 (2)	—
Sami	9.61 ± 1.24 (29)	8.01 ± 0.98 (23)	130.8 ± 11.6 (29)	132.8 ± 11.7 (24)
Euro-Americans	9.43 ± 1.05 (24)	7.55 ± 1.49 (25)	124.5 ± 6.7 (24)	131.6 ± 10.2 (25)
Australians	9.92 ± 1.01 (15)	7.06 ± 1.56 (5)	129.3 ± 8.3 (15)	141.9 ± 12.3 (5)
Jebel Sahaba	9.02—(1)	7.53 ± 0.31 (2)	117.3 ± 11.5 (12)	127.2 ± 7.4 (9)
Khoisan	7.19 ± 0.98 (18)	6.00 ± 0.77 (14)	128.5 ± 9.1 (19)	127.2 ± 10.4 (14)
African Americans	9.73 ± 1.32 (40)	8.43 ± 1.12 (30)	123.0 ± 9.2 (40)	126.4 ± 7.2 (30)
Zulu	9.83 ± 1.04 (31)	8.10 ± 1.20 (31)	121.9 ± 8.4 (31)	127.9 ± 9.9 (31)

have often been considered to be unusual in this regard, but are within the range of morphologies present in human populations (Trinkaus, 1983a; Hambücken, 1993). Our results for the comparative populations serve to further de-emphasize the distinctiveness of Neanderthal platybrachia. Many recent populations, especially populations of hunter-gatherers, show similar levels of flattening of the humeral shaft. What may be of greater interest is the observation that the populations listed in Table 2 differ greatly in the degree of sexual dimorphism between the average male and female cross-sectional shape. Neanderthals, Gravettians, Khoisan, and a variety of other populations exhibit almost no sexual dimorphism in this feature, while other populations, including the continent-wide sample of Australian Aboriginal and terminal Pleistocene people from Jebel Sahaba in northern Sudan, have marked dimorphism. If humeral shape provides a record of the direction that bending forces have been applied to a long bone, the upshot may be that Neanderthal males and females were either doing similar things or perhaps different activities that nevertheless produced similar kinds of mechanical responses in the humerus.

The correlation between index of residual strength and index of humeral midshaft shape in the pooled sample is $r = -0.17$ ($p = 0.0004$; $n = 425$). Thus, there is a weak tendency for the humeri that have the highest residual strength to also have more rounded cross-sections.

LOWER LIMB

Tables 3 and 4 present the results for the lower limb. With respect to the femur, researchers have long noted that Neanderthals tend to have a nearly round femoral midshaft with relatively little development of a femoral pilaster (Boule, 1913; McCown and Keith, 1939; Trinkaus, 1976), although the Amud 1 Neanderthal male from Israel departs from this pattern (Endo and Kimura, 1970). Analyses of the cross-sectional geometry of the femur have tended to conclude that modern humans who have elevated ratios of $I_{max/min}$ (or $I_{xx/yy}$) were more mobile or traveled longer distances than those who have more rounded midshaft cross-sections (Ruff et al., 1984; Ruff, 1992, 2000a). This interpretation is somewhat problematic with respect to the Neanderthals, who have been frequently portrayed as highly active, but perhaps engaging in patterns of movement that did not allow them to travel in a single direction for extended periods of time (comment by Trinkaus in Vandermeersch and Marks, 1992). The elevated $I_{xx/yy}$ index of the femur of St. Césaire has been interpreted as a behaviorally-driven convergence upon the greater development of a femoral pilaster present in many Upper Paleolithic anatomically modern humans, implying that St. Césaire may have had a pattern of habitual walking or running activity that was more similar to those of Upper Paleolithic moderns than earlier Mousterian Neanderthals, and thus in keeping with technological aspects of the Châtelperronian (Trinkaus et al., 1999).

These are persuasive and logically consistent interpretations, but we should not forget that we do not really know to what extent habitual activity alters cross-sectional geometry in living people (Jurmain, 1999), and only a limited number of studies have tried to investigate the question (for a review, see Pearson and Lieberman, 2004). One solution is to restrict the analysis to determining which recent populations resemble fossil groups in their residual strength or shape, then noting what factor or factors those recent populations share with the fossil groups. It is also quite possible that there are many ways to develop similar cross-sectional shapes, and until we have a better grasp of what factors cause those shapes to develop, we should perhaps err on the side of caution in not advancing interpretations beyond what we actually know.

With respect to residual femoral strength, the Neanderthals have impressively high ratios, especially those from the Middle East (i.e., predominantly individuals from Shanidar). However, the pooled sample of male Neanderthals is matched by some of the modern samples including, most prominently, Epigravettian and Gravettian males. Skhul IV and the single male from the Danish Mesolithic upon whom the index can be calculated are also quite similar to the Neanderthal male mean. The single female Neanderthal (La Ferrassie 2) also has a high ratio of residual femoral strength and is most closely approximated by the Gravettian females. The single Skhul-Qafzeh female, Qafzeh 9, has an extremely high ratio, substantially higher than Skhul IV and equal to or higher than most Neanderthal males.

The results for the pilastric index show the familiar pattern of a low index for Neanderthals and highly developed pilasters in the Skhul-Qafzeh early moderns (McCown and Keith, 1939; Endo and Kimura, 1970; Vandermeersch, 1981; Trinkaus, 1983a; Trinkaus and Ruff, 1999a). of more interest is the fact that among the many hunter-gatherer populations listed in Table 3, only the Khoisan (pastoralists and hunter-gatherers) match the Skhul-Qafzeh people in the development of the femoral pilaster. Neanderthal males do not differ significantly from many populations in the index, which is partially a function of the substantial amount of variability in the index among individuals in each population. Neanderthal females have a lower mean for pilastric index than the males, as do most of the more recent females. As was the case for

Table 3. Femur – Residual Robusticity and Pilastric Index

	Residual Robusticity		*Pilastric index*	
	Males	*Females*	*Males*	*Females*
Group	*Mean ± 1SD (n)*	*Mean ± 1SD (n)*	*Mean ± 1SD (n)*	*Mean ± 1SD (n)*
Würm European Neanderthals	17.4 ± 1.3 (4)	18.2—(1)	106.0 ± 11.4 (5)	96.6–(1)
Middle Eastern Neanderthals	21.2 ± 2.1 (3)	—	115.0 ± 6.6 (3)	96.4 ± 13.6 (2)
Neanderthals (pooled)	19.0 ± 2.5 (7)	18.4—(1)	109.3 ± 10.4 (8)	96.4 ± 9.6 (3)
Skhul-Qafzeh	18.5—(1)	24.8—(1)	129.5 ± 10.5 (4)	116.2 ± 19.8 (2)
Gravettian	18.7 ± 4.0 (10)	17.0 ± 2.2 (3)	115.0 ± 12.8 (13)	100.3 ± 6.9 (3)
Magdalenian	18.4 ± 1.9 (5)	—	115.9 ± 12.5 (5)	100.8—(1)
Epigravettian	19.1 ± 2.2 (5)	—	116.7 ± 3.7 (5)	—
Inuit – Riverine	16.1 ± 1.7 (11)	15.6 ± 1.5 (10)	113.3 ± 7.9 (11)	103.4 ± 6.2 (10)
Inuit – Coastal	17.9 ± 2.0 (36)	15.6 ± 1.7 (14)	114.5 ± 9.8 (36)	110.8 ± 6.2 (14)
Tierra del Fuego	17.6 ± 1.6 (14)	16.3 ± 1.3 (7)	111.8 ± 9.1 (15)	102.8 ± 5.2 (7)
Chinese	15.6 ± 2.0 (25)	—	109.2 ± 9.9 (25)	—
Mesolithic 1	18.9—(1)	15.6 ± 2.0 (2)	117.2 ± 23.4 (2)	103.6 ± 6.7 (2)
Mesolithic 2	16.7 ± 2.1 (4)	15.6 ± 1.3 (4)	103.9 ± 9.2 (4)	101.9 ± 9.2 (4)
Mesolithic 3	18.3—(1)	—	115.0 ± 10.5 (2)	—
Sami	15.3 ± 1.9 (29)	14.2 ± 1.9 (24)	104.8 ± 8.0 (29)	101.9 ± 7.2 (24)
Euro-Americans	16.9 ± 1.5 (24)	16.1 ± 1.8 (25)	107.2 ± 11.5 (24)	109.0 ± 8.4 (25)
Australians	16.5 ± 2.2 (15)	14.6 ± 2.4 (5)	116.1 ± 10.4 (15)	112.3 ± 10.0 (5)
Jebel Sahaba	17.9 ± 1.1 (6)	16.7 ± 1.8 (3)	116.9 ± 8.4 (12)	109.1 ± 8.1 (8)
Khoisan	16.7 ± 1.5 (14)	14.9 ± 1.5 (12)	127.7 ± 9.8 (18)	114.5 ± 6.7 (13)
African Americans	17.6 ± 1.9 (40)	16.5 ± 1.5 (30)	105.4 ± 13.9 (40)	111.8 ± 11.6 (30)
Zulu	17.5 ± 1.8 (31)	16.6 ± 1.7 (31)	111.0 ± 8.3 (31)	113.6 ± 10.9 (31)

the males, Neanderthal females do not differ significantly from a variety of recent human populations in their pilastric index, but are most similar to Gravettian, Tévic-Hoëdic, Sami, and riverine Inuit females. For the femoral indices, a final point of interest comes from the fact that the correlation between index of residual strength of the femur and the pilastric index in the pooled sample is $r = 0.19$ ($p < 0.0001$; $n = 437$).

Few Neanderthal tibiae are complete enough to provide data on the residual strength and mishaft cnemic index (Table 4). However, the few male tibiae that are sufficiently well preserved produce ratios of residual robusticity that are less elevated with respect to recent groups than was the case for the femur. The pooled group of female Neanderthals has a more elevated mean for residual tibial robusticity, but is closely approximated by Gravettian females and, somewhat less closely, by Zulu females. With respect to the midshaft cnemic index, Neanderthal males show a degree of platycnemia that is commonly found in more recent hunter gatherers and, as is also the case in modern hunter-gatherers, the index is sexually dimorphic (Ruff, 1987). Interestingly, the correlation between index of residual strength of the tibia and the midshaft cnemic index in the pooled sample is non-significant ($r = 0.06$; $p = 0.23$; $n = 408$). Unlike the case in the humerus and femur, the two indices in the tibia are not even weakly associated.

With respect to the analyses of the indices of the lower limb, it is interesting to note that in the pooled sample of fossil and recent individuals used in these analyses, the pilastric

Table 4. Tibia – Residual Robusticity and Midshaft Cnemic Index

	Residual Robusticity		*Midshaft Cnemic Index*	
	Males	*Females*	*Males*	*Females*
Group	*Mean ± 1SD (n)*	*Mean ± 1SD (n)*	*Mean ± 1SD (n)*	*Mean ± 1SD (n)*
European Würm Neanderthals	9.5 ± 0.5 (2)	7.9—(1)	66.5 ± 4.2 (4)	74.5—(1)
Middle Eastern Neanderthals	—	9.7—(1)	78.8—(1)	75.2—(1)
Neanderthals (pooled)	9.5 ± 0.5 (2)	8.8 ± 1.2 (2)	68.9 ± 6.6 (5)	74.8 ± 0.5
Skhul-Qafzeh	10.1—(1)	—	78.8 ± 2.3 (2)	—
Gravettian	10.5 ± 2.5 (7)	8.3 ± 1.1 (3)	64.1 ± 7.9 (11)	63.6 ± 1.0 (3)
Magdalenian	10.2 ± 1.0 (4)	—	63.7 ± 4.7 (4)	—
Epigravettian	10.2 ± 0.9 (4)	—	59.6 ± 4.0 (4)	—
Inuit – Riverine	8.2 ± 0.8 (11)	7.2 ± 0.8 (10)	64.4 ± 4.0 (11)	69.2 ± 5.8 (10)
Inuit – Coastal	8.6 ± 0.8 (36)	7.2 ± 0.7 (14)	67.0 ± 6.5 (36)	67.4 ± 4.7 (14)
Tierra del Fuego	9.2 ± 0.9 (14)	8.1 ± 1.1 (8)	71.4 ± 3.9 (15)	68.3 ± 5.8 (9)
Chinese	8.0 ± 0.8 (23)	—	73.8 ± 4.9 (24)	—
Mesolithic 1	10.2—(1)	7.6 ± 1.0 (2)	64.0 ± 1.0 (2)	64.4 ± 6.9 (2)
Mesolithic 2	10.3—(1)	7.1—(1)	69.6 ± 5.6 (4)	61.4 ± 7.3 (3)
Mesolithic 3	9.1—(1)	—	72.7—(1)	—
Sami	7.7 ± 0.9 (24)	7.1 ± 0.73 (22)	74.8 ± 6.8 (28)	75.4 ± 6.4 (23)
Euro-Americans	8.9 ± 1.0 (23)	7.5 ± 0.9 (25)	72.8 ± 4.8 (23)	75.9 ± 6.5 (25)
Australians	8.9 ± 1.3 (15)	8.1 ± 1.3 (5)	69.8 ± 5.5 (15)	73.5 ± 2.9 (5)
Jebel Sahaba	—	—	69.0 ± 7.1 (11)	73.3 ± 3.9 (9)
Khoisan	8.6 ± 1.1 (16)	6.9 ± 0.8 (10)	64.0 ± 5.1 (18)	69.6 ± 3.8 (11)
African Americans	9.0 ± 1.0 (40)	8.1 ± 0.8 (30)	74.2 ± 6.6 (40)	75.5 ± 8.1 (30)
Zulu	9.5 ± 1.3 (31)	8.4 ± 1.2 (31)	72.1 ± 6.8 (31)	73.2 ± 6.7 (31)

and cnemic index are, in fact, only weakly correlated ($r = -0.11$; $p = 0.02$; $n = 453$; Figure 2). Pearson and Grine (1996b) reported this for $I_{max/min}$ indices in the femur and tibia of the same individual. Stock (2004) argued that the cross-sectional properties of the tibia contain less inherent variability and are a better reflection of patterns of habitual activity than the cross-sectional properties of the femur. The pattern depicted here has been reported previously for external measurements of the femoral and tibial diaphyses (Pearson and Millones, 2003; Pearson and Cordero, 2004). Ruff and colleagues have argued that femoral cross-sectional geometry is useful for shedding light on differences in prehistoric activity patterns (e.g., Ruff et al., 1984; Ruff, 1992, 2000b) and have described cases in which variation in femoral and tibial cross-sectional geometry show concordant patterns (Trinkaus and Ruff, 1999a, b). However, Ruff et al. (2005) now argue that variation in tibia cross-sectional geometry is much more informative about "habitual activities" and less correlated with variation in body shape (bi-iliac breadth and lower limb length) than is the case for femoral midshaft shape.

If body shape affects femoral midshaft shape more than tibial midshaft shape (Ruff et al., 2005), it is possible that lumping males and females from all of the populations used in the analysis has obscured the true relationship between these diaphyseal shapes. We performed two additional sets of analyses to test this possibility. First, we subdivided the pooled recent human sample used in Figure 2 by population and sex, then re-ran the correlation in each of the resultant groups. This subdivision should control for average differences in body shape between groups and sexes and allow the relationship between the indices to be expressed more clearly. The resulting 19 correlations for groups with sample sizes of $n = 4$ or larger (average $n = 20.3 \pm 10.7$) produced an average correlation between the indices of $r = 0.030 \pm 0.257$, with a maximum of $r = 0.618$ and a minimum of $r = -0.514$ (both on small samples of $n = 4$ and $n = 8$, respectively). None of the correlations were different than $r = 0$ at the $p < 0.05$ level of significance.

Second, although subdivision by population and sex should remove average between-group differences in body shape, there is still substantial variation in body shape between individuals in a single sex in each population (Pearson et al., 2001). It is possible that this individual variation has obscured the underlying pattern. To test this possibility, we examined the relationships between pilastric index, midshaft cnemic index, and the ratio of bi-iliac breadth to maximum femoral length (as a proxy for body shape) in a sample of $n = 18$ Inuit males and $n = 34$ African American males for whom we had measurements of all of these variables. In the pooled sample of $n = 52$ males, the correlations between the ratio of bi-iliac breadth to femoral length and the pilastric and midshaft cnemic index were

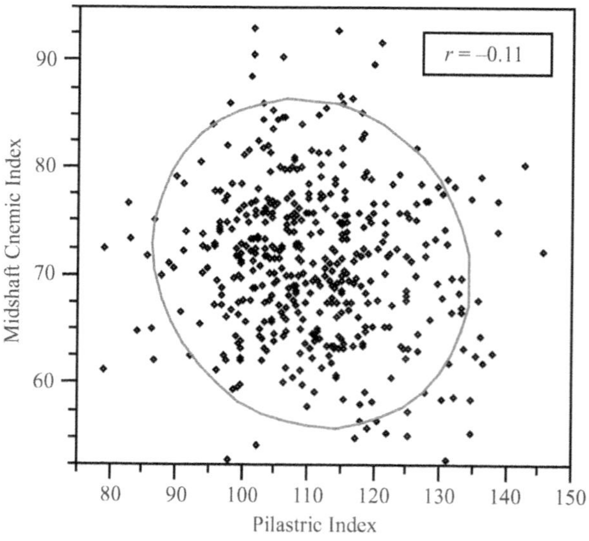

Figure 2. Plot of Pilastric index versus Cnemic index in the sample of recent humans. The ellipse depicts the 90% confidence envelope ($r = -0.11$; $p = 0.0201$; $n = 405$).

$r = 0.180$, $p = 0.201$ and $r = -0.174$, $p = 0.218$, respectively. The partial correlation for the pooled sample between the pilastric index and the midshft cnemic index with the ratio of bi-iliac breadth to maximum femoral length held constant was $r = -0.162$, which was similar to the raw pairwise correlation between the two diaphyseal shape indices of $r = -0.189$. To test for the possibility that controlling for bi-iliac breadth alone or both bi-iliac breadth and maximum femoral length (rather than their ratio) might reveal an underlying stronger correlation between midshaft femoral and tibial shapes, we also ran partial correlations with these variables held constant and found nearly identical results. These analyses should suffice to show that variation in bi-iliac breadth, femur length, or the ratio of bi-iliac breadth to femoral length cannot account for the poor correlations between midshaft femoral and tibial shapes, at least with respect to these bones' external dimensions. Perhaps we do not yet know what determines the development of femoral pilasters or platycnemic tibiae but because they are statistically nearly independent, they probably develop in response to different forces or influences.

Discussion

Our results and conclusions are necessarily based on a fairly small set of variables that comprises only some of the features that have been noted as distinctive in Neanderthal limbs (Boule, 1913; McCown and Keith, 1939; Endo and Kimura, 1970; Heim, 1982; Trinkaus, 1983a; Hambücken, 1993; Churchill, 1996). Many of these features could probably be addressed within a similar comparative framework, and some already have been (Churchill, 1994; 1996; Pearson, 1997; Holt, 1998; Trinkaus, 2000). Given that many of the features under consideration vary with body size, one set of features that could profitably be revisited would be the muscle markings and longer lever arms available to specific muscles in Neanderthal limbs noted by Trinkaus (1983a). Trinkaus (2000) has already analyzed some of these traits to see how they covary with size, but more work is needed. Musculo-skeletal stress markers (MSMs) have received a great deal of attention over the last decade (e.g., Hawkey and Merbs, 1995; Wilczak, 1998; Robb, 1998; Zumwalt et al., 2000; Weiss, 2003, 2004; Zumwalt, 2004). The etiology of observed marks is still unclear and, in most cases, still uncertainly related to specific activities or frequencies, duration, or intensity of those activities (Jurmain, 1999).

Even though the structural correlates of muscle markings have been poorly explored, there have been some interesting analyses that have pointed the way to what could be possible (e.g., Churchill, 1996; Bridges, 1997). With regard to the functional anatomy of Neanderthals, we offer one final analysis. Trinkaus (1983a) observed that Neanderthals tend to have wide insertions for their pectoralis major muscle, the major protractor of the arm. Pearson (1997) measured the breadth of the pectoralis major insertion in most of the samples used in the analyses in this paper and replicated Trinkaus' observation. However, it seemed that the ratio of pectoralis major insertion breadth to maximum humerus length was higher on average in males than in females, and in some populations (especially in Neanderthal, Mesolithic, Sami, riverine Inuit, and European American males) than others. Table 5 presents analyses of the correlations and partial correlations between pectoralis major insertion breadth, humeral torsional strength (estimated via the regression formula above), maximum humeral length, and femoral head diameter as a proxy for body mass in the pooled sample of recent and fossil humans used in these analyses.

The correlation between estimated humeral torsional strength and the breadth of the pectoralis major insertion is $r = 0.68$, while the

Table 5 Humerus – Correlations and Partial Correlations

Variable	*H6*	*J**	*F1*	*H1*
H6	—	0.68	0.58	0.30
J*	0.53	—	0.71	0.62
F1	0.30	0.25	—	0.64
H1	−0.31	0.41	0.43	—

Correlations (above the diagonal) and partial correlations with the other variables held constant (below the diagonal) among pectoralis major insertion width (H6), estimated humeral torsional strength (J*), superior-inferior femoral head diameter (F1, a proxy for body mass), and maximum humerus length (H1). Based on n 5 364 individuals from the pooled fossil and recent sample.

partial correlation after controlling for femoral and humeral head size is $r = 0.53$. This suggests that there is indeed a relationship between this muscle's insertion size and humeral torsional strength even after the other effects of size have been removed. This is a promising result for those who wish to use muscle markings to make inferences about activity. However, our analysis made no attempt to control for the effect of age nor did we subdivide our sample by sex. When the sexes are analyzed separately, the partial correlation between pectoralis major insertion size and estimated torsional second moment of area remains high for males ($r = 0.52$), but decreases in the female sample ($r = 0.38$).

In addition to new studies of muscle markings or the inter-relationships among a large variety of skeletal features, more work is needed on broad-ranging comparative analyses of the effects of specific, archaeologically visible factor in subsistence technology, terrain, or mobility (which is usually hypothesized rather than documented empirically). Ruff's (1987) analysis of tibial platycnemia and Collier's (1989, 1993) and Churchill's (1994, 1996) investigations of a series of postcranial features in populations that differed in lifestyle provide good models of what can be done to describe patterns of association between lifestyles and physical features. In a similar vein, Pearson and Cordero (2004) analyzed the effects of varying forms of subsistence practices, mobility, and the kind of terrain inhabited upon long bone residual robusticity and diaphyseal midshaft shape in most of the samples used in the present study (summarized in Table 1). In these data, variation in topography had a more noticeable effect on males than on females. Both femoral residual strength and tibial residual strength increased in more rugged terrain, replicating Ruff's (2000b) observation based on cross-sectional geometry. This increased strength was most notable in rugged terrain populations. Tibial midshaft ML/AP decreased in both males and females as terrain ruggedness increased. Female pilastric index decreased substantially as terrain increased from moderate to rugged terrain, while little change occurred between flat and moderate terrain. Males exhibited a slight increase from flat to moderate terrain, but no increase from moderate to rugged terrain. As one might suppose, male humeral residual strength increased with the presence of seafaring (rowing or paddling). Curiously, we found that tibial residual strength also increased slightly with the presence of seafaring, although this may be a statistical artifact from the limited number of populations and combinations of lifestyle factors that we were able to observe. Female humeral residual strength decreased with the presence of seafaring. It should be noted, however, that while statistically significant, most of the overall effects of the patterns found by Pearson and Cordero (2004) were weak in magnitude, generally accounting for ~10% of the observed variance in the indices under consideration.

Conclusions

Our results for the upper limb de-emphasize the distinctiveness of Neanderthals and, at least with respect to the indices we considered, situate them among a variety of recent populations. Male Neanderthals' combination of residual robusticity and midshaft shape for the humerus bear the closest resemblance to intensive foragers, such as the Epigravettian males from Italy. The sample of female Neanderthal humeri is more limited, but they too are quite similar to both groups of Inuit females, as well as to a single well-preserved Danish Mesolithic female and a single Gravettian female. In sum, both Neanderthal males' and females' humeri tend to resemble those of same-sex intensive foragers. While this does not prove that Neanderthals had the same behaviors or in the same frequencies as individuals in those groups, at a minimum, it is suggestive that something about their overall pattern of mechanical loading was similar.

The analyses of the indices of the lower limb produced similar results. Once again Neanderthal males are the most similar in terms of the combination of residual robusticity and shape to a broad spectrum of foraging peoples. Within this context, only the relatively small development of the femoral pilaster in Neanderthal males stands out as distinct. Like the males, female Neanderthals also tend to resemble a variety of intensive foraging populations most closely in their femoral indices. Due to smaller sample sizes for Neanderthals, less can be said about the relationship of their tibial indices to recent populations. However, the data indicate that Neanderthal tibiae are, if anything, even less distinctive relative to a variety of recent populations than their humeri.

The results provide no support for the hypothesis that Neanderthals "lost an evolutionary 'arms' race" to the Skhul-Qafzeh early modern humans in Israel (Gibbons, 1996), much less to the Early-Middle Paleolithic moderns in Europe. At most, the data presented here reinforce Trinkaus and Churchill's (1999) conclusion that it was the pattern of arm strength and morphology in the Skhul-Qafzeh humans that was distinctive, not that of Neanderthals.

Perhaps the findings for both the humerus and lower limb bones are most intelligible within a framework of population density and subsistence intensification. The major contrast in upper limb morphology noted between Near Eastern Neanderthal and Skhul-Qafzeh males may have to do more with the existence of a more intensive form of subsistence among the Neanderthals as proposed by Lieberman and Shea (Lieberman and Shea, 1994; Lieberman, 1993, 1998; Shea, 1998). Neanderthals were suggested to occupy their sites multi-seasonally, whereas modern human occupations tended to be for single seasons. Additionally, Shea (1998) has noted that many Neanderthal lithic assemblages contain a higher percentage of points, possibly implying a greater reliance upon intensive hunting.

Likewise, the results provide no substantiation (as also argued by Trinkaus and Churchill [1999] for the humerus) for a hypothesis that there was a universal decrease in residual robusticity from Neanderthals to the present as proposed by Ruff et al. (1993) with respect to the femur. Rather, compared to Neanderthals and many horticulturalists, some industrialized populations, and later intensive foragers, the Skhul-Qafzeh early modern humans and Early Upper Paleolithic (Gravettian) modern humans appear to have had unusually low values for residual humeral robusticity. Temporal declines in femoral and tibial residual robusticity also appear to be much less uniform than (most likely) was thought a decade ago. Finally, we should not lose sight of the fact that there are still substantial limitations to what we can interpret confidently with respect to prehistoric patterns of habitual activity, and we are still confronted by the nagging problem of equifinality – the possibility, or even likely probability, that multiple and diverse activities can produce similar types of loads and similar patterns of bone modeling.

Acknowledgements

Funding to collect the data for this research was provided by the Wenner Gren Foundation, the Boise Fund, a National Science Foundation Graduate Fellowship to O.M.P., J. C. Pearson, and the University of New Mexico. We are grateful to the numerous museum curators who facilitated access to the recent and fossil specimens under their care and to the heirs of Max Lohest who donated the Spy Neanderthal remains to science. We thank Chris Ruff, Terry Harrison, Katerina Harvati, Heather Edgar and an anonymous reviewer for useful comments on the manuscript.

References

Arsuaga, J.-L., Lorenzo, C., Carretero, J.-M., Gracia, A., Martínez, I., García, N., Bermúdez de Castro, J.-M., Carbonell, E., 1999. A complete human pelvis from the Middle Pleistocene of Spain. *Nature* 399, 255–258.

Binford, L., 1984. *Faunal Remains at Klasies River Mouth*. Academic Press, Orlando, FL.

Binford, L.R., 1985. Human ancestors: changing views of their behavior. *J. Anthropol. Arch*. 4, 292–327.

Binford, L.R., 1989. Isolating the transition to cultural adaptations: an organizational approach. In: Trinkaus, E. (Ed.), *The Emergence of Modern Humans*. Cambridge University Press, Cambridge, pp. 18–41.

Bocherens, H., Billiou, D., Mariotti, A., Toussaint, M., Patou-Mathis, M., Bonjean, D., Otte, M., 2001. New isotope evidence for dietary habits of Neandertals from Belgium. *J. Hum. Evol*. 40, 497–505.

Boule, M., 1913. L'homme fossil de la Chapelle-aux-Saints. *Ann. Paléont*. 8, 1–67.

Bridges, P.S., 1989. Changes in activities with the shift to agriculture in the Southeastern United States. *Curr. Anthropol*. 30, 385–394.

Bridges, P.S., 1997. The relationship between muscle markings and diaphyseal strength in prehistoric remains from West-Central Illinois. *Am. J. Phys. Anthropol. Suppl*. 4, 82.

Chase, P.G., 1986. *The Hunters of Combe Grenal: Approaches to Middle Paleolithic Subsistence in Europe*. British Archaeological Reports (International Series 286), Oxford.

Churchill, S.E., 1994. Human upper body evolution in the Eurasian Later Pleistocene. Ph.D. Dissertation, The University of New Mexico, Albuquerque, NM.

Churchill, S.E., 1996. Particulate versus integrated evolution in the upper body in Late Pleistocene humans: a test of two models. *Am. J. Phys. Anthropol*. 100, 559–583.

Churchill, S. E., Formicola, V., 1997. A case of marked bilateral asymmetry in the upper limbs of an Upper Palaeolithic male from Barma Grande (Liguria), Italy. *Int. J. Osteoarchaeol*. 7, 18–38.

Churchill, S.E., Pearson, O.M., Grine, F.E., Trinkaus, E., Holliday, T.W., 1996a. Morphological affinities of the proximal ulna from Klasies River Mouth Main Site: archaic or modern? *J. Hum. Evol*. 31, 213–237.

Churchill, S.E., Weaver, A.H., Niewoehner, W.A., 1996b. Late Pleistocene human technological and subsistence behavior: Functional interpretations of upper limb morphology. *Quaternaria Nova* 6, 413–447.

Clark, G.A., 2002. Neandertal archaeology – implications for our origins. *Am. Anthropol*. 104, 50–67.

Clark, G.A., Lindley, J.M., 1989. The case of continuity: observations on the biocultural transition in Europe and Western Asia. In: Mellars, P., Stringer, C. (Eds.), *The Human Revolution: Behavioral and Biological Perspectives on the Origins of Modern Humans*. Princeton University Press, Princeton, pp. 626–676.

Collier, S., 1989. The influence of economic behaviour and environment upon robusticity of the post-cranial skeleton: a comparison of Australian Aborigines and other populations. *Archaeol. Oceania* 24, 17–30.

Collier, S., 1993. Sexual dimorphism in relation to big-game hunting and economy in modern human populations. *Am. J. Phys. Anthropol*. 91, 485–504.

Debets, G.F., 1967. Skelet pozdnepaleoliticheskogo cheloveka iz pogrebeniia na Sungirskoi stoianke. *Sovetskaia Arkheologiia* 3, 160–164.

Demes, B., Stern, J.T. Jr., Hausman, M.R., Larson, S.G., McLeod, K.J., Rubin, C.T., 1998. Patterns of strain in the macaque ulna during functional activity. *Am. J. Phys. Anthropol.* 106, 87–100.

Demes, B., Qin, Y.-X., Stern, J.T. Jr., Larson, S.G., Rubin, C.T., 2001. Patterns of strain in the macaque tibia during functional activity. *Am. J. Phys. Anthropol.* 116, 257–265.

Drucker, D., Bocherens, H., 2004. Carbon and nitrogen stable isotopes as tracers of change in diet breadth during Middle and Upper Palaeolithic in Europe. *Int. J. Osteoarchaeol.* 14, 162–177.

Endo, B., 1971. Some characteristics of the deltoid tuberosity of the humerus in West-Asian and European "Classic" Neanderthals. *J. Anthropol. Soc. Jpn.* 79, 249–258.

Endo, B., Kimura, T., 1970. Postcranial skeleton of the Amud man. In: Suzuki, H., Takai, F. (Eds.), *The Amud Man and His Cave Site*. Academic Press of Japan, Tokyo, pp. 231–406.

Fizet, M., Mariotti, A., Bocherens, H., Lange-Badré, B., Vandermeersch, B., Borel, J.P., Bellon, G., 1995. Effect of diet, physiology and climate on carbon and nitrogen stable isotopes of collagen in a late Pleistocene anthropic palaeoecosystem: Marillac, Charente, France. *J. Archaeol. Sci.* 22, 67–79.

Gaudzinski, S., 1999. The faunal record of the Lower and Middle Palaeolithic of Europe: remarks on human interference. In: Roebroeks, W., Gamble, C. (Eds.), *The Middle Palaeolithic occupation of Europe*. University of Leiden, Leiden, pp. 215–233.

Gaudzinski, S., Roebroeks, W., 2000. Adults only. Reindeer hunting at the Middle Paleolithic site Salzgitter Lebenstedt, northern Germany. *J. Hum. Evol.* 38, 497–521.

Gibbons, A., 1996. Did Neandertals lose an evolutionary "arms" race? *Science 272*, 1586–1587.

Grayson, D.K., Delpech, F., 2002. Specialized early Upper Paleolithic hunters in Southwestern France? *J. Archaeol. Sci.* 29, 1439–1449.

Grine, F.E., Jungers, W.L., Tobias, P.V., Pearson, O.M., 1995. Fossil *Homo* femur from Berg Aukas, northern Namibia. *Am. J. Phys. Anthropol.* 97, 151–185.

Gross, T.S., McLeod, K J., Rubin, C.T., 1992. Characterizing bone strain distributions in vivo using three triple rosette strain gages. *J. Biomech.* 25, 1081–1087.

Hambücken, A., 1993. Révision des particularités de l'humérus des Néandertaliens européens. *C. R. Acad. Sci. Paris sér.* II 317, 109–114.

Hawkey, D.E., Merbs, C.F., 1995. Activity-induced musculoskeletal stress markers (MSM) and subsistence strategy changes among ancient Hudson Bay Eskimos. *Int. J. Osteoarchaeol.* 5, 324–338.

Heim, J.-L., 1982. Les hommes fossiles de La Ferrassie. II. Les squelettes d'adultes: squelettes des membres. *Arch. Inst. Paléont. Hum.* 38, 1–272.

Helm, J., 1993. "Always with them either a feast or a famine": living off the land with Chipewyan Indians, 1791—1792. *Arctic Anthrop.* 30, 46–60.

Holt, B.M., 1998. Biomechanical Evidence of Decreased Mobility in Upper Paleolithic and Mesolithic Europe. Ph.D. Dissertation, University of Missouri, Columbia, MO.

Hultkrantz, J.V., 1907. *Zur Osteologie der Ona- und Yahgan-Indianer des Feuerlandes: Wissenschaftliche Ergebnisse der schwedischen Expedition nach den Magellensländern 1895–97 unter Leitung von Dr. Otto Nordenskjöld*, vol. 1. Geologie, Geographie und Anthropologie, No. 5. Norstedt & Söner, Stockholm, pp. 109–173.

Hyades, P., Deniker, J., 1891. *Anthropologie et Ethnographie: Mission Scientifique du Cap Horn (1882–1883)*, Vol. 7. Gauthier-Villars et Fils, Paris.

Jurmain, R., 1999. *Stories from the Skeleton: Behavioral Reconstruction in Human Osteology*. Gordon and Breach, Amsterdam.

Kappelman, J., 1996. The evolution of body mass and relative brain size in fossil hominids. *J. Hum. Evol.* 30, 243–276.

Khrisanfova, E.N., 1980. Skelet verkhnepaleolitich-eskogo cheloveka iz Sungirya. *Vopros'i Antropologii* 64, 40–68.

Khrisanfova, E.N., 1984. *Postkranial'n'ii skelet vzrosl-ogo muzhchin'i Sungir' 1. Bedrennaya kost' Sungir' 4: Sungir' Anthropologicheskoe Issledovanie*. Izdatel'stvo "Nauka", Moscow, pp. 100–140.

Klein, R.G., 1995. Anatomy, behavior, and modern human origins. *J. World Prehist.* 9, 167–198.

Klein, R.G., 1999. *The Human Career*. 2nd Edition. University of Chicago Press, Chicago.

Klein, R.G., 2003. Whither the Neanderthals? *Science* 299, 1525–1527.

Lieberman, D.E., 1993. The rise and fall of seasonal mobility among hunter-gatherers: the case of the southern Levant. *Curr. Anthropol.* 34, 599–631.

Lieberman, D.E., 1998. Neandertal and early modern human mobility patterns: Comparing archaeological and anatomical evidence. In: Akazawa, T.,

Aoki, K., Bar-Yosef, O. (Eds.), *Neandertals and Modern Humans in Western Asia*. Plenum, New York, pp. 263–275.

Lieberman, D.E., Shea, J J., 1994. Behavioral differences between archaic and modern humans in the Levantine Mousterian. *Am. Anthropol.* 96, 300–332.

Lovejoy, C.O., Trinkaus, E., 1980. Strength and robusticity of the Neandertal tibia. *Am. J. Phys. Anthropol.* 53, 465–470.

Lieberman, D.E., Polk, J.D., Demes, B., 2004. Prediction long bone loading from cross-sectional geometry. *Am. J. Phys. Anthropol.* 123, 156–171.

Marean, C.W., 1998. A critique of the evidence for scavenging by Neandertals and early modern humans: new details from Kobeh Cave (Zagros Mountains, Iran) and Die Kelders Cave 1 Layer 10 (South Africa). *J. Hum. Evol.* 35, 111–136.

Marean, C.W., Kim, S.Y., 1998. Mousterian large-mammal remains from Kobeh Cave. *Curr. Anthropol.* 39 (Suppl.), S79-S113.

Martin, R., Saller, K., 1956. *Lehrbuch der Anthropologie*. Gustav Fischer, Stuttgart.

Matiegka, J., 1938. *Homo Predmostensis*: Fosilní Clovek z Předmostí na Morave II. Ostatní Cásti Kostrové. Nákaladem Ceské Akademie Ved e Umení, Prague.

McCown, T.D., Keith, A., 1939. *The Stone Age of Mount Carmel, II: The Fossil Human Remains from the Levalloiso-Mousterian*. Clarendon Press, Oxford.

Mellars, P., 1996. *The Neandertal Legacy: An Archaeological Perspective of Western Europe*. Princeton University Press, Princeton.

Mellars, P.A., 2004a. Reindeer specialization in the early Upper Palaeolithic: the evidence from South West France. *J. Archaeol. Sci.* 31, 613–617.

Mellars, P., 2004b. Neanderthals and the modern human colonization of Europe. *Nature* 432, 461–465.

Mulvaney, J., Kamminga, J., 1999. *Prehistory of Australia*. Smithsonian Institution Press, Washington, D.C..

Pardoe, C., 1994. Bioscapes: the evolutionary landscape of Australia. *Archaeol. Oceania* 29,182–190.

Pardoe, C., 1995. Riverine, biological and cultural evolution in southeastern Australia. *Antiquity* 69, 696–713.

Pauwels, F., 1980. *Biomechanics of the Locomotor Apparatus: Contributions on the Functional Anatomy of the Locomotor Apparatus*. Springer, Berlin.

Pearson, O.M., 1997. Postcranial Morphology and the Origin of Modern Humans. Ph.D. Dissertation, State University of New York at Stony Brook, NY.

Pearson, O.M., 2000. Activity, climate, and postcranial robusticity: Implications for modern human origins and scenarios of adaptive change. *Curr. Anthropol.* 41, 569–607.

Pearson, O.M., Cordero, R., 2004. World-wide variation in residual strength of the humerus, femur, and tibia. Am. *J. Phys. Anthropol.* Suppl. 38, 158.

Pearson, O.M., Grine, F. E., 1996a. Morphology of the Border Cave hominid ulna and humerus. *S. Afr. J. Sci.* 92, 231–236.

Pearson, O.M., Grine, F. E., 1996b. Cortical thickness and relative bending moments in human long bones: correlations among elements. *Am. J. Phys. Anthropol.* Suppl 22, 183.

Pearson O.M., Grine, F.E., 1997. Re-analysis of the hominid radii from Cave of Hearths and Klasies River Mouth, South Africa. *J. Hum. Evol.* 32, 577–592.

Pearson, O.M., Lieberman, D.E., 2004. The aging of Wolff's "Law:" Ontogeny and responses to mechanical loading in cortical bone. *Yearb. Phys. Anthropol.* 47, 63–99.

Pearson, O.M., Millones, M., 2003. Postcranial reflections of climatic adaptation and habitual activity in Tierra del Fuego. *Am. J. Phys. Anthropol.* Suppl. 36, 166.

Pearson, O.M., Millones, M. 2005. Rasgos esqueléticos de adaptación al clima y a la actividad entre los habitantes aborígenes de Tierra del Fuego. *Magallania* 33, 37–50.

Pearson, O.M., Jungers, W., Grine, F., Mowbray, K., 2001. The reliability of estimates of hominin body mass derived from bi-iliac breadth and stature. *Am. J. Phys. Anthropol.* Suppl. 32, 118.

Richards, M.P., Pettitt, P.B., Trinkaus, E., Smith, F.H., Paunovic, M., Karavanic, I., 2000. Neanderthal diet at Vindija and Neanderthal predation: The evidence for stable isotopes. *Proc. Natl. Acad. Sci. U.S.A.* 97, 7663–7666.

Robb, J., 1998. The interpretation of skeletal muscle sites: A statistical approach. *Int. J. Osteoarchaeol.* 8, 363–377.

Rosenberg, K.R., Lü, Z., Ruff, C.B., 1999. Body size, body proportions and encephalization in the Jinniushan specimen. *Am. J. Phys. Anthropol. Suppl.* 28, 235.

Ruff, C., 1987. Sexual dimorphism in human lower limb bone structure: relationship to subsistence strategy and sexual division of labor. *J. Hum. Evol.* 16, 391–416.

Ruff, C.B., 1992. Biomechanical analysis of archaeological human material. In: Saunders, S. R., Katzenberg, M. A. (Eds.), *Skeletal Biology of Past People*. Wiley-Liss, New York, pp. 41–62.

Ruff, C.B., 2000a. Body size, body shape, and long bone strength in modern humans. *J. Hum. Evol.* 38, 269–290.

Ruff, C.B., 2000b. Biomechanical analyses of archaeological human skeletons. In: Saunders, S. R., Katzenberg, M. A. (Eds.), *Skeletal Biology of Past People*. Wiley-Liss, New York, pp. 71–102.

Ruff, C.B., Hayes, W.C., 1983. Cross-sectional geometry of Pecos Pueblo femora and tibiae – a biomechanical investigation: I. Method and general patterns of variation. *Am. J. Phys. Anthropol.* 60, 359–381.

Ruff, C.B., Larsen, C.S., and Hayes, W. C., 1984. Structural changes with the transition to agriculture on the Georgia Coast. *Am. J. Phys. Anthropol.* 64, 125–136.

Ruff, C.B., Trinkaus, E., Walker, A., Larsen, C.S., 1993. Postcranial robusticity in *Homo*. I: Temporal trends and mechanical interpretation. *Am. J. Phys. Anthropol.* 91, 21–53.

Ruff, C.B., Holt, B.M., Sládek, V., Berner, M., Murphy, W.A., zur Nedden, D., Seidler, H., Recheis, W., 2005. Body size, body shape, and long bone strength of the Tyrolean ice man. *Am. J. Phys. Anthropol.* Suppl. 40, 180.

Senut, B., 1985. Computerized tomography of a Neandertal humerus from Le Régourdou (Dordogne, France): comparisons with modern man. *J. Hum. Evol.* 14, 717–723.

Shea, J.J., 1998. Neandertal and early modern human behavioral variability: A regional-scale approach to lithic evidence for hunting in the Levantine Mousterian. *Curr. Anthropol.* 39 (Suppl.), S45–S78.

Sorensen, M.V., Leonard, W. R., 2001. Neandertal energetics and foraging efficiency. *J. Hum. Evol.* 40, 483–495.

Stock, J.T., 2004. Differential constraints on the pattern of skeletal robusticity in human limbs relative to climatic and behavioral influences on morphology. *Am. J. Phys. Anthropol.* Suppl. 38, 188–189.

Stock, J.T., Pfeiffer, S.K., 2004. Long bone robusticity and subsistence behaviour among Later Stone Age foragers of the forest and fynbos biomes of South Africa. *J. Archaeol. Sci.* 31, 999–1013.

Straus, L.G., 1995. The Upper Paleolithic of Europe: an overview. *Evol. Anthropol.* 4, 4–16.

Straus, L.G., 1997. Continuity or rupture; convergence or invasion; adaptation or catastrophe; mosaic or monolith: views on the Middle to Upper Paleolithic transition in Iberia. In: Carbonell, E., Vaquero, M. (Eds.), *The Last Neandertals, the First Anatomically Modern Humans*. 1996. Fundacio Catalana per a la Recerca, Barcelona / Universitat Rovira i Virgili, Tarragona, Barcelona/Tarragona, pp. 203–218.

Trinkaus, E., 1976. The evolution of the hominid femoral diaphysis during the Upper Pleistocene in Europe and the Near East. *Z. Morph. Anthrop.* 67, 291–319.

Trinkaus, E., 1983a. Neandertal postcrania and the adaptive shift to modern humans. In: Trinkaus, E. (Ed.), *The Mousterian Legacy: Human Biocultural Changes in the Upper Pleistocene*. BAR International Series, Oxford, pp. 165–200.

Trinkaus, E., 1983b. *The Shanidar Neandertals*. Academic Press, New York.

Trinkaus, E., 1984. Western Asia. In: Smith, F. H., Spencer, F. (Eds.), *The Origins of Modern Humans: A World Survey of the Fossil Evidence*. Alan R. Liss, New York, pp. 251–293.

Trinkaus, E., 2000. The "robusticity transition" revisited. In: Stringer, C.B., Barton, R.N.E., Finlayson, J.C. (Eds.), *Neanderthals on the Edge*. Oxbow Books, Oxford, pp. 227–236.

Trinkaus, E., Churchill, S. E., 1999. Diaphyseal cross-sectional geometry of Near Eastern Middle Paleolithic humans: the humerus. *J. Archaeol.* Sci. 26, 173–184.

Trinkaus, E., Jelínek, J., 1997. Human remains from the Moravian Gravettian: the Dolní Věstonice 3 postcrania. *J. Hum. Evol.* 33, 33–82.

Trinkaus, E., Ruff, C.B., 1999a. Diaphyseal cross-sectional geometry of Near Eastern Middle Paleolithic humans: the femur. *J. Archaeol. Sci.* 26, 409–424.

Trinkaus, E., Ruff, C.B., 1999b. Diaphyseal cross-sectional geometry of Near Eastern Middle Palaeolithic humans: The tibia. *J. Archaeol. Sci.* 26, 1289–1300.

Trinkaus, E., Shipman, P., 1993. *The Neandertals*. Knopf, New York.

Trinkaus, E., Churchill, S.E., Villemeur, I., Riley, K.G., Heller, J.A., Ruff, C.B., 1991. Robusticity versus shape: the functional interpretation of Neandertal appendicular morphology. *J. Anthropol. Soc. Nippon* 99, 257–278.

Trinkaus, E., Churchill, S.E., Ruff, C.B., 1994. Postcranial robusticity in *Homo*. II: bilateral asymmetry and bone plasticity. *Am. J. Phys. Anthropol.* 93, 1–34.

Trinkaus, E., Ruff, C.B., Churchill, S.E., 1998. Upper limb versus lower limb loading patterns among Near Eastern Middle Paleolithic hominids. In: Akazawa, T., Aoki, K., Bar-Yosef, O. (Eds.), *Neandertals and Modern Humans in Western Asia*. Plenum Press, New York, pp. 391–404.

Trinkaus, E., Churchill, S.E., Ruff, C.B., Vandermeersch, B., 1999. Long bone shaft robusticity and body proportions of the Saint-Césaire 1 Châtelperronian Neanderthal. *J. Archaeol. Sci.* 26, 753–773.

Vallois, H.V., de Félice, S., 1977. Les Mésolithiques de France. Archives de l'Institute de Paléontologie Humaine, Mémoire 37.

Vandermeersch, B., 1981. *Les Hommes Fossiles de Qafzeh (Israël)*. CNRS, Paris.

Vandermeersch, B., Marks, A. E., 1992. Second day's discussion. In: Akazawa, T., Aoki, K., and Kimura, T. (Eds.), *The Evolution and Dispersal of Modern Humans in Asia*. Hokusen-sha, Tokyo, pp. 295–302.

Weiss, E., 2003. Understanding muscle markers: aggregation and construct validity. *Am. J. Phys. Anthropol.* 121, 230–240.

Weiss, E., 2004. Understanding muscle markers: Lower limbs. *Am. J. Phys. Anthropol.* 125, 232–238.

Wilczak, C.A., 1998. Consideration of sexual dimorphism, age, and asymmetry in quantitative measurements of muscle insertion sites. *Int. J. Osteoarchaeol.* 8, 311–325.

Zumwalt, A.C., 2004. A new method to quantify the 3D morphology of bone surfaces, with application to muscle enthesis rugosity. *Am. J. Phys. Anthropol.* Suppl. 38, 216.

Zumwalt, A.C., Ruff, C.B., Wilczak, C.A., 2000. Primate muscle insertions: What does size tell you? *Am. J. Phys. Anthropol.* Suppl. 30, 331.

9. Neanderthal hands in their proper perspective

W.A. NIEWOEHNER
Department of Anthropology,
California State University-San Bernardino,
San Bernardino, CA 92407, USA
wniewoeh@csusb.edu

Keywords: Hands, Neanderthal, Upper Paleolithic, carpometacarpal articulations, geometric morphometrics

Abstract

Neanderthal hand remains are usually compared to those of recent humans because recent human samples are readily available. These comparisons demonstrate that Neanderthal hand morphologies are at or beyond the range of recent human samples for traits such as: expanded distal tuberosities, rugose musculotendinous attachment sites, dorsopalmarly flat metacarpal 1 bases, relatively short thumb proximal phalanges, more parasagittally oriented capitate metacarpal 2 facets, reduced metacarpal 3 styloid processes, radioulnarly flat metacarpal 5 bases, and large, projecting carpal tubercles. Functional interpretations suggest that Neanderthal hands are adapted primarily for power during manipulation.

Neanderthal hands are rarely systematically compared to those of other Late Pleistocene humans. Metric and three-dimensional geometric morphometric analyses and qualitative observations demonstrate that Early and Late Upper Paleolithic hand remains are more similar to recent human hands than to Neanderthal hands and that the Skhūl/Qafzeh sample is similar to the Upper Paleolithic samples. However, Early and Late Upper Paleolithic hand remains are not indistinguishable from those of modern humans. Evident in the Early Upper Paleolithic sample are features reminiscent of Neanderthals, such as the presence (on some specimens) of significant muscle crests on metacarpals 1 and 5 as well as a Neanderthal-like metacarpal 1 base shape. Additionally, both the Early and Late Upper Paleolithic samples have intermediate metacarpal 2 and 3 base morphologies relative to Neanderthal and recent human samples. There are indications of increased stabilization of the mid-carpometacarpal region, the enhancement of first finger precision movements, and reductions in mechanical advantages in the Early and Late Upper Paleolithic specimens. Some features found in the Upper Paleolithic samples are likely related to increases in the frequency and sophistication of hafted tools, while functional adaptations related to more frequent precision grip usage are argued to be associated with fine finger movements.

K. Harvati and T. Harrison (eds.), Neanderthals Revisited: New Approaches and Perspectives, 157–190.

Introduction

The fossil record indicates that anatomically modern human cranio-facial and postcranial morphologies evolved in the Old World during the Late Pleistocene. Traditionally, analyses of the cranio-facial region have been the main focus of phylogenetic debates, especially regarding the "fate" of the Neanderthals. The primary concern of these debates has been whether, and to what extent, they contributed to the recent human gene pool (Stringer & Andrews, 1988; Wolpoff, 1989a, b, 1992; Bräuer, 1992; Stringer, 1992, 1994; also see contributions to this volume).

Sometimes lost in the emphasis on phylogeny are issues of human behavioral evolution, usually the subject of archaeological analyses. In general, Late Pleistocene archaeological remains indicate substantial changes in subsistence strategies, the pace of technological innovation and the spatial organization of sites, which may be indicative of an adaptive shift towards ethnohistorically documented hunting and gathering behavioral pattern (Mellars, 1994). Some paleoanthropologists have put aside the issues of phylogeny in favor of quantifying correlated shifts in postcranial morphology within the framework of the hypothesis that archaic human behavior was not simply a technologically limited version of recent hunter-gatherer behavior (see Trinkaus [1992b] for a historical perspective).

Central to this intellectual and methodological framework is the integration of Paleolithic archaeology with functional analyses of the appendicular skeleton (Trinkaus, 1976, 1977, 1978, 1983a, b, 1986a, b, 1989b, 1992a, b; Lovejoy & Trinkaus, 1980; Riley & Trinkaus, 1989; Churchill & Trinkaus, 1990; Trinkaus & Villemeur, 1991; Churchill, 1994; Niewoehner et al., 1997b; Niewoehner, 2000, 2001). These analyses indicate that Neanderthals, as the most numerous and best preserved examples of Late Pleistocene humans, are morphologically and functionally distinct from more recent humans in many developmentally plastic regions of the skeleton, including the hand. The habitual use of contrasting manipulative postures changes the positions in which articulations are peak loaded and alters the vectors of the principal joint reaction forces. Between-sample contrasts in articular facet orientation could therefore be indicative of contrasting patterns of habitual force transmission. This is a reasonable position because animal experiments demonstrate that subchondral bone has the ability to dynamically model to restore effective load transmission when load patterns are altered during growth (Frost, 1987, 1990, 1999; Llinas et al., 1993; Frost et al., 1998; Plochocki & Organ, 2003). However, the mechanisms of ontogenetic responses to loading on cortical bone and subchondral bone are complex and not well understood and it appears that the majority of skeletal responses to altered load patterns in cortical bone (and perhaps subchondral bone) occur before sexual maturity (Pearson & Lieberman, 2004). Thus, the behavioral interpretations presented later should be considered working hypotheses that, if correct, reflect sub-adult behaviors that presumably are carried into adulthood.

Why study hands? The evolution of hominid manipulative behaviors continues to generate considerable interest in paleoanthropology, given the primacy of technology in the human adaptive complex. Central to the elucidation of fossil hominid manipulative capabilities and habitual behaviors is the examination of fossil hand remains within a comparative framework. Early Pliocene hominid hands display a mosaic of ape-like and human-like traits, being separated from recent humans by a considerable morphological gap, and they appear to have not yet achieved fully human manipulative capabilities (Lewis, 1977; Susman & Creel, 1979; Marzke, 1983, 1997; McHenry, 1983; Marzke & Marzke, 1987; Marzke &

Shackley, 1987; Ricklan, 1987; Susman, 1988, 1995; Lewis, 1989; Marzke et al., 1992).

The morphologies of Neanderthal hand bones are more similar to the modern human condition than to early Pliocene hominid hand bones, so it is logical that Neanderthal hand remains are most often compared to those of recent humans, rather than to those of our Pliocene relatives. Although using recent human samples is certainly valid and necessary, the danger is that these sample contrasts might overemphasize the uniqueness of Neanderthal morpho-functional complexes relative to less frequently studied hand remains of other Late Pleistocene humans, such as those of Upper Paleolithic humans and the remains from Skhūl and Qafzeh. How do these other hand remains compare to Neanderthal and recent human hands and what, if anything, can these hand remains tell us about behavioral evolution in the Late Pleistocene? These questions are the focus of this chapter.

Methods and Samples

METHODS

Some of the information presented below consists of my qualitative observations. I also added my own data on Upper Paleolithic specimens to previously published measurements on Neanderthal and recent human hand bones in order to broaden the comparative perspective. My primary research method, though, is to collect three-dimensional landmark data on the functionally important carpometacarpal joints (of the wrist) and the second and fifth metacarpophalangeal joints and use these in a geometric morphometric shape analysis. This method is useful for determining the nature and extent of between-sample differences in joint shapes and orientations. The results, discussed later, are used to further elucidate the nature of frequency shifts in habitual manipulatory behaviors. Below, I present a brief overview of the methodology (for detailed discussions see Niewoehner, 2000, 2001, 2005).

Geometric Morphometrics

Three-dimensional landmark coordinates are used for a geometric morphometric analysis in which biological shapes are ultimately visualized in a computer software interface, Morphologika (O'Higgins & Jones, 1998), to determine how principal components and shape are related. The first step, a Prucrustes superimposition of the landmark coordinates, is followed by a principal components analysis of the now registered landmark coordinates, and finally, by a canonical discriminant function of the principal components scores.

Landmark data require the use of points that have the same meaning and locations between specimens (Bookstein, 1991). Joint surfaces have few easily identifiable natural landmarks so this problem is addressed by using a set of specially prepared slides and a slide projector to project the image of a 10×10 grid onto the joint surface. Extreme care is used in positioning the grid in the same manner on each joint surface. This procedure is detailed elsewhere (Niewoehner, 2000, 2001, 2005). Touching the specimen with a digitizing arm results in movement of the specimen and alters the original grid position, so I use photogrammetry, the extraction of 3D information from two-dimensional digitized images. The landmarks on the digital images of the joint surface are manually digitized in Photomodeler (version 3.1), a software package that calculates the three-dimensional coordinates of each digitized landmark to within $\pm$ 0.023 mm. Using the Morphologika software program, the landmark coordinates are initially registered through a generalized Procrustes analysis that removes translational and rotational differences between forms and then scales them to the best-fit configuration (Gower, 1975; Rohlf & Slice, 1990; Goodall, 1991).

The interpretation of shape variation along a principal component is accomplished solely by visualization of shape changes within the software interface. The Morphologika software accomplishes this by reconstructing hypothetical forms along any user-selected principal component. Next, specimens are assigned an *a priori* class, i.e., Neanderthal, Early Upper Paleolithic, Late Upper Paleolithic, or recent human male or female, and their principal components scores derived from the Morphologika program are used to produce a separate canonical discriminant function for each metacarpal base, carpal facet, or metacarpal head with the SAS statistical software. Only those functions with significant discriminations ($p \leq 0.05$) are discussed later in reference to joint shape and joint orientation contrasts. The Skhūl/Qafzeh specimens are inserted into the discriminant functions as unknowns and assigned to the nearest class based on the discriminant function Mahalanobis distance matrix. The results are indicative of morphological resemblance only, and they form the basis for the functional and behavioral inferences discussed later.

SAMPLES

Neanderthals

The Neanderthal specimens discussed in this chapter are identified in the literature as Neanderthals *sensu stricto*. The Neanderthal specimens that are used in the geometric morphometric analysis ($N = 13$) are listed in Table 1. I was not able to collect data on specimens from Krapina, or on Kiik-Koba 1 and Subalyuk 1. Although Kiik-Koba 1 is an important specimen, the other specimens are not critical omissions. The Krapina remains are all unassociated, none of the metacarpals has their base preserved and the single capitate has a damaged distal surface (Radovcic et al., 1988). Subalyuk 1 only has a fragmentary metacarpal 2 (Bartucz & Szabó, 1938).

All specimens are considered prime-aged adults in the sense that they have fused epiphyses. Although the determination of chronological age is difficult, it appears that Neanderthals did not survive past the age of about 45 years (Trinkaus & Thompson, 1987), so this can be considered as their maximum age. I did not divide the sample by sex for the geometric morphometric analyses, but I have nevertheless included the probable sex of the individuals taken from the literature. Excluding Le Régourdou 1, which is of indeterminate sex (Vandermeersch & Trinkaus, 1995),the sample is a little more than half (66%) presumed males. On this basis it might be appropriate to subdivide the sample by sex. However, their hand skeletons are incomplete so most analyses would not have a balanced sex ratio. The approach used is to scale for size and pool all the individuals. This procedure maximizes sample sizes, therefore increasing statistical confidence in the results.

The sample spans a great deal of geologic time and geographic space. The earliest Neanderthals include Tabun 1, and Shanidar 4 and 6, all of which may be of Last Interglacial age (Trinkaus, 1983a, 1991), and the youngest is probably Amud 1 at 40 to 50 ka (Grün & Stringer, 1991). The European Neanderthals in this sample are thought to be from the early Last Glacial (oxygen isotope stages 5 to 3b). Regardless, the Near Eastern and European Neanderthals are pooled for the analysis, ignoring the possible effects of ecogeographic patterning of body segment lengths within the sample (Holliday, 1995, 1997), or possible reductions in muscularity or changes in joint shape or orientation through time. These are interesting issues, but they cannot be addressed at present due to the lack of precise chronological ordering within the sample, and more importantly, because sample sizes are already small for most of the analyses.

Upper Paleolithic

The Upper Paleolithic specimens are, with one exception, all associated with "non-transitional" European Upper Paleolithic industries, i.e., Aurignacian, Gravettian, Magdalenian,

Table 1. The Neanderthal sample

Specimen	*Sex*	*Cultural association*	*Date/Geological age*
European Neanderthals			
La Ferrassie 1[1]	M	Charentian Mousterian[2]	Würm II/ Interpleniglacial[3]
La Ferrassie 2[1]	F	Charentian Mousterian[2]	Würm II/ Interpleniglacial[3]
La Chapelle-aux-Saints 1[4]	M	Charentian Mousterian[3]	47± 3 / 56± 4 ka[5]
Le Régourdou 1[6*]	I	Charentian Mousterian[3]	Early Würm[3]
Spy 1[7]	F	Mousterian[2]	Oxygen isotope stage 5a-3[9]
Spy 2[7]	M	Mousterian[8]	Oxygen isotope stage 5a-3[9]
Near Eastern Neanderthals			
Amud 1[10]	M	Mousterian (Tabun B)[11]	41/50 ka[5]
Kebara 2[12]	M	Mousterian (Tabun B)[13]	60 ± 4 ka[14]
Shanidar 3[15]	M	Mousterian (Zagros)[15]	ca. 50–75 ka[15]
Shanidar 4[15*]	M	Mousterian (Zagros)[15]	75–100 ka[16]
Shanidar 5[14*]	M	Mousterian (Zagros)[15]	ca. 50–75 ka[15]
Shanidar 6[15*]	F	Mousterian (Zagros)[15]	75–100 ka[16]
Tabun 1[17]	F	Mousterian (Tabun B)[11]	100/120 ka[5]

[1] Heim (1982).
[2] Bourgon (1957).
[3] Vandermeersch (1965).
[4] Boule (1911–1913).
[5] Grün and Stringer (1991).
[6] Piveteau (1963).
[7] Fraipont and Lohest (1886).
[8] Zeuner (1940).
[9] Ulrex-Claret (1990).
[10] Endo and Kimura (1970).
[11] Jelenik (1982).
[12] Vandermeersch (1991).
[13] Meignen and Bar-Yosef (1989).
[14] Valladas et al. (1987).
[15] Trinkaus (1983b).
[16] Trinkaus (1991).
[17] McCown and Keith (1939).

* Casts.

and Epigravettian. The exception is Ohalo 2, which is from the Near East, but is associated with a typologically Upper Paleolithic industry, the Kebaran (Nadel & Hershkovitz, 1991). The sample has been further sub divided into the Early ($N = 13$, Table 2) and Late Upper Paleolithic ($N = 8$, Table 3). Specimens greater than ~20 ka are included in the Early Upper Paleolithic sample, and younger than ~20 ka are included in the Late Upper

Table 2. The early Upper Paleolithic sample

Specimen	*Sex*	*Cultural association*	*Date/Geological age*
Abri Pataud 163[1]	F?	"Proto-Magdalenian"[1] (Gravettian)	21.9–19.3 ka[2]
Abri Pataud 230[1]	F	"Proto-Magdalenian"[1] (Gravettian)	21.9–19.3ka[2]
Arene Candide 1[3]	M	Gravettian[4]	> 18.6 ka[4]
Barma Grande 2[5]	M	Gravettian[6]	26,000–23,000 BP[6]
Cro-Magnon[7]	?	Aurignacian?/Gravettian[8, 9]	30 ka?[9,10]
Dolní Věstonice 13[11]	M	Eastern Gravettian[12]	27.6 ka[13]
Dolní Věstonice 14[11]	M	Eastern Gravettian[12]	27.6 ka[13]
Dolní Věstonice 15[11]	F	Eastern Gravettian[12]	27.6 ka[13]
Dolní Věstonice 16[11]	M	Eastern Gravettian[12]	27.6 ka[13]
Grotte des Enfants 4[5]	M	Late Aurignacian/ Gravettian[14]	Upper Pleniglacial[14]
Paglicci 3[15]	F	Gravettian[15]	26 ka[15]
Parabita[16]	?	Gravettian[16]	22,220 ± 360/ 22,110 ± 330 BP[16]
Pavlov 1[17]	M	Eastern Gravettian[18]	26.6–25.0 ka[19]

[1] Movius (1958).
[2] Valladas et al. (1987).
[3] Sergi et al. (1974).
[4] Bietti (1987).
[5] Verneau (1906).
[6] Formicola et al. (2004).
[7] Vallois and Billy (1965).
[8] de Sonneville-Bordes (1960).
[9] Gambier (2002).
[10] Movius (1958).
[11] Vlček (1991).
[12] Svoboda (1988).
[13] Klima (1988).
[14] Mussi (1986).
[15] Mallegni (personal communication cited in Churchill [1994]).
[16] Palma di Cesnola (1993).
[17] Vlček (1961).
[18] Klima (1959).
[19] Klima & Kukla (1963).

Paleolithic sample because a number of workers (Straus, 1983, 1990; Clark & Lindley, 1989; Clark, 2002) have argued that the behavioral changes near the Middle to Upper Paleolithic transition are not as great as those found near the end of the Upper Paleolithic. The exact date of the Early to Late transition is arbitrary, although the time period approaching the Last Glacial maximum in Europe is most clearly associated with resource intensification and the increase in the pace of technological changes associated mostly with the production of microliths used in projectile weapons for hunting dangerous and/or elusive game (Mellars, 1989, 1994; Straus, 1993).

Most of the Upper Paleolithic individuals are probably of prime age because they have fused epiphyses. The single exception is Arene Candide 1, an approximately 15 year old male (Sergi et al., 1974), in which the metacarpal heads and the bases of the proximal phalanges

Table 3. The late Upper Paleolithic sample

Specimen	*Sex*	*Cultural association*	*Date/Geological age*
Arene Candide 3[1]	?	Final Epigravettian[2]	11.8–10.9 ka[2]
Arene Candide 5[1]	M	Final Epigravettian	11.8–10.9 ka[2]
Arene Candide 12[1]	?	Final Epigravettian	11.8–10.9 ka[2]
Bruniquel 24[3]	F	Upper Magdalenian[4]	Last glacial[4]
Continenza[5]	M	Terminal Epigravettian[5]	ca. 10,000 BP[5]
Oetrange[4]	?	Magdalenian[4]	Würm[4]
Ohalo 2[6]	M	Kebaran[6]	19 ka[6]
Vado all'Arancio[7]	M	Final Epigravettian[8]	11,330 + 50 BP[8]

[1] Paoli et al. (1980).
[2] Bietti (1987).
[3] Genet-Varcin and Miquel (1967).
[4] Oakley et al. (1971).
[5] Formicola (personal communication).
[6] Nadel & Hershkovitz (1991).
[7] Minellono et al. (1980).
[8] Minellono (1985).

are partially fused. I decided not to use the metacarpal 2 or metacarpal 5 heads in the morphometric shape analysis since the epiphyseal plates are so near the metacarpal head articular margins. I did, however, decide to use the metacarpal bases and the carpals in the morphometric analysis to bolster the sample sizes.

There are a few specimens (Abri Pataud 163, Cro-Magnon, and Parabita) where the hand remains are not securely associated with diagnostic crania or postcrania for determining sex. The Abri Pataud remains consist of (probably) four hands attributed to three different individuals (Billy, 1975). One of the specimens (#163) consists of two intermixed hands from Square G. There are no duplicated bones of the same side and all are of similar size and have similar patina and are thought by Billy (1975) to be from one individual. They may be from a female, but this attribution is based on size alone. The other hand (#230) derives from a different locality (complex 22) and is more certainly part of a female skeleton found there. The Cro-Magnon metacarpals 3 and 4 are part of a large group of unnumbered, intermixed postcranial remains from the site (Oakley et al., 1971). The hand remains from Parabita (Cremonisi et al., 1972) are variably preserved. Some are heavily damaged, while others are almost perfect. They were unsorted at the time I studied them, but it is clear that there is more than one individual represented, so I sorted them based on coloration and size.

Skhūl/Qafzeh

In the Near East, the Skhūl/Qafzeh early modern humans ($N = 5$; Table 4) are associated with Mousterian lithic technologies, as are the Near Eastern Neanderthals (Jelenik, 1982; Boutié, 1989). Upon their initial discovery the Skhūl hominids were considered to be part of a late "progressive" Neanderthal group (McCown & Keith, 1939), but now both groups of hominids are generally agreed to represent an early though variable manifestation of non-Neanderthal, early modern craniofacial morphology (Klein, 1994); ~100 ka for Qafzeh (Valladas et al., 1987; Grün & Stringer, 1991), and 40 to 50 ka or 80 to 100 ka for Skhūl (McDermott et al., 1993).

Table 4. The Skhūl/Qafzeh Sample

Specimen	*Sex*	*Cultural association*	*Geological age*
Qafzeh 3[1]	F	Mousterian (Tabun C)[2]	100/120–92 ka[3,4]
Qafzeh 7[1]	?	Mousterian (Tabun C)[2]	100/120–92 ka[3,4]
Qafzeh 8[1]	M	Mousterian (Tabun C)[2]	100/120–92 ka[3,4]
Qafzeh 9[1]	F?	Mousterian (Tabun C)[2]	100/120–92 ka[3,4]
Skhūl 5[5]	M	Phase 2 Mousterian[6]	40–50 or 80–100 ka[7]

[1] Vandermeersch (1981).
[2] Boutié (1989).
[3] Grün and Stringer (1991).
[4] Valladas et al. (1987).
[5] McCown and Keith (1939).
[6] Jelenik (1982).
[7] McDermott et al. (1993).

Importantly, a number of studies have documented that the Skhūl/Qafzeh homonids have contrasting upper limb skeletal features when compared to Neanderthals (Trinkaus, 1992a; Churchill, 1994; Trinkaus et al., 1998; Trinkaus & Churchill, 1999), thus complicating simple techno-typological/morphological associations. I have previously argued (Niewoehner, 2001) that the Skhūl/Qafzeh hand anatomies are morphologically and functionally more similar to the Upper Paleolithic specimens than to Neanderthals, and some of these same arguments will be repeated later because a comparative analysis of Neanderthal and Upper Paleolithic hands would be incomplete without including this sample.

Unfortunately, the hand remains of the Skhūl/Qafzeh hominids are not well preserved, so there are only one to four specimens that can be included in different aspects of the geometric morphometric analyses. Of special note is the sex and age of the best preserved, most complete specimen, Qafzeh 9. This individual is reported as a late adolescent female by Vandermeersch (1981), although others (Rosenberg, 1988; Rak, 1990; Wolpoff, 1999) believe that Qafzeh 9 is a male. Nevertheless, even though Qafzeh 9 was most likely significantly younger than all the Neanderthals, the metacarpal epiphyses are fused, so there are probably no significant ontogenetic effects to be concerned about.

Recent Humans

Comparative data were collected on three recent human samples for the geometric morphometric analysis. The first, a modern autopsy sample of North American urban individuals ($N = 22$), was taken from the autopsied skeletal collection of the University of New Mexico's Maxwell Museum. The second, a late prehistoric Amerindian sample ($N = 24$), dates to between 1250 and 1600 AD. It consists of individuals from the Pueblo IV sites of Kuaua, Pottery Mound, and Supawe of New Mexico's central Rio Grande Valley that are curated at the University of New Mexico's Maxwell Museum. The third, a medieval cemetery sample from Mistihalj Yugoslavia ($N = 24$), is curated at the Peabody Museum, Harvard University. All of the aforementioned samples consist primarily of prime-age adults with an approximately equal number of males and females and left and right skeletal elements.

Neanderthal Hands

HISTORICAL INTERPRETATIONS OF MORPHOLOGY

Some early workers, primarily noting thumb morphology, argued that Neanderthals had limited manipulative capabilities (Boule, 1911–1913; Sarasin, 1932; Bonch-Osmolovskij, 1941; Musgrave, 1971; Vlček, 1975, 1978). For example, Neanderthal thumbs were thought to be both absolutely and relatively short (e.g., Boule, 1911–1913) which, if true, would have impaired their fine manipulative capabilities. It has been demonstrated that relative to either hand or arm lengths, Neanderthal and recent human thumbs (summed first metacarpal, first proximal phalanx, and first distal phalanx lengths) have similar proportions (Musgrave, 1971; Trinkaus, 1983b; Trinkaus & Villemeur, 1991). However, the Neanderthal proximal phalanx is short relative to the distal phalanx, and this fact has important biomechanical consequences that are discussed later.

Musgrave (1971) argued that Neanderthals may have lacked the full precision capabilities of modern humans because the non-condyloid configuration of the base of the Neanderthal first metacarpal could have restricted movements at the trapezio-metacarpal 1 joint. He also considered the role of the first dorsal interosseous muscle in the manipulative abilities of the Neanderthal hand. He argued that its powerful exertion on the metacarpophalangeal joint would have kept the first finger pronated so that the pad of the first finger could not properly oppose the pad of the thumb. Therefore, although he argued that Neanderthals would have had powerful pinch grips, they must have lacked the degree of fine control of the first finger evident in recent humans. Furthermore, Musgrave (1971) argued that the Neanderthal distal thumb phalanx shows a high degree of ulnar deviation, which would have further impaired opposition of the thumb. It is true that the Neanderthal distal phalanx deviates ulnarly, but its deviation is within the upper limits of recent human variation (Stoner & Trinkaus, 1981).

On the basis of his study of the Kiik-Koba hand remains, Vlček (1975, 1978) believed that Neanderthals and recent humans had profoundly differing manipulative capabilities, arguing that the *flexor pollicis brevis* and the *opponens pollicis* muscles in Neanderthals functioned only to adduct the thumb towards the palm of the hand. This was based on his belief that the insertion of the *opponens pollicis* on the Neanderthal metacarpal 1 shaft was shifted too far palmarly to function as a medial rotator; the lack of full medial rotation of the thumb during adduction prevented Neanderthals from forming a modern human-like precision grip. Vlček further argued that the insertion of the *opponens pollicis* must have shifted radially with the appearance of modern humans, imbuing them with greater precision.

There appears to be no anatomical or functional basis for the above claims that Neanderthal thumbs were less mobile than those of modern humans (Trinkaus, 1983b). In fact, a computer simulation of the Neanderthal thumb and first finger leaves no reason to believe that Neanderthal hands were significantly less dextrous than our own hands, at least with regards to precision movements of the thumb and index finger (Niewoehner et al., 2003).

Thus, recent analyses have exposed the methodological weaknesses of earlier studies indicating that Neanderthals were deficient in their precision capabilities. With new analyses we can still glean some significant information about Late Pleistocene human habitual manipulative behaviors through a comparative analysis of features relating to strength, mechanical advantage, and joint shape and orientation.

CRITICAL REVIEW OF CURRENT INTERPRETATIONS OF MORPHOLOGY

Features Relating to Strength

Neanderthals almost certainly possessed hypertrophied intrinsic hand musculature and habitually transmitted higher forces through their joints. Additionally, increased projection of a number of carpal tubercles and their thumb phalangeal proportions augmented mechanical advantages across the first and fifth carpometacarpal and the first metacarpophalangeal joints. This combination of hypertrophied muscles plus increased mechanical advantages without doubt imbued Neanderthals with prodigious grip strength.

The *opponens pollicis* muscle forms the bulk of the thenar eminence while the *opponens digiti minimi* forms the bulk of the hypothenar eminence. Both muscles contribute significantly to movements essential to both power and precision grips because the *m. opponens pollicis* helps in opposition of the thumb and the *m. opponens digiti minimi* flexes and radially rotates the fifth metacarpal during cupping of the palm (Kaplan & Spinner, 1984). Neanderthal hand morphology is indicative of hypertrophy of both of these muscles, indicating greater grip strength on the thumb and little finger sides of the hand. Evidence for this are prominent *opponens* crests on most Neanderthal first metacarpals and many fifth metacarpals (Trinkaus, 1983b).

Trinkaus (1983b) calculated the relative development of the *opponens pollicis* crest with the *opponens pollicis* index (maximum distal shaft breadth/minimum proximal shaft breadth × 100). Neanderthals have the largest mean values for this index (Table 5), indicating exceptional crest development in this sample. Qafzeh 9 and Skhūl V have intermediate values (122.9 and 136.5), putting them at the low end of Neanderthal values and the high end of recent human values. *Opponens pollicis* crest development in the Upper Paleolithic sample is variable (Figure 1), ranging from little to no crest development in Abri Pataud 227 and Ohalo 2 (indices[a] 100), to moderately well developed in Parabita and Grotte des Enfants 4 (indices[a] 120, 123).

A similar index for the *opponens* crest on the fifth metacarpal is not possible to calculate, since the form of the crest is not the same. However, visual inspection of the fifth metacarpals reveals that crest development here is variable within each sample. Variation

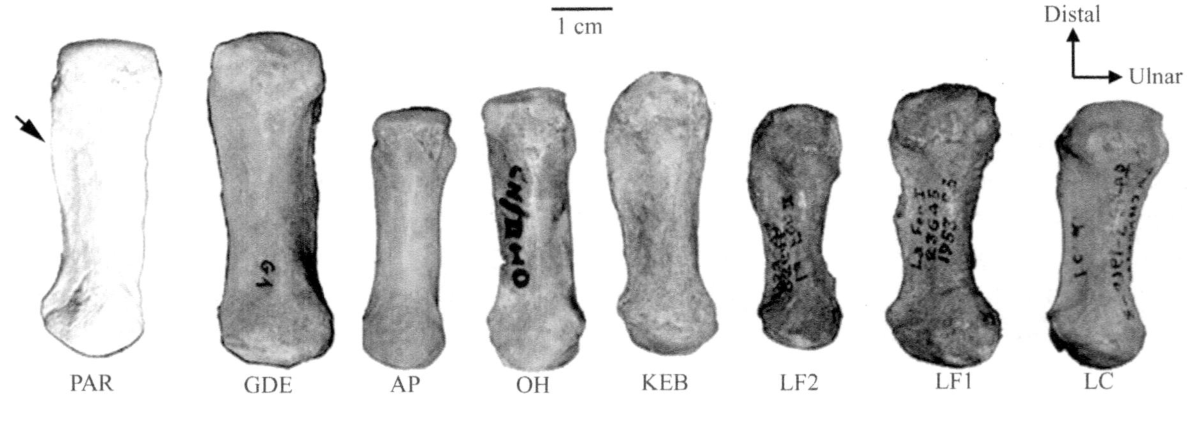

Figure 1. *Opponens pollicis* crest variation among Late Pleistocene humans. Palmar views of the first metacarpals of Parabita (PAR,) Grotte des Enfants 4 (GDE), Abri Pataud 227 (AP), Ohalo 2 (OH), Kebara 2 (KEB), La Ferrassie 2 and 1 (LF2, LF1), and La Chapelle-aux-Saints (LC). The arrow is pointing to the *opponens pollicis* crest on Parabita. As a group, the Neanderthals (KEB, LF2, LF1, and LC) have the largest crests (the crest on LF1 is damaged). Crest development in the Upper Paleolithic sample is variable, ranging from little to no crest (OH and AP) to moderately well developed (PAR, GDE). Parabita, La Ferrassie 2, and La Chapelle-aux-Saints are mirror-imaged.

Table 5. Opponens pollicis index

Sample	*N*	*Opponens pollicis index*[1]
Neanderthals[2]	8	141.1 ± 16.6[3]
Qafzeh 9[2]	1	122.9
Skhūl V[2]	1	136.5
Early Upper Paleolithic[1]	2	110.0 ± 14.1
Late Upper Paleolithic[4]	2	111.5 ± 16.2
Recent Europeans[2]	20	111.3 ± 6.1
Amerindians[2]	20	114.5 ± 10.0

[1](Maximum/minimum first metacarpal shaft breadth) × 100
[2] Data from Trinkaus (1983b).
[3] Mean ± 1 SD.
[4] Author's data.

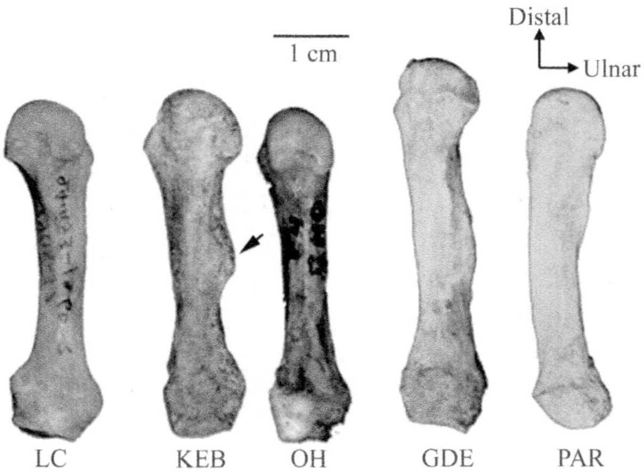

Figure 2. *Opponens digiti minimi* crest variation among Late Pleistocene humans. Palmar views of the fifth metacarpal of La Chapelle-aux-Saints (LC), Kebara 2 (KEB), Ohalo 2 (OH), Grotte des Enfants 4 (GDE), and Parabita (PAR). The arrow is pointing to the *opponens digiti minimi* crest on Kebara. The Neanderthal sample is variable. The La Chapelle-aux-Saints fifth metacarpal represents one extreme. It has small, raised ridges, without a true crest. Kebara 2 represents the other extreme. It has an extraordinarily well developed crest. The Upper Paleolithic individuals are also variable, but none examined has the crest development comparable to Kebara 2. Ohalo 2 has no crest and the shaft is quite smooth. Both Grotte des Enfants 4 and Parabita have crest development that exceeds that of most recent humans. La Chapelle-aux-Saints has been mirror-imaged.

in the Neanderthal sample ranges from barely discernable crests on the La Chapelle-aux-Saints specimen to the highly developed crests on Kebara 2 (Figure 2). However, most other Neanderthal specimens are closer to the Kebara morphology than to the La Chapelle-aux-Saints morphology. The crest morphology of the Upper Paleolithic specimens is likewise variable. For example, Abri Pataud 227 and Ohalo 2 have virtually no crest development, which is typical of most recent human fifth metacarpals. On the other hand, specimens such as Parabita and Grotte des Enfants 4 have levels of crest development that are similar to that of many Neanderthals.

The *flexor pollicis longus* muscle originates on the radius and its tendon courses up the first metacarpal shaft, crossing the thumb metacarpophalangeal and interphalangeal joints to insert on the base of the thumb distal phalanx. It acts primarily to flex the thumb joints, and is important for isometric contraction during power gripping and manipulating objects in pinch grips (Kaplan, 1965). In Neanderthals, the pit for the *flexor pollicis longus* insertion tends to be not only deeply excavated, but is also generally larger in area, taking up more of the area on the proximal half of the distal phalanx than is normally apparent in modern human samples (Trinkaus, 1983b). A large and deep insertion for the *flexor pollicis longus* tendon is likely to be indicative of hypertrophy of the muscle, but Shrewsbury et al. (2003) question this assumption. Assessing the morphology of the *flexor pollicis longus* pit is subjective, but it appears that the pits in the non-Neanderthal fossil samples, though variable, are intermediate between the recent human and the Neanderthal condition. For example, on Qafzeh 9, the pit is broad, but not as deeply excavated as in most Neanderthals, and the same can be said of Barma Grande 2, but the pit on Grotte des Enfants 4 is not very large or deep.

The prominent ridges on the proximal phalanges, marking the attachment of the flexor tendon sheaths, imply habitual increased tension on these regions in the Neanderthal hand caused, most likely, by isometric contraction of the flexor tendons during gripping. Most recent human proximal phalanges also have well defined ridges, but they are not very high and tend to be blunt. The appearance of these ridges on Neanderthals are essentially the same, but they are sharper and extend further palmarly (Trinkaus, 1983b). The flexor ridges in the Upper Paleolithic and Skhūl/Qafzeh specimens are closer to the recent human condition than to the Neanderthal condition.

Additional evidence for the inferred grip strength of the Neanderthal hand is provided by the morphology of their distal phalanges. Their pollical distal tuberosities are moderately radioulnarly expanded relative to more recent humans (Trinkaus, 1983b). Measurement of the radioulnar breadth of the pollical distal tuberosity of Early and Late Upper Paleolithic specimens and Qafzeh 7 and 9 (Table 6), indicates that they are well within the recent human range of variation. The same pattern is apparent when their remaining distal tuberosities are considered (Table 7). This radioulnar expansion, plus their generally greater dorsopalmar thickness, gives the Neanderthal finger distal tuberosities a "mushroom-like" appearance, contrasting with the dorsopalmarly short, arrow-head shape of the Qafzeh, Upper Paleolithic, and recent human second through fifth distal tuberosities (Figure 3) (Musgrave, 1973). The Neanderthal morphology is indicative of wider fingertips; the larger tuberosities providing increased attachment area for the pulp and the nail, and is most likely an adaptation for effectively transmitting large loads across the fingertips (Trinkaus, 1983b).

Table 6. Pollical distal tuberosity radioulnar breadth (mm)

Sample	*N*	*Radioulnar breadth*
Neanderthals[1]	8	12.6 ± 1.6[2]
Qafzeh 9[3]	1	10.4
Qafzeh 7[3]	1	9.0
Early Upper Paleolithic[3]	5	9.5 ± 1.0
Late Upper Paleolithic[3]	4	10.1 ± 1.0
Europeans[1]	37	10.7 ± 1.4
Amerindians[1]	20	9.3 ± 1.0

[1] Data from Trinkaus (1983b).
[2] Mean ± 1 SD.
[3] Author's data.

Features Related to Mechanical Advantages

Most Neanderthals have large and projecting trapezium, hamate, and scaphoid tubercles. Increased projection of these tubercles would have created a large carpal tunnel to accommodate hypertrophied digital flexor tendons. Increased tubercle projection also produces increased mechanical advantages for many extrinsic muscles of the wrist and for the intrinsic muscles of the thumb and fifth digit

Table 7. Finger distal tuberosity radioulnar breadth (mm)

Sample	Digit 2	Digit 3	Digit 4	Digit 5
Neanderthals[1]	9.5 ± 1.0[2]	11.4 ± 1.0	10.0 ± 1.0	7.6 ± 0.3
N	6	5	5	4
Qafzeh 9[1]	6.8	8.0	—	5.7
Early Upper Paleolithic[1]	7.0 ± 0.4	6.9 ± 1.0	6.5 ± 0.6	5.1 ± 0.6
N	4	5	5	4
Late Upper Paleolithic[1]	7.9 ± 1.0	7.7± 1.1	8.4	5.2 ± 0.7
N	3	3	1	2
Recent Europeans[3]	7.9 ± 1.0	8.7 ± 1.2	8.3± 1.2	6.5 ± 1.0
N	38	37	37	37

[1] Author's data.
[2] Mean ± 1 SD.
[3] Data from Musgrave (1973).

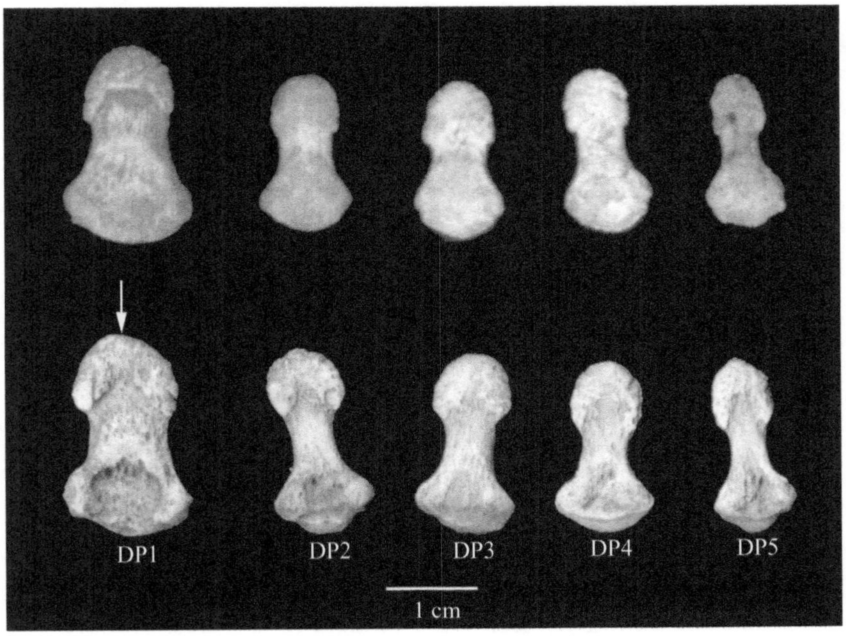

Figure 3. Distal phalanges of Kebara 2 and Barma Grande 2. Palmar view of the first through fifth distal phalanges (DP) of an Early Upper Paleolithic human, Barma Grande 2 (top row) and a Neanderthal (cast of Kebara 2). Note that all the Neanderthal distal tuberosities (arrow) are expanded radioulnarly compared to the recent human condition, but 2–5 are more markedly expanded. They also tend to be dorsopalmarly thicker. This radioulnar expansion would have provided more surface area and more effectively transmitted greater levels of force through the fingers during manipulatory behaviors.

which cross the 1st and 5th carpometacarpal articulations and contribute to flexion and rotation of the first and fifth metacarpals (Trinkaus, 1983a, b), all important movements for a variety of grip positions.

The relative projection of the trapezium tubercle is calculated as the ratio of tubercle length to metacarpal facet articular breadth (Trinkaus, 1983b). The Neanderthal sample (Table 8) has the highest mean value. The

Table 8. Trapezium tubercle projection ratio

Sample	*N*	*Tubercle projection*[1]
Neanderthals[2]	6	46.7 ± 11.4[3]
Qafzeh 9[2]	1	16.7
Skhūl V[4]	1	17.1
Early Upper Paleolithic[2]	6	17.7 ± 6.1
Late Upper Paleolithic[2]	5	17.2 ± 4.5
Recent Europeans[5]	20	27.3 ± 5.0
Amerindians[5]	20	22.6 ± 4.2

[1] (Tubercle projection/maximum articular breadth) × 100.
[2] Author's data.
[3] Mean ± 1 SD.
[4] Data from Trinkaus (1983).
[5] Data from Stoner and Trinkaus (1981).

Table 9. Relative hamulus projection of hamate

Sample	*N*	*Hamulus projection*[1]
Neanderthals[2]	9	90.9 ± 18.2[3]
Qafzeh 3,8,9	3	63.2 ± 8.2
Early Upper Paleolithic[2]	7	66.7 ± 13.3
Late Upper Paleolithic[2]	7	75.2 ± 18.7
Recent Europeans[2]	50	70.0 ± 8.4
Amerindians[2]	42	65.5 ± 7.8

[1] (Hamulus projection/maximum hamate height) × 100.
[2] Author's data.
[3] Mean ± 1 SD.

Early and Late Upper Paleolithic samples, as well as Qafzeh 9 and Skhūl V, have values that are well within the recent human range of variation. A similar ratio of hamulus projection (hamulus projection/hamate articular height) reveals that Neanderthals have, on average, increased relative palmar projection for the hamulus (Table 9). The remainder of the Late Pleistocene fossil samples, Qafzeh 3, 8, and 9, Early Upper Paleolithics, and Late Upper Paleolithics, are all within the recent human sample ranges of variation.

Neanderthal hamates can be discriminated from recent human samples based on their unique combination of more palmarly projecting, radioulnarly thicker, and proximodistally longer hamuli relative to dorsopalmarly and proximodistally short hamate bodies (Niewoehner et al. 1997a). If metacarpal 3 articular length is taken to represent the load arm of the hand, then Neanderthal and recent human samples have similar load arms since their third metacarpals are of similar length (Neanderthal mean MC 3 articular length = 64.24 mm, $N = 9$; recent human mean MC 3 articular length = 61.44 mm., $N = 92$). However, if hamulus projection plus 1/2 of the hamate body height is used to estimate the moment arm of the muscles inserting there, the sample mean for Neanderthal hamulus moment arms (18.09 mm, $N = 10$) is significantly greater than that of recent humans (mean = 15.07 mm, $N = 97$). Additionally, when expressed as a ratio of metacarpal 3 articular length to hamulus moment arm ([hamulus moment arm / MC 3 length] × 100), Neanderthals have significantly increased mechanical advantages relative to more recent humans (Neanderthal mean = 28.4, $N = 9$; recent human mean = 24.6, $N = 92$). The Qafzeh 8 and 9, and Early and Late Upper Paleolithic samples all have reduced means relative to the Neanderthal sample, although their sample ranges of variation overlap with the lower end of the Neanderthal sample range of variation (Table 10).

A comparison of Neanderthal and recent distal phalanx length to proximal phalanx length ratios indicates the average Neanderthal distal phalanx is 92% the length of the proximal phalanx. Recent humans tend to have longer proximal phalanges, their distal phalanx is only about 75% the length of the proximal phalanx (Trinkaus & Villemeur, 1991). Qafzeh 9's thumb distal phalanx is 70.5% the length of its proximal phalanx, a ratio similar to the recent human and Upper Paleolithic samples (Table 11).

Trinkaus and Villemeur (1991) investigated the biomechanical consequences of the Neanderthal's thumb phalangeal proportions. The primary flexor of the thumb, *flexor pollicis longus*, crosses and produces flexion at both the metacarpophalangeal and interphalangeal joints. The superficial head of the

Table 10. Mechanical advantage on the ulnar side of the hand

Sample	*N*	*Mechanical advantage*[1]
Neanderthals[2]	9	28.4 ± 2.8[3]
Qafzeh 8,9[4]	2	24.8 ± 1.9
Early Upper Paleolithic[4]	2	25.7 ± 2.4
Late Upper Paleolithic[4]	7	24.2 ± 4.1
Recent Humans[2]	92	24.6 ± 1.9

[1] (Hamulus moment arm/metacarpal 3 articular length) × 100.
[2] Data from Niewoehner et al. (1997b).
[3] Mean ± 1 SD.
[4] Author's data.

Table 11. Thumb phalangeal length ratio

Sample	*N*	*Thumb phalangeal ratio*[1]
Neanderthals[2]	7	92.3 ± 7.0[3]
Qafzeh 9[4]	1	70.5
Early Upper Paleolithic[4]	4	70.9 ± 6.4
Late Upper Paleolithic[4]	4	73.4 ± 3.8
Recent Europeans[2]	92	75.7 ± 5.9
Amerindians[2]	58	73.2 ± 6.3

[1] (Thumb distal phalanx length/proximal phalanx) × 100.
[2] Data from Trinkaus and Villemeur (1991).
[3] Mean ± 1 SD.
[4] Author's data.

flexor pollicis brevis flexes the proximal phalanx at the metacarpophalangeal joint and the first metacarpal at the trapezio-metacarpal 1 joint. Both of these muscles, and their actions, are important in both power and precision grips. However, strong flexion at the metacarpophalangeal joint is particularly important for executing a power grip that requires that an object be held against the palm of the hand.

When the relative shortness of the Neanderthal thumb proximal phalanx is considered in concert with their moderately dorsopalmarly taller bases, the mechanical advantage of the flexor muscles acting at the metacarpophalangeal joint are significantly increased. When the mechanical advantage of the flexors crossing the interphalangeal joint is considered, Neanderthals actually have significantly reduced mechanical advantages. These results indicate that Neanderthal thumbs were more powerful than recent thumbs during power grips, yet had reduced mechanical advantages when manipulating objects at the fingertips. However, given their general muscular hypertrophy, this decline in mechanical advantage could have been compensated for by their greater overall strength (Trinkaus & Villemeur, 1991). The reduced mechanical advantage of the thumb metacarpophalangeal joint during flexion is also apparent in the non-Neanderthal Late Pleistocene fossil samples (Table 12). Qafzeh 9's mechanical advantage is near the low end of the distribution for the recent human sample, and the Early and Late Upper Paleolithic samples are well within the modern human range of values.

Features Relating to Joint Shape and Orientation

In contrast with more recent humans, who have radioulnarly convex and dorsopalmarly concave proximal metacarpal 1 articular surfaces, Neanderthal metacarpal 1 bases are radioulnarly convex and variably dorsopalmarly flat to slightly concave or convex. The La Chapelle-aux-Saints metacarpal 1 base is extraordinarily dorsopalmarly convex, even for a Neanderthal (Figure 4), a condition also apparent in the Kiik-Koba specimen (Musgrave, 1971). Two Neanderthals, Shanidar 4 and La Ferrassie 2, have dorsopalmarly flat bases; whereas other Neanderthals have slightly concave

Table 12. Mechanical advantage at the thumb metacarpophalangeal joint

Sample	*N*	*Thumb phalangeal ratio*[1]
Neanderthals[2]	11	20.3 ± .012[3]
Qafzeh 9[4]	1	15.4
Early Upper Paleolithic[4]	5	16.6 ± .010
Late Upper Paleolithic[4]	6	17.3 ± .001
Recent Europeans[2]	60	17.7 ± .014
Amerindians[2]	79	16.9 ± .013

[1] (1/2 proximal phalanx base articular height/articular length) × 100.
[2] Data from Trinkaus and Villemeur (1991).
[3] Mean ± 1 SD.
[4] Author's data.

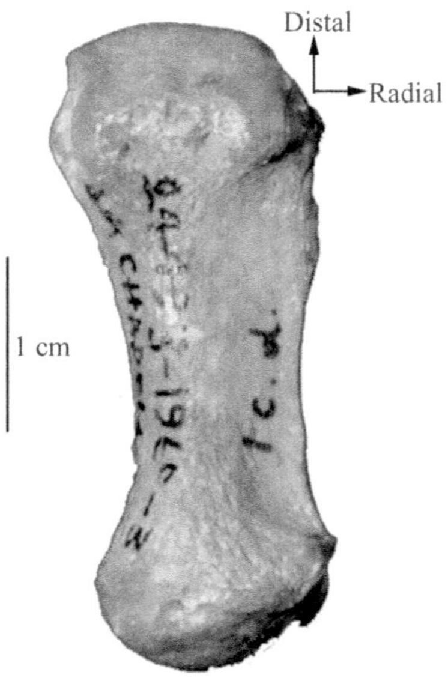

Figure 4. The La Chapelle-aux-Saints first metacarpal. Palmar view of the La Chapelle-aux-Saints first metacarpal (right side). The base is both dorsopalmarly and radioulnarly convex. This is one of the most extreme examples of Neanderthal metacarpal base dorsopalmar convexity. Most other Neanderthals have dorsopalmarly flat to slightly concave metacarpal 1 bases.

metacarpal 1 bases. In addition, the corresponding metacarpal 1 articular surface on the trapezium, although sellar, tends to be flatter than those of recent humans (Stoner & Trinkaus, 1981; Trinkaus, 1983b; Niewoehner, 1999).

Rather than arguing that Neanderthals had limited thumb mobility relative to recent humans, Stoner and Trinkaus (1981) and Trinkaus (1983 a, b) proposed that Neanderthals would have had a more mobile thumb, since primates that lack full opposability of the thumb have more tightly interlocking articular configurations. In this context, it has been argued (Trinkaus, 1989a) that the shape of the Neanderthal trapezio-metacarpal 1 articulation is an adaptive response to high levels of joint reaction forces that would have been produced by their hypertrophied musculature. This idea is based on evidence that in pinch grips, tip-to-tip, and grasping involving the thumb, most of the joint reaction forces at the first carpometacarpal joint are directed axially, parallel to the long axis of the bone, rather than perpendicular to it (Cooney & Chao, 1977) and a flatter joint surface would more evenly distribute joint reaction forces across the joint surfaces.

Riley and Trinkaus (1989) initially quantified the unusual Neanderthal mid-carpometacarpal morphology (the articulations between the capitate and the second and third metacarpals), a combination of a relatively short metacarpal 3 styloid process and a more parasagittal orientation of the facet for the capitate on the metacarpal 2 base. Niewoehner et al. (1997b) expanded the sample sizes of Neanderthals and recent humans and used a multivariate analysis of linear measurements to better quantify the morphology of this region. The univariate results confirmed Riley and Trinkaus' (1989) previous analysis; the Neanderthal sample average metacarpal 2 capitate facet angle is 22°, whereas the recent human average of 42° is significantly greater. Thus, recent human capitate facets are indeed more obliquely oriented in the coronal plane relative to the long axis of metacarpal 2. Additionally, a canonical discriminant function was performed on size corrected variables of the distal capitate facets and the metacarpal 2 and 3 bases. The results indicate that modern human capitate-to-metacarpal 2/3 morphology results from non-allometric increases in distal capitate breadth and proximal projection of the metacarpal 3 styloid process and reductions in the capitate-metacarpal 2 facet height and metacarpal 3 facet breadth and that these shape changes are associated with a significantly less parasagittal orientation of the capitate-metacarpal 2 facet.

The recent human metacarpal 5 base tends to have a sellar shape, it is normally dorsopalmarly convex and radioulnarly concave. The Neanderthal metacarpal 5 proximal articular surface, rather than being radioulnarly concave tends to be flat (Heim, 1982) and unlike recent humans, the matching articular surface on the hamate is radioulnarly flat, rather than convex. The flattening of the Neanderthal metacarpal 5 base is probably related to the same phenomenon of evenly distributing joint reaction forces as mentioned for the metacarpal 1 base.

The morphology of Neanderthal distal tuberosities adds support to the notion that Neanderthal hands are, in many ways, adapted to transmitting higher levels of force through their joints. Yet, the full ramifications (if any) of these shape changes on the kinematics of the Neanderthal trapezio-metacarpal 1 and hamate-metacarpal 5 articulations remain unclear. An additional issue that must be considered is that when the mechanical advantage of a muscle is increased it can pull with less force, thus actually reducing joint reaction forces.

Given the unique configuration of the Neanderthal metacarpal 2/3 region, which is not correlated with measures of metacarpal diaphyseal robusticity, it has been proposed that the recent human configuration is adapted for better resisting axial as well as obliquely directed joint forces. One possible behavioral interpretation is that Neanderthals relied predominantly on transverse power grips, rather than oblique power grips for most of their upper limb related subsistence tasks (Trinkaus & Villemeur, 1991; Niewoehner et al., 1997a).

If the above interpretation is correct, then there may be other morphological indications of frequency shifts in preferred manipulatory postures. One region that may be informative of a shift from a preference for power grips to a preference for precision grips is the second and fifth metacarpophalangeal joints. Recent human metacarpal 2 and 5 heads, although condyloid in shape, are asymmetrical (Susman, 1979; Lewis, 1989). The heads of the same Neanderthal metacarpals are also condyloid in shape; however, the metacarpal 2 and 5 heads appear to have reduced radioulnar asymmetry (see the geometric morphometric analysis below). These shape differences would be important given that the asymmetry of the recent human metacarpal 2 head facilitates, in part, the radial deviation of the second digit towards the thumb during metacarpophalangeal joint flexion, an important movement for precision grips. Additionally, the asymmetry of the recent human metacarpal 5 head produces an ulnar deviation of the little finger, a movement that is important in the production of both precision and power grips (Lewis, 1989).

THREE-DIMENSIONAL GEOMETRIC MORPHOMETRIC ANALYSIS OF JOINT SHAPE AND ORIENTATION

Given the above indications that Neanderthals possess some unusual carpometacarpal joint shapes and somewhat less asymmetric first and fifth metacarpal heads, I reanalyzed these regions with a geometric morphometric shape analysis to get a better idea of the nature and extent of the between-sample shape differences (Niewoehner, 2000, 2001, 2005). The number of landmarks and the number of eigenvalues used in the principal components analysis are listed in Table 13. The number of principal components and eigenvalues outputted from the Morphologika program is $N - 1$ (the number of specimens $- 1$) or km $- 1$ (k landmarks times m dimensions $- 1$), whichever is smaller. Following Jobsen (1992), the average eigenvalue of the principal components was calculated and only those principal components with eigenvalues equal to or greater than the average eigenvalue were retained for the analysis. The null hypothesis of between-sample shape equivalence is rejected for the carpal facets of the metacarpal 1, 2, 3, and 5 bases, the metacarpal facets of the capitate and hamate, and the metacarpal 2 and 5 head-shapes. These differences are presented in Table 14 as a series of contrasts between Neanderthal and recent human morphologies because these two groups are

Table 13. Number of landmarks and eigenvalues used in the geometric morphometric analysis

	Number of Landmarks	*Number of Eigenvalues*	*Percent of Total Variance*[1]
Metacarpal 1 Base	65	8	85.6
Metacarpal 2 Base	75	11	80.3
Metacarpal 3 Base	79	11	84.5
Metacarpal 5 Base	71	8	85.0
Capitate	50	8	83.6
Hamate	83	8	86.8
Metacarpal 2 Head	12	6	83.2
Metacarpal 5 Head	12	6	90.1

[1] The percent of the total sample variance represented by the eigenvalues used in.

Table 14. Neanderthal/recent human contrasts in carpometacarpal joint morphology

	Neanderthal morphology relative to the recent human morphology
Metacarpal 1 Base	Dorsopalmarly flatter bases that lack palmar beak development which tend to be torsioned and more symmetrically convex in the radioulnar direction. Neanderthal, *N* = 8; Early Upper Paleolithic, *N* = 5; Late Upper Paleolithic, *N* = 7; Recent Humans; *N* = 37; Skhūl/Qafzeh, *N* = 2.
Metacarpal 2 Base	Dorsopalmarly shorter, flatter in the radioulnar axis, and slightly more convex in the dorsopalmar axis. The facet for the capitate is proximodistally shorter and more parasagittally oriented, and the trapezium facet tends to face proximally rather than proximoradially. Neanderthal, *N* = 8; Early Upper Paleolithic, *N* = 7; Late Upper Paleolithic, *N* = 7; Recent Humans, *N* = 29; Skhūl/Qafzeh, *N* = 0.
Metacarpal 3 Base	Radioulnarly expanded capitate facets, more parasagittally oriented, dorsopalmarly flat, and proximodistally short facets for the metacarpal 2, and reduced proximal projection of the metacarpal 3 styloid process. Neanderthal, *N* = 7; Early Upper Paleolithic, *N* = 8; Late Upper Paleolithic, *N* = 7; Recent Humans, *N* = 29; Skhūl/Qafzeh, *N* = 2.
Metacarpal 5 Base	Increased proximodistal height of the facet for the metacarpal 4, a more acute orientation of the metacarpal 4 face relative to the hamate facet resulting in a more parasagittal orientation of the metacarpal 4 facet, both of which are accompanied by increases in the dorsopalmar convexity and radioulnar symmetry of the hamate facet. Neanderthal, *N* = 6; Early Upper Paleolithic, *N* = 5; Late Upper Paleolithic, *N* = 6; Recent Humans, *N* = 29; Skhūl/Qafzeh, *N* = 1.
Capitate	Metacarpal 3 facets that are dorsopalmarly taller, with metacarpal 2 facets that are proximodistally short and with reduced concavity in the dorsopalmar axis that tend to face radially rather than distally; producing a more parasagittal orientation. Neanderthal, *N* = 6; Early Upper Paleolithic, *N* = 3; Late Upper Paleolithic, *N* = 6; Recent Humans, *N* = 33; Skhūl/Qafzeh, *N* = 2.
Hamate	Dorsopalmar and radioulnar expansion of the distal facets, accompanied by both a relative decrease in the radioulnar breadth of the facet for the metacarpal 5 and a more ulnar position of the hamulus tip. Metacarpal 5 facet shape changes include decreased radioulnar convexity with increased dorsopalmar concavity. Neanderthal, *N* = 6; Early Upper Paleolithic, *N* = 6; Late Upper Paleolithic, *N* = 4; Recent Humans, *N* = 32; Skhūl/Qafzeh, *N* = 2.
Metacarpal 2 Head	Expansion of metacarpal heads on the dorsoradial and mid-dorsopalmar aspects, but the articular surfaces on the radial lobe tends to extend palmarly, rather than radially. Neanderthal, *N* = 4; Early Upper Paleolithic, *N* = 6; Late Upper Paleolithic, *N* = 4; Recent Humans, *N* = 36; Skhūl/Qafzeh, *N* = 1.
Metacarpal 5 Head	Ulnar expansion of the dorsoulnar aspect, but not the dorsal aspect, yet reduction in the degree of ulnar, palmar, and proximal projection of the ulnar lobe. Neanderthal, *N* = 4; Early Upper Paleolithic, *N* = 5; Late Upper Paleolithic, *N* = 4; Recent Humans, *N* = 34; Skhūl/Qafzeh, *N* = *1*

consistently at the extremes of the combined sample shape variation. The implications of the shape contrasts are discussed below in the context of their division into four functional regions: the trapezial-metacarpal 1 joint, the capitate-metacarpal 2/3 joint complexes, and the hamate-metacarpal 4/5 joint complex. The shape contrasts in the metacarpal 2 and 5 heads are discussed with the metacarpal 2 and 5 bases, respectively. The discussion emphasizes the morphological separation of Neanderthal and recent human samples, but as discussed later, this seeming discontinuity is often bridged by the morphologies of the Early and Late Upper Paleolithic samples.

The Trapezio-Metacarpal 1 Joint. The primary shape change identified in this anatomical complex is the degree of the development of the palmar beak of the metacarpal 1 base. The dorsopalmar flatness noted to be typical of the Neanderthal base (Musgrave, 1971; Stoner & Trinkaus, 1981; Trinkaus, 1983b) is primarily due to the lack of beak development, while the recent human metacarpal 1 base is more concave because the palmar beak is more strongly developed. Interestingly, there is considerably more shape variation in the metacarpal 1 base than in the trapezium-metacarpal 1 facet indicating that the morphology of the metacarpal 1 base is more responsive to altered force levels through the thumb than is the morphology of trapezium.

The Capitate-Metacarpal 2/3 Region. The metacarpal 2 base has to respond to a number of functional constraints because it contacts the trapezium on its radial side, the trapezoid on its proximal base surface, and the capitate and metacarpal 3 base on its ulnar side. Significant articular shape differences are found in the metacarpal 2 facet for the trapezoid. The Neanderthal sample tends to have reduced dorsopalmar concavity relative to the recent modern human samples. The most reasonable explanation for this shape difference follows the same logic proposed to explain the weak palmar beak on the Neanderthal metacarpal 1 base. Joint flattening spreads joint reaction forces more equally across the joint, thus more efficiently transmitting axially directed forces.

There are between-sample differences in the orientation of the facet for the trapezium on the radial side of the metacarpal 2 base. At the extremes of morphological variation, the Neanderthal facet for the trapezium faces more proximally, rather than proximoradially as in the recent human sample. The more proximally facing facet would be in a more effective orientation to resist axially directed forces from the metacarpal 2 diaphysis to the trapezium. Admittedly, these forces would be primarily transmitted through the larger and more centrally located trapezoid facet. However, this small facet, through its contact with the trapezium, would further increase the surface area of the base facing in approximately the same direction as the larger trapezoid facet.

The more proximal orientation of the metacarpal 2-trapezium facet could also be interpreted as an indication of a significant difference in the orientation of the Neanderthal trapezium itself. If this were true, it would mean that the trapezial-metacarpal 1 facet, as well as the thumb itself, was also radially deviated. This is somewhat similar to an idea championed by Vlček (1978), who suggested that the Neanderthal thumb was positioned more dorsally (rather than radially deviated), although his idea was based on the position of the first dorsal interosseous muscle. One has to be cautious, though, because the orientation of the trapezial-metacarpal 1 facet cannot be determined without calculating the matching trapezial-metacarpal 1 facet orientation relative to the trapezial-metacarpal 2 facet. Also, given that there is a substantial amount of variance in the position and orientation of the trapezial-metacarpal 2 facet, any changes in the orientation of the facet on the metacarpal 2 base could be compensated for by changes in the orientation of the trapezial-metacarpal 2 facet. Therefore, there is no evidence at the present time that Neanderthal and recent human thumbs differed significantly in their orientation relative to the rest of the carpus, although until this issue is explored further it remains a possibility.

On the ulnar aspect of the metacarpal 2 base is the facet for the capitate. The primary contrast between the Neanderthal and recent human samples is in the more parasagittal orientation of this facet in the coronal plane. This difference was discussed earlier in reference to the works of Riley and Trinkaus (1989) and Niewoehner et al. (1997b).

Niewoehner et al. (1997b) also noted that the recent human capitate-metacarpal 2 facet tends to face distally, rather than being oriented parasagittally in the coronal plane, mirroring the orientation of the metacarpal 2-capitate facet. This feature was also identified in this analysis and supports their argument that these features are oriented in the recent human hand in such a manner as to best transmit oblique as well as axial forces. In addition to differences in facet orientation, there are Neanderthal/recent human contrasts in the relative surface area devoted to the metacarpal 2-capitate and capitate-metacarpal 2 facets. Though more apparent in the capitate-metacarpal 2 facet, the recent human facets tend to be proximodistally expanded, thus providing more surface area for the transmission of these obliquely directed forces.

The primary contrasts in the metacarpal 3 base include a radioulnar expansion of the Neanderthal facet for the capitate. This feature can be understood by considering that on the Neanderthal capitate there is a relative reduction of joint surface devoted to the facet for the metacarpal 2 base. Similarly, the recent human capitate tends to have increased articular area devoted to the facet for the metacarpal 2 and less articular area devoted to the facet for the metacarpal 3 base. These changes are again likely indicative of adaptations to the predominance of axial forces at the Neanderthal metacarpal 3 base.

The remaining readily apparent shape contrasts on the radial aspect of the metacarpal 3 base, plus those already mentioned for the same region of the capitate and the metacarpal 2 base, are all indicative of a major shift in functional anatomy in this region. Not only does the recent human metacarpal 3 styloid process project further proximally, but a secondary effect of the recent human larger styloid process is to produce a more dorsopalmarly concave facet for the ulnar aspect of the metacarpal 2 base. This creates a deeper more rounded "cup" for the metacarpal 2 base, perhaps enhancing the conjunct rotation of the recent human metacarpal 2.

The conjunct rotation of the metacarpal 2 is important in the production of precision grips involving the first finger and thumb (Lewis, 1989; Craig, 1992). Of additional importance is the shape of the recent human metacarpal 2 head. The primary contrasts identified in this region are slight changes in the shape of the radial aspect of the metacarpal 2 head. Neanderthal metacarpal 2 heads tend to be expanded on the dorsoradial aspect; however, the radial lobe does not project as far radially as the recent human radial lobe. Admittedly, this is minor variation and most apparent at the extremes of morphological variation. Nevertheless, the evidence from this analysis for enhanced pronation of the recent human metacarpal 2, plus this orientation difference is potentially an important indicator of a "fine tuning" of the fine manipulative capabilities of the recent human hand that deserves further research.

To reiterate, the primary shape contrasts identified in the capitate-metacarpal 2/3 region are consistent with a shift from primarily axially oriented forces through this region of the Neanderthal hand to a combination of axial and oblique forces in the recent human hand. Between-sample shifts in facet orientation are apparent for the metacarpal 2-trapezium facet, the metacarpal 2-capitate facet, and the capitate-metacarpal 2 facets. This difference in facet orientations is accompanied by shifts in the relative sizes of the capitate-metacarpal 2 and 3 facets. The radioulnar flattening of the Neanderthal metacarpal 2-trapezoid facet and their non-projecting metacarpal 3 styloid process are further evidence for the predominance of axial forces in this region. There are no indications that Neanderthals lacked fine manipulative abilities due to insufficient metacarpal 2 conjunct rotation, or that the metacarpal 2 head shape is radically different from the modern human condition. Yet, significant changes in the shape of the "cup" for the ulnar aspect of the metacarpal 2 base, plus changes in the radial aspect of the metacarpal 2 head that enhance the radial deviation and supination of the first finger, are evidence for continued "fine tuning" of precision grips involving the first finger and thumb.

The Hamate, Metacarpal 5 base, and Metacarpal 5 head. One of the primary contrasts between the extremes of recent human and Neanderthal hamate facet morphology is the difference in the relative radioulnar extents of the facets for the metacarpal 5 and metacarpal 4. The tendency is for the Neanderthal hamate-metacarpal 5 facet to comprise less of the hamate's total radioulnar breadth compared to the recent human tendency of metacarpal 5 facet radioulnar expansion. Even though the Neanderthal metacarpal 5 facet tends to be relatively narrow, it nevertheless tends to be relatively taller in the dorsopalmar axis. It is possible that the recent human hamate-metacarpal 5 facet occupies a greater proportion of the distal hamate surface because the metacarpal 4 facet is reduced in its radioulnar extent relative to the Neanderthal hamate-metacarpal 4 facet. This difference in relative joint surface area, plus the fact that Neanderthals have significantly increased metacarpal 4 and 5 cortical areas (Niewoehner, 2000), implies that Neanderthals experienced primarily axial loads at both the fourth and fifth metacarpal bases, whereas recent humans are adapted to reduced axial loads at the metacarpal 4 base.

The only noticeable facet shape differences between the Neanderthal and recent human samples are found on the hamate-metacarpal 5 facet. There is a slight increase in the dorsopalmar concavity of the Neanderthal facet and a slightly more pronounced increase in the convexity of the facet in the radioulnar axis. These shape changes are more subtle than the Neanderthal/recent human differences in the shape of the metacarpal 5 base discussed below. Finally, it is not clear why the Neanderthal hamulus tends to be positioned further ulnarly than in modern humans, but it is likely that this is merely the product of increased palmar projection of the ulnarly tilted hamulus and/or their (radioulnarly) thicker hamuli.

The primary shape differences identified at the metacarpal 5 base are contrasts in the degree of concavity/convexity in the radioulnar axis of the facet for the hamate. Although among the Neanderthals this was consistently the most heavily damaged of the metacarpal bases, there is still enough of the base present to demonstrate significant shape differences between Neanderthals and recent humans. At the extreme ends of the recent human sample variation, the metacarpal 5-hamate facet is saddle-shaped, being dorsopalmarly convex and mildly radioulnarly concave. The Neanderthal metacarpal 5-hamate facet tends to lack the radioulnar concavity of most recent human specimens, and although the dorsoulnar aspect is damaged on the Neanderthal specimens, it is clear that the convexity here is generally lacking. Following the logic presented earlier for the metacarpal 1 and 2 bases, the shape contrasts in both the hamate (in terms of radioulnar flattening of the facet for the metacarpal 5) and the metacarpal 5 base are most likely further evidence of adaptations to the transmission of axial forces across the Neanderthal distal carpometacarpal row. Additionally, as in the trapezio-metacarpal 1 joint, there is more shape variation in the metacarpal base than in the matching facet on the carpal (Niewoehner, 2000).

There are more pronounced differences between the Neanderthal and recent metacarpal 5 heads that are analogous to the shape contrasts described for the metacarpal 2 head. The primary Neanderthal/recent human contrasts are in the reduced ulnar, palmar, and proximal projection of the ulnar lobe of the Neanderthal metacarpal 5 head. The functional importance of the recent human metacarpal 5 head shape has been previously detailed. The shape difference found in the Neanderthal metacarpal head also implies slightly reduced ulnar deviation and supination of the fifth proximal phalanx during finger flexion. The effect this may have had on fine manipulative abilities when objects were held at the tips of the fingers is unclear, since this shape difference is not dramatic. However, this appears to be further evidence that the recent human hand is "fine tuned" for precise manipulation.

Another interesting morphological difference is the more parasagittal orientation of the facet for the metacarpal 4 base relative to the radioulnar plane of the facet for the hamate in the Neanderthal metacarpal 5 base. It is possible that there is no functional significance to this feature, as again, it is most noticeable at the extremes of Neanderthal/recent human ranges of variation. There remains the possibility that the recent human fifth metacarpal is tilted slightly more proximoulnarly than the Neanderthal fifth metacarpal, assuming similar degrees of proximodistal curvature of the distal carpal row. This would place the recent human fifth metacarpal in a position more analogous to the first metacarpal on the radial side of the hand (however, this interpretation is the opposite of Musgrave's [1971] proposal that Neanderthal fifth metacarpals were more ulnarly deviated, giving them a wide finger span). Furthermore, the slight expansion of the articular surface on the dorsal aspect of the recent human metacarpal 5 base may be indicative of increased range of metacarpal 5 extension, and the more open configuration of the recent human hamate-metacarpal 5 facet in the dorsopalmar axis would allow for generally greater mobility in the flexion-extension axis.

One note of caution has to be sounded at this point. Although I am certain of the "reality" of the shape changes at the hamate-metacarpal 5 facet, and the relative orientation of the metacarpal 5-hamate and metacarpal 4 facets, I am less certain of the "reality" of the dorsal extension of the metacarpal 5 articular surface. First, this principal component accounts for only 2.5% of total sample variance (Niewoehner, 2000). Second, the metacarpal 5 base is difficult to digitize because it is so small and most of the Neanderthal metacarpal 5 bases have damage near the dorsal aspect. Nevertheless, slight increases in mobility of the metacarpal 5, especially in flexion, plus the metacarpal 5 head shape differences already mentioned would further enhance the fine manipulative postures of the fifth finger.

The radioulnar flattening of the Neanderthal metacarpal 5 base is analogous to the shape changes identified in the Neanderthal metacarpal 1 base. Of note is the apparent reduction of the radioulnar extent of the joint surface for the metacarpal 4 base at the extremes of recent human variation, implying reduced load transmission through the metacarpal 4 base. There is some evidence for increased range of flexion/ extension of the recent human metacarpal 5, and less certain evidence for a more pronounced proximoulnar tilt. However, the shape of the recent human metacarpal 5 head is indicative of enhanced ulnar deviation and supination of the fifth proximal phalanx.

Stasis and Change in Late Pleistocene Hand Functional Anatomy

As the preceding review has made clear, numerous features of the Neanderthal hand are consistent with hypertrophied hand musculature, indicating increased grip strength relative to recent humans. Adding to the power from hypertrophied musculature would have been increased mechanical advantages at the first and fifth carpometacarpal joints produced by their large, projecting scaphoid, trapezium, and hamate tubercles. The intrinsic hand muscles that originate on or near the trapezium tubercle and contribute to flexion, opposition, and rotation of the first metacarpal (*abductor pollicis brevis, opponens pollicis*, and *flexor pollicis brevis*) all operated with increased mechanical advantage. On the ulnar side of the hand, the extrinsic hand muscle (*flexor carpi ulnaris*) that has part of its attachment on the flexor retinaculum through the pisiform probably had increased mechanical advantage for wrist flexion and hand adduction. The intrinsic muscles on the ulnar side of the hand that originate on or near the hamulus through the flexor retinaculum are the *abductor digiti minimi, flexor digiti minimi brevis*, and *opponens digiti minimi*. Of these, the *opponens* muscle almost certainly operated with greater mechanical advantage during opposition of the fifth metacarpal while cupping the palm of the hand.

The Neanderthal hand, therefore, seems to be adapted primarily to produce increased grip strength during circumduction of the thumb at the trapezio-metacarpal 1 joint, during cupping of the palm through the opposition of the fifth metacarpal, during flexion of the wrist and adduction of the hand, and, given the previously discussed mechanical advantage of their thumb phalangeal proportions, during flexion of the thumb metacarpophalangeal joint. All of these movements are important in the production of power grips. Again, this is not to say that Neanderthals could not produce precision grips, since there are no morphological indications of limited joint movements (Stoner & Trinkaus, 1981; Trinkaus, 1983a, b; Trinkaus et al., 1991; Trinkaus & Villemeur, 1991; Niewoehner et al., 1997a; Niewoehner, 2003). Added to the previous features are the joint shape and joint orientation contrasts between Neanderthal and recent human samples that seem to be indicative of large axially directed forces being transmitted through the carpometacarpal joints.

The morphological gulf between the recent human hand and the Neanderthal hand is bridged by the other Late Pleistocene samples. With reference to joint shapes and orientations identified in the geometric morphometric analysis, there appear to be four basic patterns of morphological affinities (Figure 5). The first, evident in the lack of significant morphological separation between any of the samples, is that the trapezia and metacarpal 4 bases are essentially indistinguishable. The second pattern, evident in the capitate, hamate, metacarpal 2 and 5 heads, and the metacarpal 5 base, is one in which the Neanderthal and recent human samples are distinct, but the Early and Late Upper Paleolithic samples are morphologically indistinguishable from modern humans. The third pattern, evident in the metacarpal 2 and metacarpal 3 base morphologies, is one in which the Neanderthal and recent human samples are well separated, but the Early and Late Upper Paleolithic samples are morphologically intermediate between the Neanderthal and recent human ranges of variation. The fourth, and perhaps, the most interesting pattern, is for the first metacarpal base. Here, the Neanderthal and Early Upper Paleolithic samples are similar to each other, and as a combined sample they contrast with the recent human and Late Upper Paleolithic samples that, in turn, are indistinguishable from each other in their morphology. The Skhūl/Qafzeh

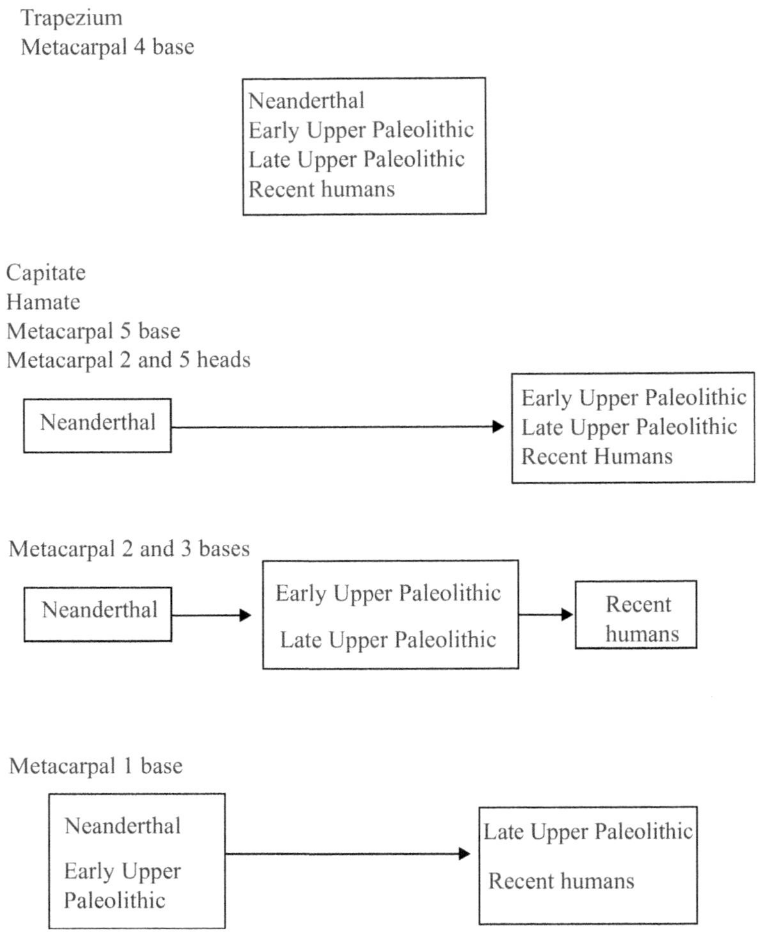

Figure 5. The mosaic nature of morphological transitions in the carpometacarpal and metacarpophalangeal regions. The samples are grouped in boxes according to their morphological similarities and arrows indicate morphological transitions.

sample is difficult to incorporate into the patterns mentioned above because of sample size problems (the largest sample size is three, see Table 14), but as has been discussed at length before (Niewoehner, 2001), in their total morphological pattern they are more similar to the combined Upper Paleolithic sample than to either Neanderthals or modern humans.

These results, coupled with qualitative observations, demonstrate that Early and Late Upper Paleolithic hand remains are more similar to recent human hands than to Neanderthal hands in features such as the morphology of the distal phalanges and in measures of carpal tubercle projection and hand/wrist mechanical advantages. The values for relative tubercle projection on both the trapezium and hamate, the thumb phalangeal ratios, and the mechanical advantage at the metacarpophalangeal joint and the ulnar side of the hand decrease rapidly with the advent of Early Upper Paleolithic humans, indicating contrasts in these measures are primarily between Neanderthals and all other samples.

Significantly though, Early and Late Upper Paleolithic hand remains are not indistinguishable from those of modern humans. Evident in the Early Upper Paleolithic sample are some features that are reminiscent of Neanderthals. These features include the presence (on some specimens) of significant muscle crests on metacarpals 1 and 5 as well as a Neanderthal-like metacarpal 1 base shape. Of special note is the fact that Early Upper Paleolithic and Neanderthal thumb carpometacarpal articulations are similar, although somewhat greater development of the metacarpal 1 palmar beak is evident in Early Upper Paleolithic humans. Evidently, Early Upper Paleolithic humans and Neanderthals shared manipulatory behaviors that produced similar levels of axial loads at the base of the thumb. However, both Late Upper Paleolithic and recent human metacarpal 1 bases are almost invariantly dorsopalmary concave, indicating reduced load levels and/or altered load orientations (Hamrick, 1996) at the base of the thumb.

Additionally, both the Early and Late Upper Paleolithic samples have intermediate metacarpal 2 and 3 base morphologies relative to Neanderthal and recent human samples. Functionally, there are indications of increased stabilization of the mid-carpometacarpal region, the enhancement of first finger precision movements, coupled with the aforementioned significant reductions in hand and wrist mechanical advantages in the Early and Late Upper Paleolithic specimens. Whereas the Neanderthal metacarpal 2 base is adapted for the transmission of primarily axially directed joint reaction forces, Early Upper Paleolithic metacarpal 2 and metacarpal 3 bases (given increased projection of their styloid process) are both adapted for increased oblique loads. These adaptations are more apparent in Late Upper Paleolithic specimens and are fully developed in recent human samples. This indicates that the Middle to Upper Paleolithic behavioral transition, insofar as manipulative behaviors are concerned, involved subtle, rather than radical frequency shifts, and the Early Upper Paleolithic humans (primarily Gravettians in this sample) employed manipulative behaviors that produced high levels of force at the base of the thumb.

Discussion

A major concern relevant to the interpretation of Neanderthal upper limb morphology is the degree to which Mousterian and Upper Paleolithic lithic technologies made use of the increased mechanical advantages of hafted tools. Microscopic indications of hafting have been reported both by Anderson-Gerfaud (1990) and Beyries (1988) for convergent and transverse sidescrapers, as well as endscrapers, but there is no indication of how frequently this may have occurred within these typological categories. Bitumen residues reported from the Umm el Tlel site in Syria (Boëda et al., 1996) are from the terminal Mousterian layer (40,000 BC), and they were found on only two pieces: a convergent scraper and a small Levallois flake. What is surprising is not that there is evidence for hafting in the Mousterian, but that the macroscopic evidence is so rare. This is a significant contrast with the more abundant evidence for hafting in the Upper Paleolithic (Kuhn, 1995). Another complicating factor is the possibility that Levantine Mousterians utilized hafted lithics (at least spear points) more frequently than European Mousterians (Shea, 1988, 1989, 1997).

Given the location and morphology of microwear traces, it is probable that single lithic elements were hafted onto the distal end of wood or bone handles. It is less likely that multiple flakes were regularly set into the lateral aspect of the shaft as is typical of some Upper Paleolithic tools. Experimental replication of the possible ways to haft a Mousterian convergent scraper (Anderson-Gerfaud, 1990) yields one plausible scenario that is in accord with functional analyses of the Neanderthal hand. Figure 6 illustrates one possible use of a scraper that has been hafted into the distal end of a wood handle and used for woodworking. The user grasps the handle transversely, rather than obliquely across the hand, requiring approximately equal flexion at each of the

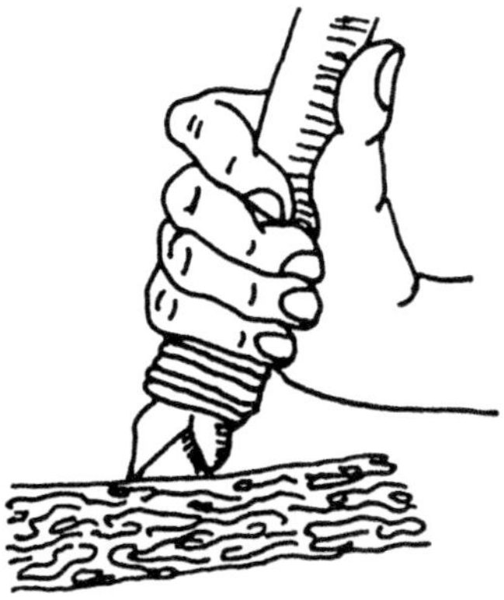

Figure 6. Hypothetical hand position for a Mousterian hafted tool. This reconstructed tool was used to scrape wood. The hand is slightly abducted and the tool placed transversely across the hand while the thumb stabilizes the shaft. Drawing by Karen Fennell of a photograph in Anderson-Gerfaud (1990).

finger metacarpophalangeal joints. The handle is stabilized in the hand not only by flexion of the fingers, but also by bracing the thumb against the handle itself. This would require powerful flexion of the fingers plus isometric contraction of flexor muscles around the thumb metacarpophalangeal and trapezio-metacarpal 1 joints. In this posture, the thumb interphalangeal joint does not have to be flexed. The increased mechanical advantage at the Neanderthal thumb metacarpophalangeal joint could reflect habitual use of this type of grasp. Also note that the hand is slightly abducted, characteristic of the most stable wrist position for power grips. Increased muscular stabilization of the wrist would be important for maintenance of this posture.

Of course, this is neither the only possible hafting method nor the only possible manipulative posture. It is merely suggestive of a possible Neanderthal positional "preference" for their use of hafted tools. Interestingly, this behavioral interpretation is also in accord with functional interpretations of the Neanderthal mid-carpometacarpal region which indicate a predominance of axial, rather than axial and oblique forces being habitually transmitted through the mid-carpometacarpal region (Riley & Trinkaus, 1989; Niewoehner et al., 1997b; Niewoehner, 2000).

The majority of non-convergent Mousterian tool types, such as many scrapers and most denticulates, probably had to be held in the hand without the benefit of a handle, since they do not show hafting marks (Anderson-Gerfaud, 1990). Neanderthal hands are better suited than other Late Pleistocene human hands for habitually and effectively using non-hafted tools. Important anatomical correlates of this are indications of generally hypertrophied hand musculature, especially of the digital flexor muscles and their (presumably) large dorsal interosseous muscles which aid in finger flexion at the metacarpophalangeal joints. Also, their expanded distal tuberosities, producing wider fingertips, would have aided their ability to firmly grasp objects against the fingertips. If the smaller tools, such as denticulates, were held in pinch grips between the thumb and the side of the index finger, the combination of a hypertrophied first dorsal interosseous muscle with powerful flexion of the thumb would have enabled Neanderthals to produce a powerful, vise-like grip.

If the use-wear studies are correct, that Mousterians spent much of their time shaping wood into various forms (Anderson-Gerfaud & Helmer, 1987; Anderson-Gerfaud et al., 1987; Beyries, 1987, 1988; Anderson-Gerfaud, 1990), a large part of their technological repertoire has not been preserved in the archaeological record. This is a reasonable inference for the Neanderthals during the early glacial period (oxygen isotope stages 5e through 5a), since even during the coldest periods (stages 5d and 5b) most of northern and western Europe still had significant patches of birch-pine forests (Zagwin, 1990). During the middle glacial period (oxygen isotope stage 4) most of northern Europe consisted of tundra (Behre, 1990;

Zagwin, 1990), severely limiting the supply of wood. However, even during the coldest periods of oxygen isotope stage 4, pollen cores from Eastern France (Le Grande Pile and Les Echets) indicate the continued presence of birch and pine in this region at least (Woillard, 1978; Beaulieu & Reille, 1984). Thus, wood was probably more or less continuously available as a raw material throughout most of Europe during the early and middle glacial periods, except the northern part of Europe during the coldest periods of oxygen isotope stage 4.

It is possible that an emphasis on woodworking, their habitual grip positions, their hafting methods, and especially their emphasis on grip strength, are all at least partially related to the modification of this archaeologically invisible component of the Neanderthal technological repertoire. The extreme reduction in Early and Late Upper Paleolithic carpal tubercle projection and most of the mechanical advantages in the hand could be indicative of a shift away from wood as the primary material and its replacement in many circumstances by alternate materials such as bone and antler, although antler and wood may have similar properties. Of course, working wood with few hafted tools and other hand held tools is not the only behavior that requires hand strength and alternate grip positions, and certainly other subsistence behaviors contributed to these functional adaptations.

It is convenient to propose a behavioral hypothesis based on archaeological "evidence" (wood) that is generally not preserved and even perhaps not universally abundant during the coldest phases of the first half of the last glaciation; yet the functional anatomy of the Neanderthal hand supports the aforementioned reliance on wood as one possible interpretation of developmentally plastic aspects of their manual anatomy. It is also possible that Neanderthals and the earliest Upper Paleolithic humans preferred different raw materials (wood vs. bone and antler), but used them for essentially the same subsistence purposes. This hypothesis is relevant to those interested in the extent and nature of Mousterian vs. Early Upper Paleolithic contrasts in subsistence behaviors and it underscores the need for functional analyses that focus on changes in mechanical effectiveness of tools as well as changes in the frequency of hafting itself during the Early Upper Paleolithic.

Unfortunately, although this scenario may make sense for the European Early Upper Paleolithic (with the reservations expressed above), it does not work at all for the Near Eastern Mousterian because it requires that the Skhūl/Qafzeh hominids be associated with Aurignacian-like bone and/or antler artifacts. Yet, it does point out that we have to begin looking at these problems in alternative ways, and develop more research strategies to investigate questions of behavioral variability (Niewoehner, 2001).

Microwear traces on Upper Paleolithic burins and some Font-Robert points (both of which are thought to have been hafted) indicate they were used to scrape, drill, and groove bone and antler (Anderson-Gerfaud & Helmer, 1987; Otte & Caspar, 1987) and the proliferation of Upper Paleolithic end scrapers is taken to be indicative of the more intensive preparation of hides (Anderson-Gerfaud, 1990). The unanswered question though, is to what degree did these shifts in tool technologies and material preparation occur in the early versus the latter stages of the Upper Paleolithic? The answer may lie, in part, in the consideration of hypothesized shifts in manipulative postures, and their morphological correlates, that would have been required to manufacture Mousterian and Upper Paleolithic tools (Table 15). Of course, there are certainly many activities beyond tool manufacture and use that could contribute to functional differences, but it is difficult to identify them archaeologically. Mousterian tools, utilized primarily to work wood, stone, skin and meat, and some bone, could have been used effectively with four basic techniques: percussion, whittling, scraping, and cutting, although other techniques listed for the Upper Paleolithic could certainly have been used at low frequencies. These four techniques all require repetitive hand motions within the same plane, most of the leverage coming from the elbow and shoulder, while the other hand holds the worked material immobile (Dennell, 1983). Functionally, these behaviors would require the maintenance of the observed upper limb muscular strength and mechanical advantages in the forearms of Neanderthals (Trinkaus, 1983a, b, 1986a; Trinkaus & Churchill, 1988; Churchill, 1994).

The manufacture and use of Upper Paleolithic tools still used basic Mousterian techniques, but had to include higher frequencies of pressure flaking, drilling, twisting, and grinding. Frequency shifts in manufacturing techniques, plus associated shifts in hand postures that were required to effectively use Upper Paleolithic tools, necessitated increased use of a number of habitual grip positions such as oblique power grips and hand/finger motions such as twisting of the fingers against the thumb for drilling, and slow precise motions of the hand for engraving (Dennell, 1983). Functionally, adaptive changes in the hand would be expected; especially those reflecting an emphasis on precision handling versus power, as well as joint configurations indicative of altered force transmission through the hand as the frequency of oblique grip use increased.

If manipulative postures changed significantly at the beginning of the Early Upper Paleolithic, one should find that Early Upper Paleolithic and Late Upper Paleolithic hand anatomies are essentially the same. In other words, we should find the same types and magnitudes of morphological contrasts in muscularity, mechanical advantages, and joint shape and orientation between Neanderthals and both Early and Late Upper Paleolithic hands. However, we do not find this. The fact that Early Upper Paleolithic and Late Upper Paleolithic hands are slightly different lends support to those who argue that the behavioral transition from the Middle to the Upper Paleolithic (at least with regards to behaviors involving manipulation such as tool use and manufacturing) was slow and gradual rather

Table 15. Probable Middle and Upper Paleolithic manufacturing techniques

Middle Paleolithic		*Upper Paleolithic*	
Technique	*Material*	*Technique*	*Material*
Percussion	Bone, Stone	Percussion	Bone, Stone
Whittling	Wood	Whittling	Wood, antler
Scraping	Wood, skin	Scraping	Wood, skin
Cutting	Wood, skin, meat	Cutting	Wood, skin, meat
		Pressure flaking	Stone
		Drilling	Bone, ivory, antler, wood, shell
		Carving	Stone, antler, bone, ivory
		Engraving	Stone, antler, bone, tooth, ivory
		Grinding, polishing	Stone, bone

Modified from Dennell (1983).

than abrupt (Clark, 2002). Finally, given the samples employed in this review it becomes clear that the most interesting and most informative hand remains are those which have yet to be discovered and analyzed, those of the earliest Aurignacian.

Acknowledgements

I thank the editors and the two anonymous reviewers for their comments and suggestions for improving this manuscript. I also thank the numerous curators and collection managers for access to their skeletal collections. Grants from the National Science Foundation, the L.S.B. Leakey Foundation, and the University of New Mexico funded this research.

References

Anderson-Gerfaud, P., 1990. Aspects of behavior in the Middle Paleolithic: Functional analysis of stone tools from Southwest France. In: Mellars, P. (Ed.) *The Emergence of Modern Humans: An Archaeological Perspective*. Edinburgh University Press, Edinburgh, pp. 389–418.

Anderson-Gerfaud, P., Helmer, D., 1987. L'emmanchement au moustérien. In: Stordeur, D. (Ed.) *La Main et l'Outil: Manches et Emmanchements Préhistoriques*. Travaux de la Maison de l'Orient, Lyon, pp. 37–54.

Anderson-Gerfaud, P., Moss, E., Plisson, H., 1987. A quoi ont-ils servi? L'apport de l'analyse fonctionelle. *Bulletin de la Société Préhistorique Française* 84, 226–237.

Bartucz, L. & Szabó, J., 1938. Der urmensch der mussolini-höhle. *Geol. Hung.* 14, 49–112.

Beaulieu, J.L.d., Reille, M., 1984. A long Upper Pleistocene pollen record from Les Echets, near Lyon France. *Boreas* 13, 111–131.

Behre, K.-E., 1990. Biostratigraphy of the last glacial period in Europe. *Quat. Sci. Rev.* 8, 25–44.

Beyries, S., 1987. *Variabilité de l'Industrie Lithique au Moustérien: Approche Functionelle sur Quelques Guisements Français*. BAR International Series, Oxford.

Beyries, S., 1988. Functional variability of lithic sets in the Middle Paleolithic. In: Dibble, H.L., Montet-White, A. (Eds) *Upper Pleistocene Prehistory of Western Asia*. The University Museum, University of Pennsylvania, Philadelphia, pp. 213–223.

Bietti, A., 1987. Some remarks on the new radiocarbon dates from the Arene Candide Cave (Savona, Italy). *Hum. Evol.* 2, 185–190.

Billy, G., 1975. *Étude anthropologique des restes humains de l'Abri Pataud*. Peabody Museum of Archaeology and Ethnology, Harvard University, Cambridge, MA.

Boëda, E., Connan, J., Dessort, D., Muhesen, S., Mercier, N., Valladas, H., Tisnérat, N., 1996. Bitumen as a hafting material on Middle Paleolithic artifacts. *Nature* 380, 336–338.

Bonch-Osmolovskij, G.A., 1941. The hand of the fossil man from Kiik-Koba (in Russian). *Paleolit Kryma* 2, 1–172.

Bookstein, F.L., 1991. *Morphometric Tools for Landmark Data: Geometry and Biology*. Cambridge University Press, Cambridge.

Boule, M., 1911–1913. L'homme fossile de La Chapelle-aux-Saints. *Ann. Paléontol.* 6, 111–172; 7, 21–56, 85–192; 8, 1–70.

Bourgon, M., 1957. Les industries moustériennes et prémoustériennes du Périgord. *Arch. Inst. Paléontol. Hum.* 9, 5–7.

Boutié, P., 1989. Étude technologique de l'industrie Moustérienne de la grotte de Qafzeh prés de Nazareth, Isreal. In: Bar-Yosef, O., Vandermeersch, B. (Eds.) *Investigations in South Levantine Prehistory*, pp. 213–229: BAR International Series 497, Oxford.

Bräuer, G.,1992. Africa's place in the evolution of *Homo sapiens*. In: Bräuer, G., Smith, F.H. (Eds.) Continuity or Replacement: Controversies in *Homo sapiens* Evolution. A.A. Balkema, Rotterdam, pp. 83–98.

Churchill, S.E., 1994. Human Upper Body evolution in the Eurasian Later Pleistocene. Ph.D. Dissertation, University of New Mexico, Albuquerque, NM.

Churchill, S.E., Trinkaus, E., 1990. Neandertal scapular glenoid morphology. *Am. J. Phys. Anthropol.* 83, 147–160.

Clark, G., Lindley, J., 1989. The case for continuity: observations on biocultural transition in Europe and Western Asia. In: Mellars, P., Stringer, C. (Eds.) *The Human Revolution: Behavioral and Biological Perspectives on the Origins of Modern Humans*. Edinburgh University Press, Edinburgh, pp. 626–676.

Clark, G.A., 2002. Neandertal archaeology – Implications for our origins. *Am. Anthropol.* 104, 50–67.

Cooney, W.P., Chao, E.Y., 1977. Biomechanical analysis of the static forces in the thumb during hand function. *J. Bone Joint Surg. Am.* 59(A)1, 27–36.

Craig, S.M., 1992. Anatomy of the joints of the fingers. *Hand Clin.* 8, 693–700.

Cremonisi, G., Parenti, R. S. R., 1972. Scheletri paleolitici della Grotta delle Veneri presso Parabita (Lecce). Atti XVa Ruin. *Sci. Ist. Ital. Preist. Protost.* Puglia, 105–117.

Dennell, R., 1983. *European Economic Prehistory: A New Approach.* Academic Press, New York.

Endo, B., Kimura, T., 1970. Postcranial skeleton of the Amud man. In: Suzuki, H., Kimura, T. (Eds.) *The Amud Man and His Cave Site.* Academic Press of Japan, Tokyo, pp. 231–406.

Formicola, V., Pettitt, P.B., Del Lucchese, A., 2004. A direct AMS radiocarbon date on the Barma Grande 6 Upper Paleolithic skeleton. *Curr. Anthropol.* 45, 114–118.

Fraipont, J., Lohest, M., 1886. La race humaine de Néanderthal ou du Candstat, en Belgique. *Bull. Acad. r. Belg. Cl. Sci.* 12, 741–784.

Frost, H.M., 1987. Bone mass and the mechanostat. A proposal. *Anat. Rec.* 219, 1–19.

Frost, H.M., 1990. Skeletal structural adaptations to mechanical usage (SATMU). *Anat. Rec.* 226, 403–422.

Frost, H.M., 1999. Joint anatomy, design, and arthroses: insights of the Utah paradigm. *Anat. Rec.* 255, 162–174.

Frost, M.H., Ferretti, J.L., Jee, W.S., 1998. Perspectives: some roles of mechanical usage, muscle strength, and the mechanostat in skeletal physiology, disease, and research. *Calcified Tissue Int.* 62, 1–7.

Gambier, H., 2002. Les fossiles de Cro-Magnon (Les Ezyies de Tayac, Dordogne): Nouvelle données sur leur position chronologique et leur attribution culturelle. *Mém. Soc. Anthropol.* Paris 14, 89–112.

Genet-Varcin, E. Miquel, M., 1967. Contribution à l'étude du squelette magdalénien de l'abri Lafaye à Bruniquel (Tarn et Garonne). *Anthropologie*, Paris 71, 467–478.

Goodall, C.R., 1991. Procrustes methods and the statistical analysis of shape. *J. R. Stat. Soc. Br.* 53, 284–340.

Gower, J.C., 1975. Generalized procrustes analysis. *Psychometrika* 40, 33–50.

Grün, R., Stringer, C.B., 1991. Electron spin resonance dating and the evolution of modern humans. *Archaeometry* 33, 153–199.

Hamrick, M.W., 1996. Articular size and curvature as determinants of carpal joint mobility and stability in strepsirhine primates. *J. Morphol.* 230, 113–127.

Heim, J.L., 1982. *Les Hommes Fossiles de La Ferrassie. Tome II. Les Squelettes Adultes.* CNRS, Paris.

Holliday, T.W., 1995. Body Size and Proportion in the Late Pleistocene Old World and the Origins of Modern Humans. Ph.D. Dissertation, University of New Mexico, Albuquerque, NM.

Holliday, T.W., 1997. Body proportions in Late Pleistocene Europe and modern human origins. *J. Hum. Evol.* 32, 423–447.

Jelenik, A.J., 1982. The Middle Paleolithic in the Levant from the perspective of Tabun Cave. In: Cauvin, J., Sanlaville, P. (Eds.) *Préhistoire du Levant.* CNRS, Paris, pp. 265–280.

Jobsen, J.D., 1992. *Applied Multivariate Data Analysis.* Springer, New York.

Kaplan, E.B., 1965. *Functional and Surgical Anatomy of the Hand.* J.B. Lippincott, Philadelphia.

Kaplan, E.B., Spinner, M., 1984. The hand as an organ. In: Spinner, M.D.M. (Ed.), *Kaplan's Functional and Surgical Anatomy of the Hand*, 3rd Edition. J.B. Lippincott, Philadelphia, pp. 3–19.

Klein, R.G., 1994. The problem of modern human origins. In: Nitecki, M.H., Nitecki, D.V. (Eds.), *Origins of Anatomically Modern Humans.* Plenum Press, New York, pp. 3–17.

Klima, B., 1959. Zur problematik des Aurignacien und Gravettian in Mittel-Europa. *Arch. Austracia* 26, 35–51.

Klima, B., 1988. A triple burial from the Upper Paleolithic of Dolní Věstonice, Czechoslovakia. *J. Hum. Evol.* 16, 831–835.

Klima, B., Kukla, J., 1963. Absolute chronological date of Czechoslovakian Pleistocene. *Conf. Int. Ass. Quatern. Res. Warsaw* 1961 1, 171–174.

Kuhn, S.L., 1995. *Mousterian Lithic Technology.* Princeton University Press, Princeton, NJ.

Lewis, O.J., 1977. Joint remodeling and the evolution of the human hand. *J. Anat.* 123, 157–201.

Lewis, O.J., 1989. *Functional Morphology of the Evolving Hand and Foot.* Oxford University Press, Oxford.

Llinas, A., McKellop, H.A., Marshall, G.J., Sharpe, F., Lu, B., Kirchin, M., Sarmiento, A., 1993. Healing and remodeling of articular incongruities in a rabbit fracture model. *J. Bone Joint Surg.* 75(A), 1508–1523.

Lovejoy, C.O., Trinkaus, E., 1980. Strength and robusticity of the Neanderthal tibia. *Am. J. Phys. Anthropol.* 53, 465–470.

Marzke, M.W., 1983. Joint function and grips *Australopithecus afarensis* hand with special reference to the capitate. *J. Hum. Evol.* 12, 197–211.

Marzke, M.W., 1997. Precision grips, hand morphology, and tools. *Am. J. Phys. Anthropol.* 102, 91–110.

Marzke, M.W., Marzke, R.F., 1987. The third metacarpal styloid process in humans. Origins and functions. *Am. J. Phys. Anthropol.* 73, 415–431.

Marzke, M.W., Shackley, M.S., 1987. Hominid hand use in the Pliocene and Pleistocene: evidence from experimental archaeology and comparative morphology. *J. Hum. Evol.* 12, 197–211.

Marzke, M.W., Wullstein, K.L., Viegas, S.F., 1992. Evolution of the power ("squeeze") grip and its morphological correlates in hominids. *Am. J. Phys. Anthropol.* 89, 283–298.

McCown, T.D., Keith, A., 1939. *The Stone Age of Mt. Carmel Vol.II: The Fossil Remains From the Levalloiso-Mousterian*. Clarendon Press, Oxford.

McDermott, F., Grün, R., Stringer, C.B., Hawkesworth, C.J., 1993. Mass spectrometric U-series dates for Israeli Neanderthal/early modern hominid sites. *Nature* 363, 252–255.

McHenry, H.M., 1983. The capitate of *A. africanus* and *A. afarensis*. *Am. J. Phys. Anthropol.* 62, 187–198.

Meignen, L., Bar-Yosef, O., 1989. Nouvelles recherches sur le Paléolithique moyen d'Isräel. In: Bar-Yosef, O., Vandermeersch, B. (Eds.), *Investigations in Southern Levant Prehistory*. BAR International Series, Oxford, pp. 169–184.

Mellars, P., 1989. Major issues in the emergence of modern humans. *Curr. Anthropol.* 30, 349–385.

Mellars, P., 1994. The Upper Paleolithic revolution. In: Cunliffe, B. (Ed.), *The Oxford Illustrated Prehistory of Europe*. Oxford University Press, Oxford, pp. 42–79.

Minellono, F., 1985. Osservazioni tecnologiche su alcune insioni paleolitiche toscane. *Museologia Scientifica* 2, 237–243.

Minellono, F., Pardini, E., Fornaciari, 1980. Le sepulture epigravettiane di Vado all'Arancio. *Riv. Sci. Preistoriche* 35, 3–44.

Movius, H.L., 1958. The Proto-Magdalénien of the Abri Pataud, Les Eyzies (Dordogne). *Ber. Vth intern. Kong. Frühgesch. (Hamburg)*, 561–566.

Musgrave, J.H., 1971. How dexterous was Neanderthal man? *Nature* 233, 538–541.

Musgrave, J.H., 1973. The phalanges of Neandertal and Upper Paleolithic hands. In: Day, M.H. (Ed.), *Human Evolution*. Taylor & Francis, London, pp. 59–85.

Mussi, M., 1986. On the chronology of the burials found in the Grimaldi Caves. *Antro. Contemp.* 9, 95–104.

Nadel, D., Hershkovitz, I., 1991. *New subsistence data and human remains from the earliest Levantine* Epipaleolithic. *Curr. Anthropol.* 631–635.

Niewoehner, W.A., 1999. A photogrammetric analysis of the MC 1 base of Late Pleistocene and recent (Holocene) humans. *Am. J. Phys. Anthropol.* Suppl. 28, 211–212.

Niewoehner, W.A., 2000. The Functional Anatomy of Late Pleistocene and Recent Human Carpometacarpal and Metacarpophalangeal Articulations. Ph. D. Dissertation, University of New Mexico, Albuquerque, NM.

Niewoehner, W.A., 2001. Behavioral inferences from the Skhūl/Qafzeh early modern human hand remains. *Proc. Natl. Acad. Sci. U.S.A.* 98, 2979–2984.

Niewoehner, W.A., 2005. A geometric morphometric analysis of Late Pleistocene human metacarpal 1 base shape. In: Slice, D. (Ed.), *Modern Morphometrics in Physical Anthropology*. Kluwer Academic Publishers, Amsterdam, pp. 285–298.

Niewoehner, W.A., Ward, C.V., Trinkaus, E., 1997a. Pliocene and Pleistocene hominid hamulus size and robusticity. *Am. J. Phys. Anthropol.* Suppl. 24, 178–179.

Niewoehner, W.A., Weaver, A., Trinkaus, E., 1997b. Neandertal capitate-metacarpal articular morphology. *Am. J. Phys. Anthropol.* 103, 219–233.

Niewoehner, W.A., Bergstrom, A., Eichele, D., Zuroff, M., Clark, J.T., 2003. Manual dexterity in Neanderthals. *Nature* 422, 395.

O'Higgins, P., Jones, N., 1998. *Morphometrika, Tools for Shape Analysis*. University College London, London.

Oakley, K.P., Campbell, B.G., Molleson, T.I., 1971. *Catalogue of Fossil Hominids. Part II: Europe*. The British Museum of Natural History, London.

Otte, M. & Caspar, J. P., 1987. Les "pointes" de la Font-Robert: outils emmanchés? In: Stordeur, D. (Ed.), *La Main et l'Outil: Manches et Emmanchements Préhistoriques*. Travaux de la Maison de l'Orient 15, Lyon, pp. 65–74.

Palma di Cesnola, A., 1993. *Il Paleolitico superiore in Italia*. Garlatti and Razzai, Florence.

Paoli, G., Parenti, R., Sergi, S., 1980. Gli Scheletri Mesolitici della Caverna delle Arene Candide (Liguria). Rome: Memorie dell 'Instituto Italiano di Paleontologia Umana, No. 3.

Pearson, O., Lieberman, D., 2004. The aging of Wolf's law: Ontogeny and mechanical responses to loading in cortical bone. *Ybk. phys. Anthropol.* 47, 63–99.

Piveteau, J., 1963. La grotte du Régourdou (Dordogne): humaine. *Ann. Paléont.* 49, 285–305.

Plochocki, J.H., Organ, J., 2003. The effects of differential mechanical loading on articular surface area in miniature swine. *Am. J. Phys. Anthropol.* (Suppl.) 36, 169.

Radovcic, J., Smith, F.H., Trinkaus, E., Wolpoff, M.H., 1988. *The Krapina Hominids: An Illustrated Catalog of Skeletal Collection*. Croatian Natural History Museum, Zagreb.

Rak, Y., 1990. On the differences between two pelvises of Mousterian context from the Qafzeh and Kebara caves, Israel. *Am. J. Phys. Anthropol.* 81, 323–332.

Ricklan, D.E., 1987. The functional anatomy of the hand of *A. africanus*. *J. Hum. Evol.* 16, 643–664.

Riley, G., Trinkaus, E., 1989. Neandertal capitate-metacarpal 2 articular morphology and Neandertal manipulative behavior. *Am. J. Phys. Anthropol.* 78, 290.

Rohlf, F., Slice, D., 1990. Extension of the Procrustes method for the optimal superimposition of landmarks. *Syst. Zool.* 39, 40–59.

Rosenberg, K., 1988. The functional significance of Neandertal pubic length. *Curr. Anthropol.* 29, 595–617.

Sarasin, F., 1932. Die Variationen im Bau des Handskeletts verschiedener Menschenformen. *Z. Morph. Anthropol.* 30, 252–316.

Sergi, S., Parenti, R., Paoli, G., 1974. Il Giovane Paleolitico della Caverna Della Arene Candide: Memorie dell 'Istituto Italiano di Paleontologia Umana, Vol. 2.

Shea, J.J., 1988. Spear points from the Middle Paleolithic of the Levant. *J. Field Arch.* 15, 441–450.

Shea, J.J., 1989. A functional study of the lithic industries associated with hominid fossils in the Kebara and Qafzeh caves. In: Mellars, P., Stringer, C. (Eds.), *The Human Revolution: Behavioural and Biological Perspectives on the Origins of Modern Humans*. Edinburgh University Press, Edinburgh, pp. 611–625.

Shea, J.J., 1997. Middle Paleolithic spear point technology. In: Knecht, H. (Ed.), *Projectile Technology*. Plenum Press, New York, pp. 79–101.

Shrewsbury, M.M., Marzke, M.W., Linscheid, R.L., Reece, S.P., 2003. Comparative morphology of the pollical distal phalanx. *Am. J. Phys. Anthropol.* 121, 30–47.

Sonneville-Bordes, D. de, 1960. Le Paléolithique supérieur en Périgord. *Delmas, Bordeaux.* I 73.

Stoner, B.P., Trinkaus, E., 1981. Getting a grip on Neandertals: Were they all thumbs? *Am. J. Phys. Anthropol.* 54, 281–282.

Straus, L.G., 1983. From Mousterian to Magdalenian: cultural evolution viewed from the Vasco-Cantabrian Spain and Pyrenean France. In: Trinkaus, E. (Ed.), *The Mousterian Legacy: Human Biocultural Evolution in the Upper Pleistocene*. BAR International Series, Oxford, pp. 73–112.

Straus, L.G., 1990. The Early Upper Paleolithic of Southwest Europe: Cro-Magnon adaptations in the Iberian peripheries, 40 000–20 000 B.P. In: Mellars, P. (Ed.), *The Emergence of Modern Humans: An Archaeological Perspective*. Cornell University Press, New York, pp. 276–302.

Straus, L.G., 1993. Upper Paleolithic hunting tactics and weapons in Western Europe. *Arch. Pap. Am. Anthropol. Assoc.* 4, 83–93.

Stringer, C.B., 1992. Replacement, continuity, and the origins of *Homo sapiens*. In: Bräuer, G., Smith, F.H. (Eds.), *Controversies in Homo sapiens Evolution*. A.A. Balkema, Rotterdam, pp. 9–24.

Stringer, C.B., 1994. Out of Africa – a personal history. In: Nitecki, M.H., Nitecki, D.V. (Eds.), *Origins of Anatomically Modern Humans*. Plenum Press, New York, pp. 149–172.

Stringer, C.B., Andrews, P., 1988. Genetic and fossil evidence for the origin of modern humans. *Science* 239, 1263–1268.

Susman, R.L., 1979. The comparative and functional morphology of hominoid fingers. *Am. J. Phys. Anthropol.* 50, 215–236.

Susman, R.L., 1988. The hand of *Paranthropus*. *Science* 240, 781–784.

Susman, R.L., 1995. Fossil evidence for early hominid tool use. *Science* 265, 1570.

Susman, R.L. & Creel, N., 1979. Functional and morphological affinities of the subadult hand (OH 7) from Olduvai Gorge. *Am. J. Phys. Anthropol.* 5, 311–332.

Svoboda, J., 1988. A new male burial from Dolní Věstonice. *J. Hum. Evol.* 16, 827–830.

Trinkaus, E., 1976. The evolution of the hominid femoral diaphysis during the Upper Pleistocene in Europe and the Near East. *Z. Morph. Anthropol.* 67, 291–319.

Trinkaus, E., 1977. A functional interpretation of the axillary border of the Neandertal scapula. *J. Hum. Evol.* 6, 231–234.

Trinkaus, E., 1978. Functional implications of the Krapina lower limb remains. In: Malez, M. (Ed.), *Krapinski Pracovjek i Evolucija Hominida*. Juvoslavenska Akademija Znanosti i Umjet, Zagreb, pp. 155–192.

Trinkaus, E., 1983a. Neandertal postcrania and the adaptive shift to modern humans. In: Trinkaus, E. (Ed.), *The Mousterian Legacy: Human Biocultural Change in the Upper Pleistocene*. BAR International Series (S164), Oxford, 165–200.

Trinkaus, E., 1983b. *The Shanidar Neanderthals*. Academic Press, New York.

Trinkaus, E., 1986a. Bodies, brawn, brains, and noses: human ancestors and human predation. In: Nitecki, M.H., Nitecki, D.V. (Eds.), *The Evolution of Human Hunting*, pp. 107–147. Plenum Press, New York.

Trinkaus, E., 1986b. The Neandertals and modern human origins. *Ann. Rev. Phys. Anthropol.* 15, 193–218.

Trinkaus, E., 1989a. Olduvai Hominid 7 trapezial metacarpal 1 articular morphology: contrasts with recent humans. *Am. J. Phys. Anthropol.* 80, 411–416.

Trinkaus, E., 1989b. The Upper Pleistocene transition. In: Trinkaus, E. (Ed.) *The Emergence of Modern Humans: Biocultural Evolution in the Later Pleistocene*. Cambridge University Press, Cambridge, pp. 42–66.

Trinkaus, E., 1991. Les hommes fossiles de la grotte de Shanidar, Irak; évolution et continuité parmi les hommes archaïques tardifs du ProcheOrient. *L'Anthropologie* (Paris) 95, 535–572.

Trinkaus, E., 1992a. Morphological contrasts between the Near Eastern Qafzeh-Skhūl and late archaic human samples: grounds for a behavioral difference? In: Akazawa, T. (Ed.), *The Evolution and Dispersal of Modern Humans in Asia*. University of Tokyo Press, Tokyo, pp. 277–294.

Trinkaus, E., 1992b. Paleontological perspectives on Neanderthal behavior. In: Toussant, M. (Ed.), *5 Millions d'Années L'Aventure Humaine*. E. R. A. U. L., Liege, pp. 151–176.

Trinkaus, E., Churchill, S., 1988. Neanderthal radial tuberosity orientation. *Am. J. Phys. Anthropol.* 75, 15–21.

Trinkaus, E., Churchill, S., 1999. Diaphyseal cross-sectional geometry of Near Eastern Middle Paleolithic humans: the humerus. *J. Archeol. Sci.* 26, 173–184.

Trinkaus, E., Thompson, D.D., 1987. Femoral diaphyseal histomorphometric age determinations for the Shanidar 3, 4, 5, and 6 Neanderthals and Neanderthal longevity. *Am. J. Phys. Anthropol.* 72, 123–129.

Trinkaus, E. & Villemeur, I., 1991. Mechanical advantages of the Neandertal thumb in flexion: a test of an hypothesis. *Am. J. Phys. Anthropol.* 84, 249–260.

Trinkaus, E., Churchill, S.E., Villemeur, I., Riley, K.G., Heller, J.A., Ruff, C.B., 1991. Robusticity versus shape: the functional interpretation of Neanderthal appendicular morphology. *J. Anthropol. Soc. Nippon* 99, 257–278.

Trinkaus, E., Ruff, C.B., Churchill, S.E., 1998. Upper limb vs. lower limb loading patterns among Near Eastern Middle Paleolithic hominids. In: Akazawa, T., Aoki, K., Bar-Yosef, O. (Eds.), *Neandertals and Modern Humans in West Asia*. Plenum Press, New York, pp. 391–404.

Ulrex-Claret, M., 1990. Le Paléolithique Moyen récent en Belgique. In: Farizy, C. (Ed.), *Le Paléolithique Moyen Récent et Paléolithique Superieur Ancien en Europe: Ruptures et Transitions*. Ass. pour la promotion de la recherche archeologique en Ile de France, Nemours, pp. 135–143..

Valladas, H., Joron, J.-L., Valladas, G., Arensburg, B., Bar-Yosef, O., Belfer-Cohen, A., Goldberg, P., Laville, H., Meignen, L., Rak, Y., Tchernov, T., Tillier, A.M., Vandermeersch, B., 1987. Thermoluminescence dates for the Neanderthal burial site at Kebara in Israel. *Nature* 330, 159–160.

Vallois, H.V., Billy, G., 1965. Nouvelles recherches sur les Hommes fossiles de l'abri de Cro- Magnon. *L'Anthropologie*, Paris 69, 47–74, 249–272.

Vandermeersch, B., 1965. Position stratographiques et chronologie relative des restes humains du Paléolithique moyen dans le Sud-Ouest de France. *Ann. Paléont.* 51, 63–126.

Vandermeersch, B., 1981. *Les Hommes Fossiles de Qafzeh (Israel)*. CNRS, Paris.

Vandermeersch, B., 1991. La ceinture scapulaire et les membres supérieurs. In: Bar-Yosef, O., Vandermeersch, B. (Eds.), *Le Squelette Moustérien de Kebara* 2. CNRS, Paris, pp. 157–178.

Vandermeersch, B., Trinkaus, E., 1995. The postcranial remains of the Régourdou 1 Neandertal: the shoulder and arm remains. *J. Hum. Evol.* 28, 439–476.

Verneau, R., 1906. *Les Grottes de Grimaldi, Monaco*. Anthropologie. Monaco Press, Monaco.

Vlček, E., 1961. Nouvelles trouvailles de l"homme du Pléistocene récent du Pavlov (CSR). *Anthropos (Brno)* 14, 141–145.

Vlček, E., 1975. Morphology of the first metacarpal of Neanderthal individuals from the Crimea. *Bull. Soc. préhist. France* 13, 257–276.

Vlček, E., 1978. Transformation of the metacarpal bones and some short hand muscles of the Neandertal man. 19th Morphological Congress Symposia Charles University (Prague), pp. 89–99.

Vlček, E., 1991. Die Mammutjäger von Dolní Věstonice: anthropologishe Bearbeitung der Skelette aus Dolní Věstonice und Pavlov. Berichte aus der Arbeit des Amtes für Museen und Archäologie des Kantons Baselland, Basel.

Woillard, G.M., 1978. Grande Pile peat bog: a continuous pollen record for the last 140,000 years. *Quatern. Res.* 9, 1–21.

Wolpoff, M.H., 1989a. Multiregional evolution: the fossil alternative to Eden. In: Mellars, P., Stringer, C.B. (Eds.), *The Human Revolution: Behavioural and Biological Perspectives on the Origins of Modern Humans*. Edinburgh University Press, Edinburgh, pp. 62–108.

Wolpoff, M.H., 1989b. The place of Neanderthals in human evolution. In: Trinkaus, E. (Ed.), The *Emergence of Modern Humans*. Cambridge University Press, Cambridge, pp. 97–155.

Wolpoff, M.H., 1992. Theories of modern human origins. In: Bräuer, G., Smith, F.H. (Eds.), *Controversies in Homo sapiens Evolution*. A.A. Balkema, Rotterdam, pp. 25–63.

Wolpoff, M.H., 1999. *Paleoanthropology*. McGraw-Hill, New York.

Zagwin, W.H., 1990. Vegetation and climate during the warmer intervals in the Late Pleistocene of western and central Europe. *Quatern. Int.* 3/4, 57–67.

Zeuner, F.E., 1940. The age of Neanderthal man. *Occ. Pap. Inst. Archeol. Univ. London* 3, 8.

10. Did Neanderthals make the Châtelperronian assemblage from La Grotte du Renne (Arcy-sur-Cure, France)?

S.E. BAILEY
Abteilung für Humanevolution
Max-Planck-Institut für Evolutionäre Anthropologie
Deutscher Platz 6
04103, Leipzig, Germany

Current address:
Center for the Study of Human Origins
Department of Anthropology
New York University
New York
New York, NY 10003, USA
sbailey@nyu.edu

J.-J. HUBLIN
Abteilung für Humanevolution
Max-Planck-Institut für Evolutionäre Anthropologie
Deutscher Platz 6
04103, Leipzig, Germany
Hublin@eva.mpg.de

Keywords: Neanderthal remains, fossil teeth, dental morphology, Late Middle Paleolithic

Abstract

Much debate has focused on the significance of the "modern" cultural elements found in European Late Middle Paleolithic (Châtelperronian, Uluzzian, and Szeletian) contexts. In light of evidence suggesting cultural interaction between the makers of these industries and the makers of the Aurignacian (presumably anatomically modern humans) it is imperative that the taxonomic affiliation of the hominins associated with these "transitional" industries be accurately identified. The fossil remains from the Châtelperronian levels (VIII-X) at the Grotte du Renne (Arcy-sur-Cure, France) comprise a series of isolated teeth, as well as a child's temporal bone. While the temporal bone has been analyzed (and identified as having Neanderthal affinity), most of the 29 teeth from these levels have not been described. The Châtelperronian dental remains from the Grotte du Renne comprise both permanent and deciduous teeth. Fortunately, most are well preserved and relatively unworn. Simple dental dimensions are not particularly helpful in attempts to differentiate between Neanderthals and anatomically modern humans. The dimensions of the postcanine teeth in these two groups overlap completely. However, Neanderthals are known to have larger anterior teeth (on average), especially relative to their postcanine tooth size. Not surprisingly, we find that the crown dimensions for the postcanine teeth from the Grotte du Renne fall within the ranges of both hominin groups. The crown dimensions of the anterior teeth, however, strongly suggest that they belong to Neanderthal individuals. The buccolingual measurements of all but one tooth fall outside the range of Upper Paleolithic modern humans and

K. Harvati and T. Harrison (eds.), Neanderthals Revisited: New Approaches and Perspectives, 191–210.

within the range of Neanderthals. Research by the first author has identified key dental morphological features that can be used to differentiate Neanderthals and anatomically modern humans. These key characters are found in the upper incisors, upper molars, P_4 and lower molars. Fortunately all but the upper molars are represented by the Châtelperronian remains at the Grotte du Renne. The strongly shoveled, labially convex lateral incisors with strong lingual tubercles, the asymmetrical P_4 with a strong, mesially placed metaconid and multiple lingual cusps, and the presence of the mid-trigonid crest on lower molars all point to a Neanderthal affinity of these individuals. In addition, the morphology of the deciduous teeth more closely resembles that of Neanderthals than it does that of anatomically modern humans. There is no single dental morphological character present exclusively in Neanderthals. Rather, it is the frequency with which certain characters occur and, more importantly, the combinations of morphological features that are important diagnostic tools. The distinctive combinations of features characteristic of Neanderthal teeth are all found in the Châtelperronian-associated teeth from the Grotte du Renne. Our analysis of both the permanent and deciduous teeth, therefore, is in agreement with the analysis of the temporal bone indicating the makers of the Châtelperronian at the Grotte du Renne were Neanderthals.

Introduction

The sites of Arcy-sur-Cure, located south-east of Paris in the Yonne department, consist of a network of caves carved out by the Cure River. These caves were excavated under the direction of André Leroi-Gourhan between 1946 and 1963 (Leroi-Gourhan, 1958, 1961). The Grande Grotte and the Grotte du Cheval are well known by Paleolithic art enthusiasts for their painted walls dating to between 24 and 32 ka. Other caves, the Grotte de l'Hyene, the Grotte du Renne and the Galerie Schoepflin, preserve evidence of Mousterian occupation, including fossils and/or artifacts.

The Grotte du Renne has been of particular interest because of the discovery of a Châtelperronian artifact assemblage, which is rich in bone tools and personal ornaments (d'Errico et al., 1998). Fourteen stratigraphic units were identified at the Grotte du Renne. The Châtelperronian artifacts are contained in three stratigraphic levels (VIII–X) that are sandwiched between an Aurignacian level (VII) and three Mousterian levels (XI–XIII). Gravettian levels (IV–VI) have also been identified (Figure 1). A child's temporal bone was recovered from the Châtelperronian level Xb, which has been dated by the ^{14}C method. If only the AMS dates are taken into consideration, the ages obtained are 33,820 ± 720 BP (OxA-3462), 34,450 ± 750 BP (OxA-8452/Ly-895) and 33,400 (OxA-9122/Ly-1055) (David et al., 2001). An older date of 38,300 ±1300 (OxA-8451/ly-894) may result from a sample inversion (David et al., 2001). Although there has been some controversy regarding dates in the Arcy sequence where conventional ^{14}C dates show evidence of contamination (David et al., 2001; White, 2001), palynological and chronostratigraphical information, together with information from other Châtelperronian sites, suggests that the Châtelperronian began at the start of the des Cottés Interstadial (Interstade des Cottés), and lasted about 5000 years, which places it generally between 38,000 and 33,000 ^{14}C years BP.

Much debate has surrounded the significance of the Châtelperronian industry at Arcy-sur-Cure. Initially, conventional thought presumed that anatomically modern humans were the makers of the Châtelperronian, as well as of other early Upper Paleolithic-like assemblages. Doubts had already been raised about this view by Leroi-Gourhan himself, who claimed some teeth from Arcy could be non-modern (Leroi-Gourhan, 1958, 1961). After the discovery of a well-preserved partial Neanderthal skeleton clearly associated with the Châtelperronian at St. Césaire (Lévêque and Vandermeersch, 1980), attention turned to hypotheses regarding the explanation of Neanderthal remains with Upper Paleolithic artifacts. Several authors have supported the view that the cultural evolution of the very last

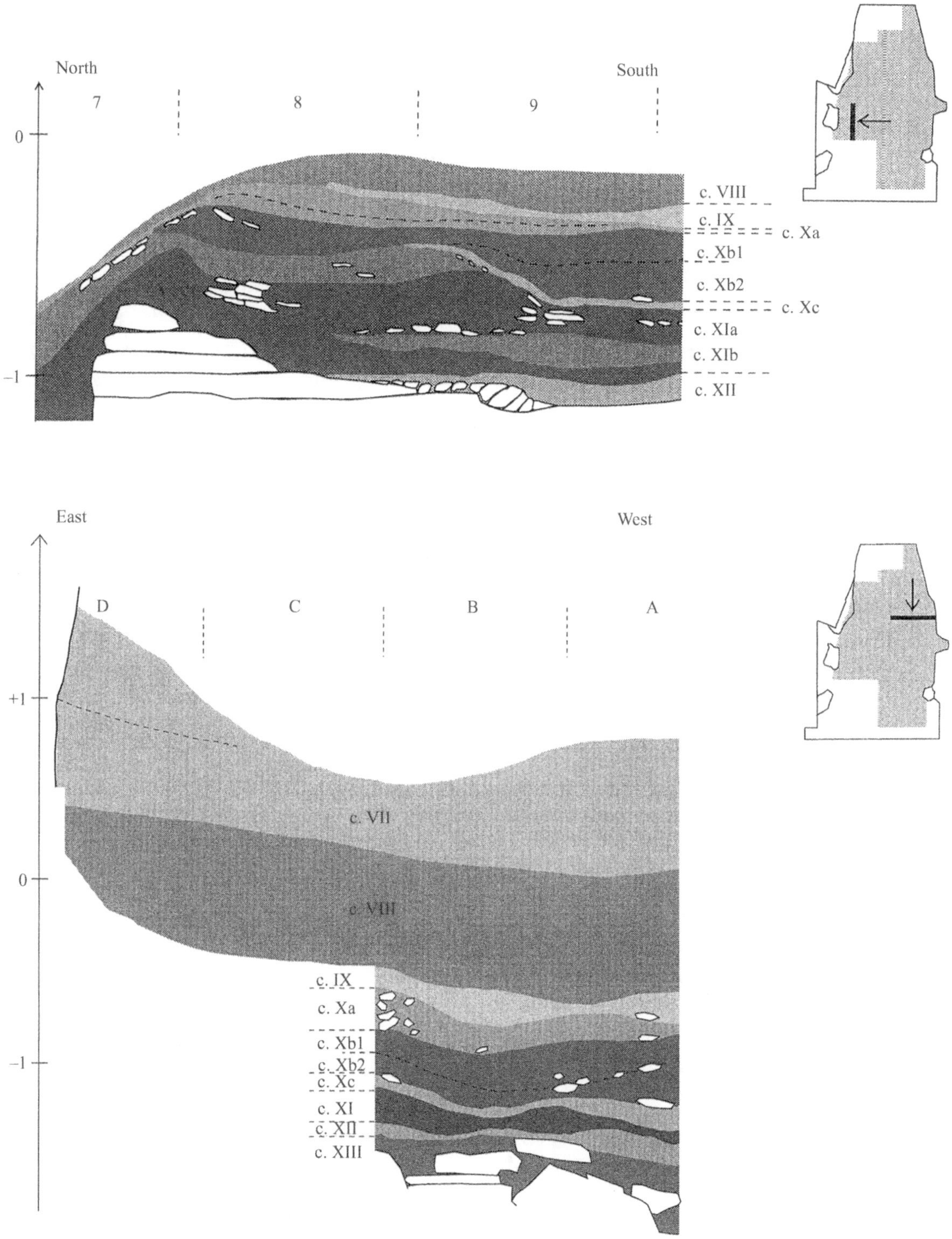

Figure 1. Profiles from the Grotte du Renne, Arcy-sur-Cure. A. Eastern aspect B. Northern aspect (Drawn by R. Humbert, taken from Connet, 2002 with permission).

Neanderthals could result from an acculturation process by the modern invaders (Demars and Hublin, 1989; Harrold, 1989; Hublin et al., 1996; Mellars, 2004). Alternatively, some have proposed an independent invention of some of the Upper Paleolithic cultural innovations by the last Neanderthals, in particular, the use of body ornaments (d'Errico et al., 1998).

Recently, the re-dating of several Upper Paleolithic sites (e.g., Vogelherd, Cro-Magnon) has led some to suggest that we do not know who the makers of any of the early or initial Upper Paleolithic assemblages were (Conard et al., 2004). One primary issue is that the human remains associated with such assemblages are generally poorly preserved, and in certain cases consist nearly wholly of teeth. For example, recently Henry-Gambier et al. (2004) claimed that the human remains associated with the early Aurignacian at Brassempouy are undiagnostic and could be either Neanderthal or anatomically modern. A critical analysis of the available data does not support this view, but rather confirms the anatomically modern nature of these remains (Bailey and Hublin, 2005).

Similar to Brassempouy, the fossil sample associated with the Châtelperronian assemblage at the Grotte du Renne consists mainly of teeth. As a result, the taxonomic affiliation of the fossils, as well as their association with the Châtelperronian artifacts, has been questioned. In addition to isolated teeth, the fossil sample consists of some skeletal remains including an infant's temporal bone. The six teeth originally recovered from the Châtelperronian levels were described as "paleoanthropic" and indistinguishable from those from lower Mousterian levels by Leroi-Gourhan (1958). More recently, Hublin et al. (1996) showed that the infant's temporal bone from Level Xb had a clear Neanderthal affinity. Coming after the Saint-Césaire discovery, this provided support that Neanderthals, not modern humans, were responsible for the assemblage.

Recently the association between the temporal bone and the Châtelperronian artifacts has been questioned by Connet (2002). This author suggests that the temporal bone derives from a part of the cave where there is the potential for disturbance resulting from the sloping of the deposits away from the cliff wall in an area where a Mousterian fossil could have "moved up" in the stratigraphy. If this were the only human fossil associated with the assemblage, it could indeed be problematic. However, subsequent to Leroi-Gourhan's original 1958 publication an additional 25 teeth mostly, but not exclusively, from level Xb have been recovered. With this expanded dental sample we observe not one, but several individuals associated with the Châtelperronian assemblage of Arcy. When plotted on a map of the site (Figure 2), it is apparent that the teeth are not limited to any particular area. An equal number of teeth come from the horizontal deposits more than one meter thick in the back of the shelter and from the thinner deposits of the slope. Although it is possible to argue, as did Connet (2002), that the latter witnessed some disturbance in relation with processes of site formation, this argument does not apply to the former.

There are two primary questions that need to be addressed with regard to the Châtelperronian dental sample from the Grotte du Renne. First, is it possible to identify the taxonomic affinity of the sample based solely on isolated teeth? And second, if the teeth are diagnosable as Neanderthal, are those teeth that exhibit *diagnostically Neanderthal* characters limited to the areas of the site where vertical displacement is a viable and likely explanation? The expanded fossil sample, as well as recent work on Neanderthal dental morphology, provides us with an opportunity to address this issue in a novel way.

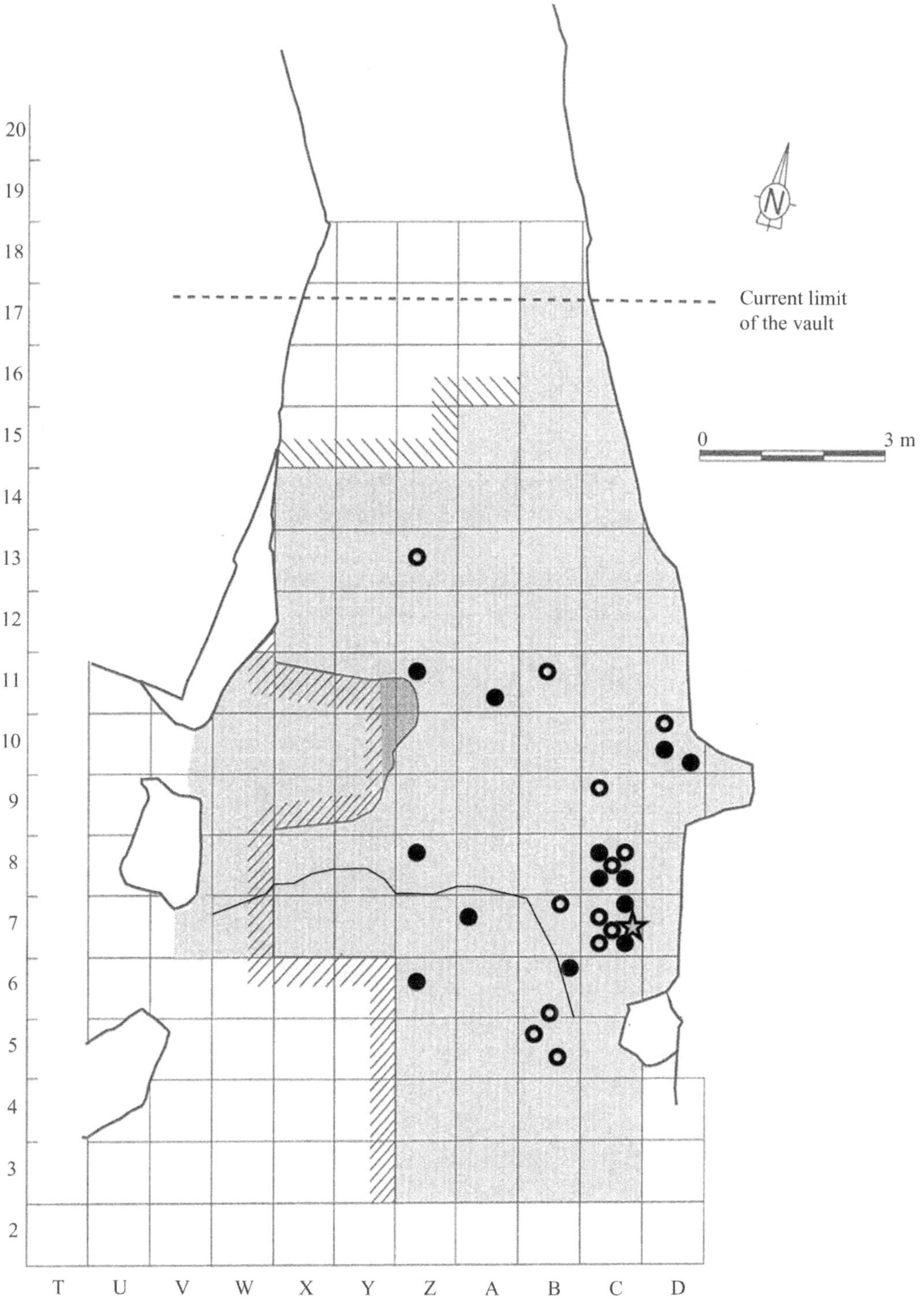

Figure 2. Distribution of the isolated teeth at the Grotte du Renne in Châtelperronian levels. Black dot: tooth with diagnostic Neanderthal features; unfilled circle: tooth consistent with (but not proving) Neanderthal morphology; grey star: child's temporal bone. Black line extending from W7 to B6 shows limit between the plateau (above) and the slope (below) in the site. All tooth positions take into account changes in the coordinate system before 1956. Four teeth (Nos. 4, 5, 6 & 7 are not plotted because of the uncertainty of their exact location (see Note 2, Table 1). Original drawn by R. Humbert, modified (with permission) from Connet, 2002.

The Dental Sample

Leroi-Gourhan (1958) originally described six human teeth recovered from the Châtelperronian levels, two of which were later identified as non-human. Subsequent excavations between 1959 and 1963 uncovered an additional 25 human teeth. The dental sample now consists of 15 permanent and 14 deciduous teeth. Most are relatively unworn and derive from young individuals. A complete description of the entire dental set can be found in Bailey and Hublin (2006).

Table 1 presents the list of Châtelperronian-associated specimens. While the 14 deciduous teeth greatly expand the fossil deciduous tooth sample, to date there has been no systematic study comparing Neanderthal and anatomically modern deciduous tooth morphology using agreed upon methods and standards. Therefore, the taxonomic assessment presented here will be based primarily on the permanent teeth.

It has sometimes been assumed that the teeth of Neanderthals and anatomically modern humans are very much alike. However, recent comprehensive analyses of the Neanderthal dentition have shown this assumption to be misconceived. While simple measurements of the postcanine teeth show complete overlap between Neanderthals and anatomically modern humans, the anterior

Table 1. List of human teeth from the Châtelperronian levels of the Grotte du Renne, Arcy-sur-Cure

Spec no	*Level*	*Specimen label*	*Tooth (side)*	*Age*	*Publication*
11	VIII	Z11 451	I_2 (L)	Subadult >8 yrs	Bailey and Hublin, 2006
4[2]	IX	IXb B7	P_4 (L)	12–18 yrs	Leroi-Gourhan, 1958
13	IX	IXc Z13	P^4 (R)	15–18 yrs	Bailey and Hublin, 2006
16[1]	IX	RIX B7	M_2 (R)	adult	Bailey and Hublin, 2006
5[2]	X	RXb A6	M_2 (R)	>15 yrs	Leroi-Gourhan, 1958
6[2]	X	RXb A6	M_3 (R)	adult	Leroi-Gourhan, 1958
7[2]	X	RXb Z8	C, (L)	12+ yrs	Leroi-Gourhan, 1958
17[1]	X	RXa C7	M_{1-2} (R)	>15 yrs	Bailey and Hublin, 2006
18	X	RXb1 D10	dm_1 (R)	4–7 yrs	Bailey and Hublin, 2006
19	X	RXb1 D10	I^2 (L)	4–6 yrs	Bailey and Hublin, 2006
20	X	RXb1 D10	P^3 (L)	5–7 yrs	Bailey and Hublin, 2006
21	X	RXb1c A11	M_2 (R)	7–9 yrs	Bailey and Hublin, 2006
22	X	RXb2 B5 1916	di^2 (R)	4–6 yrs	Bailey and Hublin, 2006
23	X	RXb2 B6 1506	I^2 (L)	6–8 yrs	Bailey and Hublin, 2006
24	X	RXb2 B6	P^3 (L)	5–7 yrs	Bailey and Hublin, 2006
25	X	RXb2 B11 3191	dm_1 (R)	5–7 yrs	Bailey and Hublin, 2006
26	X	RXb2 C7	dm^2 (R)	9–12 mo	Bailey and Hublin, 2006
27	X	RXb2 C7	di^1 (R)	<18 mo	Bailey and Hublin, 2006
28	X	RXb2 C7	di^2 (R)	7–12 mo	Bailey and Hublin, 2006
29	X	RXb2 C8	dm_2 (R)	7–15 mo	Bailey and Hublin, 2006
30	X	RXb2 C8	M_1 (R)	9–18 mo	Bailey and Hublin, 2006
31	X	RXb2 C8	dc, (R)	7–15 mo	Bailey and Hublin, 2006
32	X	RXb2 C8	di^1 (L)?	>5–6 yrs	Bailey and Hublin, 2006
33	X	RXb2 C8	dm_1 (R)	6–11 mo	Bailey and Hublin, 2006
34	X	RX C7	dm^1 (L)[1']	7–11 mo	Bailey and Hublin, 2006
35	X	RXc A7	M_1 (R)	6–9 yrs	Bailey and Hublin, 2006
36	X	RXb2 B5	di^1 (L)	birth	Bailey and Hublin, 2006
37	X	RXc Z6	dc' (R)	3–7 yrs	Bailey and Hublin, 2006
38	X	RXc C9	dc' (L)	4–8 mo	Bailey and Hublin, 2006

[1] Not used in the analysis because worn or damaged.

[2] These teeth are of uncertain location because of changes in the coordinate system before 1956. The locations of all other teeth have been checked according to the post-1956 grid.

Table 2. A list of trait frequencies that distinguish Neanderthals from Upper Paleolithic moderns and their presence or absence in the Arcy-sur-Cure sample

Tooth	*Trait*	*Trait presence (based on ASUDAS[1] and Bailey, 2002b)*	*Arcy-sur-Cure*	*Mousterian Neanderthal % present (n)*	*Upper Paleolithic Modern % present (n)*
I^2 ($n = 2$)					
	Shoveling	Grade 3+	present	93 (27)	43 (7)
	Lingual tubercles (*Tuberculum dentale*)	Grade 1+	present	100 (25)	0 (7)
	Two of the above		**present**	**100 (24)**	**0 (7)**
P^3 ($n = 2$)					
	Essential crest	Grade 1+	present	100 (19)	43 (7)
	Maxillary Premolar Accessory Ridges (MxPAR)	Grade 1+	present (1/2)	69 (16)	25 (4)
	Two of the above		**present (1/2)**	**88 (16)**	**0 (5)**
P^4 ($n = 1$)					
	Essential crest	Grade 1+	present	100 (18)	67 (6)
	Maxillary Premolar Accessory Ridges (MxPAR)	Grade 1+	present	77 (22)	50 (2)
	Two of the above		**present**	**78 (18)**	**50 (2)**
C_1 ($n = 1$)					
	Distal accessory ridge	Grade 2+	present	67 (12)	29 (7)
P_4 (n = 1)					
	Distolingual cusp	Grade 2+	present	90 (30)	39 (13)
	Transverse crest	Grade 2+	absent	77 (27)	7 (14)
	Asymmetry	Grade 1+	present	92 (25)	33 (9)
	Two of the above (distolingual cusp + asymmetry)		**present**	**91 (22)**	**9 (11)**
M_1 ($n = 2$)					
	Mid-trigonid crest	Grade 1+	present	94 (28)	0 (23)
	Cusp 6	Grade 1+	present	26 (19)	19 (21)
	Two of the above		**present (1/2)**	**57 (7)**	**0 (15)**
M_2 ($n = 2$)					
	Y-pattern	Y	present (1/2)	79 (34)	44 (25)
	Cusp 6	Grade 1+	present	55 (20)	24 (17)
	Mid-trigonid crest	Grade 1+	present	91 (24)	9 (23)
	Anterior fovea	Grade 2+	present	88 (24)	53 (19)
	Three of the above (Cusp 6+ mid-trigonid crest +anterior fovea)		**present**	**63 (20)**	**0 (13)**
M_3 ($n = 1$)					
	Four cusps		absent	0 (23)	32 (19)

[1] ASUDAS: Arizona State University dental anthropology system (Turner et al., 1991).

dentition of Neanderthals is relatively larger than that of anatomically modern humans (Bytnar et al., 1994). A recent study of tooth root lengths also indicates that the roots of several teeth are significantly longer in Neanderthals than in Upper Paleolithic modern humans (e.g., I^1, I^2, C', I_1, I_2, C, P_3, P_4 and M_2); and, for some teeth (e.g., I^1, C' and I_1) there is little or no overlap in their ranges (Bailey, 2005).

Morphologically, while no single dental morphological character is uniquely present in Neanderthals, the frequencies with which certain traits occur and, perhaps more importantly, the combination of traits in a single individual or in individual teeth has proven to be an important set of diagnostic tools (Bailey, 2002a, b, 2004; Bailey and Lynch, 2005). Table 2 provides a list of tooth traits in which frequencies distinguish Neanderthals from Upper Paleolithic modern humans and their presence or absence in the Grotte du Renne sample.

For example upper incisors show marked differences between Neanderthals and anatomically modern humans. Not only do they tend to be relatively larger in Neanderthals (especially buccolingually), but they show a distinctive combination of morphological features as well (Mizoguchi, 1985; Crummett, 1995; Bailey, 2000). Mizoguchi described Neanderthal incisors as having "extremely developed marginal ridges which run parallel to each other, a very deep lingual fossa and a

Table 3. Comparative buccolingual measurements for the Arcy-sur-Cure sample, Mousterian Neanderthals and Upper Paleolithic modern humans

	The Grotte du Renne, Arcy-sur-Cure	*Mousterian Neanderthal mean (n) range[1]*	*Upper Paleolithic modern mean (n) range[1]*
I^2	8.2, 8.8	8.2 (n = 8) 7.4–8.8	6.7 (n = 11) 5.8–8.3
P^3	11.3	10.1 (n = 17) 8.1–11.3	9.7 (n = 12) 8.7–10.6
P^4	10.5	10.2 (n = 11) 8.2–11.3	9.7 (n = 12) 8.8–10.9
I_2	7.8	7.5 (n = 7) 6.0–8.0	6.8 (n = 21) 6.0–7.5
C,	9.8	8.5 (n = 10) 5.6–9.8	8.4 (n = 16) 7.2–9.7
P_4	10.2	8.8 (n = 11) 7.6–10.5	8.4 (n = 14) 7.1–9.2
M_1	11.1	10.8 (n = 18) 9.7–11.8	10.9 (n = 28) 9.8–11.9
M_2	11.6, 11.6	10.9 (n = 16) 9.9–12.1	10.7 (n = 30) 8.6–12.3
M_3	10.8	10.8 (n = 13) 7.8–13.1	10.6 (n = 12) 7.7–12.5

[1] Bailey, unpublished data: Comparative samples include the following sites:
Mousterian Neanderthals: Arcy-sur-Cure (levels XI and XII), Ciota Ciara, Grotte Guattari, Hortus, Krapina, Kůlna, La Fate, La Quina, Melpignano, Montmaurin, Ochoz, Pontnewydd, Petit Puymoyen, Régourdou and Spy.
Upper Paleolithic moderns: Abeilles, Abri Blanchard, Abri Castanet, Abri Pataud, Aurignac, Bruniquel, Gough's Cave, Dolní Věstonice, Farincourt, Fourneau-du-Diable, Grottes d'Isturitz, La Chaud, La Ferrassie, La Gravette, La Grèze, La Linde, La Madeleine, Les Vachons, Laugerie Basse, Les Rois, Mieslingtal, St. Germain-la-Rivière and Vindija.

large lingual tubercle" ("Type 2" shoveling: Mizoguchi, 1985: 47). He clearly distinguished this form from that of modern humans and also noted that the presence of this morphology in the lateral incisors further distinguished Neanderthals from *Homo erectus*. Crummett (1995) noted that, in addition to the aforementioned characters, Neanderthal incisors are typified by marked labial convexity. Shoveling and lingual tubercles on the upper central incisor are primitive characters found in other fossil hominins as well (Mizoguchi, 1985). The degree of expression and the combination of these three characters in a single tooth, however, is distinctive of Neanderthals.

The two upper lateral incisors in the Châtelperronian sample are large. The buccolingual dimension of one falls at the upper end of the Upper Paleolithic variation, and that of the other falls above its range (Table 3). Morphologically, they show strong shoveling, marked lingual tubercles and labial convexity (Figure 3). The combination of shoveling and *lingual tubercles* in the I^2 occurs in 92% of Neanderthals and only 13% of Upper Paleolithic moderns (Table 2). In their combination and expression of these three features they clearly show affiliation to Neanderthals.

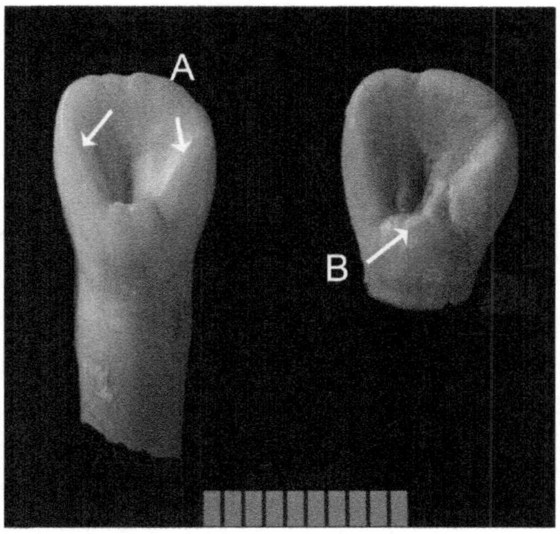

Figure 3. Two left I^2s from Level X, Grotte du Renne. Both show strong shovel shape (A), lingual tubercles (B) and labial convexity (not shown).

Upper premolars of Neanderthals are quite similar to those of other archaic humans. The three upper premolars recovered from the Grotte du Renne are, as Leroi-Gourhan first described them, "paleoanthropic". They

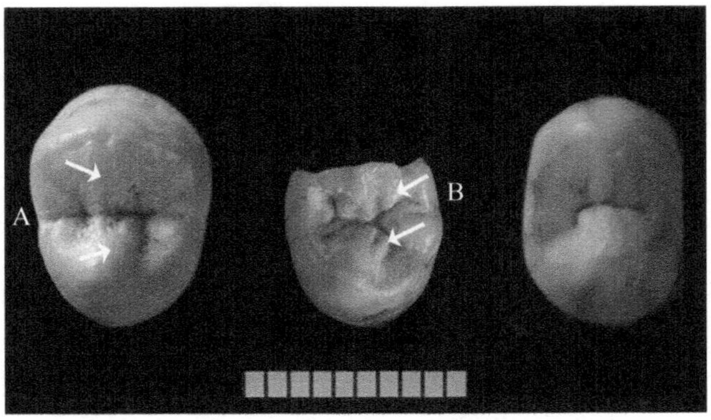

Figure 4. Upper premolars from the Grotte du Renne. Left: left P^3 (Level X). Middle: left P^3 (Level X). Right: right P^4 (Level IX). A: essential crest, B: MxPAR (maxillary premolar accessory ridges).

present strong essential (median) crests on the buccal and lingual cusps and two of the three present accessory ridges (MxPAR: Burnett, 1998) (Figure 4). The frequencies of these features are lower in Upper Paleolithic modern specimens than in Neanderthals. Separately they are not particularly useful traits for taxonomic affiliation; however, in combination they are more informative. One of the five scorable Upper Paleolithic modern P^3s presents accessory ridges (Table 2), but it does not exhibit a definite essential crest. However, a majority (88%) of the Neanderthal specimens present these traits in combination. For the P^4, again, 78% of Neanderthals show these traits in combination while only one of the two scorable Upper Paleolithic moderns shows this combination.

Metrically, the buccolingual dimension of the P^3 falls outside the range of Upper Paleolithic modern specimens and within the range of Neanderthals (Table 3), while that of the P^4 falls within the range of both Upper Paleolithic moderns and Neanderthals. The root of the P^4, however, is quite long. It falls within the range of Neanderthals and is much longer than the two Upper Paleolithic modern P^4s with measurable roots (Table 4).

The single lower incisor (an I_2) possesses archaic features, including moderate shoveling, median ridge development and a cingulum shelf (Figure 5). Its buccolingual breadth falls outside the range of Upper Paleolithic moderns and within the range of Neanderthals (Table 3). Bytnar et al. (1994) have shown that late archaic humans (Neanderthals) and early modern humans in the Near East differ significantly in I_2 buccolingual dimensions. SEB has found that European Neanderthals also have I_2s with significantly larger buccolingual dimensions than those of Upper Paleolithic moderns (see Table 3, $t = 4.34$, $p < .0001$, df = 34, Bailey unpublished data). The buccolingual dimensions of this tooth, together with its archaic morphology, suggest Neanderthal affiliation.

Like the I_2, the lower canine is archaic in its size and morphology. Leroi-Gourhan (1958) noted that its robust crown dimensions and

Table 4. Root lengths in the Arcy-sur-Cure sample compared to that of Mousterian Neanderthals and Upper Paleolithic modern humans

	The Grotte du Renne, Arcy-sur-Cure	*Mousterian Neanderthal mean (n) range*[1]	*Upper Paleolithic modern mean (n) range*[1]
P^4	16.6	17.6 (n = 10) 16.2–19.0	11.9 (n = 2) 10.5, 13.3
C,	18.0	19.7 (n = 7) 17.3–23.2	15.9 (n = 4) 13.2–19.0
P_4	17.4	18.7 (n = 7) 14.5–21.0	14.5 (n = 6) 13.0–17.1
M_1	13.6	14.3 (n = 9) 12.2–16.8	13.2 (n = 3) 11.6–14.0
M_2	14.4	15.3 (n = 6) 14.3–16.3	13.7 (n = 7) 12.6–16.8
M_3	15.0	14.3 (n = 5) 11.8–14.1	no data

[1] Bailey, unpublished data: Comparative samples include the following sites:
Mousterian Neanderthals: Krapina, Hortus, Ciota Ciara, Petit Puymoyen, Régourdou, La Quina and Spy.

Upper Paleolithic moderns: Fourneau-du-Diable, Gough's Cave, Grottes d'Isturitz, La Chaud, La Ferrassie, La Gravette, La Grèze, Les Vachons, Les Rois and St. Germain-la- Rivière.

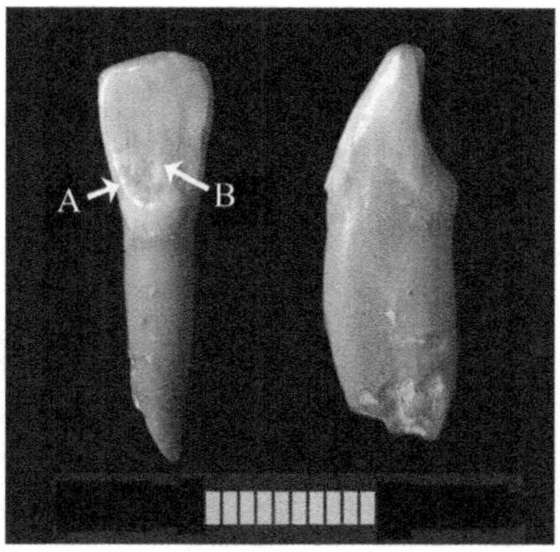

Figure 5. Left I_2 from Level VIII, Grotte du Renne with cingulum (A) and median ridge (B) development. Left: lingual view, Right: distal view.

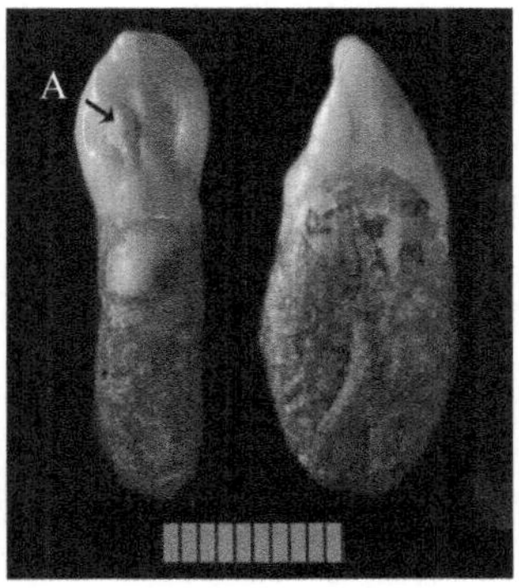

Figure 6. Left C, from Level X, Grotte du Renne showing marked distal accessory ridge (A). Left lingual view, right: mesial view.

double channeled root were similar to canines from earlier Mousterian levels. Indeed, its buccolingual dimensions are outside the range of Upper Paleolithic moderns and at the high end of the range for Neanderthals. The presence of a strong distal accessory ridge is also more common in Neanderthals than in Upper Paleolithic moderns (Table 2). Compared to other Neanderthal lower canines, the fully formed root is somewhat diminutive in length. However, the marked hypercementosis and absence of crown wear strongly suggest that the tooth was impacted (Figure 6). Therefore, we caution against using root length in the

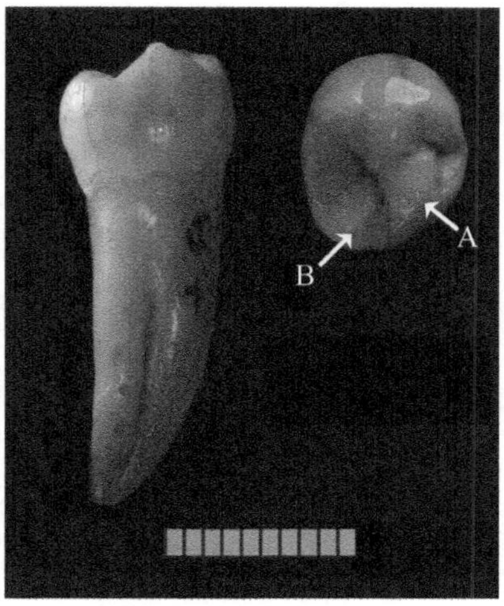

Figure 7. Left P_4 from Level IX, Grotte du Renne possessing a large mesially placed metaconid (A) and an second lingual cusp (B). Occlusal outline is markedly asymmetrical. Left: buccal view, right: occlusal view.

interpretation. In all other attributes the tooth is most closely affiliated with Neanderthals.

The single P_4 presents a markedly asymmetrical occlusal crown outline and multiple lingual cusps. It also possesses a large and mesially placed metaconid (Figure 7). The combination of asymmetry and multiple lingual cusps can be found in 91% of Neanderthals, but it is rare (~8%) in Upper Paleolithic moderns (Table 2). About 60% of Neanderthals also present a prominent continuous transverse crest connecting the metaconid and protoconid. However, while the essential crests of the buccal and lingual cusps of this tooth are markedly developed, they do not join to form a transverse crest. The buccolingual breadth of the P_4 is at the high end of the Neanderthal range of variation and two standard deviations above the range for Upper Paleolithic moderns (Table 3). Its root length is also above the Upper Paleolithic modern range. In morphology and size the affinity of this tooth is much closer to Neanderthals than to Upper Paleolithic moderns.

Seven lower molars have been recovered from the Châtelperronian levels of the Grotte du Renne. Three of these provide little or no morphological information due to their marked wear or partial preservation. Because molar length and breadth dimensions of Neanderthals and anatomically modern humans overlap extensively, there is little we can reliably infer about the taxonomic affiliation of these three worn teeth. However, the remaining teeth are less worn and more informative (Figure 8). Each of the two M_1 present a mid-trigonid crest (or epicristid: Zubov, 1992), which has a much higher frequency in Neanderthals than in Upper Paleolithic moderns (Table 2). Of the two M_2s, one presents a continuous mid-trigonid crest, while the other presents a mid-trigonid crest that is divided by a shallow groove. Both M_2s possess large hypoconulids (Cusp 5). In addition, at least one clearly possesses a *tuberculum sextum* (Cusp 6). The distal portion of the other M_2 is obscured by wear. The trait combination – mid-trigonid crest + anterior fovea + cusp 6 – is observed in 63% of Neanderthals but not in Upper Paleolithic moderns (Table 2). Thus, the combination of traits observed in the M_2 clearly suggests Neanderthal affinity. Although quite worn, the single M_3 certainly possessed more than four cusps – a characteristic found in 100% of Neanderthals and 68% of Upper Paleolithic moderns.

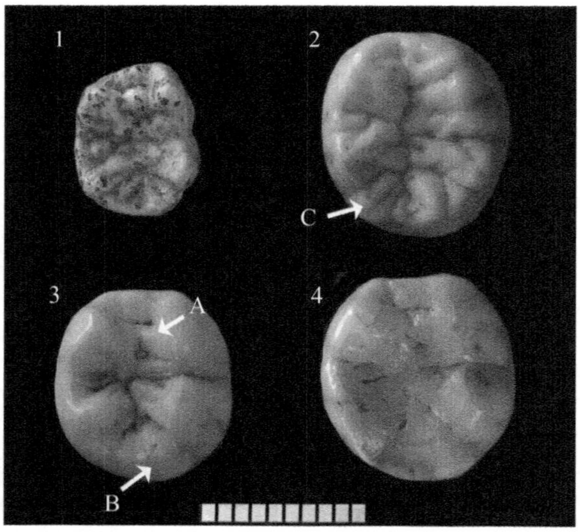

Figure 8. Unworn or slightly worn lower molars, Level X, Grotte du Renne. 1: right M1, 2: right M2, 3: right M1, 4: right M2. Traits referred to in the text: mid-trigonid crest (A), hypoconulid (B) and tuberculum sextum (C).

To summarize, in size and morphology all of the teeth from the Châtelperronian levels at the Grotte du Renne are consistent with Neanderthal affinity. Certain teeth in this sample present traits and/or trait combinations that are rare or absent in Upper Paleolithic modern humans, but occur with high frequency in Neanderthals. When we consider each of the traits for which Upper Paleolithic and Neanderthals have substantial differences in frequency (Table 2), and combine that with the available metric data, it is clear that the likelihood that these teeth come from anatomically modern humans is quite low. In fact, if we assume that the Châtelperronian assemblage represents a single population, and compare it to the combination of trait frequencies observed in Neanderthals and Upper Paleolithic moderns, we find the posterior probability that the population can be assigned to Upper Paleolithic moderns to be 0[1].

SPATIAL DISTRIBUTION

Three teeth likely belonging to a single individual between the ages of 4 and 6 years were recovered from level Xb1, Square D10 located towards the back of the shelter (Figure 9). In addition to developmental age, the color and the state of preservation all suggest the teeth derive from the same individual. The I^2 possesses the distinctive Neanderthal combination of strong shovel shape, labial convexity and a well-developed lingual tubercle. The dm_1 presents a strong crest connecting buccal and

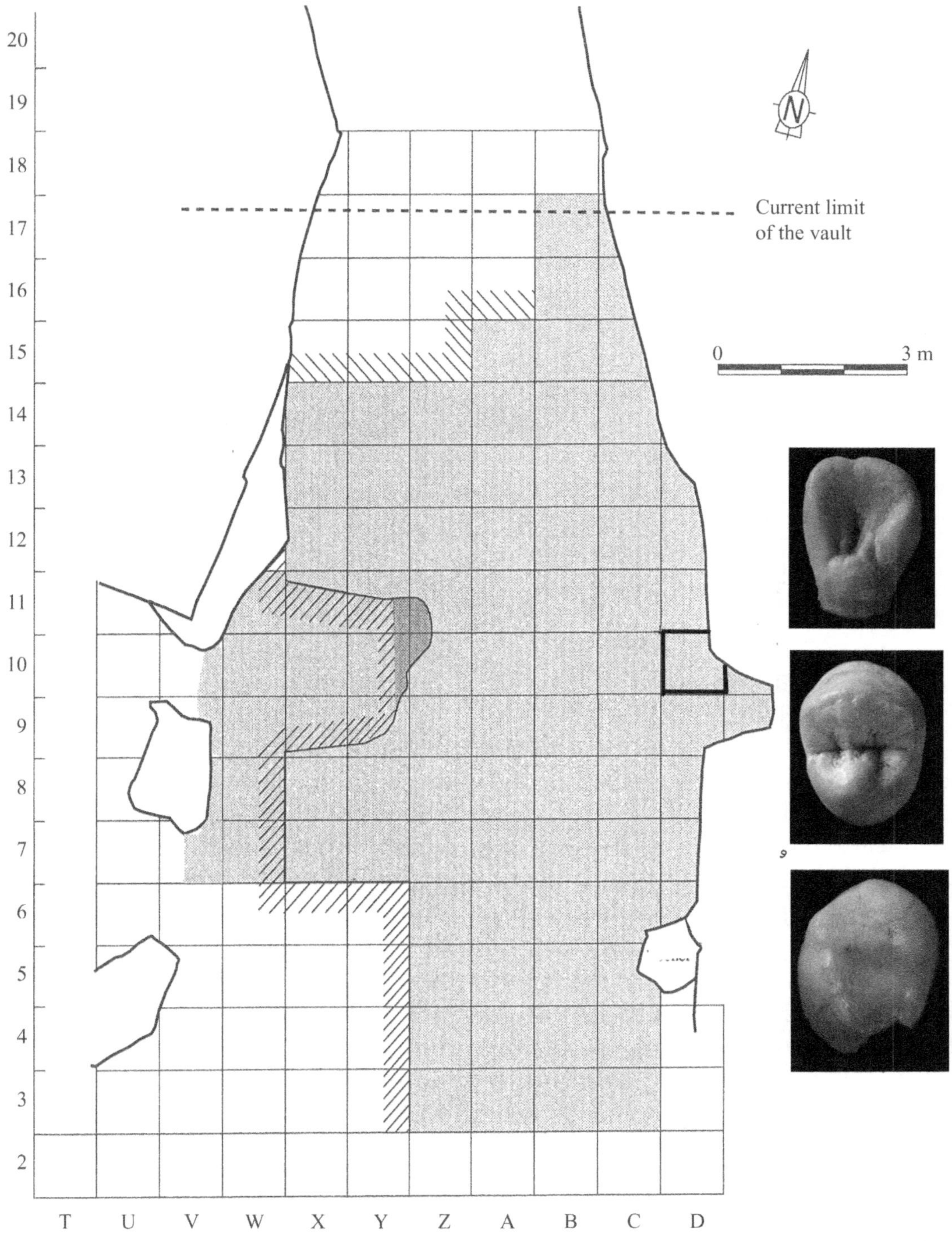

Figure 9. Location of square D10 and associated teeth (see Table 1) Original drawn by R. Humbert, modified (with permission) from Connet, 2002.

lingual cusps, which is similar to that found in the permanent P_4 of Neanderthals. The P^3 presents a strong and bifurcated essential crest, consistent with (but not proving) Neanderthal affinity. The fragmentary nature and incompleteness of this set suggest possible disturbance. However, it remains that three teeth likely belong to the same individual and come from a square that is in the back of the shelter, away from the sloping part of the deposits. This suggests some level of integrity of the layer in this particular square.

Similarly, four of the five teeth recovered from Xb2, Square C8 (Figure 10) have a high probability of belonging to the same individual with a developmental age of 9 to 18 months (dc_1, dm_1, dm_2, M_1 but not the di^1). The permanent M_1 has a high probability of belonging to a Neanderthal, with its marked mid-trigonid crest, low mesial marginal ridge and six cusps. The dm_2 presents compressed and internally placed cusps typical of Neanderthals, and the morphology of its mid-trigonid crest is nearly identical to that of the M_1. The dm_1 shows morphology similar to that seen in Neanderthal P_4s, including a prominent, continuous crest between the buccal and lingual cusps. The color, preservation and developmental age of these four specimens, together with the morphological similarity between the dm_2 and the permanent M_1, strongly suggest that they belong to the same infant. This again suggests that disturbance in this part of the cave, and this square in particular, was limited and likely not to account for the association between these Neanderthal teeth and the Châtelperronian assemblage.

Finally, an M_2 from Xb1c, Square A11 (Figure 11) presents a combination of traits found in 63% of Neanderthals, but not found in any Upper Paleolithic modern humans sampled. The occlusal complexity of the tooth, together with the morphology of the mid-trigonid crest strongly suggests it belonged to a Neanderthal. Square A11, is located in the back of the shelter where the likelihood is low that vertical displacement would explain the presence of a Neanderthal tooth in this level.

Summary and Conclusions

Our goal was to use new methods (dental anthropology) to identify the makers of the Châtelperronian at the Grotte du Renne, and to investigate the likelihood that inter-level movement of objects and/or teeth could account for the association between the fossils and the cultural remains. The use of the dental sample increases the number of individuals associated with the assemblage (MNI = 6) and confirms the conclusions derived from the study of the temporal bone (Hublin et al., 1996) that Neanderthals are the only fossils associated with the Châtelperronian at Arcy-sur-Cure.

The Grotte du Renne is indeed geologically and stratigraphically complex. There is evidence of periodic roof collapse and there is the potential for mixture of once distinct archaeological units in the section of the site formed before the threshold of the bedrock located some 8 m from the back of the shelter. However, nearly half of the Neanderthal teeth derive from areas where this type of disturbance and mixture is unlikely.

After the recent excavation of a preserved section at the Grotte du Renne located virtually at the center of the cave, David et al. (2001: 218) concluded that level X is "*a key bed in the stratigraphy of sedimentary refilling of the Grotte du Renne. It is well defined by its thickness, color, nature, structure and archeological contents*". Moreover, our investigation of the association of certain fossils shows conclusively that, although some level of disturbance is possible, spatial clusters of teeth belong to particular individuals, suggesting that movement of the remains was minimal. Finally, it should be noted that level Xb, which yielded

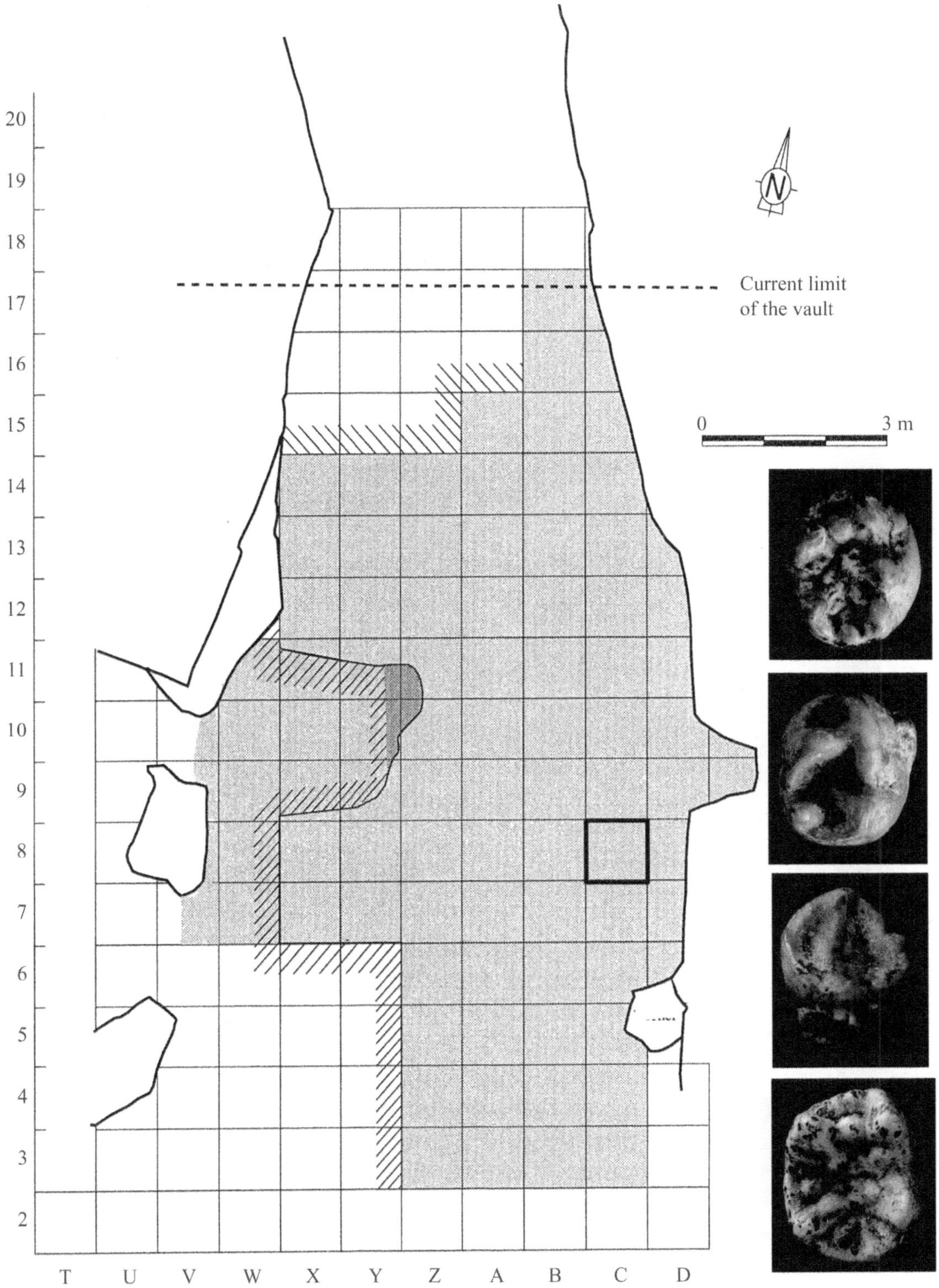

Figure 10. Location of Xb2, square C8 and associated teeth (see Table 1). Original drawn by R. Humbert, modified (with permission) from Connet, 2002.

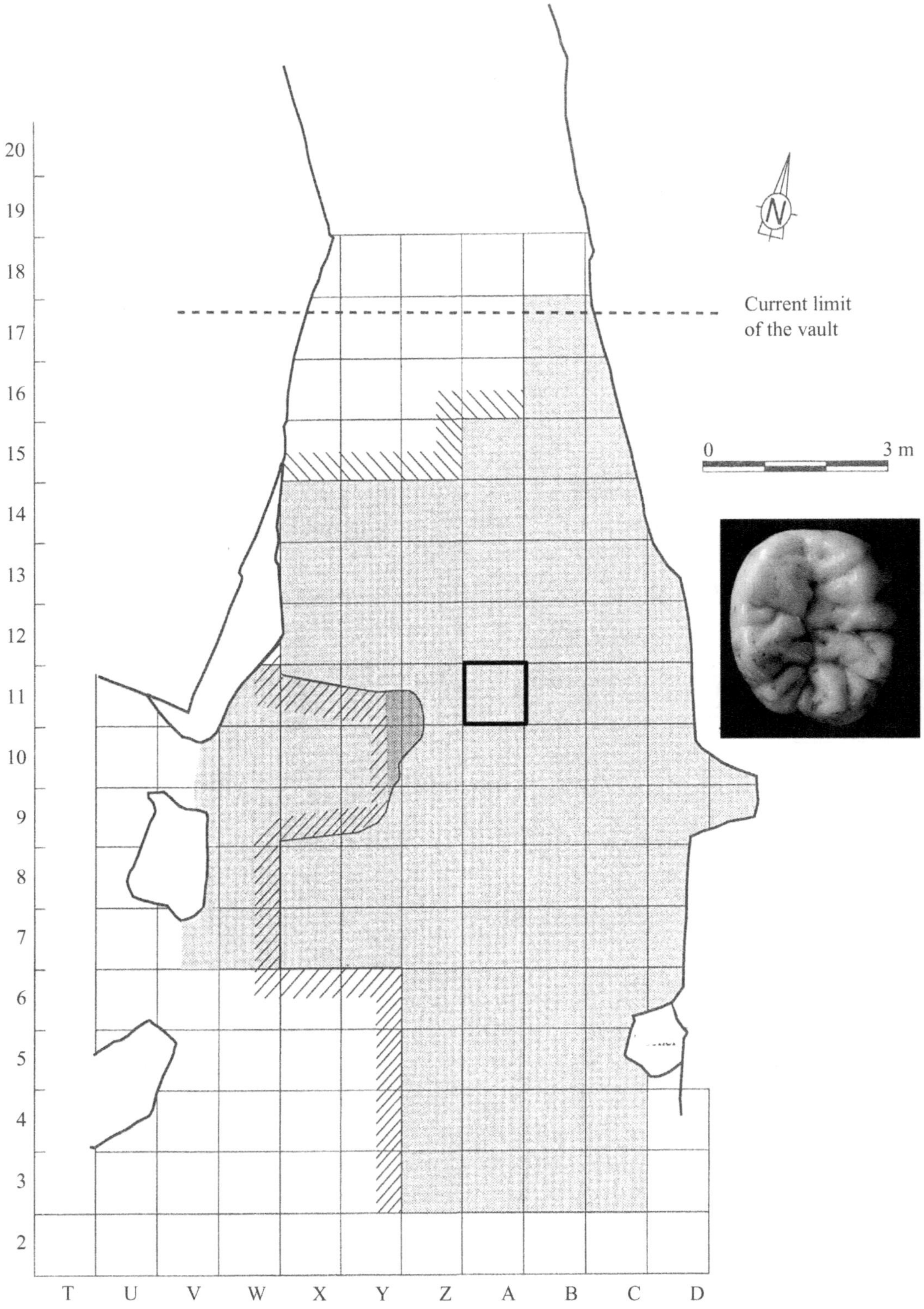

Figure 11. Location of Xb1c, Square A11 and associated tooth (see Table 1). Original drawn by R. Humbert, modified (with permission) from connet, 2002.

the majority of these remains, is not the lowermost Châtelperronian level of the site. On part of its extension it is separated from the Mousterian levels by an older Châtelperronian level (Xc). Conversely, level Xb is also separated from the overlying Aurignacian deposits by two other Châtelperronian levels (IX and VIII) reaching 50 cm and 20 cm thick in some places. The uppermost layer VIII also yielded a diagnosable Neandertal tooth, which is well above the uppermost Mousterian Layer XI. In addition, it must be noted that no Aurignacian small artifacts or modern human teeth are found in level Xb. It is, in our view, unlikely that the occurrence of body ornaments associated to the Neanderthal remains in level Xb could be explained by a selective migration of such items from level VII through levels VIII and IX (also see discussion in d'Errico et al., 1998, 2003).

Recently, the idea that there are no distinguishing characteristics of Neanderthal teeth, espoused by Boule and Vallois (1957), has been revived by Henry-Gambier et al. (2004). This idea is based on the fact that metrically there is a great deal of overlap between the groups, and morphologically there are no traits found in Neanderthals that cannot be found, at least on occasion, in anatomically modern humans. It seems to be linked to a recent trend to question the generally held assumption that the makers of the early Aurignacian were modern humans. This trend is partly the result of recent re-dating of several Aurignacian sites (e.g., Vogelherd, Cro-Magnon), which now appear to be much younger than once thought. Conard et al. (2004) have suggested that we perhaps do not know who made the Aurignacian. This suggestion has recently been echoed by Henry-Gambier et al. (2004) in their analysis of the early Aurignacian fossils from Brassempouy.

Neanderthal teeth are distinguished from those of anatomically modern humans primarily in their trait frequencies and in the combination of traits in a single tooth, not in the presence of a particular trait (the mid-trigonid crest on the M_3 may be an exception: Bailey, 2002a). However, it is misguided to conclude that taxonomic affiliation cannot be determined from isolated teeth. To date, we find no evidence to support that any hominins other than Neanderthals are associated with the Châtelperronian or that any hominins other than anatomically modern humans are associated with the early Aurignacian.

Acknowledgements

We would like to thank Katerina Harvati and Terry Harrison for organizing the Neanderthals Revisited symposium and for inviting us to participate. We also appreciate the reviewers' helpful comments and suggestions on the original manuscript. For access to the Arcy-sur-Cure fossils we thank Professor H. de Lumley and D. Grimaud-Hervé. We would also like to thank N. Connet, C. Letourneux and M. Julien for providing excavation data on the spatial distribution. Much of the comparative data presented in this study was collected while SEB was a PhD student at Arizona State University, and she is grateful to all the people who provided access to, and assistance with, the fossils and contemporary specimens examined during this time. Data collection and research for SEB's dissertation were supported by grants from the National Science Foundation (BCS-0002481), LSB Leakey foundation and the Philanthropic Educational Organization (PEO).

Note

1. The combination of features found in some of the Arcy Châtelperronian teeth was never observed in the modern human series. Assuming that the Arcy series represents a homogeneous sample, a posterior probability of this series to be modern, computed from the observed frequency in our reference populations, will inevitably give a null result.

References

Bailey, S.E., 2000. Dental morphological affinities among late Pleistocene and recent humans. *Dent. Anthropol.* 14, 1–8.

Bailey, S.E., 2002a. A closer look at Neanderthal post-canine dental morphology. I. The mandibular dentition. *New Anat.* 269, 148–156.

Bailey, S.E., 2002b. Neandertal Dental Morphology: Implications for modern human origins. Ph.D. Dissertation, Arizona State University.

Bailey, S.E., 2004. A morphometric analysis of maxillary molar crowns of Middle- Late Pleistocene hominins. *J. Hum. Evol.* 47, 183–198.

Bailey, S.E., 2005. Diagnostic dental differences between Neandertals and Upper Paleolithic humans: Getting to the root of the matter. In Żądzińska, E. (Ed.) *Current Trends in Dental Morphology Research: Proceedings of the 13th International Symposium on Dental Morphology.* Wydawnictwo Uniwersytetu Łódzkiego, Łódź, pp. 201–210.

Bailey, S.E., Hublin, J.-J., 2005. Who made the Early Aurignacian? A reconsideration of the Brassempouy dental remains. *Bull. Mém. Soc. Anthropol. Paris* 17, 115–121.

Bailey, S.E., Hublin, J.-J., 2006. Dental remains from the Grotte de Renne at Arcy-sur-Cure (Yonne). *J. Hum. Evol.* 50, 485–508.

Bailey, S.E., Lynch, J.M., 2005. Diagnostic differences in mandibular P4 shape between Neandertals and anatomically modern humans. *Am. J. Phys. Anthropol.* 126, 268–277.

Boule, M., Vallois, H.V., 1957. *Fossil Men.* Dryden Press, New York.

Burnett, S.E., 1998. Maxillary Premolar Accessory Ridges (MXPAR): Worldwide Occurrence and Utility in Population Differentiation. Masters Thesis, Arizona State University.

Bytnar, J.A., Trinkaus, E., Falsetti, A.B., 1994. A dental comparison of Middle Paleolithic Near Eastern hominids. *Am. J. Phys. Anthropol.* Suppl. 19, 63.

Conard, N.J., Grootes, P.M., Smith, F.H., 2004. Unexpectedly recent dates for human remains from Vogelherd. *Nature* 430, 198–201.

Connet, N., 2002. Le Châtelperronien: Réflexions sur l'unité et l'identite techno-économique de l'industrie lithique. L'apport de l'analyse diachronique des industries lithiques des couches Châtelperronienes de la Grotte du Renne a Arcy-sur-Cure (Yonne). Ph.D. Dissertation, Universite de Lille 1.

Crummett, T., 1995. The three dimensions of shovel-shaping. In: Moggi-Cecchi, J. (Ed.), *Aspects of Dental Biology: Palaeontology, Anthropology and Evolution.* International Institute for the Study of Man, Florence, pp. 305–313.

d'Errico, F., Zilhão, J., Julien, M., Baffler, D., Pelegrin, J., 1998. Neanderthal acculturation in Western Europe? A critical review of the evidence and its interpretation. *Curr. Anthropol.* 39, S1–S44.

d'Errico, F., Julien, M., Liolios, D., VanHaren, M. and Baffier, D. 2003. Many awls in our argument: bone tool manufacture and use in the Châtelperronian and Aurignacian levels of the Grotte de Renne at Arcy-sur-Cure. In: Zilhão, J. and d'Errico, F. (Eds.), *The Chronology of the Aurignacian and of the Transitional Technocomplexes: Dating, Stratigraphies, Cultural Implications. Proceedings of Symposium 6.1 of the 14th Congress of the UISPP*, Instituto Portugês de Arqueologia, Lisbon, pp. 247–270.

David, F., Connet, N., Girard, M., Lhomme, V., Miskovsky, J.C., Roblin-Jouve, A., 2001. Le Châtelperronien de la Grotte du Renne à Arcy-sur-Cure (Yonne). Données sédimentologiques et chronostratigraphiques. *Bull. Soc. Préhist. Fr.* 98, 207–230.

Demars, P.Y., Hublin, J.-J., 1989. La transition néandertaliens/hommes de type moderne en Europe occidentale: aspects paleontologiques et culturels. In: Otte, M. and Laville, H. (Eds.), *L'homme Néandertal 7: l'extinction.* ERAUL, Liège, pp. 29–42.

Harrold, F.B., 1989. Mousterian, Châtelperronian and early Aurignacian in Western Europe: continuity or discontinuity? In: Mellars, P. and Stringer, C.B. (Eds.), *The Human Revolution.* Princeton University Press, Princeton, pp. 677–713.

Henry-Gambier, D., Maureille, B., White, R., 2004. Vestiges humains des niveaux de l'Aurignacien ancien du site de Brassempouy (Landes). *Bull. Mém. Soc. Anthropol. Paris* 16, 49–87.

Hublin, J.-J., Spoor, F., Braun, M., Zonneveld, F.W., Condemi, S., 1996. A late Neanderthal associated with Upper Palaeolithic aftefacts. *Nature* 381, 224–226.

Leroi-Gourhan, A., 1958. Étude des restes humains fossiles provenant des Grottes d'Arcy-sur-Cure. *Ann. Paléontol.* 44, 87–148.

Leroi-Gourhan, A., 1961. Les fouilles d'Arcy-sur-Cure (Yonne). *Gallia préhistoire* 4, 3–16.

Lévêque, F., Vandermeersch, B., 1980. Découverte de restes humains dans un niveau castelperronien à

Saint-Césaire (Charente-Maritime). *C. R. Acad. Sci. Paris* 291, 187–189.
Mellars, P., 2004. Neanderthals and the modern human colonization of Europe. *Nature* 432, 461–465.
Mizoguchi, Y., 1985. *Shovelling: A Statistical Analysis of Its Morphology*. University of Tokyo Press, Tokyo.
Turner, C.G., II, Nichol, C.R., Scott, G.R., 1991. Scoring procedures for key morphological traits of the permanent dentition: The Arizona State University Dental Anthropology System. In: Kelley, M., Larsen, C. (Eds.), *Advances in Dental Anthropology*. Wiley Liss, New York, pp. 13–31.
White, R., 2001. Personal ornaments from the Grotte du Renne at Arcy-sur-Cure. *Athena Rev.* 2, 41–46.
Zubov, A., 1992. The epicristid or middle trigonid crest defined. *Dental Anthropol. Newslett.* 6, 9–10.

11. The fate of European Neanderthals: results and perspectives from ancient DNA analyses

D. SERRE
Max-Planck-Institute for Evolutionary Anthropology,
Leipzig D-04103,
Germany

Current address:
Genome Quebec Innovation Center and McGill University,
Montreal,
Canada
david.serre@mail.mcgill.ca

S. PÄÄBO
Max-Planck-Institut für Evolutionäre Anthropologie,
Leipzig D04103,
Germany
paabo@eva.mpg.de

Keywords: mtDNA, genetic diversity, population history, bottleneck, gene flow, coalescence, glaciation

Abstract

Analyses of mitochondrial DNA sequences extracted from several Neanderthal remains have provided new information on their genetic relationship with modern human individuals. However, these results have been interpreted very differently among anthropologists. Here we review these results and present additional data directly addressing the question of genetic continuity among human populations during the Late Pleistocene. An analysis of additional Neanderthal and early modern human remains from Western and Central Europe do not provide any evidence of gene flow between the two groups. We also show that under reasonable assumptions of human demography, these data rule out a major genetic contribution by Neanderthals to the modern human gene pool. Finally, we present preliminary results showing that ancient DNA studies can also contribute to unraveling aspects of Neanderthal demography. Promising avenues of research, such as the investigation of Neanderthal population genetic diversity and organization, as well as analyses of mammal populations contemporary with Neanderthals, could allow us to better understand the dynamics, and perhaps causes, of the demographic changes that occurred in Eurasia during the Late Pleistocene.

Introduction

Most researchers agree that the first hominids evolved in Africa (e.g., Campbell, 1988; Klein, 1989; Lewin, 1999) and that *Homo erectus* left Africa around two million years ago to colonize Europe and Asia as far as Indonesia (e.g., Wolpoff and Caspari, 1997; Gabunia et al., 2000; Oms et al., 2000; Wood and Richmond, 2000; Roebroeks, 2001; Balter and

K. Harvati and T. Harrison (eds.), Neanderthals Revisited: New Approaches and Perspectives, 211–220.

Gibbons, 2002; Vekua et al., 2002). However, the fate of archaic human populations that evolved regionally from this ancestral stock is much debated, especially with regards to a second wave of colonization from Africa around 100 ka. Most notably, attention has been focused on the fate of the Neanderthals,[1] the archaic humans that inhabited Europe and Western Asia during the later part of the Pleistocene (i.e., between 300 ka and 30 ka). Recent ^{14}C-dating confirms that the last Neanderthals could have co-existed with the first modern humans in Europe (Bocquet-Appel and Demars, 2000). However, it is still unclear whether this possible cohabitation influenced the gene pool of the newcomers or if, on the contrary, the Neanderthals went extinct without contributing to the gene pool of early modern humans in Europe.

In 1997, Krings extracted DNA from a humerus of the Neanderthal holotype (Krings et al., 1997). The 379 base pairs (bp) amplified from the hypervariable region of the mitochondrial (mt) genome were different from all modern human DNA sequences. Furthermore, this DNA sequence was too different from the current human sequences observed in the gene pool to be likely to be found in an individual that has not been analyzed yet. Tree reconstructions confirmed these analyses: while all human mtDNA sequences group together with a recent common ancestor,[2] the sequence retrieved from the type specimen of Neanderthal shows a much deeper separation with strong statistical support. This result is often interpreted as compelling evidence for the absence of interbreeding between Neanderthals and modern humans, or even as proof that Neanderthals and modern humans were two different species (e.g., Lindahl, 2000). However, even after the publication of two additional mtDNA sequences, very similar to that of the first individual (Krings et al., 2000; Ovchinnikov et al., 2000), many scenarios are still consistent with the data.

Two problems limit the range of the conclusions drawn from these studies: first, due to the impossibility of differentiating modern contamination from endogenous DNA sequences, a sample from a Neanderthal individual carrying a sequence similar to that of a current human could be discarded as putative contamination (Nordborg, 1998; Trinkaus, 2001). Second, the absence of early modern human DNA sequences leaves a long time span during which simple demographic processes can lead to the loss of Neanderthal sequences even with a substantial amount of admixture in the past (e.g., Relethford, 1998, 1999, 2001). Thus, the Neanderthal mtDNA could have been swamped by a continuous influx of modern human mtDNA into the Neanderthal gene pool (Enflo et al., 2001), lost by genetic drift (Nordborg, 1998), or by a population replacement much later than the Paleolithic transition, for example during the Neolithic expansion (e.g., Cavalli-Sforza et al., 1993). Here we summarize results that overcome these problems. The paper addresses the question of continuity or replacement between Neanderthals and early modern humans, as well as, more generally, aspects of what happened to the human populations during the transition from the Middle to Upper Paleolithic in Europe.

Looking for Gene Flow Between Neanderthals and Early Modern Humans

Contamination is the major problem of ancient DNA studies dealing with human remains because it is currently impossible to differentiate endogenous DNA sequences from modern contaminants present on the bones and those potentially left by excavators, curators and scientists that handled the bones. It has been shown that most ancient animal remains yield human DNA sequences when sensitive enough amplifications are used (Hofreiter et al., 2001; Wandeler et al., 2003).

This hampers the range of conclusions that can be drawn from studies of Neanderthal mtDNA, since (1) possible evidence of gene flow from modern humans to Neanderthals, such as a Neanderthal specimen yielding only a modern mtDNA sequence, could be considered a contamination artifact and therefore discarded, and (2) there are no conclusive mtDNA sequences from early modern humans that can be compared with Neanderthals. These DNA sequences would be especially informative due to their closer proximity in time to that of Neanderthals than current genetic diversity (Relethford, 1998, 1999, 2001).

However, one can investigate the genetic relationship between Neanderthals and early modern humans by making use of the fact that the Neanderthal mtDNA sequences retrieved so far are distinguishable from all current mtDNA sequences found in the human population. Thus, one way to look for gene flow between Neanderthals and early modern humans is to ask two questions: (1) Do all Neanderthal remains yield a "Neanderthal-like" mtDNA sequence? (2) Do any early modern human remains yield a "Neanderthal-like" mtDNA sequence?

As this approach relies on the presence/absence of a Neanderthal mtDNA sequence it requires some independent criteria to validate that any non-retrieval of Neanderthal mtDNA is effectively due to its absence and not to a lack of preservation of the biomolecules. We used animal remains, for which contamination is easily differentiable from endogenous DNA, to determine which state of biomolecular preservation is correlated with successful retrieval of endogenous DNA. We looked at the preservation of amino acids, the building blocks of the proteins that represent the major biomolecular component of the bone. Analyses of numerous faunal remains showed that using three independent measurements of amino acids preservation (i.e., the total amount of molecules, the ratio of two amino acids, and the chemical preservation of a particular amino acid) we could define strict criteria by which endogenous DNA from animal remains could always be successfully retrieved and amplified (Serre et al., 2004). This method also offers the advantage of being quick and largely non-destructive (less than 10 mg of bone powder is required), thus allowing screening of a large collection of material from which one can later choose only the most promising ones. We screened more than 25 Neanderthal and 40 early modern human remains for amino acid preservation. Five Neanderthal bones and five early modern humans (Table 1) fulfilled our criteria of preservation and therefore must contain retrievable endogenous DNA sequences (Serre et al., 2004; Beauval et al., 2005).

We extracted DNA from each of the ten remains and amplified it under two different conditions:

1. an amplification of mtDNA was performed under conditions where modern human and Neanderthal, as well as chimpanzee and gorilla, DNA were successfully amplified. This amplification allowed a wide screening of possible molecules present in the bones. For example, if a bone contained an mtDNA sequence different both from Neanderthal and from modern human

Table 1. Specimen included in the gene flow study

Specimen
Neanderthal remains
Vindija 77 (Vi-77) (Croatia)
Vindija 80 (Vi-80) (Croatia)
Engis 2 (Belgium)
La-Chapelle-aux-Saints (France)
Les-Rochers-de-Villeneuve (RdV 1) (France)
Early modern human remains
Mladeč 25c (Czech Republic)
Mladeč 2 (Czech Republic)
Cro-Magnon (France)
Abri Pataud (France)
La Madeleine (France)

sequences, this "unspecific" amplification could likely detect it.
2. a "Neanderthal-specific" amplification was performed. Under the conditions used, only mtDNA sequences similar to those retrieved from the previously analyzed Neanderthal remains could be amplified while the amplification did not work on modern human mtDNA sequences. This procedure allowed us to "fish out" a Neanderthal mtDNA sequence, even if it was in the presence of a much larger amount of contaminant sequences.

All remains (the five Neanderthals and the five early modern humans) analyzed yielded DNA sequences identical to contemporary human DNA sequences when amplified using the "unspecific" conditions. In 75% of the cases, more than one human mtDNA sequence was amplified from a single bone (Serre et al., 2004). This confirmed previous results that most ancient remains yield human DNA sequences when sensitive enough amplifications are used (Hofreiter et al., 2001; Wandeler et al., 2003). Additionally, all DNA sequences retrieved from the early modern human remains were identical to modern human mtDNA sequences present in DNA sequence database (http://www.ncbi.nlm.nih.gov/Genbank/). Due to ubiquitous contamination in four samples (i.e., those yielding each more than one sequence) and the fact that any DNA sequence amplified can potentially be a contaminant, it is impossible to identify the endogenous mtDNA sequence for any of the early modern human remains. In our view, this shows that it is currently impossible to trust the veracity of any ancient DNA sequence similar to the one found in the modern human gene pool.

By contrast, when the DNA amplification was performed under "Neanderthal-specific" conditions, none of the five early modern human remains yielded an amplification product. Interestingly, all five Neanderthal remains did yield an amplification product and, after sequencing, a short mtDNA sequence fragment was identified that was identical to the corresponding region of one of the four Neanderthals already sequenced (Krings et al., 1997, 2000; Ovchinnikov et al., 2000; Schmitz et al., 2002). Given that the overall state of preservation of the biomolecules is similar, this shows that the Neanderthals formed a homogenous genetic population different from that of early modern humans (Serre et al., 2004; Beauval et al., 2005). This result is supported by the mtDNA sequence of a fragment of 47bp recently retrieved from a Neanderthal from El Sidrón Cave, Spain, that is identical to the sequences from Vindija and Feldhofer 1 (Lalueza-Fox et al., 2005).

Thus, while we applied an unbiased methodology that can detect gene flow between populations, we did not find any evidence of gene flow in either direction. It is important to stress here that some of the samples analyzed in this study have been described as "transitional" between "classical" Neanderthals and early modern humans, such as the Vindija Neanderthals (Smith and Spencer, 1984; Wolpoff, 1999) and the Mladeč individuals (Frayer, 1992; Wolpoff, 1999), so they represent good candidates to reveal potential gene flow.

What is the maximum genetic contribution that might have occurred?

Our analysis of five Neanderthal remains and five early modern humans did not detect any evidence of gene flow. However, given the small sample size one might question the power of this study to detect genetic contribution. In other words, one might want to estimate the level of genetic contribution that can be statistically ruled out given the data. It is important to note that, while the former results were obtained by straight-forward analyses of the data, estimation of the maximum genetic contribution relies on a theoretical model of what we think is a fair representation of human demographic history: what were the population sizes of Neandertals and early modern humans, when did they meet each other, how long did they interact for, when and how quickly did the modern human population expanded? All these parameters need to be estimated in the model. Therefore, one should keep in mind that any results obtained using this approach are dependant on the assumptions made.

We decided to use the simplest model possible (to account for the small data set we have) and to work under the assumption that the human population is panmictic (i.e., random mating) and of constant size through time. We estimated, using this model (Tavare, 1984), that the current mtDNA gene pool had only between four and seven ancestors at 20–30 ka. This shows the limitations of using only current diversity to obtain insights about the mtDNA gene pool in the late Pleistocene. In fact, the five early modern human individuals analyzed here provide almost as much information about the mtDNA gene pool of modern humans in the late Pleistocene as would the sequencing of mtDNA sequence from all now-living humans. They also add information that could not be obtained by studying additional now-living individuals. The mtDNA ancestry of current humans is already intensively explored with respect to deep divergences, so that additional major lineages are unlikely to be discovered (Sykes, 2001). Given that all Neanderthal bones analyzed yield mtDNA sequences that are similar to each other and absent in the five early modern humans analyzed, as well as in all modern humans, we can exclude (at 95% confidence) any Neanderthal contribution to the modern human gene pool greater than 25% (Serre et al., 2004). This might seem a rather uninformative result, but it is in fact a major improvement. When Neanderthal mtDNA sequences are considered alone, only a scenario of random-mating population comprising both Neanderthals and modern humans can be excluded (Nordborg, 1998). Thus, even using a conservative model of population history we can exclude a large Neanderthal contribution to the modern human gene pool.

If we consider a more realistic scenario where the spread of modern humans (before and during their migration out of Africa and subsequent colonization of western Eurasia) was accompanied by a population growth, we can exclude a smaller Neanderthal contribution. However, the importance of the contribution that can be excluded depends critically on when and how the expansion occurred. For example, Currat and Excoffier (2004) recently estimated that under a much more complex scenario, in which an expanding modern human population spread progressively in Europe and competed with the less numerous Neanderthals, the maximal genetic contribution compatible with the data is smaller than 0.1%.

Can Ancient DNA Studies Tell us What Happened to the Neanderthals During the Middle to Upper Paleolithic Transition?

The genetic data collected so far support a scenario of no major interbreeding between the two human populations in the Late Pleistocene. Leaving aside discussions of species/sub-species status and interbreeding capacity/incapacity, we can still try to understand why Neanderthals disappeared during the transition from Middle to Upper Paleolithic. Two avenues of research are promising for this purpose: (1) the analyses of genetic diversity within Neanderthals that can lead to a greater understanding of their demographic history; and (2) the investigations of potential demographic changes in animal populations contemporary with the Neanderthals to obtain a more global understanding of the environment and its influences.

By comparing the Neanderthal mtDNA sequences of the four individuals with the most complete genetic information, we find that the Neanderthals carry a genetic diversity for the mtDNA similar to that of the current human population and approximately 5 times smaller than that of the African great apes (Krings et al., 2000; Schmitz et al., 2002). We have shown that this low diversity within Neanderthals is not an artifact, since all well-enough preserved remains yield very similar sequences (Serre et al., 2004). One commonly proposed explanation for the reduced genetic diversity in humans relative to our closest living relatives is that gorillas and chimpanzees have always lived in the African rainforest, which was not drastically modified by climatic changes (e.g., Lahr and Foley, 1998). The African great apes may, therefore, have maintained a stable population over a long period of time and accumulated a large genetic diversity. In contrast, human populations expanding in open environments were more exposed to climatic fluctuations and likely underwent a series of drastic reductions in population size followed by expansions (e.g., Takahata, 1994; Lahr and Foley, 1998; Reich and Goldstein, 1998; Zietkiewicz et al., 1998; Adams et al., 2000). The preliminary data concerning the Neanderthal population show the same general trend, and suggest a rather unstable population history. Additionally, it is interesting to note that the mtDNA sequence retrieved from the second individual of Feldhofer, Germany (Schmitz et al., 2002) carries three differences from the type specimen mtDNA sequence (Krings et al., 1997) while carrying only one difference from the Croatian Neanderthal mtDNA sequence (Krings et al., 2000). This suggests that no strong geographical clustering of mtDNA sequences was present in Neanderthals, at least in western and central Europe. It is clear that more individuals are needed in order to arrive at more definitive conclusions about the geographic organization of the Neanderthal mtDNA gene pool, but it is interesting that this preliminary observation contrasts with the picture given by some paleoanthropologists who present Neanderthals as having strong cultural or behavioral differences correlated with their geographical origins (e.g., Bahn, 1998; d'Errico et al., 1998; Stringer et al., 2000). In this context, one can note that all Neanderthal DNA analyzed so far dates from the early to middle part (~59–35 ka) of the MIS 3 interstadial. An interesting working hypothesis would be that the Neanderthals of the Saalian glaciation (MIS 6, ~195–128 ka) consisted of a metapopulation with strong phylogeographical structure, and that the MIS 3 Neanderthal population is the result of postglacial expansion of only one, or a few, surviving local population(s).

Another promising approach to better understand the history of Neanderthal and early modern human populations is to analyze faunal remains contemporary with these populations. Ancient DNA analyses of animal remains are far easier and more efficient than those of human remains because: (1) many more samples are available for analyses; and (2) contamination is not an issue.[3] In a recent pilot study we analyzed remains from cave bears, cave hyenas, and brown bears across Europe, all dated to ~70–30 ka (Hofreiter et al., 2004). In none of these data sets were we able to detect a correlation between the mtDNA sequence carried by an individual and its geographical origin (sometimes this is referred to as phylogeographic structure). This finding is striking when compared to current genetic diversity data: most species living today in Europe show a strong correlation with the mtDNA gene pool organized in two or three clades found almost exclusively in Western Europe, Eastern Europe or Southern Europe (e.g., Taberlet et al., 1998; Avise, 2000; Hewitt, 2000). This organization of current genetic diversity is believed to be the result of glacial periods when many species survived only in a few ice-free refugia (the Iberian Peninsula, the Balkans, and Italy) and spread from there across Europe at the end of the glaciation. Interestingly, while the time of the setting of this phylogeographic structure is believed to date to early in the Pleistocene (e.g., Hewitt, 2000), we find no evidence of such organization in the three species we looked at. We concluded that the setting of this phylogeographic structure possibly occurred just a couple of tens of thousands of years ago (Hofreiter et al., 2004). It will be interesting to see if this result holds when more species contemporary with the Neanderthals are analyzed. This preliminary result might indicate that many species underwent major demographic rearrangements around the time that Neanderthals became extinct. This observation is of particular interest as any event that affected the environment so drastically must have affected the human populations as well, if not directly, at least through the changes of the dietary resource availability. An understanding of the dynamics of animal populations in the Pleistocene might therefore lead to major breakthroughs in our understanding of Neanderthal extinction.

Conclusion

We have shown here that genetic analyses of Neanderthal and early modern human remains can provide information about the relationship and dynamics of these two populations. Neanderthals, at least those living in the last interglacial period, constitute a homogenous genetic population different from the early modern humans that followed them in Europe. Recent analyses of Paleolithic human remains found no evidence of gene flow between the two populations in either direction, and we can show that, if any, the genetic contribution from the Neanderthals to the modern human gene pool must have been limited. We are also beginning to obtain some information concerning the demography of the Neanderthals. Their low genetic diversity relative to that of the African great apes, and similar to that of current humans, suggests major demographic changes during the Late Pleistocene. The geographic homogeneity of the gene pool of the Neanderthals investigated so far, strikingly contrasts with their apparent cultural diversity and requires further investigation. Preliminary analyses of faunal remains contemporary with the Neanderthal suggest that major demographic changes occurred in Europe around the time when Neanderthals became extinct. Further investigations in this direction might lead to a better understanding of the context in which this disappearance occurred and perhaps to its causes. Analyses of the DNA molecules preserved in Pleistocene human bones are tedious and, unfortunately, still require the destruction of a small amount of material. Nonetheless, these analyses provide information that cannot be obtained by looking at the current genetic diversity or through morphological/archeological studies. Eight years after the publication of the first Neanderthal mtDNA sequence we have shifted the research focus towards understanding of the Neanderthal population history, and we are only beginning to reveal this fascinating period of human evolution. The conclusions are still limited, but future analyses of additional individuals will allow us to verify (or contradict) our preliminary results and offer an exciting challenge for the coming years.

Notes

1. Throughout this paper we use the term "human population" to describe both Neanderthals and early modern humans. All the results presented here deal with the population history of "modern humans" and "Neanderthals" and can be explained by demographic processes that do not necessitate reproductive isolation or any other biological criterion that can be used to define species.
2. The concept of genetic ancestry, as used throughout this paper, is not identical to the popular meaning of ancestry. In its most common meaning, the ancestors of a particular individual are his/her parents, the

parents of this individual's parents and so on. As a consequence, the number of ancestors increases continuously when one looks back in time, at least during the first generations. In contrast, if one considers a short fragment of a DNA molecule in an individual, it is inherited from only one of her/his parents, who has also inherited it from only one parent. Therefore, the number of genetic ancestors does not increase with the number of generations. Additionally, as one looks back in time, two now-living individuals will have inherited the fragment of DNA considered from a common ancestor in the *n*th generation. Working from this definition of genetic ancestry, only this last individual will be a genetic ancestor of the two now-living individuals in the *n*th generation, while all other individuals will not be (despite the fact that they are all ancestors per the popular meaning). Thus, the number of genetic ancestors decreases when one looks back in time as more and more individuals have common ancestors until, eventually, a single most recent common ancestor (MRCA) remains. It is worth noting here that this MRCA (sometimes referred to as "eve" for the mitochondrial DNA) is not an isolated individual, but the particular member of a large population that carries the fragment of DNA present in all now-living individuals (who can harbor different DNA sequences due to the accumulation of mutations).

3. It is trivial to differentiate a human DNA sequence from that of non-human animal and, additionally, animal DNA contamination is unlikely if standard laboratory procedures are followed.

References

Adams, E.J., Cooper, S., Thomson, G., Parham, P., 2000. Common chimpanzees have greater diversity than humans at two of the three highly polymorphic MHC class I genes. *Immunogenetics* 51, 410–424.

Avise, J.C., 2000. *Phylogeography*. Harvard University Press, Cambridge, MA.

Bahn, P.G., 1998. Neanderthals emancipated. *Nature* 394, 719–721.

Balter, M., Gibbons, A., 2002. Were "little people" the first to venture out of Africa? *Science* 297, 26–27.

Beauval, C., Maureille, B., Lacrampe-Cuyaubere, F., Serre, D., Peressinotto, D., Bordes, J.G., Cochard, D., Couchoud, I., Dubrasquet, D., Laroulandie, V., Lenoble, A., Mallye, J.B., Pasty, S., Primault, J., Rohland, N., Pääbo, S., Trinkaus, E., 2005. A late Neandertal femur from Les Rochers-de-Villeneuve, France. *Proc. Natl. Acad. Sci. U.S.A.* 102, 7085–7090.

Bocquet-Appel, J.-P., Demars, P.Y., 2000. Neanderthal contraction and modern human colonization of Europe. *Antiquity 74*, 544–552.

Campbell, B.C., 1988. *Humankind Emerging*. Scott, Foresman and Company, Glenview, Boston, London.

Cavalli-Sforza, L.L., Menozzi, P., Piazza, A., 1993. Demic expansions and human evolution. *Science* 259, 639–646.

Currat, M., Excoffier, L., 2004. Modern humans did not admix with Neanderthals during their range expansion into Europe. *PLoS Biol.* 2, 2264–2274.

d'Errico, F., Zilhão, J., Julien, M., Baffier, D., Pelegrn, J., 1998. Neanderthal acculturation in Western Europe? A critical review of the evidence and its interpretation. *Curr. Anthropol.* 39, S1–S44.

Enflo, P., Hawkes, K., Wolpoff, M., 2001. A simple reason why Neanderthal ancestry can be consistent with current DNA information. *Am. J. Phys. Anthropol.* 114, 62.

Frayer, D.W., 1992. Evolution at the European edge: Neanderthal and Upper Paleolithic relationships. *Prehist. Europeenne* 2, 9–69.

Gabunia, L., Vekua, A., Lordkipanidze, D., Swisher, C.C. III, Ferring, R., Justus, A., Nioradze, M., Tvalchrelidze, M., Antón, S.C., Bosinski, G., Joris, O., Lumley, M.A., Majsuradze, G., Mouskhelishvili, A., 2000. Earliest Pleistocene hominid cranial remains from Dmanisi, Republic of Georgia: taxonomy, geological setting, and age. *Science* 288, 1019–1025.

Hewitt, G. 2000. The genetic legacy of the Quaternary ice ages. *Nature* 405, 907–913.

Hofreiter, M., Serre, D., Poinar, H.N., Kuch, M., Pääbo, S., 2001. Ancient DNA. *Nat. Rev. Gene.* 2, 353–359.

Hofreiter, M., Serre, D., Rohland, N., Rabeder, G., Nagel , D., Conard, N., Munzel, S., Pääbo, S., 2004. Lack of phylogeography in European mammals before the last glaciation. *Proc. Natl. Acad. Sci. U.S.A.* 101, 12963–12968.

Klein, R.G., 1989. *The Human Career: Human Biological and Cultural Origins*. University of Chicago Press, Chicago.

Krings, M., Capelli, C., Tschentscher, F., Geisert, H., Meyer, S., von Haeseler, A., Grossschmidt, K., Possnert, G., Paunovic, M., Pääbo, S., 2000. A view of Neandertal genetic diversity. *Nat. Genet.* 26, 144–146.

Krings, M., Stone, A., Schmitz, R.W., Krainitzki, H., Stoneking, M., Pääbo, S., 1997. Neandertal

DNA sequences and the origin of modern humans. *Cell* 90, 19–30.
Lahr, M.M., Foley, R.A., 1998. Towards a theory of modern human origins: geography, demography, and diversity in recent human evolution. *Am. J. Phys. Anthropol.* Suppl. 27, 137–176
Lalueza-Fox, C., Sampietro, M.L., Caramelli, D., Puder, Y., Lari, M., Calafell, F., Martínez-Maza, C., Bastir, M., Fortea, J., Rasilla, M.D., Bertranpetit, J., Rosas, A., 2005. Neandertal evolutionary genetics: mitochondrial DNA data from the Iberian Peninsula. *Mol. Biol. Evol.* 22, 1077–1081.
Lewin, R., 1999. *Human Evolution: An Illustrated Introduction*. Blackwell Science, London.
Lindahl, T. 2000. Fossil DNA. *Curr. Biol.* 10, R616.
Nordborg, M., 1998. On the probability of Neanderthal ancestry. *Am. J. Hum. Genet.* 63, 1237–1240.
Oms, O., Pares, J.M., Martínez-Navarro, B., Agusti, J., Toro, I., Martinez-Fernandez, G., Turq, A., 2000. Early human occupation of Western Europe: paleomagnetic dates for two Paleolithic sites in Spain. *Proc. Natl. Acad. Sci. U.S.A.* 97, 10666–10670.
Ovchinnikov, I.V., Gotherstrom, A., Romanova, G.P., Kharitonov, V.M., Liden, K., Goodwin, W., 2000. Molecular analysis of Neanderthal DNA from the northern Caucasus. *Nature* 404, 490–493
Reich, D.E., Goldstein, D.B., 1998. Genetic evidence for a Paleolithic human population expansion in Africa. *Proc. Natl. Acad. Sci. U.S.A.* 95, 8119–8123.
Relethford, J.H., 1998. Genetics of modern human origins and diversity. *Ann. Rev. Anthropol.* 27, 1–23.
Relethford, J.H., 1999. Models, predictions, and the fossil record of modern human origins. *Evol. Anthropol.* 8, 7–10.
Relethford, J.H., 2001. Absence of regional affinities of Neandertal DNA with living humans does not reject multiregional evolution. *Am. J. Phys. Anthropol.* 115, 95–98.
Roebroeks, W., 2001. Hominid behaviour and the earliest occupation of Europe: an exploration. *J. Hum. Evol.* 41, 437–461.
Schmitz, R.W., Serre, D., Bonani, G., Feine, S., Hillgruber, F., Krainitzki, H., Pääbo, S., Smith, F.H., 2002. The Neandertal type site revisited: interdisciplinary investigations of skeletal remains from the Neander Valley, Germany. *Proc. Natl. Acad. Sci. U.S.A.* 99, 13342–13347
Serre, D., Langaney, A., Chech, M., Teschler-Nicola, M., Paunovic, M., Mennecier, P., Hofreiter, M., Possnert, G.G., Pääbo, S., 2004. No evidence of Neandertal mtDNA contribution to early modern humans. *PLoS Biol.2*, 313–317.
Smith, F.H., Spencer, F. (Eds.), 1984. *The Origins of Modern Humans: A World Survey of the Fossil Evidence*. Alan R. Liss, New York.
Stringer, C., Barton, N.H., Finlayson, C. (Eds.), 2000. *Neanderthals on the Edge*. Oxbow Books, Oxford.
Sykes, B., 2001. *The Seven Daughters of Eve*. W. W. Norton, New York.
Taberlet, P., Fumagalli, L., Wust-Saucy, A.G., Cosson, J.F., 1998. Comparative phylogeography and postglacial colonization routes in Europe. *Mol. Ecol.* 7, 453–464.
Takahata, N., 1994. Repeated failures that led to the eventual success in human evolution. *Mol. Biol. Evol.* 11, 803–805.
Tavare, S., 1984. Line-of-descent and genealogical processes, and their applications in population genetics models. *Theor. Popul. Biol.* 26, 119–164.
Trinkaus, E., 2001. The Neandertal paradox. In: Finlayson, C. (Ed.), *Neanderthals and Modern Humans in Late Pleistocene Eurasia*. The Gibraltar Museum, Gibraltar.
Vekua, A., Lordkipanidze, D., Rightmire, G.P., Agusti, J., Ferring, R., Maisuradze, G., Mouskhelishvili, A., Nioradze, M., De León, M.P., Tappen, M., Tvalchrelidze, M., Zollikofer, C., 2002. A new skull of early *Homo* from Dmanisi, Georgia. *Science* 297, 85–89.
Wandeler, P., Smith, S., Morin, P.A., Pettifor, R.A., Funk, S.M., 2003. Patterns of nuclear DNA degeneration over time – a case study in historic teeth samples. *Mol. Ecol.* 12, 1087–1093.
Wolpoff, M., Caspari, R., 1997. *Race and Human Evolution*. Simon & Schuster, New York.
Wolpoff, M.H., 1999. The systematics of *Homo*. *Science* 284, 1774–1775.
Wood, B., Richmond, B.G., 2000. Human evolution: taxonomy and paleobiology. *J. Anat.* 197 (Pt 1), 19–60.
Zietkiewicz, E., Yotova, V., Jarnik, M., Korab-Laskowska, M., Kidd, K.K., Modiano, D., Scozzari, R., Stoneking, M., Tishkoff, S., Batzer, M., Labuda, D., 1998. Genetic structure of the ancestral population of modern humans. *J. Mol. Evol.* 47, 146–155.

12. Selection on mitochondrial DNA and the Neanderthal problem

J. HAWKS
Department of Anthropology
University of Wisconsin–Madison
5240 Social Science Building
1180 Observatory Drive
Madison, WI 53706–1393, USA
jhawks@wisc.edu

Keywords: Positive selection, adaptation, demographic inference, molecular evolution, selective sweep

Abstract

At present, the direct evidence for Neanderthal genetic variation and gene phylogeny is limited to the control region of the mitochondrial DNA (mtDNA). Neanderthal mtDNA sequences are divergent from those of recent humans. This fact, when coupled with the assumptions of selective neutrality and a recently expanding human population, argues for the complete and utter extinction of Neanderthals without living issue. But an alternative hypothesis is that human mtDNA has recently undergone an episode of positive selection, or a "selective sweep." Five converging lines of evidence suggest that mtDNA has undergone recent positive selection: (1) mtDNA variants in living humans are associated with life history and metabolic traits that changed dramatically during recent human evolution; (2) Statistical tests show that the distribution of human mtDNA variation is clearly inconsistent with neutrality; (3) Nuclear genomic variation is not consistent with a single recent population expansion as necessary to explain human mtDNA variation; (4) A neutral mtDNA necessitates a population replacement to explain its pattern of variation, but many autosomal and X chromosomal loci show strong phylogeographic or genealogical evidence for the survival of archaic human gene lineages and therefore reject population replacement; and (5) Anatomical and archaeological evidence shows some degree of anatomical and behavioral continuity between Upper Paleolithic Neanderthals and later Europeans and likewise reject population replacement. The hypothesis of positive selection on mtDNA is in accord with recent estimates of genome-wide rates of selection and is contradicted by no known evidence. Molecular and comparative evidence further suggests that the current pattern of human mtDNA variation represents only the most recent episode of positive selection among many during human evolution. Selection on mtDNA cannot prove that other Neanderthal genomic lineages survived, although such survival may be suggested by other anatomical and genetic evidence. Nevertheless, the substantial probability of such selection renders Neanderthal mtDNA variation phylogenetically uninformative.

K. Harvati and T. Harrison (eds.), Neanderthals Revisited: New Approaches and Perspectives, 221–238.

Introduction

As Darwin recognized over a century ago, the most important force leading to the evolution of human populations has been natural selection. Molecular genetic discoveries during the past fifteen years have increasingly demonstrated the importance of adaptive evolution to the variation of human genes and their pattern of similarities and differences compared to other hominoid species. Some of the known genetic targets of selection are tabulated by Vallender and Lahn (2004), including many with roles in neurobiology, the developmental genetics of the skeleton, longevity, and other systems, that have changed substantially during human evolution.

Mitochondrial DNA (mtDNA) has a history of adaptive evolution in recent human populations (reviewed in Wallace, 2005a). This molecule is unique compared to most of the nuclear genome because of its lack of recombination, its clonal transmission, and its maternal pattern of inheritance. All of these factors reduce the expected diversity of the mtDNA within populations, and combine to ensure that any adaptive changes affect the variation of the entire molecule (Spuhler, 1989). On the other hand, the mtDNA has a relatively higher mutation rate than the nuclear genome (Wallace, 2005b). This high mutation rate tends to increase the genetic diversity within populations, but it also may increase the rate of both negative and positive selection. The mtDNA exists in each mitochondrion, meaning that there are thousands of copies in each cell (Wallace, 2005a). This high copy number makes it possible to extract mtDNA more reliably from degraded skeletal remains than nuclear DNA. But it also greatly increases the possibility of deleterious somatic mutations. From these considerations, the selective history of human mtDNA can be expected to be complex.

However, some of these same factors make genealogical reconstruction from mtDNA enormously simpler than for most nuclear genetic loci (Cann et al., 1987). It is possible to trace ancestry along the maternal line back to a single common ancestor for all humans. The high mutation rate means that close and distant relatives may be distinguished easily, even within the past few generations (Howell et al., 1996) – a fact that has enabled the growth of mtDNA genealogy testing services. Because they are clonal, these maternal lineages can be traced over thousands of years, enabling the examination of the relative maternal contribution of different ancient regions to recent populations. And the high copy number has made it possible to extract partial mtDNA sequences from the ancient skeletal remains of Neanderthals (Krings et al., 1997).

Study of these Neanderthal mtDNA sequences has shown them to be distinct from those of living humans in several ways. This distinctiveness is great enough to suggest that Neanderthal mtDNA variants probably do not survive today (Currat and Excoffier, 2004). Under the assumption that mtDNA has not been under selection across the intervening 30,000 or more years, the lack of any Neanderthal mtDNA lineages in living human populations is highly significant: it would indicate that any Neanderthal ancestry for living people must have been next to negligible (Currat and Excoffier, 2004; Weaver and Roseman, 2005). But this conclusion hinges on the assumption that mtDNA has in fact been selectively neutral during this time period.

There are good reasons to think that mtDNA has been significantly selected within the past 30,000 to 50,000 years or more. There are five lines of evidence leading to this conclusion:

1. Presently segregating mtDNA variants are associated with phenotypic variation in longevity, degenerative disease, mortality, metabolic efficiency, brain-related disorders, diet and climate.

2. Human mtDNA variation is statistically inconsistent with neutral evolution.
3. Human nuclear genomic variation is inconsistent with the pattern of variation of human mtDNA.
4. Mitochondrial DNA neutrality can only be maintained if a population replacement occurred (Currat and Excoffier, 2004; Weaver and Roseman, 2005), but many nuclear genetic loci show evidence of ancient phylogeographic structure, and therefore are inconsistent with population replacement.
5. Fossil and archaeological data from Neanderthals and later Europeans are likewise inconsistent with complete population replacement, and instead indicate some amount of Neanderthal genetic persistence.

Each of these lines of evidence points toward the same conclusion: the replacement of Neanderthal mtDNA variation was a consequence of selection rather than extinction. This chapter reviews these lines of evidence, beginning briefly with the current state of knowledge about Neanderthal mtDNA variation and its interpretation in the absence of selection.

Neanderthal mtDNA Variation

Serre et al. (2004) review the evidence of Neanderthal mtDNA variation and its relation to the mtDNA variation in recent humans. However, there are certain weaknesses in the evidence that have been reviewed elsewhere. Here, I review many of these arguments. Many of these are not central to the hypothesis of selection, but they do lend ambiguity to the interpretation of the data. I review them here as necessary cautions on the dates and samples that document ancient mtDNA variability.

For the purposes of examining the demography of the Neanderthal population and the relationship of Neanderthals to later humans, there are four salient observations on human and Neanderthal mtDNA variation:

- Relative to all living and recent humans, the known Neanderthal sequences belong to a separate clade (Knight, 2003).
- This clade is basal compared to those including living humans (although see Gutiérrez et al. [2002] concerning possible ambiguity on this position).
- The date linking this clade with that of living humans is likely between 317,000 and 741,000 years ago (Krings et al., 1999).
- The variation within the Neanderthal clade is comparable in level to that within the clade including all recent humans (Krings et al., 2000); indicating a most recent common ancestor for Neanderthal mtDNA substantially more recent than the Neanderthal-modern human mtDNA ancestor.

These observations depend mainly on the accurate estimation of the mutation rate of mtDNA and the accurate estimation of phylogeny. They do not depend strongly on the assumption of neutrality, except to the extent that the effective rate of mutations and the occurrence of homoplastic variants may have been affected by negative selection on new deleterious variants and positive selection on new adaptive variants. Estimating the possible error in such estimates of phylogeny and dates is a complicated subject in its own right (Templeton, 1993; Wills, 1995; Wise et al., 1997). However, while the dates and phylogenies may have been affected by selection, they do not themselves suffice to demonstrate it. Nor would the presence of some selection necessarily result in erroneous dates or mtDNA genealogical reconstructions. Instead, the key error is the assumption that mtDNA variants form a one-to-one correspondence with populations (Templeton, 1993).

Our confidence in the samples is a more relevant issue to the interpretation of selection versus neutrality. Clearly, if some Neanderthal-like mtDNA variant were to be found in a post-Neanderthal European, it would be clear evidence that the mtDNA replacement was discordant with the boundaries between Neanderthal and later populations, as currently defined. Such a late-persisting Neanderthal mtDNA variant has not yet been found. Serre et al. (2004) failed to amplify Neanderthal-like mtDNA sequences in a sample of five early modern European fossils. Likewise, the amplification of mtDNA from two Upper Paleolithic Italian fossils resulted in only modern variants (Caramelli et al., 2003). But this sample is too small and too late to establish clearly when the mtDNA type of recent humans initially arrived in Europe, or the mode by which it replaced the earlier Neanderthal mtDNA type. On the one hand, the modern variant may have arrived within Neanderthal populations as a consequence of gene flow from elsewhere; on the other, Neanderthal variants may have survived into later populations at not-yet-detectable frequencies. At present, it can be said only that some recent European mtDNA haplogroups (specifically, I, HV, and U) originated earlier than 25,000 years ago (U may be as old as 50,000), and could have arrived in Europe during this time period (Richards, 2003). The contemporaneity of this haplogroup with the transition from Neanderthals to modern humans in Europe suggests an association, although the hypothesis of an exclusive association of recent human mtDNA types with modern human anatomy would require further samples from early modern Europeans.

The other major caveat about the fossil sample is the possibility of contamination. Where an mtDNA variant is amplified that is clearly not present in known samples of living people, contamination is unlikely (Serre et al., 2004). But the amplification of sequences that are present in samples of recent humans may not be verifiably endogenous. Multiple controls may reduce the likelihood of contamination to some extent, but even known Neanderthal fossils result in a high proportion of modern human contaminants. This problem may mitigate against the temporal resolution of the mtDNA discontinuity: although a large sample of later humans can indicate the lack of Neanderthal-like mtDNA variants, a similarly large sample may never demonstrate the absence of humanlike variants within Neanderthals.

The interpretation of Neanderthal mtDNA variation in terms of population replacement depends very strongly on the assumption of neutrality. This is because in the absence of selection, the likelihood of loss of mtDNA variants is related to the size of a population. Within a very small population of a few thousand individuals, the disappearance of an initially widespread mtDNA variant would be conceivable over the course of a few thousand generations. Indeed, Nordborg (1998) used the assumption of a small effective population size to show that the odds of Neanderthal mtDNA disappearance would be as high as 50 percent. But by 30,000 years ago, the global human population was not small. With a geographic distribution spanning the Old World, the ancestors of living humans must have numbered in the hundreds of thousands or more (Biraben, 2003).

Nor was the European population after 30,000 years ago stable in size; it was growing (Harpending et al., 1993; Hawks, 1999; Stiner et al., 2000). In a growing population, the likelihood is very small that a neutral allele will be extinguished by genetic drift – especially an allele once possessed by a substantial proportion of the European population (Manderscheid and Rogers, 1996). As demonstrated by Currat and Excoffier (2004), the clear implication is that if Neanderthals contributed their mtDNA type to later modern Europeans *and if mtDNA variation was neutral*, then we should observe Neanderthal-like mtDNA variants in the gene pool of present-day Europeans (also discussed by Weaver and Roseman, 2005). The fact that we observe no such variants is a strong refutation of one of these assumptions.

Selection and Recent mtDNA Variation

Based on current samples, 167 fixed amino acid substitutions separate the mtDNA genomes of humans and chimpanzees (Kivisild et al., 2006). This number does not include any of the 414 amino acid sites that are polymorphic in humans, some of which may have functional effects, nor does it count the possibility of repeated substitutions at hotspots. Only around half this number occurred on the human lineage (the other half occurred on the chimpanzee lineage), and the vast majority of these substitutions were probably neutral. But the number is large enough to indicate a substantial possibility of functional change.

The hypothesis of past selection on human mtDNA depends on two observations: that mtDNA variants may have advantageous functions for the individuals who carry them, and that the pattern of variation in humans is inconsistent with neutral evolution.

Adaptive Evolution of mtDNA in Recent Human Populations

Within humans, there is abundant evidence that mtDNA variants have phenotypic expressions related to health and lifestyle. Human mtDNA variants have been found to be associated with chronic diseases of aging (Wallace et al., 2001; Wallace and Lott, 2002), brain disorders (Wallace et al., 1999, 2001; Zhu et al., 2004), performance in athletes (Niemi and Majamaa, 2005), and longevity itself (Niemi et al., 2005). The present pattern of variation also appears to be correlated with climate (Ruiz-Pesini et al., 2004), and may affect dietary energetics and insulin metabolism (Lowell and Shulman, 2005).

Not only are mtDNA variants expressed in phenotypes, but there is also substantial evidence that they have been adaptively evolving in human populations. Ruiz-Pesini et al. (2004) proposed that human mtDNA variants might mediate a tradeoff between increased energy production and increased longevity. In their hypothesis, specific mtDNA lineages were proposed to promote health and survival due to a decrease in the production of radical oxygen species (ROS). But this decrease in ROS was a side-effect of a decrease in ATP production. Thus, they viewed this balance as a reflection of selection for greater heat production in cold climates: decreasing the efficiency of ATP production tends to increase heat at the same time it decreases oxidative damage to cellular function. They found that these variants occur repeatedly in macrohaplogroups common at higher latitudes.

Wallace (2005a) reviews evidence for ancient adaptive mutations of the mtDNA in humans, which today result in regional specificity of certain macrohaplogroups. This evidence includes regional correlations for mutations in several mtDNA genes (Mishmar et al., 2003), with different patterns among *ATP6*, cytochrome *b* and cytochrome oxidase I. Some haplogroups are more frequent at northern latitudes and others are restricted to Africa, and these restrictions correspond to an increase in the variation of *ATP6* in northern populations. This increase is especially suggestive of selection, since *ATP6* is highly conserved in other mammals. Additional evidence of adaptive mutations comes from the strong effect of some common variants on neuropathological diseases of aging (Wallace, 2005a), which may represent a changed tradeoff between longevity itself and metabolic production in the wake of recent increases in human longevity (Caspari and Lee, 2004).

Kivisild et al. (2006) found that nonsynonymous variants were generally more common in recent mtDNA clades, which had a nonsynonymous-to-synonymous (NS/S) ratio of 0.62, compared to older clades that have a ratio of 0.37. The predominance of nonsynonymous variants in younger clades may represent a weaker degree of purifying selection in recently expanding populations. But according to Kivisild et al. (2006), the NS/S ratio even in the oldest human mtDNA clades is still significantly higher than the interspecific value between humans and chimpanzees (ca. 0.20). This difference would require purifying selection to have been more effective on the human lineage as a whole than it has been during the recent worldwide diversification of humans. This seems very unlikely considering that a global increase in effective population size should have heightened purifying selection, rather than decreasing it. An alternative hypothesis that is more consistent with the recent expansion of the human population is that the NS/S ratio in recent human populations has been inflated by positive selection on present mtDNA variants in humans. In support of this idea, Kivisild et al. (2006) suggest that changes in human diet may have favored the threonine and valine codon replacements that are observed more commonly among living humans than in other primate lineages.

Additionally, there is evidence for substantial changes in the frequency of some mtDNA variants in recent European prehistory. Ancient DNA extracted from Neolithic skeletons of the *Linearbandkeramik* culture found a high proportion (ca. 25 percent) of the N1a haplotype, currently carried by only one in five hundred people (Haak et al., 2005). The radical decrease in the frequency of this variant over the past 7000 years is analogous to the disappearance of Neanderthal-like variants in Europe over a much longer time period. Haak et al. (2005) found that the decrease was inexplicable as a result of genetic drift alone.

Tests of Neutrality

As suggestive as recent adaptive mtDNA mutations may be, ancient mtDNA substitutions within the human population as a whole must be substantiated by demonstrating a departure from neutrality for the global human population. The proportion of rare nucleotide sites for the mtDNA is far higher than expected under the hypothesis of neutrality (Merriwether et al., 1991). This has been shown not only for the noncoding D-loop but also for functional genes (Wise et al., 1997, 1998). The pattern of geographic diversity of human mtDNA, with a sequence of apparent expansions early in Africa and later in Asia and Europe (Harpending et al., 1993) is consistent with a recent selective sweep originating in Africa and spreading from there to the rest of the Old World (Hawks et al., 2000a). A natural inference from these data would be that a recent episode of positive selection did in fact occur (Hawks and Wolpoff, 2001).

A weaker argument is that positive selection on mtDNA during human evolution must have been rare, because the ratio of synonymous to nonsynonymous site differences between humans and chimpanzees is consistent with neutrality. In the absence of selection, this ratio within species is expected to be equal to the same ratio between species (McDonald and Kreitman, 1991). But this test is very weak, particularly in the presence of purifying selection (Fay et al., 2001): it would take many repeated instances of positive selection on the human lineage to refute the hypothesis of neutrality using this test. Even so, where this test has been applied, human mtDNA has been shown to be significantly nonneutral. Wise et al. (1998) has shown a departure from neutrality for the *NADH2* mitochondrial gene, inferring the possibility that both purifying and positive selection may have occurred during human evolution.

Population Expansions

There is one kind of *neutral event* that could explain the violation of mutation-drift equilibrium in human mtDNA without recourse to selection: a massive population expansion. A number of geneticists have estimated the magnitude of such an expansion under the assumption that mtDNA is neutral. The extent of disequilibrium of human mtDNA leads to the estimate of an extreme demographic event: an expansion of a hundred-fold or greater from an initial effective population size of fewer than 10,000 individuals to many millions (Harpending et al., 1993, 1998; Sherry et al., 1994). This event is proposed to have occurred anytime from 40,000 years ago to as much as 150,000 years ago or longer – although the data indicate that it must have occurred earlier in Africa and later in Europe and East Asia (Sherry et al., 1994). The appearance of a series of expansions may be consistent with the progressive dispersal of humans from an initially East African source (Sherry et al., 1994; Tishkoff et al., 1996; Macaulay et al., 2005).

The hypothesis of population expansion can be tested by looking at other regions of the genome. If the ancient human population actually expanded massively in size, then all neutral regions of the genome should show evidence of the *same* pattern of increasing population size. However, this hypothesis has been very difficult to test. Because of the unique aspects of mtDNA inheritance, and its uniquely high mutation rate, nuclear and mtDNA loci are expected to show somewhat different patterns of variation after a population expansion. Also problematic is the fact that some nuclear genomic loci have themselves been subject to recent positive selection (Enard et al., 2002), meaning that the overall pattern of variation includes sites and loci that have not evolved neutrally (Harpending and Rogers, 2000). Only recently has the study of nuclear genomic loci begun to show the inconsistency of nuclear and mtDNA variation (Eswaran et al., 2005). The most parsimonious explanation for this inconsistency is selection on mtDNA.

Multilocus Genomic Comparisons

The demographic history that has been reconstructed to explain human mtDNA variation is a massive expansion from a single small population, beginning approximately 50,000 to 100,000 years ago (Harpending et al., 1993; Sherry et al., 1994; Harpending et al., 1998). This expansion is estimated to have increased the population size by anywhere from a hundredfold to a thousandfold from an initial effective population size of around 10,000 individuals. The pattern of demographic growth is different in different regions of the world, and apparently oldest within Africa and youngest in Europe (Sherry et al., 1994).

Of nuclear genomic approaches, the results that come closest to consistency with mtDNA have been derived from human microsatellite loci. The largest study of such loci is that of Zhivotovsky et al. (2003). This study examined four different geographically dispersed samples: African hunter-gatherers, African farmers, Eurasia, and East Asia. All except the hunter-gatherers were consistent with substantial recent population expansions, with dates ranging from 35,000 years ago in African farmers to 17,600 years ago in East Asia. The population sizes before this expansion were estimated as 1883 in African farmers, 1760 in Eurasia, and 1688 in East Asia. Estimates from African hunter-gatherers diverged from this pattern, with expansion from only 4300 years ago and a slightly larger initial size of 2609 individuals. These estimates undercut the *most recent* estimates from mtDNA by a factor of around threefold (comparing like geographic regions).

At the other extreme, Ptak and Przeworski (2002) examined the variation of more than 400 nuclear genetic loci. They found that the main demographic assumption that had not previously been considered in studies of variation was population structure. They showed an increase in the skew toward low-frequency variants in samples that combined many different human populations, while samples of single populations showed little or no evidence of population growth. The conclusion of this study was that there is no substantial evidence of population growth from the nuclear genome. This conclusion mirrored earlier results that found individual genetic loci to be inconsistent with a global population expansion (Hawks, 1999; Przeworski et al., 2000; Wall, 2000).

Williamson et al. (2005) examined a similar sample of 301 genes in geographically dispersed populations, but considered an alternative model in which the variation of some of these loci could represent selection in addition to population expansion. They conclude that many of the low-frequency variants in humans are actually slightly deleterious mutations and therefore do not reflect demographic expansion. From this sample of genes, the study estimated a global population expansion at 18,200 years ago, from an initial effective size of approximately 8000 to a final size of around 50,000.

Marth et al. (2004) present estimates of population size changes based on single nucleotide polymorphisms drawn from across the genome. These estimates suggest an ancient population expansion for African populations, at around 150,000 years ago from an initial size of 10,000 individuals to a final size of 18,000. In contrast, European and Asian populations are suggested to have undergone a bottleneck, with a duration of between 10,000 and 12,000 years, with expansions at 60,000 and 64,000 years ago. The pre-expansion sizes for both populations are estimated as 10,000 individuals, to post-expansion sizes at 20,000 and 25,000 individuals, and the bottleneck sizes are estimated at 2000 and 3000 individuals in Europe and Asia, respectively.

Unlike the other studies of nuclear genes, the dates estimated by Marth et al. (2004) are within the time range estimated for mtDNA expansion. But these expansions are not tenfold or more; they are threefold or less. And these SNPs indicate not a simple expansion, but a *bottleneck*. Can this bottleneck be consistent with the diversity of mtDNA? Fay and Wu (1999) showed that certain kinds of bottlenecks might be consistent with both mtDNA disequilibrium and nuclear DNA equilibrium. But the bottleneck estimated by Marth et al. (2004) is much less severe than the simulations of Fay and Wu (1999). For example, Fay and Wu (1999) model a 20-fold reduction in population size with a duration of 1500 generations. For an initial *Ne* of 10,000 and generation time of 20 years, this would approximate a bottleneck to 500 individuals lasting 30,000 years to match the pattern of mtDNA and nuclear genetic diversity in humans. In contrast, Marth et al. (2004) estimates a much more shallow reduction from an initial *Ne* of 10,000 to 2000 effective individuals for only 500 generations in Europe; 3000 effective individuals for 600 generations in Asia; and no bottleneck at all in Africa. Fay and Wu (1999) did not run simulations with such shallow values, but the mildest bottlenecks they *did* simulate, with a 10-fold reduction in population size, did not have the strong effect on mtDNA diversity that is observed today. This result is confirmed by the simulations presented by Ambrose (1998), who tested whether a short bottleneck associated with the Toba volcanic event 71,000 years ago might be consistent with human mtDNA variation. These simulations found that such a bottleneck could not be excluded by mtDNA variation, but that the bottleneck could not by itself *explain* the pattern of mtDNA variation. Instead, a more ancient reduction in population size must have occurred, if mtDNA is neutral. Such a reduction appears to be ruled out by nuclear genomic data: if it occurred, then the nuclear genomic loci should preserve much less variability than they do.

These large nuclear genomic analyses are significantly different not only from the variation of mtDNA, but also from each other. For example, the initial population sizes of the microsatellite estimates and the SNP estimates are different by an order of magnitude. These estimates are not merely different, but strongly inconsistent because they engender different expectations for the distribution of variation across genes. For example, the estimate of long-term effective size from human SNP variation is 10,000. This estimate comes from observations about the diversity of nuclear loci, including the observation that the coalescence ages of these loci have an extensive range of variation – from as little as 200,000 years (or less) to as ancient as 3 million years (or older). A wide range of dates is a normal consequence of stochastic genetic drift: this wide range is exactly what would be expected for an effective size of 10,000.

But the range of coalescent ages from different genes is simply inconsistent with a long-term effective size of 1000–2000, as estimated from microsatellite variation. The microsatellite estimates arrive at this small estimate of long-term effective size because of a *uniformity* of allele length variance among different microsatellite loci (Kimmel et al., 1997; Zhivotovsky et al., 2000). Apparently, human microsatellites have had different evolutionary dynamics than the genome-wide SNPs. The likely reason for this inconsistency is that natural selection has acted to limit the allele length variance of microsatellite loci. In any event, the two estimates are significantly at variance with each other. If we accept the SNP estimate of large long-term effective population size, we cannot explain the coalescence age and variation of mtDNA without selection. If we accept the small long-term effective size of microsatellites, we cannot explain the wide range of coalescence ages of nuclear loci, the age of the mtDNA coalescent, or the geographic spread of modern humanity before 20,000 years ago.

Resolving the Inconsistency

This discrepancy among nuclear datasets is not crucial to the comparison of nuclear and mitochondrial genomes. Whether microsatellites, SNPs or haplotype data are considered, large samples of loci from the nuclear genome have not yet produced an estimate of population expansion consistent with those produced for mtDNA. But the discrepancies among datasets raise a vital point: perhaps all genetic estimates depend on evolutionary models that exclude important factors. This alternative cannot at present be excluded. Some demographic hypothesis, involving complex interactions of ancient populations, might yet be found that would match all the genetic data. But considering the failure of any simple demographic model to it the data, we must recognize that there is already a hypothesis that resolves the inconsistencies: the hypothesis of selection.

The issue here is not whether a population expansion occurred in the Late Pleistocene and Holocene. It certainly did. Since the development of agricultural and pastoral subsistence patterns in the early to mid-Holocene, the human population has expanded by as much as a thousand-fold, from an initial population of between 4 and 10 million people (Coale, 1974; Biraben, 1979, 2003). Within the Late Pleistocene, it is probable that the global population grew by tenfold or more, because of increases in geographic range and population density. Most obviously, during the past 50,000 years, human populations colonized areas that previously had no human inhabitants (Hawks, 1999). These areas include the northern tier of Eurasia, the Arctic, Beringia and the Americas, the Melanesian archipelago, Japan, the Philippines, Australia, and many others. Archaeological evidence substantiates an increase in population density in other inhabited regions, where ancient people utilized a broader spectrum of faunal resources (Klein, 1999; Stiner et al., 2000), and developed new techniques of resource extraction such as nets, projectile weapons, and harpoons (Yellen et al., 1995; Bar-Yosef, 2002; Soffer, 2004). All told, during the past 50,000 years the human population expanded by as much as 10,000 times.

But this expansion by itself cannot explain the unique pattern of variation of human mtDNA, which ultimately depends on even earlier events. Neither does the pattern of genetic variation across the nuclear genome support the unique parameters necessary to explain mtDNA variation as a consequence of demography. These different sources of data are significantly inconsistent with each other in their evidence concerning the degree and extent of possible population growth during the Late Pleistocene and Holocene. While it will take additional work to reconcile the nuclear genomic and archaeological records, it appears evident that this resolution will be in the *opposite* direction from the extreme mtDNA population expansion model.

Phylogeography and Genetic Variation

The demographic models discussed in the previous section have all assumed that human genes represent panmictic populations, without considering their geographic structure or phylogeographic relations (e.g., Fay and Wu, 1999; Marth et al., 2004). But the geographic locations and frequencies of genetic variants have increasingly been shown to be important evidence of their evolutionary origin. Nuclear genetic variation has shown that human mtDNA variation is inconsistent with the same demography as other genomic loci. Human genetic phylogeography has shown that gene lineages from archaic Eurasians not only persist in living humans, but comprise a substantial portion of our current genetic complement.

Eswaran et al. (2005) suggest that the survival of archaic gene lineages explains the signature of a bottleneck in nuclear SNPs. They find that such a pattern is the expected result of the fusion of Eurasian and African populations, under the assumption of substantial population dispersal from Africa into other regions. This pattern contrasts strongly with the expected signature of population expansion under a replacement scenario of modern human origins. They conclude that archaic assimilation is more consistent with the pattern of genomic SNP data than replacement.

Templeton (2002) examined the phylogeographic variation of 6 autosomal and 2 X chromosomal loci. The study concluded that although mtDNA and Y chromosomal loci are consistent with recent out-of-Africa population movement, other nuclear loci indicate much more ancient patterns of population movement out of Africa. This pattern is inconsistent with a recent population replacement from Africa: if such a replacement happened, then autosomal genes should not show evidence for yet more ancient events. Templeton (2002) has argued that the pattern of mtDNA (and Y chromosomal) variation may represent one recent migration among many out of Africa; but the observations are also consistent with a recent selective sweep of mtDNA. These data conclusively refute the hypothesis that no archaic gene lineages survived into living human populations.

Templeton's studies have been critiqued on the grounds that they may not actually distinguish survival of archaic gene lineages *outside* Africa from survival of archaic lineages *inside* Africa. In other words, it has been claimed (Pearson, 2003; Eswaran et al., 2005) that ancient population structure *within* Africa might mimic the survival of gene lineages from outside Africa. Eswaran et al. (2005) show that this scenario is likely not the case, as nuclear genomic data apparently reflect the survival of archaic non-African lineages. But this equivocation bears little importance to the explanation of mtDNA: even the survival of archaic lineages from *within* Africa challenges the idea that mtDNA variation reflects the expansion of one small African population and the displacement of others. Instead, survival of archaic African lineages suggests that the ancient population size of Africans was effectively much larger (perhaps many orders of magnitude larger) than a neutral mtDNA hypothesis would admit. Together Templeton's work and the analysis of Eswaran et al. (2005) indicate that the majority of genomic loci preserve allelic variation that originally characterized archaic human populations.

In addition to these phylogeographic comparisons of many genetic loci, recent work has uncovered individual genes that appear to show strong evidence of archaic human genetic persistence. These genes cannot be accommodated within the framework of a recent mtDNA expansion and replacement of archaic humans. These genes include (but are not limited to) the region around Xp21.1 (Garrigan et al., 2005), the Xp/Yp and 12q telomeric regions (Baird et al., 2000), and an inversion on 17q21.31 (Stefansson et al., 2005). The first of these genes suggests possible modern-archaic admixture within Africa, a reflection of ancient population structure that would be inconsistent with the mtDNA evidence for recent expansion from a single small African population. The other loci are more significant in suggesting ancient Eurasian gene lineages. In particular, Hardy et al. (2005) suggest that the MAPT locus preserves evidence of a Neanderthal introgressive allele into recent Europeans.

In short, genomic data are not consistent with mtDNA neutrality, and a growing number of detailed studies have documented loci that represent the survival (and proliferation) of archaic human gene lineages. It remains possible that none of these loci preserve evidence of specifically Neanderthal gene lineages, as opposed to archaic human lineages from other regions of the world. But the idea that all non-African gene lineages were replaced by African gene lineages during the Late Pleistocene is rejected by these observations. Simply put, *some* Eurasian archaic humans were ancestral to living people. The proportion of these ancestors that were Neanderthals is not yet known, but there is no reason to think that the proportion was zero. Whether these ancestors were Neanderthals or not, the survival of archaic nuclear genomic lineages is inconsistent with the hypothesis that mtDNA evolution in humans has been neutral.

Fossil and Archaeological Observations

Human genetic loci are not the only sources of evidence about human origins. Nor were they the initial evidence suggesting an African origin for recent human populations (Protsch, 1975; Bräuer, 1984). Instead, the out-of-Africa hypothesis was initially based on the morphological resemblance of recent humans compared to archaic humans (Howells, 1942) and the early Late Pleistocene appearance of certain features in African and Levantine human fossils (Stringer and Andrews, 1988).

The fossil origins of the out-of-Africa model have often been ignored in discussions of the relevance of the genetic evidence (Stringer and Bräuer, 1994). When fossil and genetic interpretations have been conflated, widespread confusion about the predictions of different hypotheses of human origins has been the result. The out-of-Africa hypothesis, as described by Bräuer et al. (2004: 702), is not synonymous with the hypothesis of complete genetic replacement of archaic humans.

As we have pointed out elsewhere (Stringer & Bräuer, 1994; Bräuer & Stringer, 1997; Stringer, 2002), the idea of complete replacement without any interbreeding is one variant of the Out of Africa hypothesis, often identified with a particular interpretation of the mitochondrial DNA data, dubbed the Eve theory (Frayer et al., 1993). Complete replacement does not feature in all versions of the Out of Africa hypothesis, and therefore we contend that it would not be falsified by demonstration of some modern-archaic human gene flow outside of Africa. Both before and after the pioneering study by Cann et al. (1987), proponents of the out of-Africa hypothesis accepted the possibility of a certain amount of gene flow between the migrating early anatomically modern humans and the non-African archaic groups.

When construed in this way, the out-of-Africa hypothesis and its alternatives do not differ on whether "some modern-archaic human gene flow outside of Africa" occurred. Instead, they differ on the evolutionary significance of such gene flow. But for a neutral genetic locus "evolutionary significance" is a moot point: neutral genes have no adaptive role. Their only evolutionary significance is to mark demographic changes or population movements (Harpending et al., 1993; Cann et al., 1987), a fact which resulted in the linkage of mtDNA evidence to the out-of-Africa hypothesis (Stringer and Andrews, 1988; Wilson and Cann, 1992). Templeton (1993) argued that demographic inferences from mtDNA were nevertheless consistent with a multiregional origin of modern humans, because the mutation rate of mtDNA remained uncertain, and low genetic diversity by itself is a poor indicator of population history. Even so, the low mtDNA diversity of humans, the relatively higher diversity within Africans, and the lack of Neanderthal mtDNA variants in living Europeans all have been used as arguments in favor of a recent out-of Africa population replacement (Vigilant et al., 1991; Jorde et al., 1998; Foley, 1998). But the pattern of mtDNA variation – if neutral – can only be explained by a relatively extreme demographic interpretation of expansion from an extremely small initial population (Harpending et al., 1998), followed by no detectable archaic intermixture whatsoever (Currat and Excoffier, 2004; Weaver and Roseman, 2005). It is true that many may hold a less extreme view of the putative out-of-Africa expansion (Bräuer et al., 2004). But the extreme variant is the only one for which mtDNA variation has any relevance; as demonstrated well by recent work (Manderscheid and Rogers, 1996; Currat and Excoffier, 2004; Weaver and Roseman, 2005), almost any less extreme model should predict the survival of *some* archaic mtDNA into present populations.

This extreme variant of the out-of-Africa hypothesis has been rejected repeatedly by consideration of fossil evidence (Frayer et al., 1993; Frayer, 1993; Duarte et al., 1999; Hawks et al., 2000b; Wolpoff et al., 2001). These studies do not indicate the level of recent human ancestry attributable to Neanderthals and other archaic humans, but their conclusions are inconsistent with the hypothesis that no genetic survival of archaic humans occurred. Simply put, if the morphological variation of early modern humans has *any* informative value regarding the origins of those humans, the hypothesis of a complete population replacement has long been falsified. This evidence has recently been reviewed by Trinkaus (2005: 218), who called the replacement scenario "intellectually dead." Yet a neutral mtDNA is inconsistent with softer hypotheses that incorporate some amount of modern-archaic human gene flow (Currat and Excoffier, 2004).

It should be noted that not all specialists in fossil human morphology agree with this assessment of the evidence (Lahr, 1996; Tattersall and Schwartz, 1999; Bräuer et al., 2004). A central argument is that the features that apparently reflect archaic contributions to modern populations may be phylogenetically uninformative (Lahr, 1996; Bräuer et al., 2004). One possibility is that these morphological features do not indicate ancestry, for example if they mostly have evolved in parallel in distantly related archaic human populations. Another possibility is that the extant sample of African fossil humans from the late Middle Pleistocene is insufficient exclude the possibility that the features of archaic Eurasians actually occurred there at low frequency. Either of these conditions might mean that features shared between Neanderthals and early modern Europeans, for example, were homoplasies or plesiomorphies and therefore do not provide evidence of relationship. These objections, however, would cast doubt on whether *any* extant morphological evidence can provide evidence of modern human origins. If the sample of fossils is really too small to reject hidden plesiomorphies, and if parallelism is really so pervasive as to explain features like the mandibular foramen form or lambdoidal flattening, then morphological features will not test hypotheses of modern human origins until the sample becomes much stronger than it is today.

The known pattern of anatomical features in European Neanderthals and early modern Europeans is consistent with the genetic contribution of Neanderthals to later populations (Frayer, 1993; Duarte et al., 1999; Wolpoff et al., 2001; Trinkaus et al., 2003). At the same time, the morphological evidence for massive change suggests strong extra-European influences on the post-Neanderthal population. Genetic evidence for greater variation within Africa suggests that Africa may have been the ultimate source for much of this morphological change, although intervening populations may have lived for some time in western Asia or even northern Asia before substantially contributing genes to populations in Europe. The fossil evidence is inconsistent with the hypothesis of mtDNA neutrality, because the lack of Neanderthal mtDNA variants can only be explained by the near total replacement of Neanderthal mtDNA gene pool (Currat and Excoffier, 2004). If this fossil evidence is accepted, it must support the hypothesis of selection on mtDNA, under which the Neanderthal mtDNA gene pool was replaced by an adaptively superior mtDNA variant, and not by a neutral population replacement.

Discussion

These five lines of evidence all point to the same conclusion: human mtDNA has been a target of adaptive substitution during human evolution, and it is this positive selection rather than population replacement that explains the disappearance of archaic mtDNA variants, like those of the Neanderthals. This is not a new idea. In their review of genetic evidence for modern human origins, Stringer and Andrews (1988) caution that a history of selection may invalidate the interpretation of population history, although they averred that mtDNA neutrality "does appear approximately valid" (Stringer and Andrews, 1988: 1265). Spuhler (1989) raised the hypothesis of positive selection, or a "selective sweep" of mtDNA based on the complete linkage of the mtDNA molecule. Others have suggested the hypothesis of selection as an alternative to mtDNA neutrality since that time (e.g., Wise et al., 1998; Kreitman, 2000; Hawks and Wolpoff, 2001).

But arriving at this conclusion has required both time and additional research into human mtDNA and nuclear genomic variability. In a 2000 review, Kreitman (2000: 553) wrote:

> Given that the mitochondrial genome has a smaller effective population size (being maternally inherited and effectively haploid) than the nuclear genome, the conflicting portraits of polymorphism in the two genomes may be consistent with a population bottleneck. The exciting possibility of a selective sweep in the modern mitochondrial genome remains, unfortunately, an unresolved issue.

Only within the past few years has a resolution begun to be possible. During this time, researchers have reported direct evidence of adaptive mtDNA variants in living human populations (Ruiz-Pesini et al., 2004; Wallace, 2005a, b). It has also become clear that the pattern of nuclear genomic variation does not support the kind of population expansion or bottleneck that would be necessary to retain the hypothesis of mtDNA neutrality (Hawks, 1999; Kreitman, 2000; Przeworski et al., 2000; Marth et al., 2004).

Recently, evidence for positive selection has been found for many human genes Vallender and Lahn (2004). The logic supporting the inference of selection for other genes can be directly applied to the mtDNA. Consider the case of *FoxP2*. Enard et al. (2002) proposed that this gene had undergone a selective sweep within the past 200,000 years in humans, and Klein and Edgar (2002) made it the centerpiece of their argument that language had evolved recently at the origin of modern humans. Like mtDNA, *FoxP2* is strongly out of mutation-drift equilibrium. But human *FoxP2* shows *many fewer* amino acid substitutions compared to chimpanzees than does human mtDNA (2 for *FoxP2*, 167 for mtDNA), meaning that the possibility of selection on human mtDNA should be *greater*, not less. And *FoxP2* is only one functional gene; mtDNA contains 13 peptide-coding regions (Arnason et al., 1996), selection on any one of which would affect the entire molecule. And while a rare dysfunction of *FoxP2* affects language skill, many *common* variants of mtDNA directly affect mortality rates.

So what exactly is the difference that leads the same people to say that *FoxP2* is selected and mtDNA is not? There is no statistical test to support such a difference of interpretation. Nor is there any functional argument to support such a difference. No known demographic hypothesis can make either locus consistent with other genomic evidence.

The simplest hypothesis is that a similar pattern of variation in each case was generated by the same mechanism: positive selection. All five lines of evidence considered here point to this conclusion. Taken together, they are consistent with no other. If mtDNA has been positively selected during recent human evolution, then its distribution among fossil humans may not inform us about their demography or phylogenetic relationships.

What was the target of selection on human mtDNA? The mitochondria are chiefly involved in the energy metabolism of the cell. Certain tissues, such as the muscles of performance athletes, the sensory cells of the retinas, and the brain demand higher energy output than others either constantly or at intervals. Mitochondria tend to degrade somewhat in performance over the course of an individual's lifespan, with older people facing lower metabolic performance and occasional disorders known as mitochondriopathies. Mitochondrial dysfunction has been implicated in several neurodegenerative diseases, including Parkinson's disease, Alzheimer's disease, Huntington's disease, and amyotrophic lateral sclerosis (ALS) (Zhu et al., 2004). It is a reasonable conjecture that selection on mtDNA was related to evolutionary change in the metabolic requirements of the brain coupled with changes in longevity. An increase in longevity has been associated with the Upper Paleolithic in Europe (Caspari and Lee, 2004), and this evolutionary change may explain the disappearance of Neanderthal mtDNA. It is also possible that selection related to changing dietary energetics occurred (Lowell and Shulman, 2005), or that selection related to climate may explain some of the recent pattern of selection (Ruiz-Pesini et al., 2004).

But the most recent selective sweep of human mtDNA was almost certainly not the only one. The fact that Neanderthal sequence variation is so similar to that of recent humans is highly suggestive. The pattern found in these two groups is different from that within most hominoid species or subspecies. A reasonable hypothesis is that ancient humans underwent several episodes of positive selection on their mitochondrial genomes. A test of this hypothesis would be the observation of mtDNA sequences in non-European archaic humans. If ancient humans were restricted in mtDNA variation due to recurrent positive selection, then these ancient humans should also be very similar to the Neanderthals, or close to the root of the Neanderthal-modern human mtDNA divergence. Such a situation may be true of the chromosome 11 numt insertion, which lies on the human side of this divergence, but is basal to all recent human mitochondrial sequences (Zischler et al., 1995).

Determining whether Neanderthals – as opposed to other archaic groups – contributed genetic material to the living human population is a challenge. Even if clear evidence of archaic lineages is found, it is difficult to substantiate that these lineages were found in a particular region of Europe over 40,000 years ago. Yet, substantial evidence of archaic lineages has been found. There is no question that some – perhaps most – human genes preserve allelic variation from archaic human populations. The morphological and archaeological evidence suggest strongly that Neanderthal genetic lineages survived into later Upper Paleolithic populations. Ultimately, the genetic test of Neanderthal survival may be carried out by finding nuclear DNA sequences from Neanderthal fossils themselves. Until that time, we can say only that some Neanderthal contribution to the modern human nuclear gene pool is consistent with the known evidence.

References

Ambrose, S.H., 1998. Late Pleistocene human population bottlenecks, volcanic winter and differentiation of modern humans. *J. Hum. Evol.* 34, 623–652.

Arnason, U., Gullberg, A., Janke, A., Xu, X., 1996. Pattern and timing of evolutionary divergences among hominoids based on analyses of complete mtDNAs. *J. Mol. Evol.* 43, 650–661.

Baird, D.M., Coleman, J., Rosser, Z.H., Royle, N.J., 2000. High levels of sequence polymorphism and linkage disequilibrium at the telomere of 12q: implications for telomere biology and human evolution. *Am. J. Hum. Genet.* 66, 235–250.

Bar-Yosef, O., 2002. The Upper Paleolithic revolution. *Ann. Rev. Anthropol.* 31, 363–393.

Biraben, J.-N., 1979. Essai sur l'evolution du nombre des hommes. *Population* 1, 13–25.

Biraben, J.-N., 2003. Lévolution du nombre des hommes. *Population et Sociétés* 394, 1–4.

Bräuer, G., 1984. A craniological appoach to the origin of anatomically modern *Homo sapiens* in Africa and implications for the appearance of modern Europeans. In: Smith, F.H., Spencer, F. (Eds.), *The Origins of Modern Humans: A World Survey of the Fossil Evidence*. Alan R. Liss, New York, pp. 327–410.

Bräuer, G., Stringer, C., 1997. Models, polarization, and perspectives on modern human origins. In: Clark, G.A., Willermet, C.M. (Eds.), *Conceptual Issues in Modern Human Origins Research*. Aldine de Gruyter, New York, pp. 191–201.

Bräuer, G., Collard, M., Stringer, C., 2004. On the reliability of recent tests of the Out of Africa hypothesis for modern human origins. *Anat. Rec.* 279A, 701–707.

Cann, R.L., Stoneking, M., Wilson, A.C., 1987. Mitochondrial DNA and human evolution. *Nature* 325, 31–36.

Caramelli, D., Lalueza-Fox, C., Vernesi, C., Lari, M., Casoli, A., Mallegni, F., Chiarelli, B., Dupanloup, I., Bertranpetit, J., Barbujani, G., Bertorelle, G., 2003. Evidence for a genetic discontinuity between Neandertals and 24,000-year-old anatomically modern Europeans. *Proc. Natl. Acad. Sci. U.S.A.* 100, 6593–6597.

Caspari, R., Lee, S.-H., 2004. Older age becomes common late in human evolution. *Proc. Natl. Acad. Sci. U.S.A.* 101, 10,895–10,900.

Coale, A.J., 1974. The history of the human population. *Sci. Am.* 231, 40–52.

Currat, M., Excoffier, L., 2004. Modern humans did not admix with Neanderthals during their range expansion into Europe. *PLoS Biol.* 2, e421.

Duarte, C., Maurício, J., Pettitt, P.B., Souto, P., Trinkaus, E., van der Plicht, H., Zilhão, J., 1999. The early Upper Paleolithic human skeleton from the Abrigo do Lagar Velho (Portugal) and modern human emergence in Iberia. *Proc. Natl. Acad. Sci. U.S.A.* 96, 7604–7609.

Enard, W., Przeworski, M., Fisher, S.E., Lai, C.S., Wiebe, V., Kitano, T., Monasco, A.P., Pääbo, S., 2002. Molecular evolution of *FOXP2*, a gene involved in speech and language. *Nature* 418, 869–872.

Eswaran, V., Harpending, H., Rogers, A.R., 2005. Genomics refutes an exclusively African origin of humans. *J. Hum. Evol.* 49, 1–154.

Fay, J.C., Wu, C.-I., 1999. A human population bottleneck can account for the discordance between patterns of mitochondrial versus nuclear DNA variation. *Mol. Biol. Evol.* 16, 1003–1005.

Fay, J.C., Wyckoff, G.J., Wu, C.-I., 2001. Positive and negative selection on the human genome. *Genetics* 158, 1227–1254.

Foley, R.A., 1998. Genes, evolution and diversity: yet another look at the problem of modern human origins. *Evol. Anthropol.* 6, 191–193.

Frayer, D.W., 1993. Evolution at the European edge: Neanderthal and Upper Paleolithic relationships. *Préhistoire Européenne* 2, 9–69.

Frayer, D.W., Wolpoff, M.H., Smith, F.H., Thorne, A.G., Pope, G.G., 1993. The fossil evidence for modern human origins. *Am. Anthropol.* 95, 14–50.

Garrigan, D., Mobasher, Z., Severson, T., Wilder, J.A., Hammer, M.F., 2005. Evidence for archaic Asian ancestry on the human X chromosome. *Mol. Biol. Evol.* 22, 189–192.

Gutiérrez, G., Sánchez, D., Marín, A., 2002. A reanalysis of the ancient mitochondrial DNA sequences recovered from Neandertal bones. *Mol. Biol. Evol.* 19, 1359–1366.

Haak, W., Forster, P., Bramanti, B., Matsumura, S., Brandt, G., Tänzer, M., Villems, R., Renfrew, C., Gronenborn, D., Alt, K.W., Burger, J., 2005. Ancient DNA from the first European farmers in 7500-year-old Neolithic sites. *Science* 310, 1016–1018.

Hardy, J., Pittman, A., Myers, A., Gwinn-Hardy, K., Fung, H.C., de Silva, R., Hutton, M., Duckworth, J., 2005. Evidence suggesting that *Homo neanderthalensis* contributed the H2 *MAPT* haplotype to *Homo sapiens*. *Biochem. Soc. Trans.* 33, 582–585.

Harpending, H., Rogers, A., 2000. Genetic perspectives on human origins and differentiation. *Ann. Rev. Genomics Hum. Genet.* 1, 361–385.

Harpending, H.C., Sherry, S.T., Rogers, A.R., Stoneking, M., 1993. The genetic structure of ancient human populations. *Curr. Anthropol.* 34, 483–496.

Harpending, H.C., Batzer, M.A., Gurven, M., Jorde, L.B., Rogers, A.R., Sherry, S.T., 1998. Genetic traces of ancient demography. *Proc. Natl. Acad. Sci. U.S.A.* 95, 1961–1967.

Hawks, J., Wolpoff, M.H., 2001. Paleoanthropology and the population genetics of ancient genes. *Am. J. Phys. Anthropol.* 114, 269–272.

Hawks, J., Hunley, K., Lee, S.-H., Wolpoff, M.H., 2000a. Bottlenecks and Pleistocene human evolution. *Mol. Biol. Evol.* 17, 2–22.

Hawks, J., Oh, S., Hunley, K., Dobson, S., Cabana, G., Dayalu, P., Wolpoff, M.H., 2000b. An Australasian test of the recent African origin theory using the WLH-50 calvarium. *J. Hum. Evol.* 39, 1–22.

Hawks, J.D., 1999. The evolution of human population size: A synthesis of fossil, archaeological, and genetic data. Ph.D. Dissertation, University of Michigan, Ann Arbor, MI.

Howell, N., Kubacka, I., Mackey, D., 1996. How rapidly does the human mitochondrial genome evolve? *Am. J. Hum. Genet.* 59, 501–509.

Howells, W.W., 1942. Fossil man and the origin of races. *Am. Anthropol.* 44, 182–193.

Jorde, L.B., Bamshad, M., Rogers, A.R., 1998. Using mitochondrial and nuclear DNA markers to reconstruct human evolution. *BioEssays* 20, 126–136.

Kimmel, M., Chakraborty, R., King, J., Bamshad, M., Watkins, W., Jorde, L.B., 1997. Signatures of population expansion in microsatellite repeat data. *Genetics* 148, 1921–1930.

Kivisild, T., Shen, P., Wall, D.P., Do, B., Sung, R., Davis, K.K., Passarino, G., Underhill, P.A., Scharfe, C., Torroni, A., Scozzari, R., Modiano, D., Coppa, A., deKnjiff, P., Feldman, M.W., Cavalli-Sforza, L.L., Oefner, P.J., 2006. The role of selection in the evolution of human mitochondrial genomes. *Genetics* 172, 373–387.

Klein, R., 1999. *The Human Career: Human Biological and Cultural Origins*. 2nd Edition. University of Chicago Press, Chicago.

Klein, R., Edgar, B., 2002. *The Dawn of Human Culture*. John Wiley and Sons, New York.

Knight, A., 2003. The phylogenetic relationship of Neandertal and modern human mitochondrial DNAs based on informative nucleotide sites. *J. Hum. Evol.* 44, 627–632.

Kreitman, M., 2000. Methods to detect selection in populations with applications to the human. *Ann. Rev. Genom. Hum. Genet.* 1, 539–559.

Krings, M., Stone, A., Schmitz, R.W., Krainitzid, H., Stoneking, M., Pääbo, S., 1997. Neandertal DNA sequences and the origin of modern humans. *Cell* 90, 1–20.

Krings, M., Geisert, H., Schmitz, R.W., Krainitzki, S.P., 1999. DNA sequence of the mitochondrial hypervariable region ii from the Neandertal type specimen. *Proc. Natl. Acad. Sci. U.S.A.* 96, 5581–5585.

Krings, M., Capelli, C., Tachentacher, F., Geisert, H., Meyer, S., von Haeseler, A., Grossschmidt, K., Possnert, G., Paunovic, M., Pääbo, S., 2000. A view of neandertal genetic diversity. *Nat. Genet.* 26, 144–146.

Lahr, M.M., 1996. *The Evolution of Modern Human Diversity: A Study of Cranial Variation*. Cambridge Studies in Biological Anthropology, Vol. 18, Cambridge University Press, Cambridge.

Lowell, B.B., Shulman, G.I., 2005. Mitochondrial dysfunction and Type 2 diabetes. *Science* 307, 384–397.

Macaulay, V., Hill, C., Achilli, A., Rengo, C., Clarke, D., Meehan, W., Blackburn, J., Semino, O., Scozzari, R., Cruciani, F., Taha, A., Shaari, N.K., Raha, J.M., Ismail, P., Zainuddin, Z., Goodwin, W., Bulbeck, D., Bandelt, H.-J., Oppenheimer, S., Torroni, A., Richards, M., 2005. Single, rapid coastal settlement of Asia revealed by analysis of complete mitochondrial genomes. *Science* 308, 1034–1036.

Manderscheid, E.J., Rogers, A.R., 1996. Genetic admixture in the Late Pleistocene. *Am. J. Phys. Anthropol.* 100, 1–5.

Marth, G.T., Czabarka, E., Murvai, J., Sherry, S.T., 2004. The allele frequency spectrum in genome-wide human variation data reveals signals of differential demographic history in three large world populations. *Genetics* 166, 351–372.

McDonald, J.H., Kreitman, M., 1991. Adaptive protein evolution at the *Adh* locus in *Drosophila*. *Nature* 351, 652–654.

Merriwether, D., Clark, A.G., Ballinger, S.W., Schurr, T.G., Soodyall, H., Jenkins, T., Sherry, S.T., Wallace, D.C., 1991. The structure of human mitochondrial DNA variation. *J. Mol. Evol.* 33, 543–555.

Mishmar, D., Ruiz-Pesini, E., Golik, P., Macaulay, V., Clark, A.G., Hosseini, S., Brandon, M., Easley, K., Chen, E., Brown, M.D., Sukernik, R.I., Olckers, A., Wallace, D.C., 2003. Natural selection shaped regional mtDNA variation in humans. *Proc. Natl. Acad. Sci. U.S.A.* 100, 171–176.

Niemi, A.-K., Majamaa, K., 2005. Mitochondrial DNA and ACTN3 genotypes in Finnish elite endurance and sprint athletes. *Eur. J. Hum. Genet.* 13, 965–969.

Niemi, A.-K., Moilanen, J.S., Tanaka, M., Hervonen, A., Hurme, M., Lehtimäki, T., Arai, Y., Hirose, N., Majamaa, K., 2005. A combination of three common inherited mitochondrial DNA polymorphisms promotes longevity in Finnish and Japanese subjects. *Eur. J. Hum. Genet.* 13, 166–170.

Nordborg, M., 1998. On the probability of Neanderthal ancestry. *Am. J. Hum. Genet.* 63, 1237–1240.

Pearson, O., 2003. Has the combination of genetic and fossil evidence solved the riddle of modern human origins? *Evol. Anthropol.* 13, 145–159.

Protsch, R.R., 1975. The absolute dating of Upper Pleistocene sub-Saharan fossil hominids and their place in human evolution. *J. Hum. Evol.* 4, 297–322.

Przeworski, M., Hudson, R.R., DiRienzo, A., 2000. Adjusting the focus on human variation. *Trends Genet.* 16, 296–302.

Ptak, S.E., Przeworski, M., 2002. Evidence for population growth in humans is confounded by fine-scale population structure. *Trends Genet.* 18, 559–563.

Richards, M., 2003. The Neolithic invasion of Europe. *Ann. Rev. Anthropol.* 32, 135–162.

Ruiz-Pesini, E., Mishmar, D., Brandon, M., Procaccio, V., Wallace, D.C., 2004. Effects of purifying and adaptive selection on regional variation in human mtDNA. *Science* 303, 223–226.

Serre, D., Langaney, A., Chech, M., Teschler-Nicola, M., Paunovic, M., Mennecier, P., Hofreiter, M., Possnert, G., Pääbo, S., 2004. No evidence of Neandertal mtDNA contribution to early modern humans. *PLoS Biol.* 2, 313–317.

Sherry, S.T., Rogers, A.R., Harpending, H., Soodyall, H., Jenkins, T., Stoneking, M., 1994. Mismatch distribution of mtDNA reveal recent human population expansions. *Hum. Biol.* 66, 761–775.

Soffer, O., 2004. Recovering perishable technologies through use wear on tools: Preliminary evidence for Upper Paleolithic weaving and net making. *Curr. Anthropol.* 45, 407–413.

Spuhler, J.N., 1989. Evolution of mitochondrial DNA in human and other organisms. *Am. J. Hum. Biol.* 1, 509–528.

Stefansson, H., Helgason, A., Steinthorsdottir, G.T.V., Masson, G., Bernard, J., Baker, A., Jonasdottir, A., Ingason, A., Gudnadottir, V.G., Desnica, N., Hicks, A., Gylfason, A., Gudbjartsson, D.F., Jonsdottir, G.M., Sainz, J., Agnarsson, K., Birgisdottir, B., Ghosh, S., Olafsdottir, A., Cazier, J.-B., Kristjansson, K., Frigge, M.L., Thorgeirsson, T.E., Gulcher, J.R., Kong, A., Stefansson, K., 2005. A common inversion under selection in Europeans. *Nat. Genet.* 37, 129–137.

Stiner, M.C., Munro, N.D., Surovell, T.A., 2000. The tortoise and the hare: Small-game use, the broad-spectrum revolution, and Paleolithic demography. *Curr. Anthropol.* 41, 39–73.

Stringer, C., 2002. Modern human origins: Progress and prospects. *Phil. Trans. R. Soc. Lond. B* 357, 563–579.

Stringer, C.B., Andrews, P., 1988. Genetic and fossil evidence for the origin of modern humans. *Science* 239, 1263–1268.

Stringer, C.B., Bräuer, G., 1994. Methods, misreading and bias. *Am. Anthropol.* 96, 416–424.

Tattersall, I., Schwartz, J.H., 1999. Hominids and hybrids: The place of Neanderthals in human evolution. *Proc. Natl. Acad. Sci. U.S.A.* 96, 7117–7119.

Templeton, A., 1993. The "Eve" hypothesis: a genetic critique and reanalysis. *Am. Anthropol.* 95, 51–72.

Templeton, A.R., 2002. Out of Africa again and again. *Nature* 416, 45–51.

Tishkoff, S.A., Dietzsch, E., Seed, W., Pakstis, A.J., Kidd, J.R., Cheung, K., Bonné-Tamir, B., Santachiara-Benerecetti, A.S., Moral, P., Krings, M., Pääbo, S., Watson, E., Risch, N., Jenkins, T., Kidd, K.K., 1996. Global patterns of disequilibrium at the CD4 locus and modern human origins. *Science* 271, 1380–1387.

Trinkaus, E., 2005. Early modern humans. *Ann. Rev. Anthropol.* 34, 207–230.

Trinkaus, E., Milota, S., Rodrigo, R., Mircea, G., Moldovan, O., 2003. Early modern human cranial remains from the Peştera cu Oase, Romania. *J. Hum. Evol.* 45, 245–253.

Vallender, E.J., Lahn, B.T., 2004. Positive selection on the human genome. *Hum. Mol. Genet.* 13, R245–R254.

Vigilant, L., Stoneking, M., Harpending, H., Hawkes, K., Wilson, A.C., 1991. African populations and the evolution of human mitochondrial DNA. *Science* 253, 1503–1507.

Wall, J.D., 2000. Detecting ancient admixture in humans using sequence polymorphism data. *Genetics* 154, 1271–1279.

Wallace, D.C., 2005a. The mitochondrial genome in human adaptive radiation and disease: on the road to therapeutics and performance enhancement. *Gene* 354, 169–180.

Wallace, D.C., 2005b. A mitochondrial paradigm of metabolic and degenerative diseases, aging, and cancer: A dawn for evolutionary medicine. *Ann. Rev. Genet.* 39, 359–407.

Wallace, D.C., Lott, M.T., 2002. Mitochondrial genes in degenerative diseases, cancer, and aging. In: Rimoin, D.L., Connor, J.M., Pyeritz, R.E., Korf, B.R. (Eds.), *Emery and Rimoin's Principles and Practice of Medical*. Churchill Livingstone, London, pp. 299–409.

Wallace, D.C., Brown, M.D., Lott, M.T., 1999. Mitochondrial DNA variation in human evolution and disease. *Gene* 238, 211–230.

Wallace, D.C., Lott, M.T., Brown, M.D., Kerstann, K., 2001. Mitochondria and neuro-opthalmological diseases. In: Scriver, C.R., Beaudet, A.L., Sly, W.S., Valle, D. (Eds.), *The Metabolic and Molecular Basis of Inherited Disease*, Vol. 2. McGraw-Hill, New York, pp. 2425–2512.

Weaver, T.D., Roseman, C.C., 2005. Ancient DNA, late Neandertal survival, and modern-human-Neandertal genetic admixture. *Curr. Anthropol.* 46, 677–683.

Williamson, S.H., Hernandez, R., Fledel-Alon, A., Zhu, L., Nielsen, R., Bustamante, C.D., 2005. Simultaneous inference of selection and population growth from patterns of variation in the human genome. *Proc. Natl. Acad. Sci. U.S.A.* 102, 7882–7887.

Wills, C., 1995. When did Eve live? An evolutionary detective story. *Evolution* 49, 593–607.

Wilson, A.C., Cann, R.L., 1992. The recent African genesis of humans. *Sci. Am.* 266, 68–73.

Wise, C.A., Sraml, M., Rubinsztein, D.C., Easteal, S., 1997. Comparative nuclear and mitochondrial genome diversity in humans and chimpanzees. *Mol. Biol. Evol.* 14, 707–716.

Wise, C.A., Sraml, M., Easteal, S., 1998. Departure from neutrality at the mitochondrial *NADH* dehydrogenase subunit 2 gene in humans, but not in chimpanzees. *Genetics* 148, 409–421.

Wolpoff, M.H., Hawks, J., Frayer, D.W., Hunley, K., 2001. Modern human ancestry at the peripheries: a test of the replacement theory. *Science* 291, 293–297.

Yellen, J.E., Brooks, A., Cornelissen, E., Mehlman, M., Stewart, K., 1995. A Middle Stone Age worked bone industry from Katanda, Upper Semiliki Valley, Zaire. *Science* 268, 553–556.

Zhivotovsky, L.A., Bennett, L., Bowcock, A.M., Feldman, M.W., 2000. Human population expansion and microsatellite variation. *Mol. Biol. Evol.* 17, 757–767.

Zhivotovsky, L.A., Rosenberg, N.A., Feldman, M.W., 2003. Features of evolution and expansion of modern humans, inferred from genomewide microsatellite markers. *Am. J. Hum. Genet.* 72, 1171–1186.

Zhu, X., Smith, M.A., Perry, G., Aliev, G., 2004. Mitochondrial failures in Alzheimer's disease. *Am. J. Alzheimer's Dis. Other Dementias* 19, 345–352.

Zischler, H., Geisert, H., von Haeseler, A., Pääbo, S., 1995. A nuclear "fossil" of the mitochondrial D-loop and the origin of modern humans. *Nature* 378, 489–492.

13. Reliability of cranial morphology in reconstructing Neanderthal phylogeny

K. HARVATI
Max-Planck-Institut für Evolutionäre Anthropologie
Leipzig D-04103
Germany
harvati@eva.mpg.de

T.D. WEAVER
Max-Planck-Institut für Evolutionäre Anthropologie
Leipzig D-04103
Germany
weaver@eva.mpg.de

Keywords: craniofacial morphology, population history, genetics, climate, phylogeny, Neanderthals

Abstract

The usefulness of cranial morphology in reconstructing the phylogeny of closely related taxa is often questioned due to the possibility of convergence or parallelism and epigenetic response to the environment. However, it has been suggested that different cranial regions preserve phylogenetic information differentially. Some parts of the face and neurocranium are thought to be relatively developmentally flexible, and therefore to be subject to the epigenetic influence of the environment. Other parts are thought to be particularly responsive to selection for adaptation to local climate. The basicranium, on the other hand, and in particular the temporal bone, is thought to be largely genetically determined and has been argued to preserve a strong phylogenetic signal with little possibility of homoplasy. Here we test the hypotheses that cranial morphology is related to population history among recent humans, and that different cranial regions reflect population history and local climate differentially. Morphological distances among ten recent human populations were calculated from the face, vault and temporal bone using three-dimensional geometric morphometrics methods. The distance matrices obtained were then compared to neutral genetic distances and to climatic differences among the same or closely matched groups. Results indicated a stronger relationship of the shape of the vault and the temporal bone with neutral genetic distances, and a stronger association of facial shape with climate. Vault and temporal bone centroid sizes were associated with climate and particularly temperature; facial centroid size was associated with genetic distances. Temporal bone shape was more successful in tracking older population history than vault shape. Of the three cranial regions, it is therefore most appropriate for phylogenetic reconstructions among fossil humans. Analysis of temporal bone shape of both recent and Middle-Late Pleistocene humans showed Neanderthals to be morphologically very distant from both recent and fossil modern humans, indicating that Neanderthals represent a distinct evolutionary lineage.

K. Harvati and T. Harrison (eds.), Neanderthals Revisited: New Approaches and Perspectives, 239–254.

Introduction

Among the major problems in phylogenetic reconstruction from skeletal morphology is the potential unreliability of morphological characters in reflecting phylogeny (among species) and population history (within species). Convergence, parallelism, reversals and epigenetic changes are often thought to overwhelmingly influence craniofacial anatomy and to erase any phylogenetic information it might have contained (see e.g., Lieberman et al., 1996; McCollum, 1999). This criticism has been leveled most recently by Collard and Wood (2000, 2001), who found that diverse cranio-dental datasets failed to reproduce molecular phylogenies in hominoids and papionins. These authors concluded that cranial morphology cannot be used to successfully reconstruct primate and human phylogenetic relationships.

This view, however, is not universally accepted, and some researchers have proposed a differential preservation of phylogenetic information in different cranial anatomical regions. Olson (1981) suggested that the basicranium is the most genetically determined and evolutionarily conservative aspect of the cranium, and as such should be highly phylogenetically informative. This view was echoed by Wood and Lieberman (2001), who also proposed that different cranial regions reflect phylogenetic information differentially. Since the basicranium develops from cartilaginous, rather than intramembranous, origin, they suggested that its development is genetically determined, so the resulting adult morphology is only minimally influenced by environmental factors. On this basis they argued that the basicranium is more phylogenetically informative and more appropriate for phylogenetic reconstruction than the facial or cranial vault regions, which are thought to be more developmentally plastic. Recent analyses of the complex three-dimensional shape of the petrous portion of the temporal bone have provided tentative support for these hypotheses. Using three-dimensional geometric morphometrics, Harvati (2001) found that temporal bone shape tracks relationships among recent human populations better than the occipital and parietal regions of the skull. Lockwood et al. (2004) also analyzed three-dimensional temporal bone landmark coordinates to reconstruct the hominoid phylogeny, with results that closely matched the molecularly derived relationships.

Even if a cranial region reliably reflects underlying genetic variation, it will not be useful for phylogenetic reconstruction if it is particularly responsive to selection for adaptation to climate, other aspects of the local environment or behaviors. The face, in particular, has previously been linked to climatic adaptation (e.g., Coon et al., 1950; Roseman and Weaver, 2004) and to dietary and masticatory practices (e.g., Hylander, 1977; Rak, 1986; Smith, 1983), probably through a combination of epigenetic responses and genetic adaptation. The shape of the vault has also been linked to climatic adaptation (e.g., Beals, 1983; Roseman, 2004).

Here we tested the reliability of morphological evidence from three regions of the cranium – face, temporal bone and vault – in tracking population history by comparing morphological distances among recent human groups to those derived from a large number of microsatellites (neutral genetic loci, Rosenberg et al., 2002, see below). Ten globally distributed recent human groups represented in the genetic database (or their close neighbors) were also represented in our three-dimensional cranio-facial landmark database (two African, two Asian, two European, two Australasian, one Middle Eastern and one New World Arctic, see Table 1). Mahalanobis squared distance matrices (hereafter Mahalanobis D^2), corrected for unequal sample sizes, were calculated among the recent human groups based on landmark coordinates from each of the three cranial regions. The

Table 1. Morphological and genetic samples

Morphological samples	*Vault*	*Face*	*Temp.*	*Genetic samples*	*n*
W. African Dogon	33	32	33	Yoruba, Nigeria	25
S. African Khoi-San	30	29	30	San, Namibia	7
Australian, S. Australia	31	29	31	Papuan, New Guinea	17
Melanesian, New Britain	28	28	28	Melanesian, Bougainville	19
Italian	6	6	6	Italy	13
Greek	5	5	5	Sardinian, Italy	28
Syrian	20	16	18	Palestinian, Israel	51
Chinese, North China	20	17	19	Han, China	34
Thai	20	18	20	Cambodian	11
Inugsuk, Greenland	30	30	30	Yakut, Siberia	25
Total	**223**	**210**	**220**	**Total**	**230**

delta mu squared genetic distance (*Ddm*) was also calculated based on the microsatellite data to create a genetic distance matrix. The morphological distance matrices were then compared to the genetic distance matrix for the matched recent human groups using a Mantel test of matrix correspondence (Mantel, 1967; Smouse et al., 1986; Sokal and Rohlf, 1995). Importantly, the Mantel Test compares pairwise distances, so it does not assume a tree-like model for recent human population history. The morphological distance matrices were also compared to latitude, mean temperature and mean vapor pressure distance matrices for the location of origin of each recent human sample, in order to test the hypothesis that the morphology of these cranial regions reflects climatic adaptation.

We hypothesized that (a) cranial morphology reflects population history (as reflected by neutral genetic distances) in recent humans, (b) the temporal bone reflects population history best, and (c) the face also reflects adaptation to climate. Based on these hypotheses, we predicted that the morphological distances between our recent human samples would be significantly correlated with the genetic distances between the same groups; that the temporal bone distances would show the highest correlation coefficients with the genetic distances; and that the facial distances would also be significantly correlated with climatic differences. Finally, the implications of the recent human analysis were applied to the problem of Neanderthal phylogenetic relationships.

Materials and Methods

SAMPLES

This analysis included ten globally distributed recent human populations for which both morphological and genetic data were available. Exact matching of the morphological and genetic samples was not always possible due to the limitations of both the morphological and the genetic datasets. Therefore, matching between populations that were not identical but instead relatively close geographic neighbors was allowed in order to preserve a meaningful number of samples in the analysis (Table 1). The matching was loosest in two cases. The Australian morphological sample was not represented in the genetic dataset and was matched with a sample from New Guinea, which is geographically the closest group included in the genetic samples. The Greenland Inugsuk morphological sample was matched with a Siberian population in the genetic dataset. These two samples match closely in terms of latitude and climatic conditions. Furthermore, previous work has shown that Siberian and Mongolian populations approach New World populations, including Inuit groups, in their cranial morphology (Howells, 1989: 66–79).

DATA

Morphological Data

Morphological data were collected in the form of three-dimensional coordinates of osteometric landmarks on the cranium using a portable Microscribe 3DX digitizer and following the definitions of Howells (1973). All measurements were collected by Harvati. In geometric morphometrics landmarks are defined as homologous points that can be reliably and repeatedly located in all specimens under study (Bookstein, 1990; Valeri et al., 1998). Here they mostly represented standard osteometric points. Other landmarks were also included (their definitions are given in Table 2). The temporal bone dataset comprised thirteen landmarks from the right temporal bone; the facial dataset also comprised thirteen landmarks, both bilateral and midline; finally the vault dataset included eight bilateral and midline landmarks (Table 2). The three datasets overlapped minimally. Asterion was included in both the temporal bone and vault datasets, and glabella in both the vault and the face datasets. Where fossil specimens were included, minimal reconstruction was allowed during data collection for specimens with very little damage in a particular area of interest. Additionally, landmarks preserved only on one side were reconstructed by least-squares superimposing the specimens with their reflections using the Morpheus geometric morphometric software package (Slice, 1994–1999). The coordinates for each of the missing landmarks were then substituted from the fitted homologous landmark in the reflection.

Table 2. Landmarks included in the three morphological datasets

Temporal bone landmarks
1. Asterion, 2. Stylomastoid foramen, 3. Most medial point of the jugular fossa, 4. Most lateral point of the jugular fossa, 5. Lateral origin of the petro-tympanic crest, 6. Most medial point of the petro-tympanic crest at the level of the carotid canal, 7. Porion, 8. Auriculare, 9. Parietal Notch, 10. Mastoidale, 11. Most inferior point on the juxtamastoid crest (following Hublin,1978), 12. Deepest point of the lateral margin of the articular eminence, 15. Most inferior point on the entoglenoid process
Vault landmarks
1. Inion, 2. Lambda, 3. Bregma, 4. Glabella, 5–6. Asterion right and left, 7–8. Anterior pterion right and left
Facial landmarks
1. Glabella, 2. Nasion, 3. Prosthion, 4–5. Frontomalare temporale right and left, 6–7. Infraorbital foramen right and left, 8–9. Suture between the temporal and zygomatic bones on the superior aspect of the zygomatic process, right and left, 10–11. Suture between palatine pyramidal process and pterygoid plate of the sphenoid, right and left, 12–13. Malar root at alveolus, right and left.

The landmark coordinate data were processed using Generalized Procrustes Analysis, which superimposes the landmark configurations of the specimens and scales them for size, so that the differences they exhibit are due to "shape" (Rohlf, 1990; Rohlf and Marcus, 1993; Dryden and Mardia, 1998; O'Higgins and Jones, 1998). Multivariate methods based on Procrustes-aligned specimens have been shown to have the highest statistical power among alternative geometric morphometric approaches (Rohlf, 2000). Superimposition was performed using the software Morpheus (Slice, 1994–1999). Specifically, specimen configurations were translated to a common origin, scaled to unit centroid size (the square root of the sum of squared distances of all landmarks to the centroid of the object), the measure of size used here, and rotated according to a least-squared best-fit criterion. Procrustes superimposition leads to points that lie in a hemispherical variant of Kendall's shape space (Kendall, 1984; Rohlf, 1999; Slice, 2001). The present analysis was undertaken on differences among populations in the superimposed coordinates themselves.

Because size is an important aspect of morphology which can be both phylogenetically informative and related to environmental factors (e.g., Shea et al., 1993), an analysis of centroid size of the three cranial regions examined was also undertaken. Centroid size, which was removed from the coordinate data during Procrustes superimposition, was analyzed separately for its relationship with neutral genetics and climatic factors.

Genetic Data

The genetic data consisted of an expanded set of the data analyzed by Rosenberg et al. (2002) and Zhivotovsky et al. (2003). They comprised 784 microsatellite loci from 230 individuals representing 10 populations, which closely matched the populations for the morphological datasets (Table 1). The individuals in these populations are a subset of those used in the Human Genome Diversity Project–CEPH cell line panel (Cann et al., 2002). The samples were typed by the Mammalian Genotyping Service (Marshfield panel 10–52; http://www2.marshfield-clinic.org/RESEARCH/GENETICS).

Climate Data

For the climate data, we first estimated approximate latitudes and longitudes for the populations in the study. Then, based on the latitude and longitude, we were able to obtain estimates of mean yearly temperature and mean yearly vapor pressure (a proxy for humidity) from the global climate dataset published by New et al. (1999, 2000). The global climate dataset was constructed by interpolating observations collected at thousands of climate stations spread throughout the world to obtain estimates for each cell in a 0.5° latitude by 0.5° longitude grid (New et al., 1999, 2000). These two variables, as well as latitude, were used here as climatic indicators.

ANALYSES

Morphology

The morphological distances among the modern human samples were estimated using Mahalanobis D^2. This method represents the morphological variation among groups, scaled by the pooled within-group variation and accounting for covariance between variables (Neff and Marcus, 1980). Unlike other distance measures used with landmark data, such as Procrustes distance, Mahalanobis D^2 accounts for the covariation among landmark coordinates that is pervasive in biological datasets by weighting the distance by the inverse of the pooled within-group covariance matrix (see also Klingenberg and Monteiro, 2005). Additionally, by standardizing by the pooled within-group variation, Mahalanobis D^2 can be directly related to expected rates of morphological divergence predicted by population genetic theory for neutral evolution. This is because the neutral rate of morphological evolution is expected to be proportional to the within-population variation (Lynch, 1990).

For each of the three morphological datasets a principal components analysis was conducted on the superimposed coordinates. The principal components representing approximately 90% of the total variance were used as variables in calculating a Mahalanobis D^2 matrix of the recent human samples for each cranial region. Because the morphological samples used were not of equal size, a correction for unequal sample sizes was used (Marcus, 1993).

Centroid size was analyzed separately from shape information, and separately for each cranial region. The mean centroid sizes for the ten population samples were calculated. A squared distance matrix was created by calculating the squared difference in mean centroid size for all possible population pairs.

Genetics (Genetic Distance)

The genetic distances among the samples were calculated using the delta mu squared

(*Ddm*) statistic (Goldstein et al., 1995b). This distance measure is specifically designed for microsatellites, because it takes into account details of their step-wise mutation process. Under mutation-drift equilibrium, *Ddm* is expected to be linear with time, with a slope equal to twice the neutral mutation rate (Goldstein et al., 1995a, b). *Ddm* is a suitable distance for comparison with morphological Mahalanobis D^2, because both these distances measure the squared pairwise differences between populations. Other genetic distances such as F_{ST}, or the R_{ST} for microsatellites (Slatkin, 1995), measure variation both among and within the population pairs and are thus not directly comparable to morphological Mahalanobis D^2. The latter does take into account within population variation, but this is pooled-across all the populations in the sample, so there is a constant scaling of all the pairwise population comparisons.

Comparisons

The distance matrices were compared using a Mantel test of matrix correlation with NTSYSpc (Rohlf 1986–2000). This test measures the degree of relationship between two distance matrices. A permutation test is performed to assess if the relationship between the two matrices is significantly different from no relationship. Traditional tests of significance do not apply, because the matrix entries are not independent of each other. For the permutation test, one matrix is held rigid while the other is randomly permuted many times (here 10,000 times). The distribution of the matrix associations generated by the permutations can be used to construct a null distribution for tests of significance. It is also possible to compare three matrices for a partial Mantel test, which is analogous to a partial correlation among three variables (Mantel, 1967; Smouse et al., 1986; Sokal and Rohlf, 1995).

To examine the distance matrices in more detail, we conducted metric multidimensional scaling on the genetic and morphological distances using MATLAB (The MathWorks, Natick, MA). Multidimensional scaling arranges "objects" (in our case populations) in a space with a particular number of dimensions (in our case 2), so as to reproduce the observed distances as closely as possible in a low-dimensional coordinate system (Johnson and Wichern, 1998). This allowed us to explore the structure of distance matrices from the different datasets on two-dimensional plots. Unlike tree-based methods of distance comparison, the use of multidimensional scaling does not assume a bifurcating branching pattern, which makes this method more appropriate for comparisons among recent human populations where both bifurcations and reticulation may have occurred.

Results

The results of the Mantel matrix correlation tests are reported in Table 3. Out of the three cranial shape datasets, the vault and the temporal bone shape distances were significantly associated with neutral genetic distances, although none of the correlations were very strong. Contrary to our predictions, the vault distances showed a stronger relationship to genetic distances than did the temporal bone distances. The weakest correlation with the genetic distance matrix, not reaching statistical significance, was found with the facial shape distance matrix. Neither temporal bone nor vault distances were correlated with any of the three climatic variables. As predicted, facial distances showed a relationship with climatic factors. They were significantly correlated with both latitude and mean temperature, but not with mean vapor pressure. The fact that facial distances were significantly correlated with both latitude and temperature is not surprising given that latitude and temperature are strongly associated with each other (Mantel test $r = 0.91$, $p = 0.001$). The relationship between distances based on facial shape and genetic distances, however, was found to be stronger and statistically significant once the effects of latitude or temperature were adjusted for in a partial Mantel test (Table 3).

The results of the analysis of centroid size differed from the shape analyses. Although distances based on facial shape were not strongly associated with neutral genetic distances, facial size was found to be significantly correlated with the genetics distance matrix. It also did not show a relationship with any of the climatic variables. Facial size remained significantly associated with genetics even when the effects of the three climatic variables were adjusted for in partial Mantel tests. Temporal bone and vault centroid sizes were not found to be associated with genetic distances, but instead showed a significant relationship with climate. These results were the inverse of what was found in the shape analysis. Vault centroid size was significantly

Table 3. Mantel test comparisons results. Statistically significant values are indicated by asterisks

Mantel tests	*Genetics*	*Latitude*	*Mean Temperature*	*Mean Vapor Pressure*
Facial Shape	$r = 0.2988^*$ $p = 0.0904^*$	$r = 0.4556^*$ $p = 0.0148^*$	$r = 0.4515^*$ $p = 0.0151^*$	$r = 0.0116$ $p = 0.4577$
Temporal Bone Shape	$r = 0.4879^*$ $p = 0.0266^*$	$r = 0.0335$ $p = 0.3535$	$r = 0.1079$ $p = 0.2400$	$r = -0.0890$ $p = 0.6021$
Vault Shape	$r = 0.5512^*$ $p = 0.0012^*$	$r = 0.0137$ $p = 0.4712$	$r = 0.0281$ $p = 0.4468$	$r = 0.0789$ $p = 0.3306$
Face Centroid Size	$r = 0.5254^*$ $p = 0.0270^*$	$r = 0.0703$ $p = 0.2455$	$r = 0.1459$ $p = 0.1526$	$r = -0.1199$ $p = 0.5746$
Temporal Bone Centroid Size	$r = 0.2857$ $p = 0.1211$	$r = 0.3627$ $p = 0.0793$	$r = 0.5076^*$ $p = 0.0410^*$	$r = 0.0315$ $p = 0.3587$
Vault Centroid Size	$r = -0.0672$ $p = 0.5698$	$r = 0.5286^*$ $p = 0.0154^*$	$r = 0.6612^*$ $p = 0.0051^*$	$r = 0.2451$ $p = 0.1050$
Partial Mantel tests		*Adjusted for Latitude*	*Adjusted for Mean Temperature*	*Adjusted for Mean Vapor Pressure*
Facial Shape *vs.* Genetics		$r = 0.3836^*$ $p = 0.0316^*$	$r = 0.4098^*$ $p = 0.0169^*$	$r = 0.3060$ $p = 0.0838$
Temporal Bone Shape *vs.* Genetics		$r = 0.4932^*$ $p = 0.0236^*$	$r = 0.5110^*$ $p = 0.0175^*$	$r = 0.4816^*$ $p = 0.0216^*$
Vault Shape *vs.* Genetics		$r = 0.5548^*$ $p = 0.0011^*$	$r = 0.5609^*$ $p = 0.0011^*$	$r = 0.5768^*$ $p = 0.0006^*$
Face Centroid Size *vs.* Genetics		$r = 0.5353^*$ $p = 0.0216^*$	$r = 0.5573^*$ $p = 0.0167^*$	$r = 0.5159^*$ $p = 0.0213^*$
Temporal Bone Centroid Size *vs.* Genetics		$r = 0.3432^*$ $p = 0.0282^*$	$r = 0.4182^*$ $p = 0.0067^*$	$r = 0.2964$ $p = 0.0998$
Vault Centroid Size *vs.* Genetics		$r = -0.0228$ $p = 0.5798$	$r = 0.0349$ $p = 0.4008$	$r = -0.0240$ $p = 0.5264$

correlated with temperature and latitude, while temporal bone centroid size was significantly correlated with temperature only. When the effects of latitude and temperature (but not vapor pressure) were adjusted for in partial Mantel tests, however, temporal bone size also became significantly correlated with the genetic distances.

In order to compare the pattern of genetic distances to those shown by the three morphological distance datasets, multidimensional scaling plots were made for each of the distance matrices (Figure 1). Mean centroid sizes were also plotted by group for each of the three cranial regions (Figure 2). All of these were compared to the genetics multidimensional scaling plot (Figure 1a). The genetic distances showed a strong differentiation of the two African groups from the rest of the modern human populations along the first axis, with the South African Khoi-San being the most distinct population. The Eurasian samples fell close to each other on both axes, with two tight clusters representing the European/Near Eastern samples and the Asian populations, grouping also with the Greenland sample (matched with the Siberian group in the genetics dataset). Australians (matched with New Guineans) and Melanesians clustered together less tightly, and separated along the second axis from the Eurasian/New World samples, but not from the African groups.

The multidimensional scaling plot of facial shape showed a different pattern (Figure 1b). The first axis here clearly separated the two African and the Australian and Melanesian samples from the remaining groups, with the Greenland population further differentiated along the second axis. This pattern differed considerably from that shown by the genetic distances. It was consistent with a climatic influence on facial shape, as was found in the Mantel tests. Populations from tropical and subtropical climates (as defined by annual maximum and minimum temperatures) were found on the left of the first axis, while temperate – cold climate groups (again defined by annual maximum and minimum temperatures) were found on the right side. The only exception was the Thai group, a tropical population, which showed an intermediate position, but closest to the Chinese sample. It is noteworthy that the present population of Thailand, which this sample represents, migrated south from China very recently in historical times. Facial shape therefore seems to reflect a combination of climate and population history.

The temporal bone multidimensional scaling plot (Figure 1c) separated the two African samples from all other groups along the first axis, with the Khoi-San being the most distinct. These features were consistent with the major African *vs.* non-African dichotomy shown by the genetic distances. However, temporal bone shape did not produce the same clusters within the remaining populations as found in the genetic distances: the Near Eastern sample here grouped with the Asian populations rather than the Europeans; Australians and Melanesians grouped together but were not separate from the Eurasian samples; and the Greenland Inugsuk were distinct from all other groups along the second axis, rather than clustering with the Asian samples. Temporal bone shape, therefore, successfully reflected the deepest separation found in the genetic distances, that between Africans and non-Africans, but not more recent population history.

The multidimensional scaling plot of the vault shape distances (Figure 1d) showed yet another pattern. Unlike the genetic pattern, the two African groups did not separate from the other samples, although the West African Dogon were most distinct along the second axis. Vault shape, however, did separate Australians and Melanesians from the Eurasian samples. It also showed the European and Near Eastern groups clustering tightly, and the Asian and Greenland samples falling relatively close to each other. Vault shape, therefore, unlike the temporal bone, appeared to reflect recent

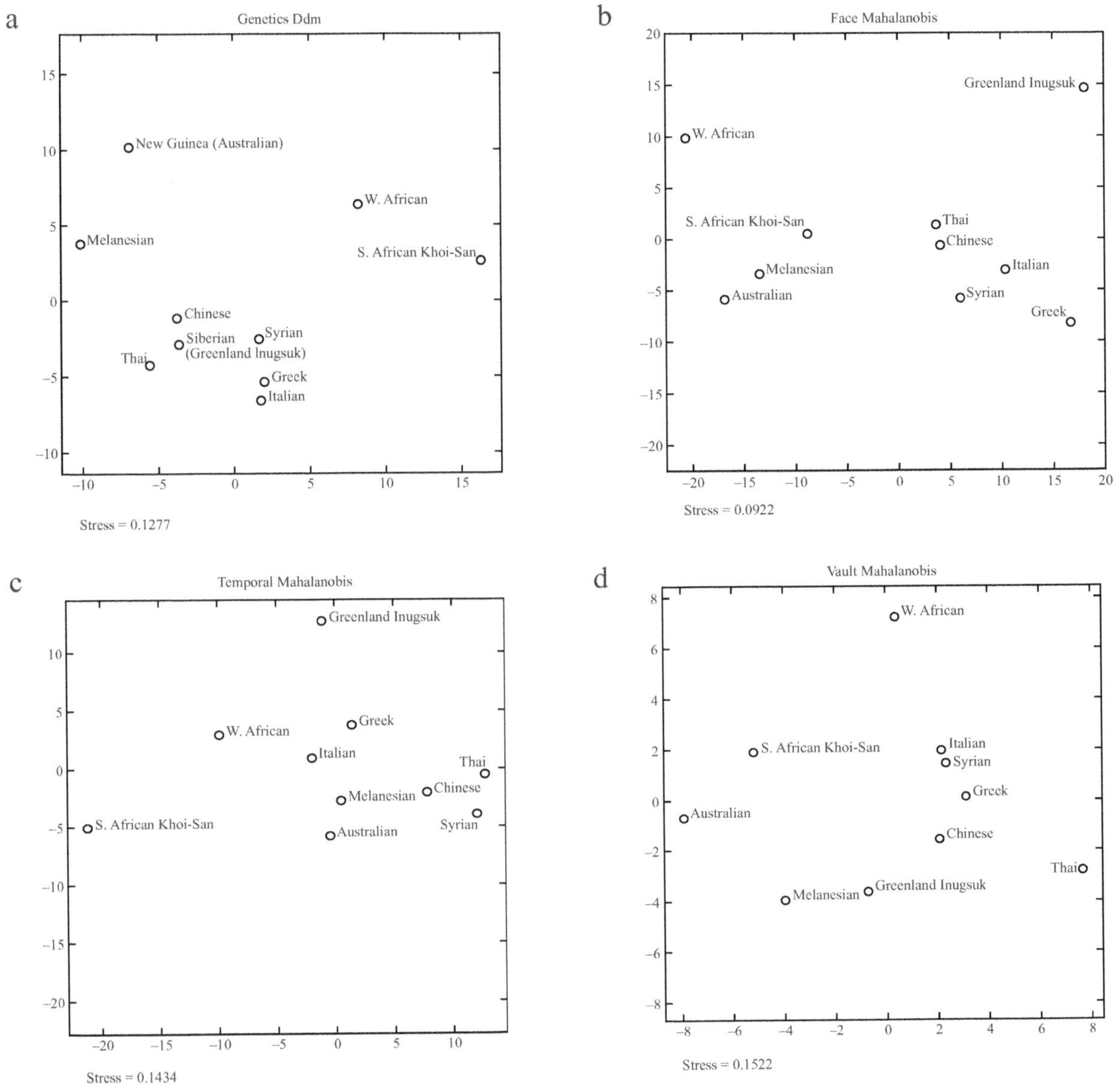

Figure 1. Multidimensional scaling plots of (a) genetic distance, in parenthesis shown matched groups from the morphological datasets, (b) face shape Mahalanobis D^2, (c) temporal bone shape Mahalanobis D^2, and (d) vault shape Mahalanobis D^2 among recent human samples.

population history among modern humans, rather than older events.

Face mean centroid size differences were found to be strongly correlated with genetic distances. When plotted by population (Figure 2a), they showed a separation between the two African samples and all other groups, with the Khoi-San being the smallest sample. West Africans were the second smallest, but very close to the other samples. The remaining samples were not differentiated in a meaningful way. Since the Khoi-San were by far the sample with the smallest faces, they could be driving the association. Only with the addition of other small populations will it be possible to answer this question. Temporal bone centroid size again showed the African groups as the smallest, but less different than the others,

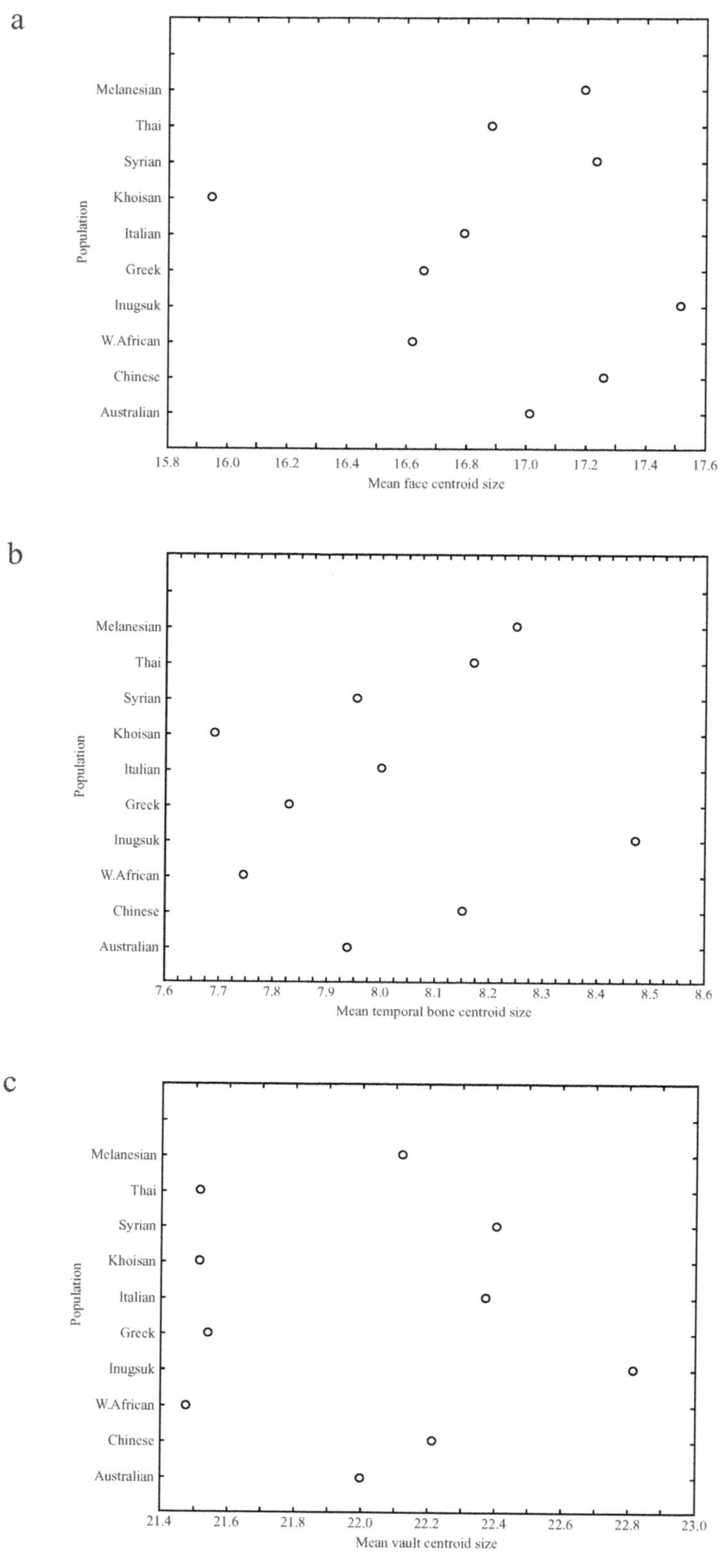

Figure 2. Mean centroid sizes for (a) face, (b) temporal bone and (c) vault, plotted by population.

and the Greenland sample as largest, with the remaining groups intermediate (Figure 2b). Vault centroid size also showed the Greenland group being the largest, and the African, Thai and Greek samples the smallest (Figure 2c).

Discussion

The results of this analysis agreed with the predictions of our hypotheses in most cases, but also differed in some important ways. As predicted, facial shape distances showed the weakest relationship to genetics, and were instead more strongly associated with climate. When the effects of climate were adjusted for, however, the relationship between facial shape and genetics became stronger and statistically significant, suggesting an interaction of genetic and climatic factors influencing human facial form. Vault and temporal bone shape distances were both significantly correlated with genetic distances, but, contrary to predictions, the highest association was with vault shape. Vault shape differences among recent human groups, however, showed a different pattern from that found in the temporal bone shape, suggesting that these two cranial regions reflect different aspects of population history. The analysis of the vault shape was most successful in tracking recent events among recent humans, grouping together populations such as the Europeans and Near Easterners, and the Chinese and Thai. It failed to show the deep split between African and non-African groups observed in the genetic data. This major dichotomy was shown in the temporal bone shape analysis, which, however, was less successful in grouping samples with a recent common history. Centroid size for the three cranial regions was found to be correlated with genetic distances in the face, and in the temporal bone after accounting for the effects of climate. This result may be driven by the small size of the Khoi-San, one of the two African groups included here and the smallest group in the recent human sample. However, it may also suggest that size may be phylogenetically informative and should not be *a priori* ignored in phylogenetic analyses. An extension of the present analysis to include additional small-bodied recent human populations is necessary to confirm this result. Finally, vault and temporal bone centroid size were found to be related to temperature (and the former also to latitude). This finding is consistent with previous studies indicating that cold-climate populations show larger brains on average than warm-climate groups among recent humans (Beals et al., 1983). Again, an extension of this analysis to include more cold-climate populations would be necessary to confirm this finding.

These results suggest that the temporal bone morphology tracks older events in population history more faithfully, while the vault reflects more recent affinities. This interpretation is consistent with the suggestion of greater developmental stability for the basicranium, which would necessarily result in slower change of this region's morphology, and for greater developmental plasticity for the vault, which would enable faster rates of morphological change (Wood and Lieberman, 2001). It is also consistent with the successful recovery of the hominoid molecular phylogeny using temporal bone three-dimensional shape (Lockwood et al., 2004). We therefore tentatively interpret our findings as indicating that temporal bone shape preserves old population history/phylogenetic signals, while vault shape preserves a more recent signal. The temporal bone would, therefore, appear to be the most appropriate of these three cranial regions for use in reconstructions of the phylogenetic relationships of Neanderthals and Pleistocene humans.

In light of this interpretation, the analysis of temporal bone shape was repeated with fossil human specimens from Europe, Africa and the Near East dated to the Middle and Late Pleistocene (see also Harvati, 2002, 2003a,

2004). The fossil samples included fourteen Neanderthal and seven Upper Paleolithic specimens, as well as Kabwe, Skhul 5 and Qafzeh 9 (Table 4). Mahalanobis squared distances were calculated among these samples, and are plotted in a multidimensional scaling plot in Figure 3. In this plot, Neanderthals fell on one end of the x-axis and recent humans on the opposite end, reflecting the great difference in temporal bone shape between these two hominin taxa. The Upper Paleolithic sample was very close to the recent human groups. Kabwe was placed close to the center, as might be expected for an older specimen that may represent the common ancestral taxon to both Neanderthals and modern humans. This finding is in agreement with previous results of both temporal bone and overall morphology (e.g., Stringer, 1974; Harvati, 2002, 2003a, b, 2004). The position of the two early anatomically modern human specimens was more difficult to interpret. While this sample showed a very small morphological distance to the Upper Paleolithic specimens, its distances to the recent human groups were as great as that to the Neanderthals and Kabwe. These large distances may be due to the very small size of this sample, and may reflect the increased influence of error and deformation on small samples of fossils. Similar distances have been

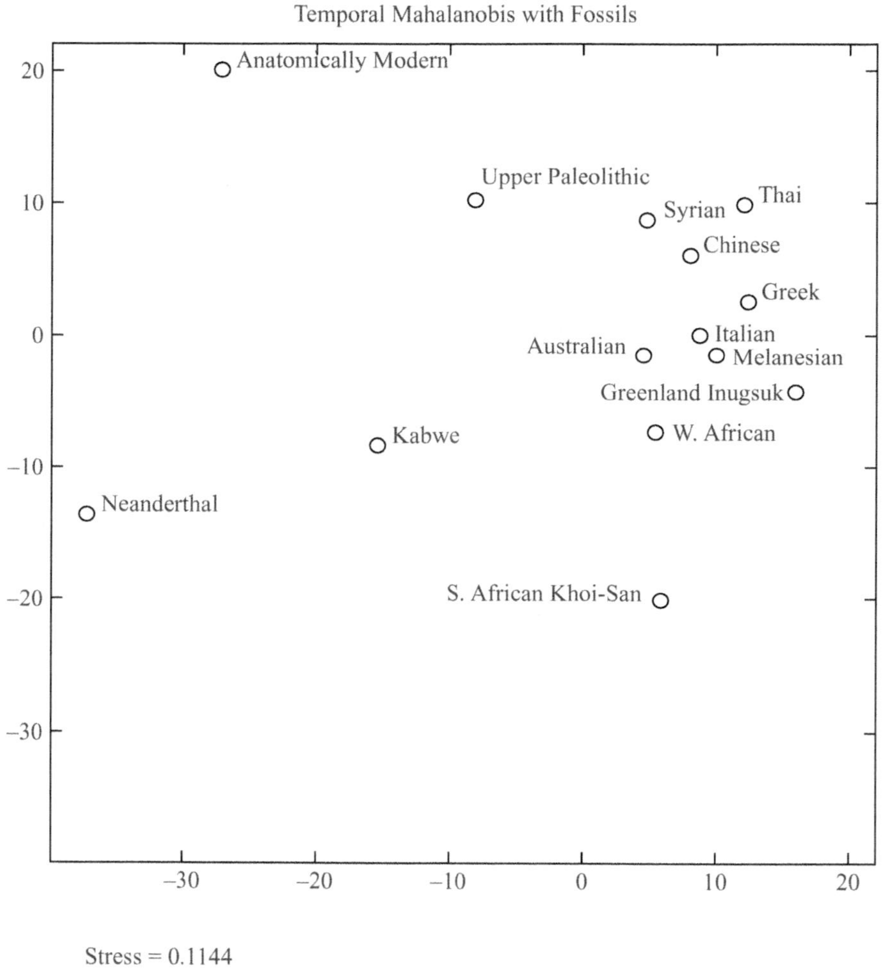

Figure 3. Multidimensional scaling plot of the temporal bone Mahalanobis D^2 among recent and Middle-Late Pleistocene humans.

Table 4. Fossil samples used

Neanderthal	Saccopastore 2, La Chapelle, La Ferrassie 1, La Ferrassie 2, Spy 1, Spy 2, La Quina 5, La Quina 27, Circeo 1, Gibraltar 1, Krapina C, Krapina 39-1, Amud 1, Shanidar 1[1]
Upper Paleolithic	Cro Magnon 1, Abri Pataud, Předmostí 3[1], Předmostí 4[1], Mladeč 2, Mladeč 5[1], Ein Gev
Early Anatomically Modern	Skhul 5, Qafzeh 9
Middle Pleistocene	Kabwe

[1] Specimens for which casts from the American Museum of Natural History Department of Anthropology were used.

found previously in studies of temporal bone morphology, as well as analyses based on other cranial and postcranial measurements (Stringer, 1974, 1992; Bräuer and Rimbach, 1990; Bräuer, 1992; Kidder et al., 1992; Pearson, 2000; Harvati, 2001, 2002, 2003a, b, 2004). They have been interpreted in terms of retention of primitive features in these specimens (Stringer, 1992).

Conclusions

The results of this analysis indicate that cranial morphology does preserve population history/ phylogenetic information, but that it does so differentially, as previously suggested by various researchers. Both vault and temporal bone shape were significantly correlated with neutral genetics (albeit not very strongly) while facial shape seemed to be affected both by climatic factors and population history. Contrary to expectations, the association between morphological and genetic distances was strongest in the shape of the vault, rather than that of the temporal bone. However, temporal bone shape appeared to be preserving older population history/phylogenetic signals, while vault shape seemed to register more recent events. Centroid size was also found to be correlated with genetic distances in some cases, and may also be phylogenetically important. It must be pointed out, however, that these results are based on ten recent human groups only, and may be highly dependent on the individual properties of these groups. Further analysis is planned to include a greater number of population samples in order to further test these hypotheses.

Due to its greater preservation of older population history/phylogenetic information, we conclude that temporal bone shape is more appropriate for use in reconstructing the phylogeny of fossil humans. Analysis of temporal bone shape in a sample of fossil and recent humans showed Neanderthals to be distinct from recent humans and Upper Paleolithic specimens. The latter were very close to recent humans. No evidence for a close phylogenetic link between Neanderthals and any recent human group or Neanderthals and the Upper Paleolithic sample used here was found, even though limited interbreeding may have been possible (see Holliday, this volume; Serre and Pääbo, this volume). This result is consistent with other analyses of Neanderthal craniodental and postcranial morphology, development, genetics and behavior (e.g., Hublin, 1978; Holliday, 1997; Krings et al., 1997; Pearson, 2000; Ponce de León and Zollikofer, 2001; Rak et al., 2002; Klein, 2003; Bailey, 2004; Harvati et al., 2004; Mellars, 2004; Serre and Pääbo, this volume; Tattersall and Schwartz, this volume), and indicates that Neanderthals are best considered a distinct evolutionary lineage.

Acknowledgments

We thank Mark Stoneking, David Serre, Bob Franciscus, Terry Harrison, Charles Lockwood, James Rohlf, Charles Roseman and Kieran McNulty for their helpful comments and suggestions. We are grateful to many curators and collections managers in various institutions in Europe, Israel and the US, for allowing study of the fossil and recent material in their care, and to the family of Prof. Max Lohest. This research was supported by the National Science Foundation, the Care Foundation for Archeological Research in Israel, the American Museum of Natural History and the Max-Planck Society. This is NYCEP morphometrics contribution number 19.

References

Bailey, S.E., 2004. A morphometric analysis of maxillary molar crowns of Middle-Late Pleistocene hominins. *J. Hum. Evol.* 47, 183–198.

Beals, K.L., Smith, C.L., Dodd, S.M., 1983. Climate and the evolution of brachycephalization. *Am. J. Phys. Anthropol.* 62, 425–437.

Bookstein, F.L., 1990. Introduction to methods for landmark data. In: Rohlf, F.J., Bookstein, F.L. (Eds.), *Proceedings of the Michigan Morphometrics Workshop.* The University of Michigan Museum of Zoology, Ann Arbor, pp. 216–225.

Bräuer, G., 1992. Africa's place in the evolution of *Homo sapiens*. In: Bräuer, G., Smith, F.H. (Eds.), *Continuity or Replacement: Controversies in Homo sapiens* Evolution. A. A. Balkema, Rotterdam, pp. 83–98.

Bräuer, G., Rimbach, K.W., 1990. Late archaic and modern *Homo sapiens* from Europe, Africa and Southwest Asia: Craniometric comparisons and phylogenetic implications. *J. Hum. Evol.* 19, 789–807.

Cann, H.M., de Toma, C., Cazes, L., Legrand, M.-F., Morel, V., Piouffre, L., Bodmer, J., Bodmer, W.F., Bonne-Tamir, B., Cambon-Thomsen, A., Chen, Z., Chu, J., Carcassi, C., Contu, L., Du, R., Excoffier, L., Friedlaender, J.S., Groot, H., Gurwitz, D., Herrera, R.J., Huang, X., Kidd, J., Kidd, K.K., Langaney, A., Lin, A.A., Mehdi, S.Q., Parham, P., Piazza, A., Pistillo, M.P., Qian, Y., Shu, Q., Xu, J., Zhu, S., Weber, J.L., Greely, H.T., Feldman, M.W., Thomas, G., Dausset, J., Cavalli-Sforza, L.L., 2002. A human genome diversity cell line panel. *Science* 296, 261–262.

Collard, M., Wood, B., 2000. How reliable are human phylogenetic hypotheses? *Proc. Natl. Acad. Sci. U.S.A.* 97, 5003–5006.

Collard, M., Wood, B., 2001. Homoplasy and the early hominid masticatory system: inferences from analyses of extant hominoids and papioni. *J. Hum. Evol.* 41, 167–194.

Coon, C.S., Garn, S.M., Birdsell, J.B., 1950. *Races: a study of the problems of race formation in man.* Charles C. Thomas, Springfield, IL.

Dryden, I.L., Mardia, K.V., 1998. *Statistical Shape Analysis.* John Wiley, New York.

Goldstein, D.B., Linares, A.R., Cavalli-Sforza, L.L., Feldman, M.W., 1995a. An evaluation of genetic distances for use with microsatellite loci. *Genetics* 139, 463–471.

Goldstein, D.B., Ruiz Linares, A., Cavalli-Sforza, L.L., Feldman, M.W., 1995b. Genetic absolute dating based on microsatellites and the origin of modern humans. *Proc. Natl. Acad. Sci. U.S.A.* 92, 6723–6727.

Harvati, K., 2001. The Neanderthal problem: 3-D geometric morphometric models of cranial shape variation within and among species. Ph.D. Dissertation, City University of New York, New York.

Harvati, K., 2002. Models of shape variation between and within species and the Neanderthal taxonomic position: a 3D geometric morphometrics approach based on temporal bone morphology. BAR International Series 1049, Oxford, pp. 25–30.

Harvati, K., 2003a. Quantitative analysis of Neanderthal temporal bone morphology using 3-D geometric morphometrics. *Am. J. Phys. Anthropol.* 120, 323–338.

Harvati, K., 2003b. The Neanderthal taxonomic position: models of intra- and inter-specific morphological variation. *J. Hum. Evol.* 44, 107–132.

Harvati, K., 2004. 3-D geometric morphometric analysis of temporal bone landmarks in Neanderthals and modern humans. In: Elewa, A.M.T. (Ed.), *Morphometrics, Applications in Biology and Paleontology*. Springer, Berlin, pp. 245–258.

Harvati, K., Frost, S.R., McNulty, K.P., 2004. Neanderthal taxonomy reconsidered: Implications of 3D primate models of intra- and inter-specific differences. *Proc. Natl. Acad. Sci. U.S.A.* 101, 1147–1152.

Holliday, T.W., 1997. Body proportions in Late Pleistocene Europe and modern human origins. *J. Hum. Evol.* 32, 423–447.

Howells, W.W., 1973. Cranial variation in Man: A study by multivariate analysis of patterns of difference among recent human populations. Papers of the Peabody Museum of Archaeology and Ethnology, Harvard University, Vol. 67.

Howells, W.W., 1989. Skull Shapes and the Map: Craniometric Analyses in the Dispersion of Modern *Homo*. Papers of the Peabody Museum of Archaeology and Ethnology, Harvard University, Vol. 79.

Hublin, J.-J., 1978. Quelques caractères apomorphes du crâne néanderthalien et leur interprétation phylogénique. *C. R. Acad. Sci. Paris* 287, 923–926.

Hylander, W.L., 1977. The adaptive significance of Eskimo craniofacial morphology. In: Dahlberg, A.A.T., Graber, M. (Eds.), *Orofacial Growth and Development*. Mouton, Paris, pp.129–170.

Johnson, R.A., Wichern, D.W., 1998. *Applied Multivariate Statistical Analysis*. Prentice Hall, Upper Saddle River, NJ.

Kendall, D.G., 1984. Shape manifolds, Procrustean metrics and complex projective spaces. *Bull. Lond. Math. Soc.* 16, 81–121.

Kidder, J.H., Jantz, R.L., Smith, F.H., 1992. Defining modern humans: a multivariate approach. In: Bräuer, G., Smith, F.H. (Eds.), *Continuity or Replacement: Controversies in Homo sapiens Evolution*. A.A. Balkema, Rotterdam, pp. 157–177.

Klein, R.G., 2003. Whither the Neanderthals? *Science* 299, 1525–1527.

Klingenberg, C.P., Monteiro, L.R., 2005. Distances and directions in multidimensional shape spaces: implications for morphometric applications. *Syst. Biol.* 54, 678–688.

Krings, M., Stone, A., Schmitz, R.W., Krainitzki, H., Stoneking, M., Pääbo, S. 1997. Neandertal DNA sequences and the origin of modern humans. *Cell* 90, 19–30.

Lieberman, D.E., Wood, B., Pilbeam, D.R., 1996. Homoplasy and early *Homo*: an analysis of the evolutionary relationships of *H. habilis sensu stricto* and *H. rudolfensis*. *J. Hum. Evol.* 30, 97–120.

Lockwood, C.A., Kimbel, W.H., Lynch, J.M., 2004. Morphometrics and hominoid phylogeny: support for a chimpanzee–human clade and differentiation among great ape subspecies. *Proc. Natl. Acad. Sci. U.S.A.* 101, 4356–4360.

Lynch, M., 1990. The rate of morphological evolution in mammals from the standpoint of neutral expectation. *Am. Nat.* 136, 727–741.

Mantel, N.A., 1967. The detection of disease clustering and a generalized regression approach. *Cancer Res.* 27, 209–220.

Marcus, L.F., 1993. Some aspects of multivariate statistics for morphometrics. In: Marcus, L.F., Bello, E., A. García-Valdecasas, A. (Eds), *Contributions to Morphometrics*. Monografias Museo Nacional de Ciencias Naturales, Madrid, pp. 99–130.

McCollum, M.A., 1999. The robust australopithecine face: A morphogenetic perspective. *Science* 284, 301–305.

Mellars, P., 2004. Neanderthals and the modern human colonization of Europe. *Nature* 432, 461–465.

Neff, N.A., Marcus, L.F., 1980. *A Survey of Multivariate Methods for Systematics*. American Museum of Natural History, New York.

New, M., Hulme, M., Jones, P., 1999. Representing twentieth-century space-time climate variability. Part I: development of a 1961–90 mean monthly terrestrial climatology. *J. Climate* 12, 829–856.

New, M., Hulme, M., Jones, P., 2000. Representing twentieth-century space-time climate variability. Part II: development of 1901–96 grids of terrestrial surface climate. *J. Climate* 13, 2217–2238.

O'Higgins, P., Jones, N., 1998. Facial growth in *Cercocebus torquatus*: an application of three-dimensional geometric morphometric techniques to the study of morphological variation. *J. Anat.* 193, 251–272.

Olson, T.R., 1981. Basicrania and evolution of the Pliocene hominids. In: Stringer, C.B. (Ed.), *Aspects of Human Evolution*. Taylor and Francis, London, pp. 99–128.

Pearson, O.M., 2000. Postcranial remains and the origin of modern humans. *Evol. Anthropol.* 9, 229–247.

Ponce de León, M.S., Zollikofer, C.P.E., 2001. Neanderthal cranial ontogeny and its implications for late hominid diversity. *Nature* 412, 534–538.

Rak, Y., 1986. The Neanderthal: A new look at an old face. *J. Hum. Evol.* 15, 151–164.

Rak, Y., Ginzberg, A., Geffen, E., 2002. Does *Homo neanderthalensis* play a role in modern human ancestry? The mandibular evidence. *Am. J. Phys. Anthropol.* 119, 199–204.

Rohlf, F.J. (1986–2000 Copyright) NTSYSpc version 2.10t. Exeter Software, Setauket, NY.

Rohlf, F.J., 1990. Rotational fit (Procrustes) methods. In: Rohlf, F.J., Bookstein, F.L. (Eds.), *Proceedings of the Michigan Morphometrics Workshop*. University of Michigan Museum of Zoology, Ann Arbor, MI, pp. 227–236.

Rohlf, F.J., 1999. Shape statistics: Procrustes superimpositions and tangent spaces. *J. Classification* 16, 197–223.

Rohlf, F.J., 2000. Statistical power comparisons among alternative morphometric methods. *Am. J. Phys. Anthropol.* 111, 463–478.

Rohlf, F.J., Marcus, L.F., 1993. A revolution in morphometrics. *Trends Ecol. Evol.* 8, 129–132.

Roseman, C.C., 2004. Detection of interregionally diversifying natural selection on modern human cranial form by using matched molecular and morphometric data. *Proc. Natl. Acad. Sci. U.S.A.* 101, 12824–12829.

Roseman, C.C., Weaver, T.D., 2004. Multivariate apportionment of global human craniometric diversity. *Am. J. Phys. Anthropol.* 125, 257–263.

Rosenberg, N.A., Pritchard, J.K., Weber, J.L., Cann, H.M., Kidd, K.K., Zhivotovsky, L.A., Feldman, M.W., 2002. Genetic structure of human populations. *Science* 298, 2381–2385.

Shea, B.T., Leigh, S.R., Groves, C.P., 1993. Multivariate craniometric variation in chimpanzees: Implications for species identification. In: Kimbel, W.H., Martin, L.B. (Eds.), *Species, Species Concepts and Primate Evolution*. Plenum Press, New York, pp. 265–296.

Slatkin, M., 1995. A measure of population subdivision based on microsatellite allele frequencies. Genetics. 139, 457–462.

Slice, D.E., 1994–1999 Morpheus et al.: Software for Morphometric Research. Department of Ecology and Evolution, State University of New York, Stony Brook, NY.

Slice, D.E., 2001. Landmark coordinates aligned by Procrustes analysis do not lie in Kendall's shape space. *Syst. Biol.* 50, 141–149.

Smith, F.H., 1983. Behavioral interpretations of changes in craniofacial morphology across the archaic/modern *Homo sapiens* transition. In: Trinkaus, E. (Ed.), *The Mousterian Legacy: Human Biocultural Change in the Upper Pleistocene*. BAR International Series, Oxford, pp. 141–163.

Smouse, P.E., Long, J.C., Sokal, R.R., 1986. Multiple regression and correlation extensions of the Mantel Test of matrix correspondence. *Syst. Zool.* 35, 627–632.

Sokal, R.R., Rohlf, F.J., 1995. *Biometry: The Principals and Practice of Statistics in Biological Research*. W.H. Freeman, New York.

Stringer, C.B., 1974. Population relationships of later Pleistocene hominids: a multivariate study of available crania. *J. Archaeol. Sci.* 1, 317–142.

Stringer, C.B., 1992. Reconstructing recent human evolution. *Phil. Trans. Biol. Sci.* 337, 217–224.

Valeri, C.J., Cole, T.H. III, Lele, S., Richtsmeier, J.T., 1998. Capturing data from three-dimensional surfaces using fuzzy landmarks. *Am. J. Phys. Anthropol.* 107, 113–124.

Wood, B., Lieberman, D.E., 2001. Craniodental variation in *Paranthropus boisei*: a developmental and functional perspective. *Am. J. Phys. Anthropol.* 116, 13–25

Zhivotovsky, L.A., Rosenberg, N.A., Feldman, M.W., 2003. Features of evolution and expansion of modern humans, inferred from genomewide microsatellite markers. *Am. J. Hum. Genet.* 72, 1171–1186.

14. Non-metric variation in recent humans as a model for understanding Neanderthal-early modern human differences: just how "unique" are Neanderthal unique traits?

J.C.M. AHERN
Department of Anthropology, University of Wyoming
Dept. 3431, 1000 E. University Ave,
Laramie, WY 82071, USA
jahern@uwyo.edu

Keywords: Neanderthals, Species, Systematics, Apomorphies, Non-Metrics, Human Evolution, Modern Human Origins

Abstract

Using living humans as an extant referent, this paper examines the probability that the frequency differences in Neanderthal "unique" non-metric traits observed between Neanderthals and Upper Paleolithic modern humans could be sampled from two major populations of the same species. Neanderthal-like features occur in very low frequencies in living humans, if present at all. Rather, other features distinguish major human populations. The population frequency differences of these features are used as a model by which the Neanderthal – Upper Paleolithic frequency differences are assessed using a resampling simulation. This methodological approach tests the null hypothesis that the observed Neanderthal – Upper Paleolithic differences are not greater than what can be sampled from between two major human populations (Amerindians and Euroamericans). Results of the analysis fail to falsify this null hypothesis. Implications of these results for Neanderthal taxonomy are examined.

Introduction

The taxonomic position of Neanderthals has been the longest ongoing debate in paleoanthropology (cf., Trinkaus and Shipman, 1993). Since the late 1980s, this debate has been framed within the contemporary controversy surrounding modern human origins. In many respects, the taxonomy of Neanderthals has become the lynchpin of the modern human origins debate (Wolpoff et al., 2000). If Neanderthals were not one of us and were rather "*Homo neanderthalensis*," Multiregional Evolution could be regarded as incorrect, at least for Europe, while if Neanderthals were *Homo sapiens*, Recent African Evolution could be regarded as incorrect.

Greater morphometric difference between Neanderthals and modern humans than among living human populations and non-human

K. Harvati and T. Harrison (eds.), Neanderthals Revisited: New Approaches and Perspectives, 255–268.

primate species and subspecies has been interpreted as evidence that Neanderthals were a separate species (e.g., Harvati et al., 2004). Such an argument is based upon the assumption that fossil species should be comparable to extant species in their range of variation. Although there is good reason to doubt that such an extant referent "yardstick" can ever effectively *falsify* a single species hypothesis in fossil contexts (Kimbel and Rak, 1993; Ahern et al., 2005), the degree of variation in extant referents could potentially be used to *support* either a single or multiple species interpretation. Rigid hypothesis testing is preferable to such a probabilistic approach (i.e., where hypotheses are "supported" or not "supported" rather than "falsified" or "tentatively accepted"), yet it is clear that many multiple species controversies in the hominid fossil record lack the evidentiary resolution to effectively falsify a single species hypothesis (i.e., Frayer et al., 1993; Kramer et al., 2001; Henneberg and De Miguel, 2004; Wolpoff et al., 2004; Hawks, this volume; but see Ponce de León and Zollikofer, this volume; Rosas et al., this volume; Tattersall and Schwartz, this volume; Zollikofer and Ponce de León, this volume; for assumption of the contrary). Despite a relatively extensive late hominid fossil record, sufficient evidence has not yet accumulated to falsify a hypothesis of Neanderthal and modern human conspecificity. Thus, we are faced with either ignoring the issue and assuming a single species, or attempting to assess the probability that Neanderthals and modern humans represent more than one species.

NEANDERTHAL "AUTAPOMORPHIES"

A variety of non-metric traits have been posited as uniquely-derived (autapomorphic) for Neanderthals (see Table 1; Santa Luca, 1978; Stringer et al., 1984). Applying either the Phylogenetic Species Concept (Cracraft, 1989; cf. Rak, 1993) or the Morphospecies Concept (Cronquist, 1978; cf., Tattersall and Schwartz, 1998), the presence of unique derived features would be consistent with a separate species designation for Neanderthals. Yet, recent studies (e.g., Frayer, 1992a, b; Franciscus and Trinkaus, 1995; Quam and Smith, 1998; Stefan and Trinkaus, 1998; Jabbour et al., 2002; Trinkaus et al., 2003; Wolpoff and Frayer, 2005; Cartmill and Smith, in prep.) have reported that many of

Table 1. A list of purported Neanderthal non-metric cranial autapomorphies[1]

Trait	*Reference*[2]
Mandibular	
Asymmetrical Mandibular Notch	Rak, 1998; Rak et al., 2002
Horizontal-Oval Mandibular Foramen	Stringer et al.,1984
Medial Crest of the Mandibular Notch	Rak, 1998
Retromolar Space	Stringer et al., 1984
Cranial	
Anterior Mastoid Tubercle	Santa Luca, 1978; Stringer et al., 1984
Occipital Bun	Stringer et al., 1984
Suprainiac Fossa	Santa Luca, 1978
Large Occipitomastoid Crest	Stringer et al., 1984

[1] Other non-metric Neanderthal craniomandibular traits, that have been called autapomorphic, were excluded from the present analysis because of limited published data and/or unclear and contentious definitions. See Materials and Methods.
[2] References for the data used in this study are given in Table 2.

Table 2. Samples and summary counts and percentages: Neanderthals and Upper Paleolithic moderns

	The Replaced				D_{NUP}
	Neanderthals		*Upper Paleolithic*		*Percentage*
Trait	*Present (%)*	*n*	*Present (%)*	*n*	*difference*
Anterior Mastoid Tubercle[1]	8 (34.8)	23	4 (11.4)[1]	44	23.4
Asymmetrical Mandibular Notch[2]	10 (71.4)	14	2 (11.8)	17	59.6
H-O Mandibular Foramen[3]	10 (52.6)	19	6 (11.5)	52	41.1
Medial CMN	11 (84.6)	13	1 (50)	2	34.6
Occipital Bun (liberal)[3, 4]	9 (81.8)	11	9 (60)	15	21.8
Occipital Bun (conservative)[4, 5]	9 (81.8)	11	0 (0)	15	81.8
Retromolar Space[6]	22 (78.6)	28	4.5 (28.1)	16	50.5
Suprainiac Fossa (liberal)[7]	23 (100)	23	19 (29.7)	64	70.3
Suprainiac Fossa (conservative)[8]	23 (100)	23	1 (1.6)	64	98.4
Occipitomastoid (liberal)[9]	11 (78.6)	14	1 (5.3)	19	73.3
Occipitomastoid (conservative)[10]	11 (100)	11	0 (0.0)	19	100

[1] Frayer (1992a).
[2] Cartmill & Smith (in prep.).
[3] Trinkaus & LeMay (1982). Assumes that the Upper Paleolithic "occipital buns" are homologous to those of Neanderthals, albeit different in form.
[4] Count excludes Stetten and Zlatý Kůn, which were included by Trinkaus & LeMay (1982) but have since been shown to be very recent.
[5] From Trinkaus & LeMay (1982) but assumes that the Upper Paleolithic "occipital buns" are not homologous to those of Neanderthals, albeit different in form.
[6] Franciscus & Trinkaus, 1995.
[7] After Frayer (1992a, b). The typical Upper Paleolithic suprainiac depression is scored as a suprainiac fossa.
[8] Hublin (1978, 1980) contends that the Neanderthal suprainiac fossa is not homologous to that seen in the Upper Paleolithic. Wolpoff et al. (2004) report that one Upper Paleolithic specimen, Mladeč 6, exhibits a Neanderthal-pattern suprainiac fossa.
[9] Assumes that the Upper Paleolithic specimen Mladeč 5 exhibits one (Wolpoff et al., 2004) and the Neanderthal specimens Saccopastore 1, Saccopastore 2, and La Quina 5 lack them.
[10] Assumes that no Upper Paleolithic specimens exhibit one. Also, the two Saccopastore specimens are eliminated from the Neanderthal sample, leaving only one Neanderthal that lacks a larger occipitomastoid crest, La Quina 5.

these traits are either not ubiquitous among Neanderthals or are also found in other populations. Especially interesting is the presence of many of these traits among Upper Paleolithic modern humans in Europe, since they were the immediate successors to Neanderthals (see Table 2). Frayer (1992a, b), Trinkaus and Zilhão (2002), Wolpoff et al. (2004) and others (e.g., Hawks, this volume) have argued that such persistence of Neanderthal features in post-Neanderthals indicates some level of genetic contribution of the former to the latter and, thus, Neanderthals and modern humans should be regarded as the same species. Yet, the frequencies of these traits can differ dramatically between Neanderthals and modern humans. The *pattern* of traits seems to indicate conspecificity, but is the *degree* of trait frequency difference between Neanderthals and early modern humans greater than what we would expect to see between populations of the same species? Are the features in question species markers or population markers?

A MODEL OF INTRASPECIFIC REPLACEMENT

An ideal extant referent model for assessing the difference between Neanderthals and the modern humans who succeeded them would be one that compares a replaced population with the population that replaced it. During the past 500 years, North American Amerindian populations have been largely replaced and/or assimilated by European colonists, as well as by African and

Asian immigrants. Recent estimates for the 1492 AD Amerindian population size north of Mexico, range from 1,894,280 (Ubelaker, 1988) to 18,000,000 (Dobyns, 1983) with a reasonable estimate of 7,000,000 given by Thornton (1987, 1997). By 1900 AD, the Amerindian population of this region had dwindled to 375,000 (Thornton, 1997). Depending on the estimated population size at contact, the annual rate of decline was −0.28% to −0.97% (Thornton, 1997). While the Amerindian percentage of the total population was 100% prior to contact, it was only 0.5% of the total north-of-Mexico population by 1900. Of course, Amerindians have not been fully replaced by any means. However, although Amerindian numbers have increased during the 20th century, self-identified Amerindians (including people who identified themselves as only part Amerindian) made up only 0.9% of the total 2000 U.S. census (Oswalt, 2006). Like other contemporary American population classifications, the biological meaning of "Amerindian" has already significantly lessened. Only 9.5% of the North American Amerindian samples surveyed by Post et al. (1968) showed no European admixture, while 65% of the samples exhibited 5% or higher admixture (see also Szathmary and Auger, 1983; Crawford, 1998; Williams et al., 2000). Admixture rates for some eastern North American groups are as high as 50% (Pollitzer et al., 1967; Szathmary and Auger, 1983). Thus, North American Amerindians, as a biological population, have been largely assimilated and replaced by immigrant populations. Although the mechanisms and processes of Amerindian replacement and assimilation are certainly different from those by which Upper Paleolithic modern humans replaced Neanderthals (cf., Diamond, 1997), the North American analogy can potentially be used as a model of intraspecific human population replacement. For example, a scenario of intraspecific population replacement in Pleistocene Europe would be supported if the differences between Upper Paleolithic moderns and Neanderthals were found to be less than the differences between Euroamericans and Amerindians.

The null hypothesis tested is: the differences, between Neanderthals and Upper Paleolithic moderns in terms of Neanderthal non-metric cranial traits, are not significantly greater than those seen between Amerindians and Euroamericans for traits that characterize Amerindians. If the null hypothesis is falsified, the results of this study would lend support to a morphospecies designation for Neanderthals. If the null hypothesis is tentatively accepted, the results of this study would lend tentative support to classifying Neanderthals as *Homo sapiens*, since it would mean that the differences between Neanderthals and Upper Paleolithic modern humans are no greater than what is seen between two populations of living humans. The use of the Amerindian-Euroamerican model reinforces the null hypothesis compared to the use of many other possible examples of recent human population replacements, since it is not a complete replacement and there has been considerable admixture between Amerindians and Euroamericans. Yet, it is still useful since most of the population that has largely succeeded Amerindians in North America lacks significant Amerindian ancestry.

Materials and Methods

Data for frequencies and counts of Neanderthal traits among Neanderthals and Upper Paleolithic moderns were drawn from the sources cited in the footnotes of Table 2. Neanderthal traits were chosen if: (1) they were non-metric, (2) appropriate sources of data were available, and (3) they had been identified as autapomorphic for Neanderthals. Some cranial traits that have been reported as non-metric Neanderthal autapomorphies were not included since their frequency among Upper Paleolithic moderns has not been

Table 3. Samples and summary counts and percentages: Amerindians and Euroamericans

	The Replaced		*The Replacer*		
	Amerindians		*Euroamericans*		
Trait	*Present (%)*	*n*	*Present (%)*	*n*	*Percentage difference*
Angled Zygomaticomax. Suture	97 (74)	131[1]	42 (35)	120[2]	39
Elliptic Palate	32 (42.7)	75[3]	2 (2.9)	68[4]	39.8
Not Sharp Nasal Sill[3]	46 (97.9)	47[5]	20 (31.2)	64[6]	66.7
Shoveled Incisors[4]	1180 (85)	1388[7]	168 (8.4)	2000[8]	76.6
Straight Palatine Suture[5]	37 (49.3)	75[3]	15 (12.6)	119[4]	36.7

[1] Holborow, 2002. Plains (n = 61), Southwest (n = 27), Peru (n = 43).
[2] Holborow, 2002. Northwest Plains frontier and forensic (n = 9), Terry Collection (n = 107).
[3] Rawlings, 2002. Northwest Plains and Smithsonian North American.
[4] Rawlings, 2002. Northwest Plains frontier and forensic, Terry Collection, and Maxwell Museum.
[5] Willson, 2004. Northwest Plains.
[6] Willson, 2004. Northwest Plains frontier and forensic (n = 12), Terry Collection (n = 52).
[7] Wissler, 1931. Southwest.
[8] Hrdlička, 1920. U.S. Whites.

established (e.g., the medial pterygoid tubercle, Rak et al., 1994; Quam and Smith, 1998) or whose anatomical variation has been shown to be more complex than previously thought (e.g., internal nasal specializations, Schwartz and Tattersall, 1996; Franciscus, 1999; and the medial pterygoid tubercle, Richards and Plourde, 1995; Antón, 1996). Two features that were used, mandibular notch form and position of the crest of the mandibular notch (Rak et al., 1994; Rak, 1998) were used despite work that has posited different interpretations of these features' variations (Quam and Smith, 1998; Stefan and Trinkaus, 1998; Jabbour et al., 2002; Wolpoff and Frayer, 2005). Frequency data for the Amerindian traits were drawn from the literature cited in the footnotes of Table 3. These traits have been documented as particularly useful in the assessment of ancestry from skeletal remains (Gill and Rhine, 1990) and help define Amerindian and Euroamerican populations. Frequencies and counts for these traits among Amerindian and Euroamerican samples are given in Table 3. As with most of the anatomy that comprises direct data from prehistoric populations, the genetic basis to all of the traits used in this study, Amerindian and Neanderthal, is far from clear. There is no reason to think that any of the Amerindian traits are any less heritable than the Neanderthal traits and vice versa.

A basic assumption of this analysis is that characteristics that distinguish one pair of intraspecific populations can be compared with non-homologous characteristics that distinguish another pair of populations. For testing the null hypothesis, there would be little utility in comparing the frequencies of Neanderthal traits among recent Amerindians and Euroamericans, or any other pair of extant human populations. Neanderthal traits purportedly distinguish Neanderthals from other human groups, and only some of these traits (e.g., midfacial projection, Wolpoff et al., 2004) still vary significantly between human populations. What characterizes population differences has changed over time and space, not to mention the fact that both Neanderthals *and* Upper Paleolithic moderns are extinct as definable populations. Thus, this analysis focuses upon comparing the past distribution of traits that characterized Neanderthals with the recent distribution of traits that characterize a recent human population, Amerindians.

Another assumption of this analysis is that "Amerindians," "Euroamericans," "Neanderthals," and "Upper Paleolithic modern humans" are roughly comparable types of populations. In terms of the samples that are used to represent these populations (see footnotes to Table 3 for details about the Amerindian and Euroamerican samples), the population groups have significant differences. The Amerindian population, as used in this study, is largely North American, although data for one trait (angled zygomaticomaxillary suture) include specimens from Peru. As far as can be gleamed from the literature, most, if not all, of the Amerindian samples used are from the past 2,000 years (see references cited in Table 3). Thus, the Amerindian "population" is represented by mostly recent North American samples. Nevertheless, such samples might be expected to exhibit regional subpopulation differences. Thus, if only one subpopulation of Amerindians is sampled, then the actual variation for a trait for Amerindians as a whole might be greatly underestimated. The samples that comprise the Euroamerican dataset suffer from the same limitations: they likely do not sample the full European colonizing population adequately, both in terms of subpopulations and time. Thus, how useful are these "Amerindian" and "Euroamerican" "populations?" Also, how do the limitations of these recent "populations" differ from those of the fossil populations "Neanderthals" and "Upper Paleolithic modern humans?" Both of the fossil samples have been greatly affected by discovery and preservation bias. For example, most of the well-preserved Neanderthal specimens are male and from Western Europe (cf. Wolpoff, 1999). Does such sampling adequately reflect a larger "Neanderthal" population? The unfortunate reality is that none of the "populations" that are usually used in comparative skeletal studies are adequately sampled. The present study is far from alone in this respect (e.g., Frayer, 1992a, b; Schwartz and Tattersall, 1996; Antón, 1996; Harvati et al., 2004). It is the unfortunate reality of skeletal studies, and it is a reality that must be recognized as one of the potential pit-falls of any such work.

The Neanderthal – Upper Paleolithic trait frequency differences given in Table 2 can readily be compared to those given for Amerindians and Euroamericans given in Table 3. Yet, the small fossil samples confound the interpretation of such a comparison. In order to statistically assess the probability of finding as much percentage difference between Amerindians and Euroamericans as is observed in limited samples of Neanderthals and Upper Paleolithic moderns, a resampling procedure was employed. This procedure consisted of the following steps for each Neanderthal trait:

(1) Calculate:

$$P_{\mathrm{n}} - P_{\mathrm{up}} = D_{\mathrm{nup}} \tag{1}$$

where P_{n} is the observed frequency of the trait in the Neanderthal sample and P_{up} the frequency of the trait in the Upper Paleolithic sample.

(2) For each Amerindian trait:

(a) Draw a sample, X_{ai}, from the Amerindian sample with $n = n_{\mathrm{n}}$, where n_{n} is the Neanderthal sample size.

(b) Draw a sample, X_{ea}, from the Euroamerican sample with $n = n_{\mathrm{up}}$, where n_{up} is the Upper Paleolithic sample size.

(c) Calculate:

$$P_{\mathrm{ai}} - P_{\mathrm{ea}} = D_{\mathrm{x}} \tag{2}$$

where P_{ai} is the frequency of the Amerindian trait in X_{ai} and P_{ea} is the frequency of the trait in X_{ea}.

(d) Store D_{x} in a bin, Z.

(e) Repeat steps 2a – 2d until Z contains 10,000 D_{x} values.

(f) Calculate:

$$P_x = n_{Dx} \div 10{,}000 \quad (3)$$

where n_{Dx} is the total number of D_x values in Z that are larger than D_{nup}, the frequency difference between the Neanderthal and Upper Paleolithic samples.

One problem with using frequency difference (i.e., D_{nup} and D_x) is that it emphasizes difference when a trait frequency is high. For example, if one sample has a frequency for a trait of 0.90 and the other sample has a frequency of 0.45, the frequency difference is 0.45. However, another trait could be in lower frequencies in the two samples: 0.20 versus 0.10 yielding a difference of 0.10. In both cases the ratio between the samples' frequencies is 0.50. Using the ratio would be a possible alternative, yet it is more problematic precisely because it would equate a 45% difference with a 10% difference. A run of simulated data demonstrates that the ratio approach results in a higher rate of Type II error than the frequency difference approach. Thus, frequency difference is used as the primary statistic in the resampling procedure.

The end result for each Neanderthal trait is a set of P_x values, each based upon an Amerindian- Euroamerican trait frequency difference model. P_x values are deemed significant if they are <0.05. A lower level of significance (i.e., as determined from Bonferroni correction [Hochberg, 1988]) might be preferable in different contexts because the many univariate tests will likely result in higher Type I error by chance alone. However, as noted below, a weak null hypothesis is preferable to an overly strong one in this analysis. If all of the P_x values for a Neanderthal trait are less than 0.05, the null hypothesis is deemed falsified for that trait. In other words, it is highly improbable that one would find as much difference between Amerindians and Euroamericans as one would between Neanderthals and Upper Paleolithic modern humans.

The assumptions and methodologies of this analysis increase the probability of incorrectly rejecting the null hypothesis that the Neanderthal – Upper Paleolithic difference is no greater than the Amerindian – Euroamerican difference (Type I error). This study assumes that if Neanderthals and Upper Paleolithic moderns are not different morphospecies we should not find any greater differences between them than found between two extant human populations for population characterizing traits. The selection of characteristic Amerindian traits that do not differ greatly between Amerindians and Euroamericans will weaken the null hypothesis while the selection of traits that show great difference still should not be greater than what would be expected between two morphospecies. The selection of Neanderthal traits is much more critical, since selecting traits that do not reportedly characterize Neanderthals from Upper Paleolithic moderns will increase the chance of Type II error. Thus, only Neanderthal traits that have been described as "autapomorphies" have been included in this study.

Another problem facing this study is the definition of the characters and how they are scored. The suprainiac fossa is a perfect example of this. Both Neanderthals and many modern humans, especially those in the European Upper Paleolithic, exhibit depressions just superior to inion (Bräuer and Brög, 1998). Frayer (1992a) scores both such structures as suprainiac fossae, while Hublin (1978, 1980, pers. comm.) contends that they are not homologous. Further clouding the definition of the suprainiac fossa as a Neanderthal autapomorphy is the presence of broad suprainiac fossae that meet all of Santa Luca's (1978) and Hublin's (1978) definitions of the Neanderthal form on the later Middle Pleistocene African specimen, Eyasi 1 (Trinkaus, 2004) and the Late Pleistocene African specimen, ADU-VP-1/3 (Haile-Selassie et al., 2004). For traits, like the suprainiac fossa, whose anatomy and distribution are controversial, both liberal (e.g., Frayer, 1992a, b) and conservative (e.g., Hublin, 1980; Bräuer and Brög, 1998) data were used. The best approach to dealing with such variation in character scoring would be to provide extensive character descriptions accompanied by ontogenetic and functional analyses of each trait. Although this is highly recommended, it is beyond the scope of this paper to resolve such character scoring controversies. By including both liberal and conservative data, it is hoped that this study will provide more than just a one-sided story.

Results

The results of the resampling procedure are given in Table 4. The null hypothesis was falsified for only one of the cranial traits, the conservative data set for the suprainiac fossa. In this data, only one Upper Paleolithic specimen (Mladeč 6) was scored as possessing a suprainiac fossa while all of the Neanderthals were scored as exhibiting them. If this is the accurate interpretation of suprainiac fossa anatomy, then the frequency difference between Neanderthals and their immediate modern successors for at least one Neanderthal "autapomorphy" is significantly greater than what we would see between Amerindians and Euroamericans. Yet, all of the Neanderthal – Upper Paleolithic differences for all of the other Neanderthal traits, whether scored liberally or conservatively, could potentially be sampled from two intraspecific populations.

Discussion

The results of this study fail to refute the null hypothesis in all instances but one. Whether or not the Upper Paleolithic anatomy is homologous to the Neanderthal anatomy is a significant issue, as demonstrated by the results for the occipital bun, occipitomastoid crest, and

Table 4. Resampling Procedure Results[1]

	Referent Model: Amerindian – Euroamerican Differences				
	Elliptic Palate	*Curved ZM Sut.*	*Sharp Nas. Sill*	*Shovel. Incisors*	*Straight Pal. Sut.*
Mandible					
H-O Mand. For.	0.452	0.410	0.999	>0.999	0.385
Asym. Mand. Notch	0.063	0.098	0.692	0.932	0.071
Medial CNM	0.999	0.996	0.923	0.999	>0.999
Retromolar Space	0.125	0.202	0.891	0.992	0.139
Cranial[2]					
Ant. Mastoid Tub.	0.946	0.903	>0.999	>0.999	0.874
Suprainiac Fossa 1	**0.002**	**0.001**	0.333	0.797	**0.002**
Suprainiac Fossa 2	**<0.001**	**<0.001**	**<0.001**	**<0.001**	**<0.001**
Occipital Bun 1	0.856	0.821	>0.999	>0.999	0.807
Occipital Bun 2	**0.006**	**0.003**	0.093	0.427	**0.006**
Occipitomastoid 1	**0.006**	**0.009**	0.343	0.670	**0.010**
Occipitomastoid 2	**0.001**	**0.000**	**0.006**	0.123	**0.001**

[1] The values given are the proportions of Amerindian – Euroamerican resampling distribution that exhibited greater percentage difference than observed between the Neanderthal and Upper Paleolithic samples (see Methods for full description). Values in **bold** are $P < 0.05$ and are considered significant.

[2] The results for Suprainaic Fossa 1, Occipital Bun 1, and Occipitomastoid 1 are based upon liberal estimates of the frequency of Neanderthal traits in the Upper Paleolithic (e.g., Frayer, 1992), while Suprainianc Fossa 2, Occipital Bun 2, and Occipitomastoid 2 are based upon conservative estimates (see Table 2).

suprainiac fossa. When the Upper Paleolithic anatomy is assumed to not be homologous, the probabilities of sampling the Neanderthal – Upper Paleolithic difference from the referent model becomes much lower (although still possible with the exception of the suprainiac fossa). Although more explicit definitions and scorings are needed for these traits, the fact that only one of the conservative datasets falsified the null hypothesis, means that the Neanderthal – Upper Paleolithic difference in terms of Neanderthal characteristic nonmetrics traits is not great.

INDIVIDUAL TRAITS OR COMBINATION OF TRAITS?

This study examined the likelihood of sampling as much Amerindian-Euroamerican difference as observed between Neanderthals and Upper Paleolithic moderns. In doing so, traits were examined univariately. It is possible that the most significant way that Neanderthals differed from Upper Paleolithic moderns was in their *combination* of traits, as opposed to the presence or frequency of individual traits. A probabilistic assessment of such trait combination differences was not made in this study, although this would be an interesting avenue for future work. A fundamental, practical problem with this approach would be the lack of specimens, both fossil and contemporary, that preserve multiple features. In a survey of forty Amerindian crania in the University of Wyoming Skeletal Repository, only fourteen preserved four of four traits examined (nasal sill, zygomaticomaxillary suture, palate shape, and transverse palatine suture). The forty crania were selected since they preserved at least one of the traits surveyed. Of thirteen Neanderthal specimens for which individual data was either available from the literature or could be scored directly from originals or casts, nine (Biache 1, Forbes' Quarry, Guattari 1, La Chapelle, La Ferrassie 1, La Quina 5, Saccopastore 1, Spy 1, Spy 2) preserved four of the four traits surveyed (occipital bun, occipitomastoid crest, anterior mastoid tubercle, and suprainiac fossa). Yet, unlike the survey of the Amerindian collection, there is a bias in which Neanderthal specimens are reported on in the literature or are available as casts. Well-preserved specimens tend to be better represented both in individual observations and in availability of casts. The reality of the human skeletal record, either fossil or recent, is that very few specimens are well preserved. Furthermore, this preservation is not random, especially in the case of Neanderthals. Far more well-preserved Neanderthals are known from Western Europe, than from Central or Eastern Europe. Of the nine that preserve four of the four traits surveyed, all were from Western Europe. Since Western Europe would have been the portion of the Neanderthal range farthest from other human populations, it would be expected that Western European Neanderthals would be the most different from other humans (including Upper Paleolithic invaders from the East) if Neanderthals were a population of a larger human species.

When only Neanderthal and Amerindian individuals that preserved four of the four traits are examined, six of nine (66.7%) Neanderthals exhibit the Neanderthal variant for all four traits and six of fourteen (42.9%) Amerindians exhibit the Amerindian variant for all four traits. Given how small these samples are, it is not surprising that the frequency of "all-Neanderthal" Neanderthals is not significantly different from the frequency of "all-Amerindian" Amerindians ($p = 0.40$, Fisher's Exact Test). Fundamentally, a meaningful assessment of just how common Neanderthals with a combination of multiple Neanderthal traits are will have to wait until sufficient comparative samples are acquired. Of course, more well-preserved Neanderthals from across their range would be useful (Stefan and Trinkaus, 1998), although sample sizes may never be sufficient to know how homogeneous Neanderthals were in terms of trait combinations except in comparison with extant referents. Even if sufficient samples were available, would a significantly greater frequency of Neanderthals with multiple Neanderthal traits than Upper Paleolithic specimens with multiple traits demonstrate Neanderthal specificity? Such a degree of difference could perhaps justify their classification as a different morphospecies. Yet, would such a classification even be evolutionarily meaningful?

SPECIES

The issue of how past variation compares with extant variation, although relevant, is not necessarily the same as whether or not Neanderthals and Upper Paleolithic moderns were the same species. In their morphometric analysis of distance between Neanderthals and modern humans, Harvati et al. (2004; see also Harvati, 2003) use the morphospecies concept in the form of yardsticks developed from extant species variation, although they overstate its acceptance in the field (cf. Kimbel and Martin, 1993). They state: "Species in paleontology should be equivalent to living ones in the variation that they accommodate" (Harvati et al., 2004: 1147). Harvati et al. (2004) conclude that, since the Mahalanobis Distance (D^2) between their Neanderthal and Upper Paleolithic samples was statistically greater than almost all of their intraspecific and most of their interspecific comparisons of cercopithecoids and African great apes, Neanderthals likely represented a different species from modern humans. The Harvati et al. (2004) study and the current analyses are applicable to a test of a single morphospecies, yet are morphospecies evolutionarily meaningful units? The morphospecies concept is pragmatically strong but ontologically weak, since it potentially constructs species units that do not have evolutionary meaning (Kimbel and Rak, 1993; Mayden, 1997). Temporal variation within a phyletic lineage, whether directional or stochastic, has the potential to be misinterpreted as multiple species when the morphospecies approach is taken (Ahern et al., 2005; but see Harvati et al., 2005).

In his recent review and analysis of species and species concepts in human evolution, Holliday (2003, this volume) offers a refreshing take on the issue of whether or not Neanderthals should be designated their own species. He notes that many closely related, yet readily distinguished, "species" mate and produce fertile offspring. Although a liberal application of the biological species concept would necessitate assigning these "species" to the same species, Templeton (1989), as Holliday (2003) notes, argues that they should be identified as separate species because they maintain separation through cohesion. Such a view lessens the importance of autapomorphies for species identification, since some "autapomorphic" traits can become homoplasies (or at least homoplasy-like) through introgression (Holliday, 2003). Holliday (2003) thus labels Neanderthals as their own species, *Homo neanderthalensis*, but allows for the sharing of traits through introgression. In order for this explanation to be valid, within Holliday's framework, Neanderthals and modern humans should retain separate cohesiveness. How this cohesiveness should be defined is difficult, however. This is especially the case when we compare one group, Neanderthals, with the group that succeeded them, Upper Paleolithic moderns. This is a very different situation than the papionin (Jolly, 2001), deer (Cathey et al., 1998), and canid examples cited by Holliday (2003) to illustrate interspecific gene flow. The main lines of evidence for these cases are the presence of hybrids, hybrid zones and evidence of genetic introgression. Neanderthals are usually compared to a potentially post-hybridization population: Upper Paleolithic modern humans. We lack any F1 (or F10, for that matter) Neanderthal – modern human hybrids and probably always will because of the limitations of the fossil record. What we do seem to have, in the case of the Upper Paleolithic moderns, is evidence of genetic admixture, since many "uniquely" Neanderthal features persist into the Upper Paleolithic. That the present study indicates that the frequency in these traits does not differ between Neanderthals and the Upper Paleolithic moderns more than we would expect for two populations of the same species is a further demonstration of some degree, perhaps significant, admixture.

While extreme viewpoints on the fate of the Neanderthals, such as overall regional continuity (Coon, 1962) or complete replacement (Stringer and Andrews, 1988) are explicit regarding Neanderthal specificity, the well-occupied intermediate ground (e.g., Bräuer, 1989; Smith et al., 1989; Stringer, 1992; Stringer and Gamble, 1993; Hawks and Wolpoff, 2001; Trinkaus and Zilhão, 2002) is far from unified regarding the issue. Although seemingly more true to the intermediate ground and the amorphous reality of species boundaries, Holliday's (2003) and others' (e.g., Jolly, 2001) interpretations have not resulted in consensus on the issue of Neanderthal systematics. Both Holliday (2003) and Wolpoff (1999) adhere to the evolutionary species concept (Wiley, 1981), yet Holliday sees a unique origin and fate and evolutionary tendency for Neanderthals (thus, *Homo neanderthalensis*), while Wolpoff does not. Wolpoff contends that the only identifiable evolutionary lineage is the human one that appears approximately 2 million years ago and is still alive today (thus, Neanderthals represent an extinct population). The question, whether or not the degree of admixture and the degree of "cohesiveness" of the Neanderthals are sufficient to falsify a single species hypothesis for Neanderthals and modern humans, remains unanswered. Even more fundamentally, any measure of species boundaries is dependent upon the species concept being employed.

Summary

This study failed to refute the hypothesis that the Neanderthal trait frequency differences between Neanderthals and Upper Paleolithic moderns are not significantly greater than the Amerindian trait frequency differences between Amerindians and Euroamericans. Although the statistical results are not unanimous, their overall pattern is consistent with a tentative acceptance of the single morphospecies hypothesis based upon the data examined. Additional trait data and/or an analysis of trait combinations could potentially falsify the single morphospecies hypothesis. A better understanding of the variation, ontogeny, and function of Neanderthal non-metric traits will be important for future work. Although this study's results are relevant to whether or not Neanderthals and Upper Paleolithic moderns were different morphospecies, morphospecies are not necessarily evolutionarily meaningful units.

Acknowledgements

Thanks to Drs. Harvati and Harrison for editing this volume and for organizing an excellent, productive, and insightful conference. A special acknowledgement goes to Dr. D. W. Frayer for his pioneering work on the persistence of Neanderthal traits among post-Neanderthal Europeans (Frayer 1992a, b). I would also like to thank Drs. F.H. Smith, M.H. Wolpoff, G. Gill, M. Harkin, and M. Kornfeld for information and insights that helped in the production of this paper.

References

Ahern, J.C.M., Hawks, J.D., Lee, S-H., 2005. Neanderthal taxonomy reconsidered ... again: a response to Harvati et al. (2004). *J. Hum. Evol.* 48, 647–652.

Antón, S., 1996. Tendon-associated bone features of the masticatory system in Neandertals. *J. Hum. Evol.* 31, 391–408.

Bräuer, G., 1989, The evolution of modern humans: a comparison of the African and non-African evidence. In: Stringer, C.B., Mellars, P. (Eds.), *The Human Revolution*. Edinburgh University Press, Edinburgh, pp. 121–154.

Bräuer, G., Brög, H., 1998. On the degree of Neanderthal-modern continuity in the earliest Upper Palaeolithic crania from the Czech Republic: evidence from non-metrical features. In: Omoto, K., Tobias, P.V. (Eds.), *Origins and Past of Modern Humans: Towards Reconciliation*, World Scientific, Singapore, pp. 106–125.

Cartmill, M., Smith, F.H., Brown, K.B., in press. *The Human Lineage*, Wiley-Liss, New York.

Cathey, J.C., Bickham, J.W., Patton, J.C., 1998. Introgressive hybridization and nonconcordant evolutionary history of maternal and paternal lineages in North American deer. *Evolution* 52, 1224–1229.

Coon, C.S., 1962. *The Origin of the Races*. Knopf, New York.

Cracraft, J., 1989. Speciation and its ontology: the empirical consequences of alternative species concepts for understanding patterns and processes of differentiation. In: Otte, D., Endler, D. (Eds.), *Speciation and its Consequences*. Sinauer Associates, Sunderland, MA, pp. 28–59.

Crawford, M.H., 1998. *The Origins of Native Americans: Evidence from Anthropological Genetics*. Cambridge University Press, Cambridge.

Cronquist, A., 1978. Once again, what is a species? In: Knutson, L.V. (Ed.), *BioSystematics in Agriculture*. Allenheld Osmun, Montclair, NJ, pp. 3–20.

Diamond, J., 1997. *Guns, Germs, and Steel: The Fates of Human Societies*. W.W. Norton, New York.

Dobyns, H.F., 1983. Disease transfer at contact. *Ann. Rev. Anthropol.* 22, 273–291.

Franciscus, R.G., 1999. Neanderthal nasal structures and upper respiratory tract "specialization". *Proc. Natl. Acad. Sci. U.S.A.* 96, 1805–1809.

Franciscus, R., Trinkaus, E., 1995. Determinants of retromolar space presence in Pleistocene *Homo* mandibles. *J. Hum. Evol.* 28, 577–595.

Frayer, D.W., 1992a. The persistence of Neanderthal features in post-Neanderthal Europeans. In: Bräuer, G., Smith, F.H. (Eds.), *Continuity or Replacement: Controversies in Homo sapiens Evolution*. AA Balkema, Rotterdam, pp. 179–188.

Frayer, D.W., 1992b. Evolution at the European edge: Neanderthal and Upper Paleolithic relationships. *Préhistoire Européenne 2*, 9–69.

Frayer, D.W., Wolpoff, M.H., Thorne, A.G., Smith, F.H., Pope, G.G., 1993. Theories of modern human origins: the paleontological test. *Am. Anthropol.* 95, 14–50

Gill, G.W., Rhine, S. (Eds.), 1990. Skeletal Attribution of Race: Methods for Forensic Anthropology. Anthropol. papers Maxwell Mus. Anthropol. 4, 47–53.

Haile-Selassie, Y., Asfaw, B., White, T.D., 2004. Hominid cranial remains from Upper Pleistocene deposits at Aduma, Middle Awash, Ethiopia. *Am. J. Phys. Anthropol.* 123, 1–10.

Hawks, J., Wolpoff, M.H., 2001, Brief communication: paleoanthropology and the population genetics of ancient genes. *Am. J. Phys. Anthropol.* 114, 269–272.

Harvati, K., 2003. The Neanderthal taxonomic position: models of intra- and inter-specific craniofacial variation. *J. Hum. Evol.* 44, 107–132.

Harvati, K., Frost, S.R., McNulty, K.P., 2004. Neanderthal taxonomy reconsidered: implications of 3D primate models of intra- and interspecific differences. *Proc. Natl. Acad. Sci. U.S.A.* 101, 1147–1152.

Harvati, K., Frost, S.R., McNulty, K.P., 2005. Neandertal variation and taxonomy – a reply to Ackerman (2005) and Ahern et al. (2005). *J. Hum. Evol.* 48, 653–660.

Henneberg, M., De Miguel, C., 2004. Hominins are a single lineage: brain and body size variability does not reflect postulate taxonomic diversity of hominins. *HOMO* 55, 21–37.

Hochberg, Y., 1988. A sharper Bonferroni procedure for multiple tests of significance. *Biometrika* 75, 800–803.

Holborow, A.A., 2002. The zygomaticomaxillary suture: a study of the frequencies of suture shapes within various major populations of *Homo sapiens*. M.A. Thesis, University of Wyoming.

Holliday, T.W., 2003. Species concepts, reticulation, and human evolution. *Curr. Anthropol.* 44, 653–673.

Hrdlička, A., 1920, Shovel-shaped teeth. Am. J. Phys. Anthropol. 3, 429–465.

Hublin, J.-J., 1978. Quelques caractères apomorphes du crâne néandertalien et leur interprétation phylogénique. *C. R. Acad. Sci., Paris* 287, 923–926.

Hublin, J.-J., 1980. La Chaise Suard, Engis 2, et La Quina H 18: développement de la morphologie externe chez l'enfant prénéandertalien et néandertalien. *C. r. Acad. Sci., Paris* 291, 669–672.

Jabbour, R.S., Richards, G.D., Anderson, J.Y., 2002. Mandibular condyle traits in Neanderthals and other *Homo*: a comparative, correlative, and ontogenetic study. *Am. J. Phys. Anthropol.* 119, 144–55.

Jolly, C.J., 2001. A proper study for mankind: analogies from the papionin monkeys and their implications for human evolution. *Yrbk. Phys. Anthropol.* 44, 177–204.

Kimbel, W.H., Martin, L.B. (Eds.), 1993. *Species, Species Concepts, and Primate Evolution.* Plenum, New York.

Kimbel, W.H., Rak, Y., 1993, The importance of species in paleoanthropology and an argument for the phylogenetic concept of the species category. In: Kimbel. W.H., Martin, L. (Eds.), *Species, Species Concepts, and Primate Evolution.* Plenum, New York.

Kramer, A., Crummett, T.L., Wolpoff, M.H., 2001. Out of Africa and into the Levant: replacement or admixture in Western Asia? *Quatern. Int.* 75, 51–63.

Mayden, R.L., 1997. A hierarchy of species concepts: the denouement in the saga of the species problem. In: Claridge, M.F., Dawah, H.A., Wilson, M.R. (Eds.), *Species: The Units of Biodiversity.* Chapman and Hall, London.

Oswalt, W.H., 2006. *This Land Was Theirs.* Oxford University Press, New York.

Pollitzer, W.S., Phelps, D.S., Waggoner, R.E., Leyshon, W.C., 1967. Catawba Indians: Morphology, genetics, history. *Am. J. Phys. Anthropol.* 26, 5–14.

Post, R.H., Neel, J.V., Schull, W.J., 1968. Tabulations of phenotype and gene frequencies for 11 different genetic systems studied in American Indians. In: *Biomedical Challenges Presented by the American Indians*, Pan American Health Organization, Washington, D.C., pp. 141–185.

Quam, R., Smith, F.H. 1998. A reassessment of the Tabun C2 mandible. In: Akazawa, T., Aoki, K., Bar-Yosef, O. (Eds.), *Neanderthals and Modern Humans in Western Asia*, Plenum, New York, pp. 405–421.

Rak, Y., 1993. Morphological variation in *Homo neanderthalensis* and *Homo sapiens* in the Levant: a biogeographic model. In: Kimbel, W.H., Martin, L. (Eds.), *Species, Species Concepts, and Primate Evolution.* Plenum, New York, pp. 523–536.

Rak, Y., 1998. Does any Mousterian cave present evidence of two hominid species? In: Akazawa, T., Aoki, K., Bar-Yosef, O. (Eds.), *Neanderthals and Modern Humans in Western Asia.* Plenum, New York, pp. 353–366.

Rak, Y., Kimbel, W.H., Hovers, E., 1994. A Neanderthal infant from Amud Cave, Israel. *J. Hum. Evol.* 26, 313–324.

Rak, Y., Ginzburg, A., Geffen, E., 2002. Does *Homo neanderthalensis* play a role in modern human ancestry? The mandibular evidence. *Am. J. Phys. Anthropol.* 119, 199–204.

Rawlings, K.J., 2002. Racial variation in palate shape and form of the transverse palatine suture. M.A. Thesis, University of Wyoming.

Richards, G.D., Plourde, A. M., 1995. Reconsideration of the "Neandertal" infant, Amud-7. *Am. J. Phys. Anthropol.* 20 (Suppl.), 180–181 (Abstract).

Santa-Luca, A., 1978. A re-examination of presumed Neanderthal-like fossils. *J. Hum. Evol.* 7, 619–636.

Schwartz, J. H., Tattersall, I., 1996. Significance of some previously unrecognized apomorphies in the nasal region of *Homo neanderthalensis. Proc. Natl. Acad. Sci. U.S.A.* 93, 10852–4.

Smith, F.H., Falsetti, A.B., Donnelly, S.M., 1989. Modern human origins. *Yrbk. Phys. Anthropol.* 32, 35–68.

Stefan, V.H., Trinkaus, E., 1998. Discrete trait and dental morphometric affinities of the Tabun 2 mandible. *J. Hum. Evol.* 34, 443–468.

Stringer, C.B., 1992. Replacement, continuity and the origin of *Homo sapiens.* In: Bräuer, G. Smith, F.H. (Eds.), *Continuity or Replacement: Controversies in Homo sapiens Evolution.* AA Balkema, Rotterdam, pp. 9–24.

Stringer, C.B., Andrews, P., 1988. Genetic and fossil evidence for the origin of modern humans. *Science* 239, 1263–1268.

Stringer, C.B., Gamble, C., 1993. *In Search of the Neanderthals.* Thames and Hudson, London.

Stringer, C.B., Hublin J.J., Vandermeersh, B., 1984. The origin of anatomically modern humans in Western Europe. In: Smith, F.H., Spencer, F. (Eds.), *The Origins of Modern Humans: a World Survey of the Fossil Evidence.* Alan R. Liss, New York, pp. 51–135.

Szathmary, E.J.E., Auger, F., 1983. Biological distances and genetic relationships within Algonkians. In: Steegman, A.T. (Ed.), *Boreal Forest Adaptations.* Plenum, New York, pp. 289–315.

Tattersall, I., Schwartz, J., 1998. Morphology, paleoanthropology, and Neanderthals. *Anat. Rec.* 253, 113–117.

Templeton, A.R., 1989. The meaning of species and speciation: a genetic perspective. In: Otte, D., Endler, J.A. (Eds.), *Speciation and Its Consequences. Sinauer Associates*, Sunderland, MA.

Thornton, R., 1987. *American Indian Holocaust and Survival: A Population History Since 1492.* University of Oklahoma Press, Norman.

Thornton, R. (1997) Aboriginal North American population and rates of decline, ca. A.D. 1500–1900. *Curr. Anthropol.* 38, 310–315.

Trinkaus, E., 2004. Eyasi 1 and the Suprainiac Fossa. *Am. J. Phys. Anthropol.* 124, 28–32.

Trinkaus, E., LeMay, M., 1982. Occipital bunning among later Pleistocene hominids. *Am. J. Phys. Anthropol.* 57, 127–135.

Trinkaus, E., Shipman, P., 1993. *The Neanderthals: Changing the Image of Mankind.* Knopf, New York.

Trinkaus, E., Zilhão, J., 2002. Phylogenetic implications. In: Zilhão, J., Trinkaus, E. (Eds.), *Portrait of the artist as a child. The Gravettian human skeleton from the Abrigo do Lagar Velho and its archeological context.* Trabalhos Arqueologia 22. Instituto Portugues de Arqueologia, Lisbon, pp. 497–518.

Trinkaus, E., Milota, S., Rodrigo, R., Mircea, G., Moldovan, O., 2003. Early modern human cranial remains from the Peştera cu Oase, Romania. *J. Hum. Evol.* 45, 245–253.

Ubelaker, D.H., 1988. North American Indian population size, A.D. 1500 to 1895. *Am. J. Phys. Anthropol.* 77, 289–294.

Wiley, E.O., 1981. *Phylogenetics: The Theory and Practice of Phylogenetic Systematics.* New York: John Wiley and Sons.

Williams, R.C., Long, J., Hanson, R.L., Sievers, M.L., Knowler, W.C., 2000. Individual estimates of European genetic admixture associated with lower body-mass index, plasma glucose, and prevalence of type 2 diabetes in Pima Indians. *Am. J. Hum. Genet.* 66, 527–538.

Willson, G.F., 2004. Variation in the lower mid-face of three American skeletal populations. M.A. Thesis, University of Wyoming.

Wissler, C., 1931. Observations on the face and teeth of North American Indians. *Anthropol. Papers Am. Mus. Nat. Hist.* 33, 22–23.

Wolpoff, M.H., 1999. *Paleoanthropology.* McGraw-Hill, New York.

Wolpoff, M.H., Frayer, D.W., 2005. Unique ramus anatomy for Neanderthals? *Am. J. Phys. Anthropol.* 128, 245–251.

Wolpoff, M.H., Hawks, J.D., Caspari, R., 2000. Multiregional, not multiple origins. *Am. J. Phys. Anthropol.* 112, 129–136.

Wolpoff, M.H., Mannheim, B., Mann, A., Hawks, J., Caspari, R., Rosenberg, K., Frayer, D.W., Gill, G.W., Clark, G., 2004. Why not the Neanderthals? *World Archaeol.* 36, 527–546.

15. Earliest Upper Paleolithic crania from Mladeč, Czech Republic, and the question of Neanderthal-modern continuity: metrical evidence from the fronto-facial region

G. BRÄUER
Abteilung für Humanbiologie, Universität Hamburg
Allende-Platz 2, 20146 Hamburg, Germany
guenter.braeuer@uni-hamburg.de

H. BROEG
Abteilung für Humanbiologie, Universität Hamburg
Allende-Platz 2, 20146 Hamburg, Germany

C.B. STRINGER
Department of Palaeontology,
Natural History Museum
London SW7 5BD, UK
c.stringer@nhm.ac.uk

Keywords: Neanderthal, early modern humans, Out-of-Africa, replacement, Mladeč, regional continuity, Upper Palaeolithic Europeans

Abstract

Direct AMS radiocarbon dates of around 31 ka BP (Wild et al., 2005) for several well preserved crania and other human specimens from Mladeč, Czech Republic, confirm their association with the Aurignacian. This material, which thus represents the earliest modern European remains with archaeological associations, has long featured in discussions of regional continuity or gene flow from Neanderthal into early Cro-Magnon populations. Here, the four most complete Mladeč crania are compared with Neanderthal fossils in metrical characters of the fronto-facial region. Both univariate and multivariate analyses show no evidence of Neanderthal affinities, and thus of Neanderthal-derived genes.

Introduction

In a commentary on the evidence from molecular biology, Gibbons (2001: 1052) stated that no-one can rule out the possibility that some of us have inherited nuclear DNA from Neanderthals, but detection of such archaic lineages is so difficult that many geneticists despair that they will ever be able to prove or disprove whether the genetic replacement of archaic people outside of Africa was complete. A population geneticist (Rosalind Harding) is cited in this article as saying, "we're going to have to let the fossil people

K. Harvati and T. Harrison (eds.), Neanderthals Revisited: New Approaches and Perspectives, 269–280.

answer this one." However, since the effects of genetic exchange are difficult to estimate with polygenic morphological features, this question is also a challenge for fossil experts. For example, there is continuing disagreement whether the presence of cranial features such as a bun-like morphology of the occiput or a supranuchal depression found in some early Cro-Magnons is the result of gene flow from Neanderthals. Also, the identification of a possible Neanderthal – Cro-Magnon hybrid from Lagar Velho (Zilhão and Trinkaus, 2002) is disputed and alternative explanations have been suggested (Tattersall and Schwartz, 1999; Stringer, 2002a; Bräuer, 2006).

Several different perspectives can currently be distinguished regarding the Neanderthal-modern transition in Europe (Bräuer, 2006): (1) an ancestor-descendant relationship as proposed by the classic Multiregional Evolution model, which sees a considerable Neanderthal ancestry for modern Europeans (Frayer et al., 1993; Wolpoff et al., 2001); (2) modern humans moved into Europe accompanied by significant assimilation of Neanderthals (Churchill and Smith, 2000; Trinkaus, 2005; Smith et al., 2005); (3) the Out-of-Africa replacement view, which allows for gene flow but sees little evidence for interbreeding in the fossil record (Bräuer and Stringer, 1997; Bräuer, 2001; Stringer, 2002b); and (4) the complete replacement view, which excludes any gene flow between Neanderthals and dispersing modern humans (e.g., Tattersall, 2003).

In order to reach further agreement on the extent of gene flow at the archaic-modern interface in Europe we feel that it is important to carefully examine all suggested indications of regional continuity (Bräuer and Stringer, 1997: 197). With this intention, the present paper examines aspects of fronto-facial morphology in the earliest anatomically modern cranial remains from the Czech Republic, and their affinities to Neanderthals. This material represents the best early modern sample from Central Europe, and if there was either regional continuity or significant Neanderthal-derived gene flow into such a population we should expect to see traces in the morphology of these specimens. Indeed, it has been claimed by Wolpoff et al. (2001) that such traces can be recognized and even quantified.

Material and Methods

The early modern sample from the Czech Republic examined here includes the four well-preserved crania Mladeč 1 (assumed ♀), 2 (assumed ♀), 5 (assumed ♂), 6 (assumed ♂), and the maxillary fragment Mladeč 8 (assumed ♂). Recent direct AMS dating of several Mladeč specimens including Mladeč 1 and 2, as well as Mladeč 8, yielded ages of about 31 ka BP (Wild et al., 2005) which are in agreement with the Aurignacian artifacts (Vlček, 1995) and previous AMS dates for associated calcite deposits (Svoboda et al., 2002). This confirms the Mladeč assemblage as the oldest directly dated substantial assemblage of modern human remains in Europe (Wild et al., 2005). In addition to this early sample, the three somewhat more recent Brno specimens 1 (assumed ♀), 2 (assumed ♂) and 3 (assumed ♀) have been examined. A direct AMS date for the Brno 2 skeleton, associated with the Moravian Gravettian, yielded a date of 23,680 ± 200 yrs BP (Pettitt and Trinkaus, 2000). The probable female calvaria from Zlatý Kůn, formerly thought to date to the Aurignacian or Szeletian (Jelínek, 1978) has now been redated by AMS to about 13 ka BP (Svoboda et al., 2002). This specimen was also included in our Upper Paleolithic comparative sample. With the exception of Mladeč 6 and Brno 3, of which only casts survive, the originals were measured by one of us (HB).

The comparative material (Table 1) includes Neanderthals, early modern humans from Africa and the Levant, additional Upper Paleolithic Europeans, and the terminal

Table 1. Comparative material used

Site	*abbr.*	*sex*	*Site*	*abbr.*	*sex*
			European Neanderthals		
European Upper Palaeolithic			La Chapelle	LACP	(♂)
Abri Pataud 1	ABRI	(♀)	La Ferrassie 1	LAF1	(♂)
Arène Candide 1	ARCA	(♂)	La Quina 5	LAQ5	(♀)
Barma Grande 5	BAG5	(♂)	Le Moustier	LEMR	(♂)
Bruniquel 24	BRUN	(♀)	Monte Circeo	MOCI	(♂)
Chancelade 1	CHAN	(♂)	Neandertal	NEAN	(♂)
Combe Capelle 1	COMB	(♂)	Spy 1	SPY1	(♀)
Cro-Magnon 1	CRO1	(♂)	*Near Eastern Neanderthals*		
Cro-Magnon 2	CRO2	(♀)	Amud 1	AMU1	(♀)
Dolní Věstonice 3	DOL3	(♀)	Shanidar 1	SHA1	(♂)
Duruthy (Sorde) 3	DUR3	(♀)	Shanidar 5	SHA5	(♂)
Grottes des Enfants 4	GRE4	(♂)	Tabun 1	TAB1	(♀)
Kostenki Markina Gora 2	KOS2	(♂)	*Early modern specimens (Near East)*		
Oberkassel 1	OKA1	(♂)	Qafzeh 6	QAF6	(♂)
Oberkassel 2	OKA2	(♀)	Qafzeh 9	QAF9	(♀)
Paderborn 1	PADB	(♂)	Skhul 4	SKH4	(♂)
Pavlov 1	PAV 1	(♂)	Skhul 5	SKH5	(♂)
Předmosti 3	PRD3	(♂)	Skhul 9	SKH9	(♂)
Předmosti 4	PRD4	(♀)	*Early modern specimens (Africa)*		
Předmosti 9	PRD9	(♂)	Border Cave 1	BOR1	(♂)
Předmosti 10	PR10	(♀)	Omo Kibish 1	OMO1	(♂)
St. Germain-La-Riviére 4	STG4	(♀)	Dar-es-Soltane 5	DAR5	(♂)
Urtiaga 1	URT1	(♂)	Nazlet Khater	NAZK	(♂)
			Wadi Kubanniya 1	WAKU	(♂)

In addition, 23 specimens from Afalou-bou-Rhummel and Taforalt including 4 ♀♀ (AFA3, AF29, TO8C, T171) were included.

Pleistocene Afalou/Taforalt series from northern Africa. The metrical data were derived from different published sources (Heim, 1976; Howells, 1975; Sergi, 1974; Suzuki and Takai, 1970; Trinkaus, 1983, 1987; Vandermeersch, 1981; Bräuer and Rimbach, 1990), from the present authors, and also kindly provided by D. Ferembach, D. Frayer, W. Henke, F. Smith, and F. Wendorf.

The metrical variables used to describe aspects of fronto-facial morphology follow Howells (1973) and Bräuer (1988). Both univariate and multivariate comparisons were conducted. For Principal Components Analysis (PCA) of Neanderthal and modern groups, individual size was eliminated by using log-shape data (cf., Darroch and Mosiman, 1985; Simmons et al., 1991).

Results

In spite of individual variation, a receding flat frontal squama is a typical plesiomorphous feature of the Neanderthals (e.g., Stringer and Trinkaus, 1981; Delson et al., 2000). The results with regard to the Frontal Angle (FRA) show clear differences between European and western Asian Neanderthals on the one hand, and the Czech sample (this study), other Upper Paleolithic Europeans, Skhul/Qafzeh and the Afalou/Taforalt series on the other (Figure 1). There is a slight overlap between the ranges of variation of Neanderthals and Afalou/Taforalt. Mladeč 1 and 5 show the greatest differences from Neanderthals among all the modern samples included. This is especially remarkable since Mladeč 1 is sexed as

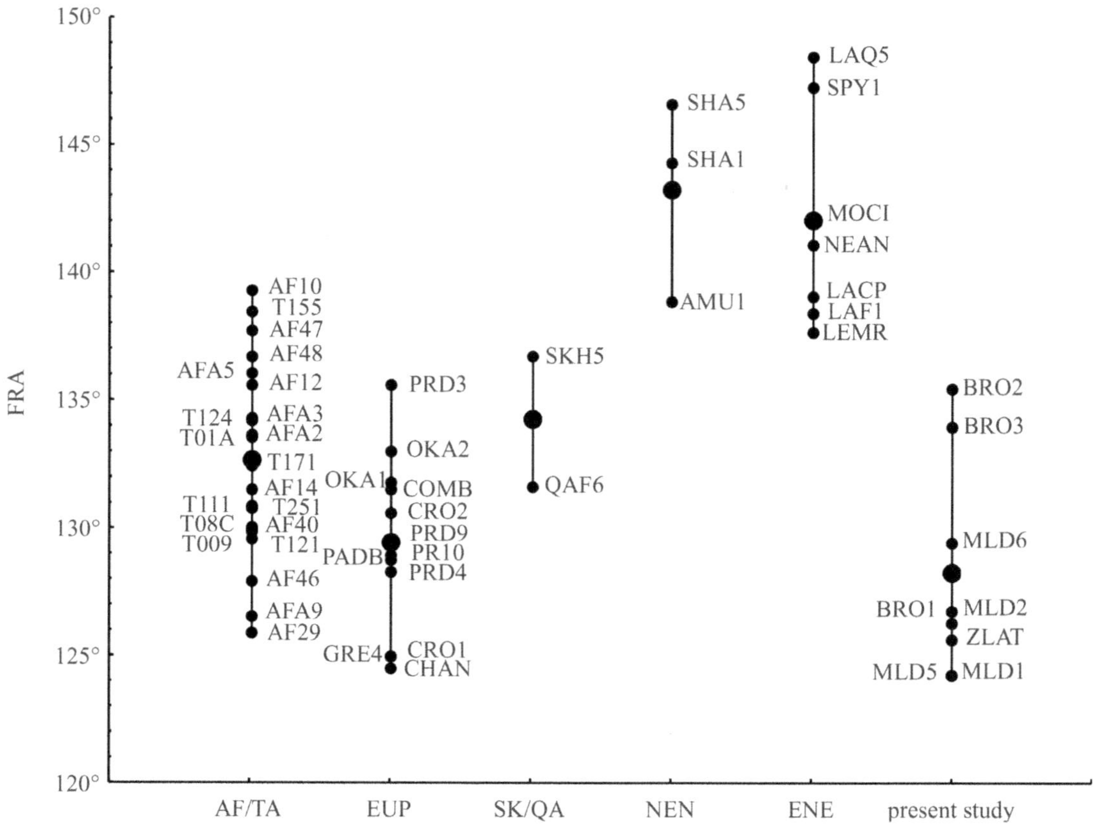

Figure 1. Frontal Angle (FRA). Comparison of early modern specimens from Czech Republic to European Neanderthals (ENE), Near Eastern Neanderthals (NEN), Skhul/Qafzeh (SK/QA), early Upper Palaeolithic Europeans (EUP) and the Afalou-Taforalt (AF/TA) sample (for abbr. see Table 1).

female and Mladeč 5 as male. Zlatý Kůn and Mladeč 2 also differ strongly from Neanderthals, as does Mladeč 6, which is close to several male and female specimens from Předmostí. The more recent Brno frontals show considerable variation even among the two females 1 and 3, but fall within the Upper Paleolithic as well as the Afalou/Taforalt ranges of variation.

Figure 2 presents the results of a PCA based on log-shape data of eight mid-sagittal frontal variables (see Table 2). Most variables, including the subtenses, have high loadings on PC1 (Table 2), which separates Neanderthals rather well from the diverse modern groups. Some Afalou/Taforalt specimens, as well as Skhul 5, show marginal affinities to the Neanderthal frontal curvature. The early modern Czech specimens again deviate most strongly from the Neanderthals, especially Mladeč 2 and 5, even approaching the very divergent Border Cave 1 specimen from South Africa. Mladeč 1 and 6, as well as Zlatý Kůn and Brno 1, also differ greatly from the Neanderthals. The robust specimen Brno 2 shows similarities to Předmostí 3 for FRA (see Figure 1), while Brno 3 exhibits a rather isolated position.

Projection of the midfacial region is one of the features in which Neanderthals have a clearly derived morphology (e.g., Stringer and Trinkaus, 1981; Stringer, 1989; Frayer, 1986, 1992). In the present study, facial morphology was analysed using the Nasio-Frontal Angle

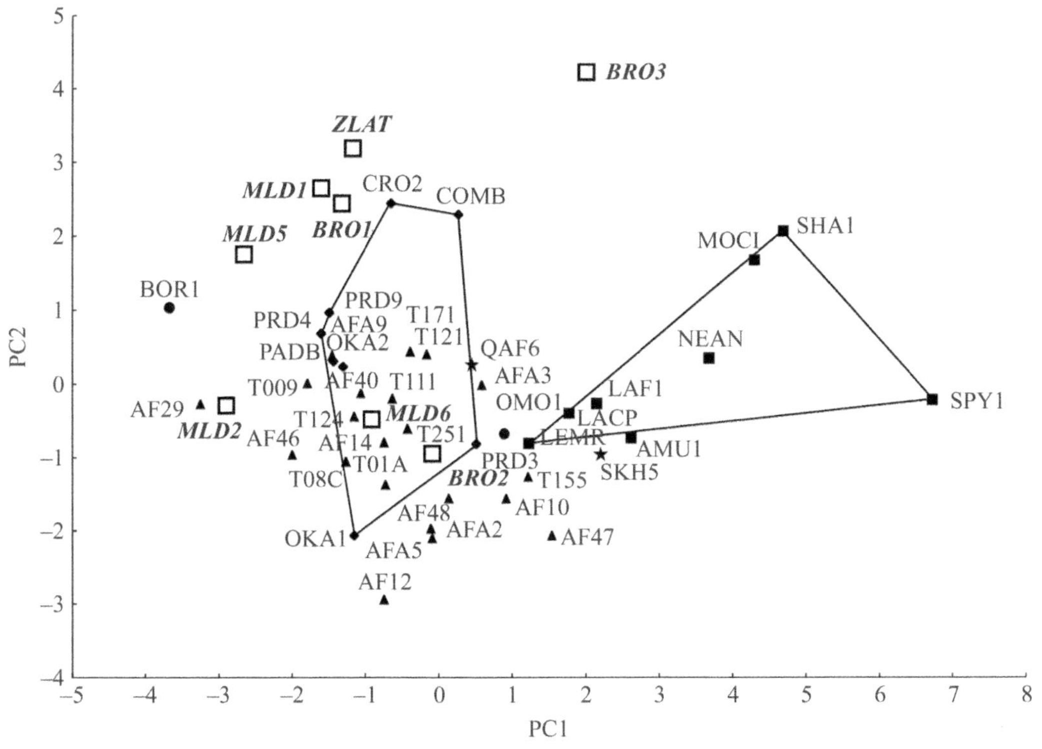

Figure 2. Principal Components Analysis based on eight frontal variables (log-shape data). PC1 accounts for 52,7% and PC2 for 28,2% of the total variance. Outlines indicate range of variation of the Neanderthal and European Upper Palaeolithic comparative samples.

Table 2. Principal Components Analysis based on log-shape data of eight mid-sagittal frontal variables.

Variable	*PC1*	*PC2*
Nasion-bregma arc (26)	0.51402	0.67050
Frontal subtense (FRS)	−0.88243	0.36641
Glabella-bregma chord (29d)	0.86574	0.38555
Glabella-subtense fraction (29f)	0.65579	−0.61420
Nasion-bregma chord (29)	0.73474	0.39576
Nasion-subtense fraction (FRF)	0.54410	−0.70677
Glabella-bregma arc (26a)	0.56439	0.65404
Glabella-bregma subtense (29e)	−0.91720	0.24823

Abbreviations after Howells (1973), numbers after Bräuer (1988).

(NFA), Subspinale (Zygomaxillary) Angle (SSA) and the major dimensions of the nasal aperture, as well as by PCA. The group means for NFA (Figure 3) show that, as expected, Neanderthals differ from the various modern samples by having smaller values, i.e., a more projecting nasion. The variation of the different groups shows some overlap between Neanderthals and the Upper Paleolithic (including Brno) and Afalou/Taforalt samples. However, the earliest Czech specimens are not only outside the Neanderthal range of variation but also show great deviations from them. Large differences between the Neanderthals and the Czech specimens Mladeč 1 and Brno 3 are also evident with regard to SSA (Figure 4) where there is no overlap between the Neanderthal and modern samples for this feature.

A large and broad nasal aperture is another plesiomorphous Neanderthal feature (e.g., Frayer, 1992; Stringer and Gamble, 1993; Delson et al., 2000) and was measured here by Nasal Breadth and Height. Only specimens for which both measurements were available were included. The Neanderthals are quite well separated from the modern groups

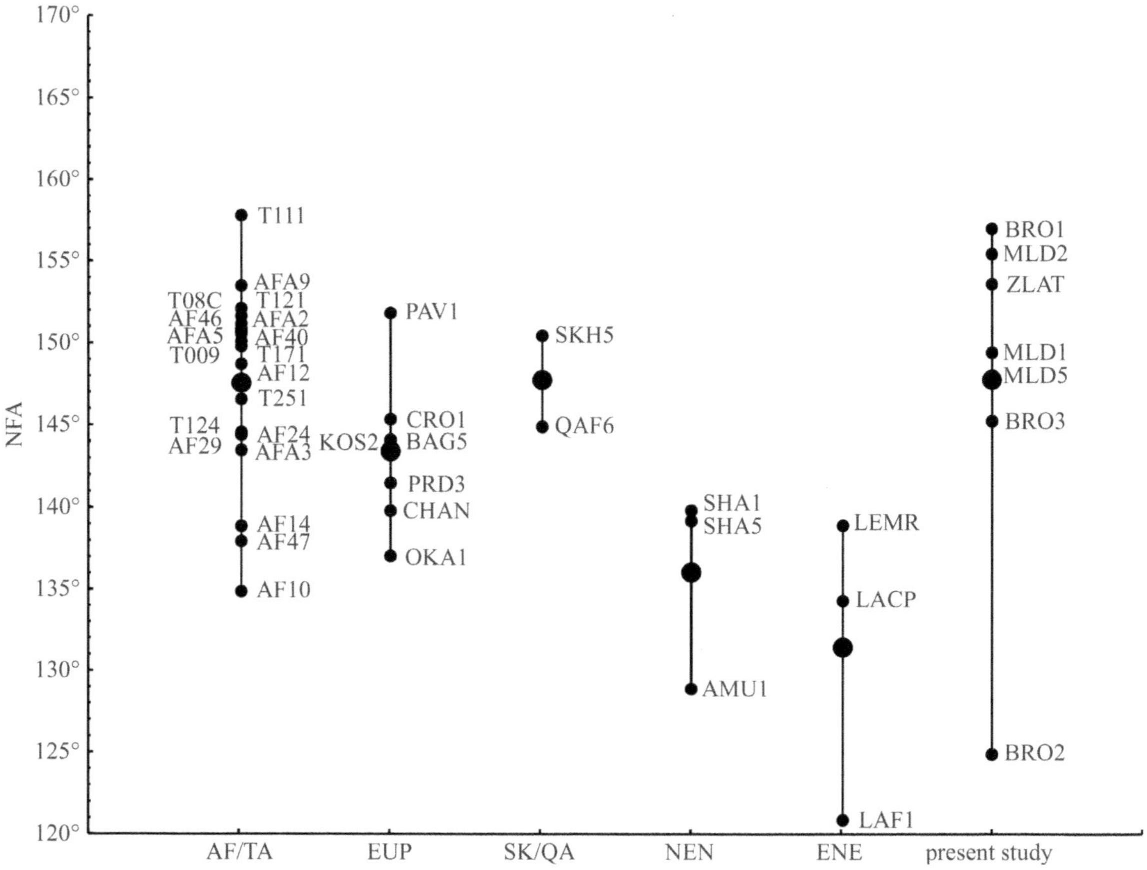

Figure 3. Nasio-frontal Angle (NFA). Comparison of early modern specimens from Czech Republic to European Neanderthals (ENE), Near Eastern Neanderthals (NEN), Skhul/Qafzeh (SK/QA), early Upper Palaeolithic Europeans (EUP) and Afalou-Taforalt (AF/TA).

(Figure 5) although some overlap is seen with regard to the western Asian Neanderthals and early moderns (Shanidar 1, Qafzeh 6). The earliest modern Czech specimens Mladeč 1 and 2 are quite distinct from Neanderthal dimensions, as is Brno 3. However, one of the Mladeč specimens (Mladeč 8), which only consists of a maxillary fragment and thus could not be included in this analysis, has a rather broad nasal aperture measuring c. 32 mm. Although this could be construed as a Neanderthal-like feature (cf., Frayer, 1992), it is also identical with the value of the early modern Qafzeh 6 (Figure 5).

Table 3. Principal Components Analysis based on log-shape data of nine facial variables

Variable	*PC1*	*PC2*
Upper facial height (48)	0.43648	0.84885
Basion-nasion length (5)	0.87654	0.19819
Basion-prosthion length (40)	0.79764	−0.31509
Bifrontal breadth (FMB)	0.90246	−0.28832
Bimaxillary breadth (ZMB)	0.69527	−0.49982
Bimaxillary subtense (SSS)	−0.83884	−0.06643
Nasio-frontal subtense (NAS)	−0.79401	0.04355
Nasal breadth (54)	0.41062	−0.38795
Nasal height (55)	0.62365	0.72629

Abbreviations after Howells (1973), numbers after Bräuer (1988)

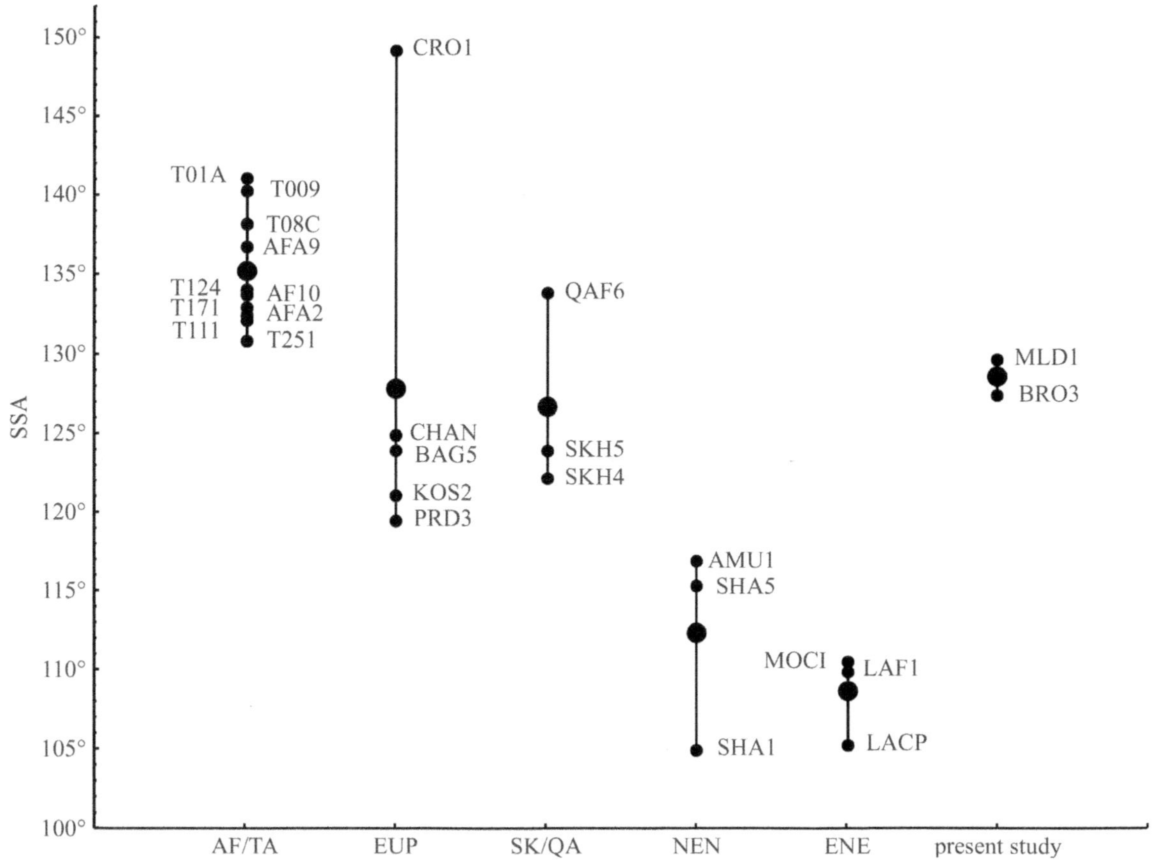

Figure 4. Zygomaxillary Angle (SSA). Comparison of early modern specimens from Czech Republic to different Neanderthal and anatomically modern samples (see Figure 3).

In a more complex approach, facial shape and projection were also analysed by PCA using log-shape data of nine variables (Table 3). These describe the dimensions of the mid-sagittal facial triangle and the nasal aperture, as well as upper facial and midfacial breadths and projections. Due to the lack of complete data sets, relatively few specimens could be included here (Figure 6). Nearly all variables have high loadings on PC1 (Table 3) and separate the three Neanderthals well from the modern specimens. Mladeč 1, the only early Czech cranium for which all variables were determinable, differs markedly from the Neanderthals, and Brno 3 also falls well within the modern group.

Conclusions

The analyses presented here suggest that the early moderns from the Czech Republic show no affinities to Neanderthals with regard to their frontal curvature. This statement holds true for the two robust (male) specimens Mladeč 5 and 6: Mladeč 5 exhibits the most strongly curved frontal measured, even showing close affinities to the probable early modern South African cranium Border Cave 1. With regard to facial morphology, no particular affinities to the Neanderthals could be found in facial projection (NFA and SSA) and the dimensions of the nasal aperture, although the maxillary fragment Mladeč 8 does exhibit

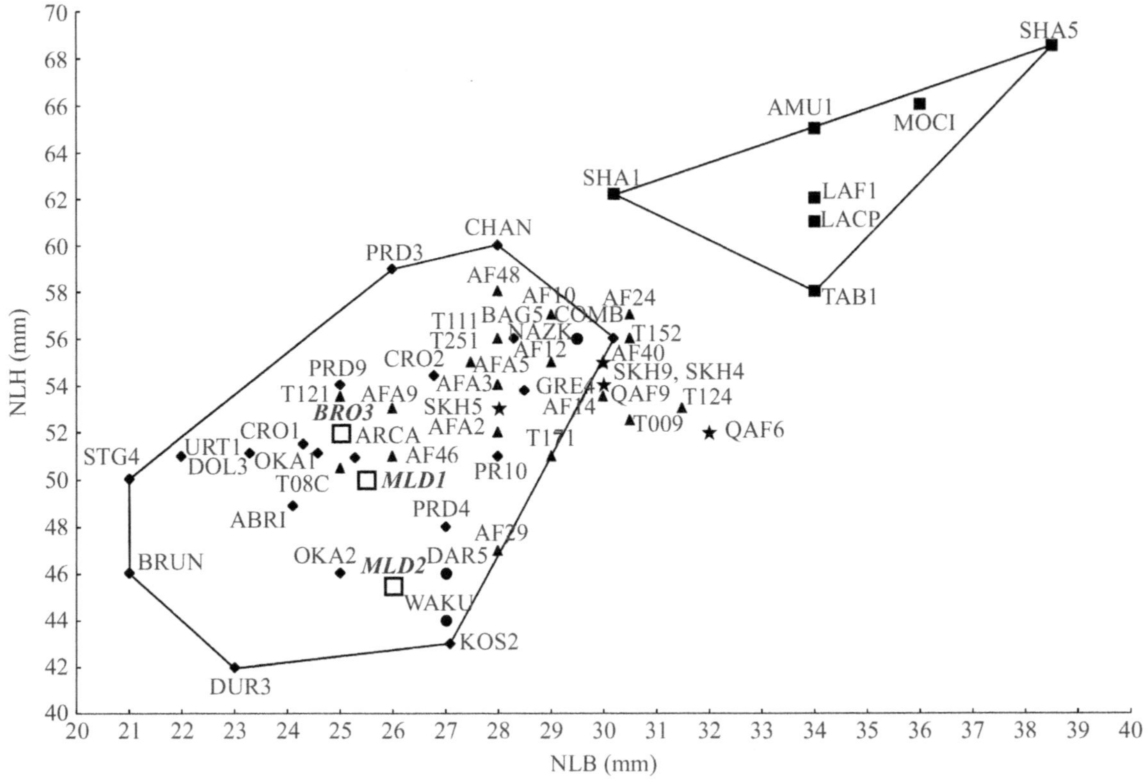

Figure 5. Bivariate scattergram for Nasal breadth and Nasal height. Comparison of early modern specimens from Czech Republic to Neanderthals and different anatomically modern samples (for abbr. see Table 1). Outlines indicate range of variation of the Neanderthal and European Upper Palaeolithic comparative samples.

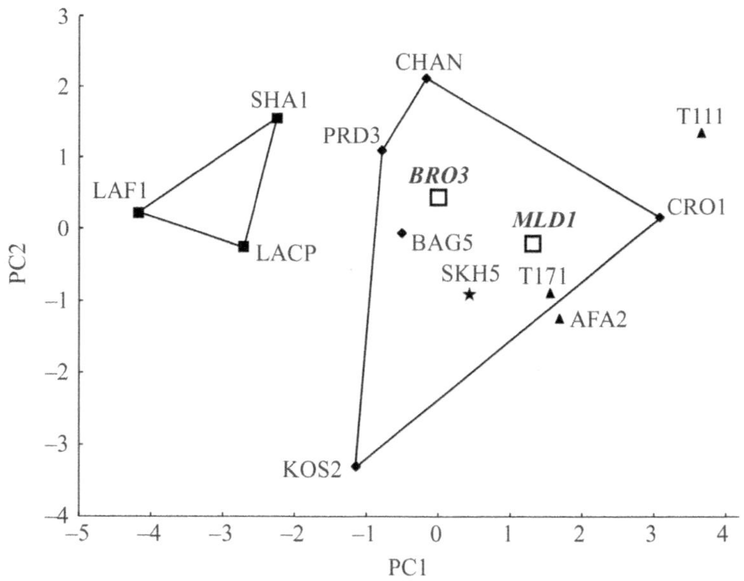

Figure 6. Principal Components Analysis based on nine variables (log-shape data) of facial shape and projection. PC1 accounts for 53.2% and PC2 for 20.9% of the total variance. Outlines indicate range of variation of the Neanderthal and European Upper Palaeolithic comparative samples.

a rather broad (Neanderthal-like or plesiomorphous?) aperture. Finally, the PCA of facial shape and projection reveals that there are no similarities between Mladeč 1 and the Neanderthals.

It can be concluded from the present study that major plesiomorphous and derived fronto-facial aspects which generally distinguish Neanderthals from early moderns do not indicate any clear affinities between the oldest modern crania from the Czech Republic and the Neanderthals. This result supports a recent re-analysis of this early Czech material that examined possible Neanderthal or Neanderthal-reminiscent non-metrical features (Bräuer and Broeg, 1998). Of 10 relevant cranial traits examined in that study, not a single character indicative of Neanderthal ancestry was found in any of these Upper Paleolithic specimens. Moreover, it emerged that most of the proposed regional continuity features of the skull (e.g., Frayer, 1992) are either highly problematic or untenable (Bräuer and Broeg, 1998: 127). This result is inconsistent with claims that the Neanderthals could have been the ancestors of these early modern Europeans (e.g., Frayer, 1992). It also contradicts the recent findings of Neanderthal-derived features in the Mladeč crania made using Pairwise Difference Analysis (Wolpoff et al., 2001). Recent reviews of this latter study demonstrated that the claimed Neanderthal affinities of the two Mladeč specimens analysed (Mladeč 5 and 6) are largely based on inadequate assessment of features, the use of traits of dubious phylogenetic utility, the selectivity employed in excluding the most complete Mladeč crania from analysis, and the inappropriate method of pairwise difference analysis used (Collard and Franchino, 2002; Bräuer et al., 2004).

For example, there does not seem to be a generally accepted definition of the suprainiac fossa even among multiregionalists (see Frayer, 1986; Caspari, 1991). Other features used in Wolpoff et al.'s (2001) analysis of the Mladeč specimens are problematic: metric traits were divided into two alternative conditions without clear justification, e.g., long frontal (glabella-bregma length > 113 mm) or thick parietal at asterion (> 9 mm); "mastoid-supramastoid crests well separated" or "fronto-nasal suture arched" cannot be properly assessed without a clear scoring system and are of dubious phylogenetic relevance. Thus it is not surprising that the use of problematic data led to confusing results, as, for example, the minimum number of differences between Skhul 4 and Mladeč 5 and the maximum number of differences between Skhul 5 and Mladeč 5. As outlined in more detail elsewhere (Bräuer et al., 2004; Bräuer, 2006) we do not believe that Wolpoff et al. (2001) provided convincing evidence for a significant Neanderthal contribution to the early modern Europeans. Our conclusions are further supported by a recent metrical study of craniofacial and cranial variation (Harvati, 2003). This study in which specimens from Mladeč were also included did not provide evidence for close similarities between Neanderthals and Upper Paleolithic Europeans nor for a Neanderthal contribution to the evolution of modern Europeans.

Based on the present study as well as on other recent analyses (e.g. Bräuer, 2006), we conclude that there is little or no clear evidence for gene flow or continuity between these early modern Central Europeans and the Neanderthals. We do not wish to deny the possibility that further studies, using other characters or other fossils, such as the early modern Oase material from Romania (Trinkaus et al., 2003, 2006), might detect indications of gene flow from Neanderthals. However, no *significant* gene flow is indicated from the study of the Mladeč material or from the current evidence of other early modern European remains.

Acknowledgements

We thank the late Jan Jelínek and Marta Dočkalová (Brno), Maria Teschler-Nicola and

Johann Szilvássy (Vienna), and the late Emanuel Vlček and Vitězslav Kuželka (Prague) for permission to study fossil specimens, and for their kind support. We would also like to thank David Frayer, Winfried Henke, Fred Smith and Fred Wendorf who generously provided data to GB that were also included in the present comparisons. Thanks go to Hermann Müller for his support with the analysis and figures. We appreciated the comments of the reviewers and editors. Finally, we thank Katerina Harvati and Terry Harrison for inviting one of us to the New York meeting, and all of us to contribute to this volume.

References

Bräuer, G., 1988. Osteometrie. In: Knussmann, R. (Ed.), *Anthropologie*. G. Fischer 4th Edition. Stuttgart, pp. 160–232.

Bräuer, G. 2001. The "Out-of Africa" Model and the question of regional continuity. In: Tobias, P., Raath, M., Moggi-Cecchi, J., Doyle, G. (Eds.), *Humanity from African Naissance to Coming Millennia*. Witwatersrand University Press, Johannesburg, pp. 183–197.

Bräuer, G., 2006. Searching for morphological evidence of Neandertal gene flow in early modern humans. In: Conard, N. J. (Ed.), *When Neanderthals and Modern Humans Met*. Kerns Verlag, Tübingen, pp. 87–103.

Bräuer, G., Broeg, H., 1998. On the degree of Neandertal-modern continuity in the earliest Upper Palaeolithic crania from the Czech Republic: Evidence from non-metrical features. In: Omoto, K., Tobias, P. (Eds.), *The Origins and Past of Modern Humans – Towards Reconciliation*. World Scientific, Singapore, pp. 106–125.

Bräuer, G., Rimbach, K., 1990. Late archaic and modern Homo sapiens from Europe, Africa and Southwest Asia: Craniometric comparisons and phylogenetic implications. *J. Hum. Evol.* 19, 789–807.

Bräuer, G., Stringer, C., 1997. Models, polarization, and perspectives on modern human origins. In: Clark, G., Willermet, C. (Eds.), *Conceptual Issues in Modern Human Origins Research*. Aldine de Gruyter, New York, pp. 191–201.

Bräuer, G., Collard, M., Stringer, C.B., 2004. On the reliability of recent tests of the Out of Africa Hypothesis for modern human origins. *Anat. Rec.* 279 A, 701–707.

Caspari, R.E., 1991. The evolution of the posterior cranial vault in the central European Upper Pleistocene. Ph.D. dissertation. University of Michigan, Ann Arbor.

Churchill, S., Smith, F., 2000. Makers of the early Aurignacian of Europe. *Yrbk. Phys. Anthropol.* 43, 61–115.

Collard, M., Franchino, N., 2002. Pairwise difference analysis in modern human origins research. *J. Hum. Evol.* 43, 323–352.

Darroch, J., Mosiman, J., 1985. Canonical and principal components of shape. *Biometrika* 72, 241–252.

Delson, E., Tattersall, I., van Couvering, J., Brooks, A., 2000. *Encyclopedia of Human Evolution and Prehistory*. 2nd Edition. Garland, NewYork.

Frayer, D., 1986. Cranial variation at Mladeč and the relationship between Mousterian and Upper Paleolithic hominids. *Anthropos* (Brno) 23, 243–256.

Frayer, D., 1992. Evolution at the European edge: Neanderthal and Upper Paleolithic relationships. *Préhistoire Européenne* 2, 9–69.

Frayer, D., Wolpoff, M., Thorne, A., Smith, F., Pope, G., 1993. Theories of modern human origins: the paleontological test. *Am. Anthropol.*95, 14–50.

Gibbons, A., 2001. Modern men trace ancestry to African migrants. *Science* 292, 1051–1052.

Harvati, K., 2003. The Neanderthal taxonomic position: models of intra-and inter-specific craniofacial variation. *J. Hum. Evol.* 44, 107–132.

Heim, J.-L., 1976. *Les hommes fossiles de La Ferrassie*. Archives l'Institute de Paléontologie Humaine, Mémoires 35.

Howells, W.W., 1973. Cranial Variation in Man. Peabody Museum Papers 67.

Howells, W.W., 1975. Neanderthal Man: facts and figures. In: Tuttle, R. (Ed.), *Paleoanthropology*. Mouton, The Hague, pp. 389–407.

Jelínek, J., 1978. Earliest *Homo sapiens sapiens* from Central Europe (Mladeč, Czechoslovakia). Paper at Xth International Congress of Anthropological and Ethnological Sciences, Delhi, India.

Pettitt, P., Trinkaus, E., 2000. Direct radiocarbon dating of the Brno 2 Gravettian human remains. *Anthropologie* (Brno) 38, 149–150.

Sergi S., 1974. *Il cranio Neandertaliano del Monte Circeo*. Accademia Nazionale dei Lincei, Rome.

Simmons, T., Falsetti, A., Smith, F., 1991. Frontal bone morphometrics of southwest Asian Pleistocene hominids. *J. Hum. Evol.* 20, 249–270.

Smith, F.H., Jankovic, I., Karavanic, I., 2005. The assimilation model and Neandertal-early modern human interactions in Europe. *Quat.Intern.* 137, 7–19.

Stringer, C., 1989. The origin of early modern humans: a comparison of the European and non-European evidence. In: Mellars, P., Stringer, C. (Eds.), *The Human Revolution: Behavioural and Biological Perspectives on the Origins of Modern Humans*. Edinburgh University Press, Edinburgh, pp. 232–244.

Stringer, C.B., 2002a. New perspectives on the Neanderthals. *Evol. Anthropol., Suppl.* 1, 58–59.

Stringer, C.B., 2002b. Modern human origins: progress and prospects. *Phil. Trans. R. Soc. Lond.* (B) 357, 563–579.

Stringer, C., Gamble C., 1993. *In Search of the Neanderthals*. Thames and Hudson, London.

Stringer, C., Trinkaus, E., 1981. The Shanidar Neandertal crania. In: Stringer, C. (Ed.), *Aspects of Human Evolution*. Taylor and Francis, London, pp. 129–165.

Suzuki, H., Takai, F., 1970. *The Amud Man and his Cave Site*. Academic Press of Japan, Tokyo.

Svoboda, J., Van der Plicht, J., Kuelka, V., 2002. Upper Palaeolithic and Mesolithic human fossils from Moravia and Bohemia (Czech Republic): some new ^{14}C dates. *Antiquity* 76, 957–962.

Tattersall, I., 2003. *The Monkey in the Mirror: Essays on the Science of What Makes Us Human*. Harcourt Brace, New York.

Tattersall, I., Schwartz, J.H., 1999. Hominids and hybrids: the place of Neandertals in human evolution. *Proc. Natl. Acad. Sci. U.S.A.* 96, 7117–7119.

Trinkaus, E., 1983. *The Shanidar Neandertals*. Academic Press, New York.

Trinkaus, E., 1987. The Neandertal face: evolutionary and functional perspectives on a recent hominid face. *J. Hum. Evol.* 16, 429–443.

Trinkaus, E., 2005. Early modern humans. *Ann. Rev. Anthropol.* 34, 207–230.

Trinkaus, E., Moldovan, O., Milota, S., Bilgar, A., Sarcina, L., Athreya, S., Bailey, S. E., Rodrigo, R., Mircea, G., Higham, T., Bronk Ramsey, C., Van der Plicht, J., 2003. An early modern human from the Peştera cu Oase, Romania. *Proc. Natl. Acad. Sci. U.S.A.* 100, 11231–11236.

Trinkaus, E., Zilhão, J., Rougier, H., Rodrigo, R., Milota, Ş., Gherase, M., Sarcina, L., Moldovan, O., Băltean, I., Codrea, V., Bailey, S.E., Franciscus, R.G. Ponce de León, M., Zollikofer, C.P.E., 2006. The Peştera cu Oase and early modern humans in southeastern Europe. In: Conard, N. J. (Ed.), *When Neanderthals and Modern Humans Met*. Kerns Verlag, Tübingen, pp. 145–164.

Vandermeersch, B. 1981. *Les Hommes de Qafzeh (Israel)*. CNRS, Paris.

Vlček, E., 1995. Evolution of human populations in the European Pleistocene. In: Ullrich, H. (Ed.), *Man and Environment in the Palaeolithic*. Etudes et Recherches Archéologique de l'Université de Liège 62, pp. 167–179.

Wild, E. M., Teschler-Nicola, M., Kutschera, W., Steier, P., Trinkaus, E., Wanek, W., 2005. Direct dating of Early Upper Palaeolithic human remains from Mladeč. *Nature* 435, 332–335.

Wolpoff, M., Hawks, J. Frayer, D., Hunley, K., 2001. Modern human ancestry at the peripheries: a test of the replacement theory. *Science* 291, 293–297.

Zilhão, J., Trinkaus, E., 2002. Portrait of the artist as a child: The Gravettian human skeleton from the Abrigo do Lagar Velho and its archaeological context. *Trabalhos de Arqueologia* 22, Lisboa.

16. Neanderthals and modern humans: an example of a mammalian syngameon?

T.W. HOLLIDAY
Department of Anthropology
Tulane University
New Orleans, Louisiana 70118, USA
thollid@tulane.edu

Keywords: hybridization, species concepts, modern human origins, divergence times

Abstract

Lotsy (1925) suggested that hybridizing plant species be grouped into larger interbreeding taxa that he named "syngameons." Such hybridizing taxa have long been well-documented among plants, but zoologists have traditionally downplayed the role of hybridization in animal evolution. Templeton (1989), however, has recently suggested that mammalian species which freely hybridize should also be grouped into syngameons. A literature survey suggests that the ability of any two mammalian species to hybridize successfully (i.e., produce viable, fertile hybrid offspring) is negatively correlated with time since phylogenetic divergence. In this regard, the genus *Homo* is a prime candidate for the presence of syngameons since the genus *Homo* (*sensu stricto* Wood and Collard, 1999) only emerged ca. 2.0 million years ago. The Late Pleistocene paleospecies *Homo neanderthalensis* is morphologically quite distinct from *H. sapiens*. The marked morphological (and genetic) distance between these two members of the genus *Homo* has led many human paleontologists to infer that these two taxa are separate species. From a current systematic perspective, such a position is justified, since in almost all species concepts species are defined by characters, of which the ability to interbreed is only one. In fact, the ability to interbreed is a plesiomorphic character, and as such we should not be surprised if two sister taxa, such as *H. neanderthalensis* and *H. sapiens*, retain this ability. There is, however, a relative dearth of paleontological evidence for such interbreeding – a somewhat surprising finding that warrants further exploration.

Introduction

The marked morphological differences between the Neanderthals, especially the European Neanderthals, and their modern or nearly-modern human contemporaries (and subsequent modern human populations) are so great as to convince many paleoanthropologists today that Neanderthals should be placed in a species separate from *Homo*

K. Harvati and T. Harrison (eds.), Neanderthals Revisited: New Approaches and Perspectives, 281–298.

sapiens (Tattersall, 1992; Howell, 1994; Franciscus, 1996; Rak, 1998; Wood and Richmond, 2000; Schillaci and Froehlich, 2001). Specifically, in addition to the long-recognized distinctive morphology of the prognathic Neanderthal midface (Rak 1986, Trinkaus, 1987; Hublin, 1998), and the plesiomorphically long, low cranium, with an apparently apomorphic suprainiac fossa and "*en bombe*" shape (Hublin, 1998), the Neanderthals also possess a suite of seemingly autapomorphic postcranial characters that distinguish them from their non-Neanderthal contemporaries (Trinkaus, 1995; Churchill, 1996; Franciscus and Schoenebaum, 2000). If Neanderthals are a separate species, then *Homo neanderthalensis* King, 1864 has taxonomic priority as the specific nomen for these hominins. As a taxon the Neanderthals appear to have evolved sometime in the Middle Pleistocene in that peninsular cul-de-sac of Eurasia known as Europe, and were largely, if not completely, isolated from non-Neanderthal hominins in Africa and Asia. Later (during OI Stage 4) these Neanderthals spread into parts of western Asia, but the degree of interaction (if any) between them and the nearly-modern humans who had been, or may have continued to be living there at the time, remains uncertain (Holliday, 2000; Shea, 2003).

The statement that *Homo neanderthalensis* is a separate species from *H. sapiens* means different things to different researchers, due to the fact that debate rages among biologists as to how species should be circumscribed. As a case in point, Mayden (1997) suggested that there are 22 species concepts in the current scientific literature, but he has more recently argued that this figure should be increased to at least 27 or 28 (Mayden, pers. comm.). While not all of these species concepts are applicable to fossil taxa, some of them nonetheless cast a long shadow over the debate concerning the place of Neanderthals in the origin(s) of modern humans. Primary among these concepts, and the best known of them all, is the Biological Species Concept (BSC) of Ernst Mayr (1942, 1963, 2000). While this concept cannot be applied to fossil taxa, for reasons to be discussed below, it almost always underlies the debate surrounding the fate of the Neanderthals.

Mayr defined species as "groups of interbreeding natural populations that are reproductively isolated from other such groups" (Mayr, 2000: 17). Reproductive isolation can be present in many forms, but at its core, it is said to be present whenever reproductive isolating mechanisms are in place. These reproductive isolating mechanisms can either be pre-mating (e.g., mate recognition) or post-mating (e.g., zygote inviability) mechanisms, but once present, the following is the case: if two taxa are reproductively isolated from each other, they either (1) no longer recognize members of the other taxon as potential mates, or (2) they are no longer capable of cross-taxic mating to produce viable (i.e., consistent with life) and/or fertile (i.e., capable of reproduction) offspring. A simple accounting of morphological differences between modern humans and Neanderthals is insufficient to answer the question of whether they would have been capable of cross-taxic mating to produce viable, fertile offspring. In spite of this fact, at times it has been argued that as separate species, Neanderthals and modern humans would have been reproductively isolated from each other (Stringer and Andrews, 1988; Shreeve, 1995; Tattersall, 1999). This implies that the species concept being used to distinguish *H. neanderthalensis* and *H. sapiens* is Mayr's Biological Species Concept – a concept that cannot be tested with fossil taxa.

One must keep in mind, however, that reproductive isolating mechanisms tend to evolve over long periods of time. Therefore, for some time after initial divergence, reintroduced taxa, which through isolation have become morphologically, ecologically, and behaviorally differentiated to such a great extent that they could be considered separate species, may in fact remain capable of some degree of genic exchange (i.e., hybridization). In this light, there are species concepts that do not require that species be reproductively isolated – and most of these species concepts have the added benefit of being applicable to fossil taxa. Primary among these, fossil taxa can be delimited via either the Evolutionary Species Concept (ESC) of Simpson (1961) or Wiley (1981), or with some version of the Phylogenetic Species Concept (PSC), originally put forward by Cracraft (1983). With regard to evolutionary species, Simpson was a paleontologist, and so the ESC is particularly applicable to paleoanthropological questions. According to Simpson (1961), an evolutionary species is "a lineage (an ancestral-descendant sequence of populations) evolving separately from others and with its own unitary evolutionary role and tendencies" (Simpson, 1961:153). Yet Simpson was careful not to insist that such independent entities be reproductively isolated from each other: "two species may interbreed to some extent without losing their distinction in evolutionary roles" (Simpson, 1961:153).

The different versions of the phylogenetic species concept (de Queiroz [1998] identifies three) have all arisen from the original PSC (Cracraft, 1983). Cracraft defined a phylogenetic species as "an irreducible (basal) cluster of organisms, diagnosably distinct from other such clusters, and within which there is a parental pattern of ancestry and descent" (Cracraft, 1989: 34–35). All versions of the PSC are intimately tied to cladistic (phylogenetic taxonomic) methodology, but only one (that of Mishler and Theriot, 2000) explicitly allows for genic exchange between members of different species. These workers maintain that the ability to interbreed is only a single character, just like any other character to be included in a phylogenetic analysis. But more importantly, the ability to interbreed, when shared by sister taxa, is merely a symplesiomorphy, and for this reason is not phylogenetically informative. All sister taxa share at least some plesiomorphic characters; it is only the apomorphies (i.e., synapomorphies) they share that are useful as valid grouping criteria.

A final species concept that is relevant to the issue of potential genic exchange between *Homo sapiens* and *H. neanderthalensis* is the Cohesion Species Concept (CSC) of Alan Templeton (1989). Templeton defined cohesive species as "the most inclusive population of individuals having the potential for phenotypic cohesion through intrinsic cohesion mechanisms" (Templeton, 1989:12; and see discussion in Holliday, 2003). His model differs from the BSC in that he finds fault with the BSC's emphasis on reproductive isolating mechanisms. He is especially critical of the role of pre-mating behavioral isolating mechanisms, which are irrelevant under a model of allopatric speciation, since they (usually) emerge as a secondary consequence of geographic isolation. In contrast, his CSC, rather than focusing solely on isolation, focuses on mechanisms (such as gene flow, genetic drift and selection) that maintain both genetic and phenotypic cohesion in those populations that are recognized as species. He also models cohesive species as lineages, and in this regard, his concept is most similar to the ESC. In fact, it is so similar to the ESC that some have cogently argued that it *is* the ESC (Wiley and Mayden, 2000; Wolpoff, 2003). Yet, with regard to the issue at hand, the most important contribution to the debate regarding potential genic exchange between Neanderthals and modern humans is Templeton's revisiting of the notion of hybridizing taxa called *syngameons*. The term "syngameon" was first used by Lotsy (1925) to describe groups of plant species that are capable of interspecific hybridization. Zoologists have traditionally argued that hybridization is less important in animal evolution than it is among plants (Mayr, 1963; White, 1973), and in fact, even among the vertebrates there is evidence that interspecific mating leading to the production of viable offspring is rarer among mammals than among reptiles or birds (Zeh and Zeh, 2000). Zeh and Zeh (2000) note that among birds, hybrids have been produced between species thought to have been diverged for up to 58 million years. In contrast, they documented no living hybrid offspring between mammalian species that had been diverged more than 8 million years. These researchers argue that this pattern is due to viviparity, since by carrying a fetus in the mother's body, rather than laying an egg, immunological incompatibility is much more likely to arise and cause spontaneous abortion of the hybrid fetus. Despite this temporal limitation (which likely includes the entire evolutionary history of the Hominini), Templeton (1989) argued that there may be many mammalian syngameons that have yet to be recognized – a prediction that Holliday (2003) investigated from a theoretical perspective with regard to the human fossil record.

Examples of mammalian syngameons discussed by Templeton include the three canid taxa *Canis lupus* (wolves), *C. latrans* (coyotes), and *C. familiaris* (dogs). Coyotes and wolves in particular are sympatric in some regions of North America, and within these regions they hybridize to produce viable and fertile offspring (Hall, 1978; Hilton, 1978; Lehman et al., 1991; Wayne et al., 1992; Roy et al., 1994; Brownlow, 1996). Nonetheless, the two taxa are easily distinguished morphologically. Genetic differences (at least in mitochondrial DNA) are also readily apparent (Templeton, 1989; Lehman et al., 1991; Wayne and Jenks, 1991; Wayne et al., 1992). Coyotes and wolves are also quite different behaviorally; wolves are pack animals with complex social networks, while coyotes tend to be more solitary. A large portion of the diet of wolves is larger prey (wolves can take moose, bison or musk oxen), while coyotes subsist primarily on smaller animals such as rodents and lagomorphs (Hall, 1978). As such, wolves and coyotes are only sympatric in relatively limited portions of their range, as they tend to live in different habitats (although anthropogenic factors almost certainly have played a role in reducing the range of the wolf – Wayne et al., 1992). These behavioral/ecological differences, combined with differences in body size between the species, limit the frequency and types of cross-specific matings that are likely to occur (Lehman et al., 1991).

According to Nowak (1978), fossil evidence from western North America indicates that the coyote and wolf lineages had split by the terminal Pliocene (ca. 2 Ma). Molecular data suggest divergence times ranging from 1.5 Ma (Wayne et al., 1991) to 2.5 Ma (Bininda-Emonds et al., 1999). This would support the position that despite millennia of interbreeding, coyotes and wolves represent cohesive evolutionary lineages/species that have maintained their own identities across space and through time (*per* the ESC or CSC). However, strict adherence to the BSC would of course relegate coyotes, wolves (and dogs) to a single species (Wolpoff, 1999:42).

A second, more recently discovered mammalian syngameon is found in the cervid genus *Odocoileus*. Mule deer (*Odocoileus hemionus*) and white-tailed deer (*O. virginianus*) are among the most widely-distributed mammals in North America. The two species (*O. hemionus* and *O. virginianus*) are ecologically differentiated, with *O. hemionus* preferring more open habitats than *O. virginianus*, and, as such, the two species are sympatric in only a small portion of their range. They are, however, distinct in morphology, behavior, and allozymes (Cathey et al., 1998). Yet studies of allozymes and mtDNA of wild mule deer and white-tailed deer in regions of west Texas, where the two species are sympatric, suggest that hybridization has been occurring for multiple generations (Ballinger et al., 1992). In particular, some individuals phenotypically identified as mule deer were molecularly aligned with white-tailed deer

(Ballinger et al., 1992). It also appears that there are relatively few F_1 individuals in the west Texas hybrid zone, but many backcrossed individuals (Ballinger et al., 1992; Cathey et al., 1998).

A third potential mammalian syngameon has only very recently come to light – the documented hybridization between *Felis* (*Lynx*) *rufus* (bobcat) and *F.* (*L.*) *canadensis* (Canadian lynx). These two species are morphologically distinct, and inhabit different ecozones. Specifically, the lynx inhabits northern boreal forests, which tend to have deep snow for much of the year. For this reason, lynx have larger paws for successfully negotiating snow, while the smaller bobcat, which has a somewhat wider North American distribution, is generally not found in areas with heavy snow cover (Schwartz et al., 2004). However, the northernmost extent of the bobcat's range does overlap with that of the lynx in northern parts of Minnesota, Michigan, Wisconsin, and Maine, and in portions of British Columbia, Manitoba, Ontario and Québec (Schwartz et al., 2004). Schwartz et al. (2004) found that in a sample of 20 lynx from northern Minnesota, three were F_1 hybrids with lynx mtDNA (indicating a lynx mother and bobcat father). Unfortunately, their tests thus far are only robust in detecting F_1 hybrids, so it is uncertain as to whether the hybrids are fertile. However, such an unexpectedly large number of F_1 hybrids is consistent with interfertility of the two species, although the time since divergence for these two species as indicated by molecular data (3.1 Ma; Bininda-Emonds et al., 1999), is consistent with either interfertility or interspecific sterility (see below).

A final example of mammalian syngameons is found among savannah baboons (genus *Papio*) and gelada baboons (genus *Theropithecus*). These examples are discussed at length in Jolly (2001) and Holliday (2003), and therefore will not be discussed in detail here. As Jolly (2001) notes, *Papio* and *Theropithecus* are distinct in the fossil record for 4–5 million years, and yet remain interfertile (there is some evidence that Haldane's Rule is in effect; Jolly et al., 1997, and see below). Additionally, despite marked morphological and ecological differences, the different species (or subspecies) within the genus *Papio* itself remain completely interfertile; these taxa are thought to have diverged ca. 1.7 Ma (Jolly, 2001).

All of the above-cited examples of mammalian syngameons suggest that interfertility among mammalian species may be more common than previously thought. However, the evidence given thus far is somewhat anecdotal; a more systematic and quantitative approach to mammalian hybridization is warranted in order to assess the possibility of hybridization between Neanderthals and modern humans. The current work will delve into this latter issue by using data gathered on a large sample of extant mammalian hybrids as a starting point. The specific questions to be addressed are: (1) how long does it take for complete reproductive isolation to evolve between mammalian sister taxa, (2) is there a correlation between time since divergence and the degree of fertility between mammalian species, and (3) in light of these data on extant mammalian hybrids, would successful hybridization (i.e., the production of viable, fertile hybrid offspring) between *Homo sapiens* and *H. neanderthalensis* have been likely or unlikely?

Materials and Methods

Data on mammalian hybrids were gathered from the literature. Most examples were culled from *Mammalian Hybrids. A Checklist with Bibliography* by Gray (1972); two examples of mammalian hybridization noted since this book's publication were added. While the author did not always agree with Gray's taxonomic assignments, in order to be conservative it was decided to follow her taxonomy, and only collect data on those examples of hybridization that she considered to be interspecific. Particular attention was paid to two major issues: (1) whether viable hybrid offspring were produced, or in contrast, whether hybrid offspring tended to be spontaneously aborted or died in infancy, and (2) whether both male and female hybrids were fertile, or if only female hybrids were fertile. This second issue relates to Haldane's (1922) Rule, who noted that if only one sex of a hybrid cross is sterile, it is always the heterogametic sex (which in mammals are males). The observation of Haldane's Rule in an interspecific cross indicates the early emergence of post-zygotic reproductive isolating mechanisms.

Divergence times (both fossil and molecular) for mammalian taxa were culled from multiple sources in the literature. In cases where multiple molecular divergence dates were obtained, a mean was computed and used for the analyses. In order to test for a relationship between these time-since-divergence data and interfertility, the hybridization data were codified into a 5-point scheme to reflect different levels of interfertility. A hybridizing pair was scored as "0" if it was well documented that most, if not all, hybrid offspring were inviable. Pairs in which viable, but infertile offspring were documented were scored as "1". Those hybridizing pairs for which hybrid offspring were viable, but likely, albeit not certain, infertile were scored as "2". Cross-specific matings out of which male offspring tended to be sterile, while female offspring tended to be fertile (i.e., Haldane's Rule is in effect) were scored as "3". For those crosses in which the viable offspring of both sexes were very likely to be fertile (albeit without certainty), the interspecific pair was scored as "4". Finally, those cross-specific matings for which viable, fertile offspring of both sexes were well-documented were scored as "5". Due to the non-parametric nature of the fertility data, divergence times (both fossil and molecular) were correlated with fertility scores using the rank-order Spearman's Rho statistic, which is a non-parametric equivalent of Pearson's r. Before analysis began, however, it was decided not to include those species pairs scored as "0," since it was thought that due to the great time since divergence for most of these species pairs, it would bias the analysis to find a significant correlation between the level of interfertility and time since divergence. Finally, in order to understand the range of variability in the relationship between interfertility and time since divergence, the minimum amount of time needed to establish complete post-zygotic reproductive isolation in the mammalian taxa sampled was noted, as was the maximum divergence time for which interfertility was still documented.

Results

Three hundred and twenty-eight cases of mammalian hybridization (from Gray, 1972, and supplemented with other sources) were found in which viable hybrid offspring were produced – no small number. Unfortunately, for nearly half of these crosses (156, or 47.6%), there were not enough data to ascertain whether the hybrids are fertile or sterile. This leaves 172 (52.3%) cases of interspecific crosses in which interfertility (or lack thereof) can be assessed. Of these 172 crosses, 110, or 64%, produced offspring in which fertile offspring are likely to be produced (although in at least 27 of these cases it appears that Haldane's Rule is in effect). In 62, or 36%, of the cross-specific matings, hybrid offspring were sterile. Thus, even if we take the extremely conservative approach and assume that all of the interspecific crosses for which sterility/fertility is unknown are in fact sterile, ca. one-third (and perhaps a much higher percentage) of all mammalian interspecific crosses result in fertile offspring.

The question remains, however, as to whether the degree of fertility is correlated

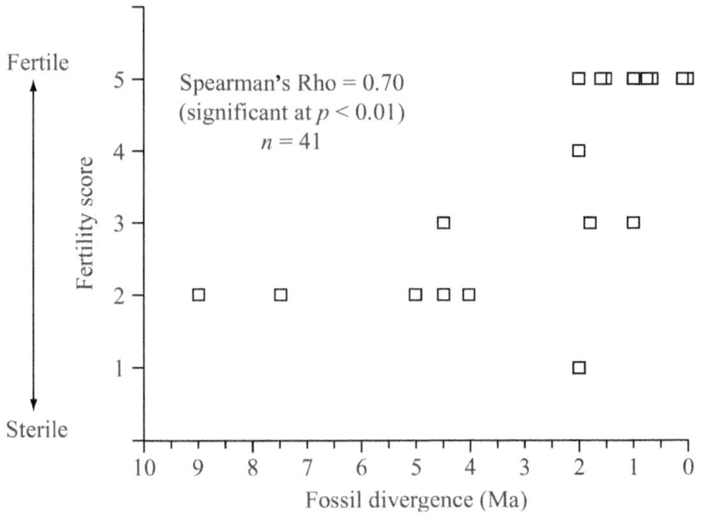

Figure 1. Scatter plot of fertility scores of mammalian interspecific hybrids on time since the hybridizing species diverged as estimated by the fossil record. Spearman's Rho for the relationship is indicated on the plot.

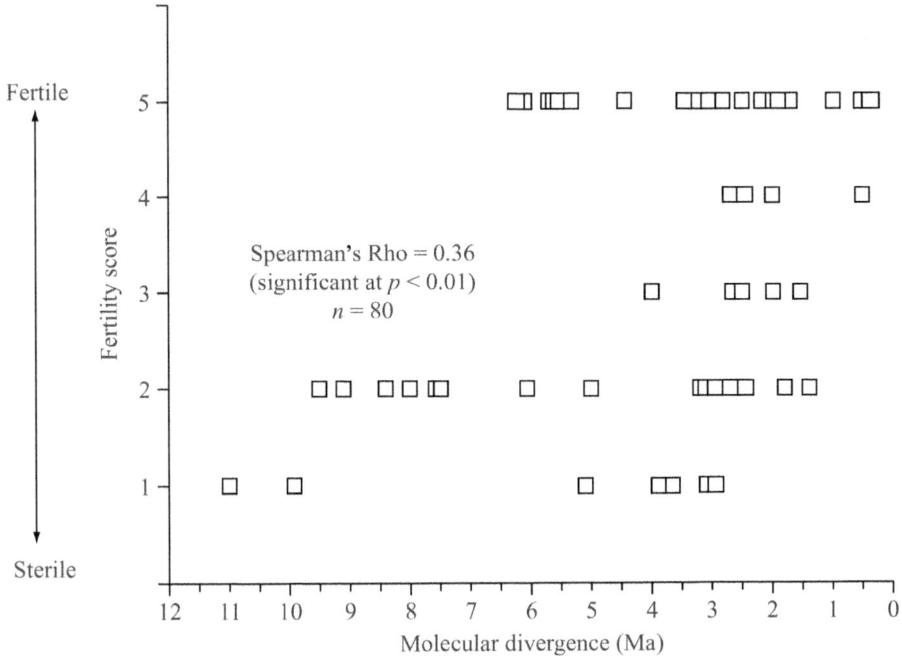

Figure 2. Scatter plot of fertility scores of mammalian interspecific hybrids on time since the hybridizing species diverged as estimated by molecular data. Spearman's Rho for the relationship is indicated on the plot.

with time since divergence. This question was evaluated via the use of Spearman's non-parametric correlation analysis of the fertility scores. The results of the Spearman's Rho tests are shown in Figures 1 and 2. Figure 1 reports the results of the relationship between divergence times as estimated by the fossil record with the fertility scores. For the 41 interspecific crosses for which paleontological divergence times are available, Spearman's

Rho with the interfertility score is 0.7, which is significant at the 0.01 level. Thus, the deeper in time two species diverged, the less likely they are to be interfertile. An examination of the graph reveals an inflection point in the relationship such that species separated more than two million years according to the fossil record are much less likely to be interfertile, whereas those species separated less than two million years (as indicated by the fossil record) tend to remain interfertile.

As evident in Figure 2, the relationship between molecular divergence times and the interfertility scores remains significant at the 0.01 level (Spearman's Rho = 0.39), such that species are less likely to be interfertile the farther back in time they diverged. There is, however, much more temporal overlap among the 80 groups for which molecular divergence times were available than was evident for the divergence times drawn from fossil data. In fact, with regard to molecular divergence, if one finds two species taxa for which molecular data suggest a divergence somewhere between 2 and 5 million years ago, then the two species could exhibit any pattern from complete interfertility to complete postzygotic reproductive isolation.

While the general trend toward decreasing fertility with increasing time since divergence is interesting (even expected), the extremes in either direction, i.e., the earliest onset of postzygotic reproductive isolation on the one hand, and the longest retention of interfertility despite a long time since divergence, on the other, are also of interest. With regard to fossil divergence, the earliest manifestation of interspecific sterility observed is that of horses vs. zebras and donkeys (genus *Equus*), which according to the fossil record are hypothesized to have split in a rapid adaptive radiation ca. 2 Ma (Vilà et al., 2001). In terms of morphology, horses are the more apomorphic of these equids, and appear to have been the first to diverge from the zebra/donkey common ancestor (MacFadden, 1992; Xu et al., 1996; Oakenfull et al., 2000; Vilà et al., 2001). This would imply that the observed sterility of donkey/zebra hybrids emerged somewhat later than 2 million years ago. However, neither the timing nor the order of the evolutionary events producing the extant members of the genus *Equus* is agreed upon by all researchers, and the phylogenetic structure of *Equus* is perhaps best considered a polytomy (Xu et al., 1996; Oakenfull and Clegg, 1998).

The earliest manifestation of Haldane's Rule, according to divergence times drawn from the fossil record, is that between member species of the genus *Bos* versus member species of the genus *Bison*, which according to the fossil record diverged ca. 1 Ma (Ritz et al., 2000). The male offspring of this transgeneric cross tend to be sterile, while females are usually fertile (Gray, 1972). By way of contrast, *Papio* and *Theropithecus* are identified as separate lineages in the fossil record for 4 to 5 million years, yet remain interfertile today, although Haldane's Rule may be in effect (Jolly et al., 1997).

With regard to the molecular divergence data, the earliest manifestation of "complete" reproductive isolation is that of *Gazella rufifrons* (red-fronted gazelles) versus *Gazella thomsoni* (Thomson's gazelles), which produce sterile hybrids and diverged (according to molecular estimates) ca. 1.4 Ma (Brashares et al., 2000). The earliest manifestation of Haldane's Rule, according to molecular data, appears to be the *Panthera leo* (lion)/ *P. tigris* (tiger) cross, two species which molecular data suggest diverged ca. 1.55 Ma (Wayne et al., 1991; Purvis and Bromham, 1997). At the opposite extreme, red deer (*Cervus elaphus*) and axis deer (*Axis axis*) remain completely interfertile, despite the fact that molecular data suggest they last shared a common ancestor ca. 6.23 Ma (Meijaard and Groves, 2004; Pitra et al., 2004).

It should be noted that one cannot rule out the possibility that reproductive isolation has evolved more rapidly between mammalian taxa that are not sampled here. There are approximately 5,000 extant mammalian species; only 328 interspecific crosses that resulted in viable offspring were considered in the current study. The current sample is, however, sufficient and appropriate for two reasons. First, it is extremely difficult to test for reproductive isolation between the thousands of mammalian sister taxa, since failure to document interbreeding does not mean it does not occur. Second, the current sample is biased toward greater representation of large-bodied mammals (arguably the more appropriate group to which humans should be compared) over smaller mammals. Most mammals are small in body size (of the 5,000 mammalian species there are over 2,000 species of rodents, over 900 species of bats, and ca. 400 species of insectivores), and it is likely that reproductive isolation evolves more rapidly among them than among larger-bodied mammals, given their tendency toward smaller home ranges and shorter generation lengths.

A final complicating issue involves the fact that unless two hybridizing species were completely geographically separated by a geological event whose date is known, it is difficult, if not impossible, to estimate the amount of gene flow that has occurred between mammalian species since their initial divergence. Low levels of gene flow between the species would in all likelihood increase the amount of time required for complete reproductive isolation to evolve. Yet in order to assess the level of gene flow between species that have not been completely geographically separated, at a minimum one would have to have a firm understanding of the species' ranges throughout the Holocene and Pleistocene (if not Pliocene) – a difficult chore indeed!

Discussion

The degree of interfertility between mammalian species decreases with time, although the rate at which postzygotic isolating mechanisms evolve is quite variable. Nonetheless, by looking at the shortest time at which hybrid sterility appears in the current mammalian sample, one can make an assessment as to whether *Homo sapiens* and *H. neanderthalensis* would have likely been reproductively isolated. Estimates for the divergence between *H. sapiens* and *H. neanderthalensis* range from ca. 700 ka (Bermúdez de Castro et al., 1997) to ca. 450 ka (Hublin, 1998) to ca. 250 ka (Lahr and Foley, 1998). These divergence times are compared to those of the most rapidly evolved postzygotic isolating mechanisms in the extant mammalian sample in Table 1. Note that even at its greatest time depth, the Neanderthal/modern human split occurred only half as long ago as the most rapid onset of postzygotic isolating mechanisms (*Gazella rufifrons* vs. *G. thomsoni*, 1.4 Ma). If one

Table 1. Estimated divergence times for Homo neanderthalensis *and* H. sapiens *compared with divergence times for those mammalian hybrid crosses for which the emergence of reproductive isolation has been the most rapid*

Estimated divergence time of *H. neanderthalensis* and *H. sapiens*:	
Lahr and Foley (1998)	250 ka
Hublin (1998)	ca. 450 ka
Bermúdez de Castro et al. (1997)	ca. 700 ka
Estimated divergence times of reproductively isolated mammals:	
Gazella thomsoni × *G. rufifrons*	1.4 Ma (Generation length ≈ 4 yrs)
Horses × Zebras/Donkeys (Genus *Equus*)	2.0 Ma (Generation length ≈ 4 yrs)

prefers fossil over molecular data, the split among the extant species of the genus *Equus* is dated to ca. 2.0 Ma, almost three times as long as the Neanderthal/modern human split. These differences in time depth of reproductively isolated species are all the more impressive when one considers that for both the *Equus* and *Gazella* species, generation length is ca. 4–5 years. Therefore, in terms of number of generations, the Neanderthals are not even close to matching the time required to evolve complete reproductive isolation. For this reason, it is safest to assume that if *H. sapiens* and *H. neanderthalensis* met and mated, their offspring would be both viable and fertile.

In this light, there are scholars who argue for morphological evidence of genetic continuity between the Neanderthals and their early modern successors in Europe. For example, Smith (1984, 1985, 1992; Churchill and Smith, 2000) has long argued that the earliest modern humans in central Europe evince characters found in the Neanderthals from the same region. Likewise, Frayer (Frayer, 1992; Frayer et al., 1993) has argued for the persistence of Neanderthal features in early modern humans across Europe, and more recently, Trinkaus and colleagues have posited that the 24.5 ka Lagar Velho juvenile skeleton from Portugal evinces features reminiscent of Neanderthals (Duarte et al., 1999; Trinkaus and Zilhão, 2002). Yet despite these claims, to many scholars (e.g., Bräuer, 1992; Holliday, 1997; Stringer, 2002) there remains a large morphological gap between the Neanderthals and their early modern successors in Europe, while still others see no evidence for interbreeding at all (e.g., Tattersall, 1999; Harvati, 2003; Klein, 2003; Harvati et al., 2004). The fact that these latter two viewpoints are so prevalent is not surprising – indeed, if anatomical manifestations of interbreeding between Neanderthals and "Cro-Magnons" were abundant, then the Neanderthal question would not still be the longest-running controversy in human evolution. Why, then, is there relatively little fossil evidence of Neanderthal-modern human hybridization? There are several potential answers to this question.

The first possibility is that there was significant admixture between the paleospecies, but we do not detect it. Given the mammalian examples cited above, this is perhaps unsurprising. As discussed earlier, in most hybrid zones there are few F_1 hybrid individuals, but many backcrossed individuals, who usually do not appear intermediate or evince obvious phenotypic signs of hybridization. In fact, for many cases (e.g., *Odocoileus* spp., *Felis* [*Lynx*] spp., *Canis* spp.), hybridization is only discovered when animals phenotypically identified as one species evince the mtDNA of another. Another reason we may not detect evidence of hybridization is the fact that the number of traits which we sample in the fossil record is quite small. We tend to forget that only a fraction of the phenotype is preserved in fossils; we have little knowledge of soft tissue differences between specimens, for example. An analogy for this situation is found in our sister discipline of paleopathology. The fact that a skeleton recovered in archaeological context appears to have been "healthy" is countered by the obvious fact that the owner of the skeleton is dead, and only those diseases that affect bone (which represent a fraction of all diseases) are bioarchaeologically detectable. Could there, not therefore have been some traits passed from *H. neanderthalensis* to *H. sapiens* that were paleontologically invisible? In this light, Barton (2001) has noted that adaptive traits can introgress out of hybrid zones, back into the parental populations, *even if there is selection against hybrid individuals*. Could the depigmented skin of Europeans be such a trait (Jolly, 2001)? It is likely that the ancestors of Neanderthals evolved depigmented skin in the cold climes of Pleistocene Europe in order to produce enough Vitamin D_3. This essential vitamin is produced in the skin via the action of ultraviolet (UV) radiation, and importantly, less Vitamin D_3 is produced in darkly pigmented individuals, since melanin absorbs much of UV light (Jablonski, 2004). Neanderthals would have had little of their skin exposed to the sun, since they would have been covered in some type of clothing. Therefore, what little skin they did have exposed to the sun would likely need to have reduced melanin content in order to facilitate production of sufficient levels of Vitamin D_3. In one sense, then, introgression is a more parsimonious explanation for the origin of the relatively depigmented skin of Europeans than is its independent evolution in immigrant populations of *Homo sapiens*, since the trait need only evolve once.

What, then, of the mtDNA evidence extracted from fossils? Recovered Neanderthal mtDNA includes sequences that are not found among modern humans (Krings et al., 1997, 2000; Ovchinnikov et al., 2000; Serre et al., 2004). Yet the fact that we do not find these "Neanderthal" sequences among people today is not surprising, given the many millennia that have passed, coupled with the high probability of lineage extinction in a (usually) maternally-inherited genome. The ancient mtDNA sequence recovered from the Australian anatomically modern Lake Mungo 3 specimen, for example, has no modern analogs in the current human mitochondrial genome (Adcock et al., 2001). What has been recently argued, however, is that because we do not find "Neanderthal" sequences among the so-called "Cro-Magnons," that there was no Neanderthal mtDNA contribution to subsequent modern European populations (Caramelli et al., 2003; Serre et al., 2004). This extreme interpretation of the data is unwarranted, given that mtDNA has thus far been extracted from only seven early modern Europeans – and perhaps ironically, a Neanderthal mtDNA sequence may have been already drawn from an early modern human. Specifically, mtDNA extracted from the Mezmaiskaya juvenile skeleton from southern Russia has a Neanderthal sequence (Ovchinnikov et al., 2000). However, as Hawks and Wolpoff (2001) note, while the skeleton itself is AMS dated to 29 ka, it was found in much older (AMS dated to > 45 ka) Mousterian deposits. Even the initial Upper Paleolithic layer, which overlies the Mousterian at the site, is apparently older than the Mezmaiskaya skeleton; it is AMS dated to 32 ka (Hawks and Wolpoff, 2001). Hawks and Wolpoff (2001) note that only a few features of the skeleton align it with Neanderthals, and all of these features are observed in some anatomically modern humans. Thus, given its radiometric date, they suspect Mezmaiskaya is an intrusive modern human burial into Mousterian deposits (but see Skinner et al., 2005). If this is the case, Mezmaiskaya would provide the first mtDNA evidence for a Neanderthal genetic contribution to subsequent human populations.

A second explanation for why we have such little evidence for *H. sapiens/ H. neanderthalensis* hybridization is that prezygotic behavioral isolating mechanisms were in place that prevented or made it less likely that Neanderthals and modern humans would recognize each other as mates. After all, many of the hybridization examples cited in Gray (1972) and used in the current analysis took place in zoos; the animals involved likely would have never mated in the wild. Some researchers have recently argued that *H. sapiens* was the only "symbolic" species of the two (Henshilwood and Marean, 2003; Klein, 2003; Mellars, 2005). This factor implies that modern humans were more intelligent than Neanderthals, and that this difference in intelligence could have served as a prezygotic isolating mechanism. Ultimately, however, this explanation rings hollow. Based on endocasts, there are no obvious differences in brain morphology manifest between the two species (Holloway, 1985). In absolute brain size, Neanderthals have larger brains on average than modern humans, although relative to body size *H. neanderthalensis* appears to have been somewhat less encephalized than our own species (Kappelman, 1996; Ruff et al., 1997). Closer examination of these data reveal, however, that while Neanderthals do consistently fall below the *H. sapiens* brain size/body size regression line, there are some modern human individuals who deviate more negatively from the regression line than do the Neanderthals (Figure 3). For this reason, it would be unwise to assume a significant difference in mental capacity based on Neanderthals' relative brain size.

In addition to a lack of morphological evidence for a cognitive difference between the two species, the archaeological record lends support to the view of Neanderthals as symbolic animals. Specifically, the fact that the late Neanderthal makers of the Châtelperronian and similar industries made beads out of pierced shell and animal teeth (d'Errico et al., 1998) or that there is now evidence for the even earlier use of manganese crayons in the French Mousterian (d'Errico and Soressi, 2002) suggests that Neanderthals were as capable of symbolic thought as their *H. sapiens* cousins. It does, however, appear that the frequency of this type of symbolic behavior among the Neanderthals was lower than that displayed by anatomically modern Europeans (Mellars, 2005). Nonetheless, as Henshilwood and Marean (2003) have recently argued, the frequency with which many "modern" behaviors are likely to be manifest in the archaeological record is heavily influenced by factors such as population density and local environmental variability. If Neanderthals were living at significantly lower population densities than their modern human cousins, then perhaps we should not be surprised that they exhibit lower levels of symbolic, "modern" behaviors such as art. It should also be noted that while sociocultural differences among recent human groups tend to reduce the amount of gene flow between them, some

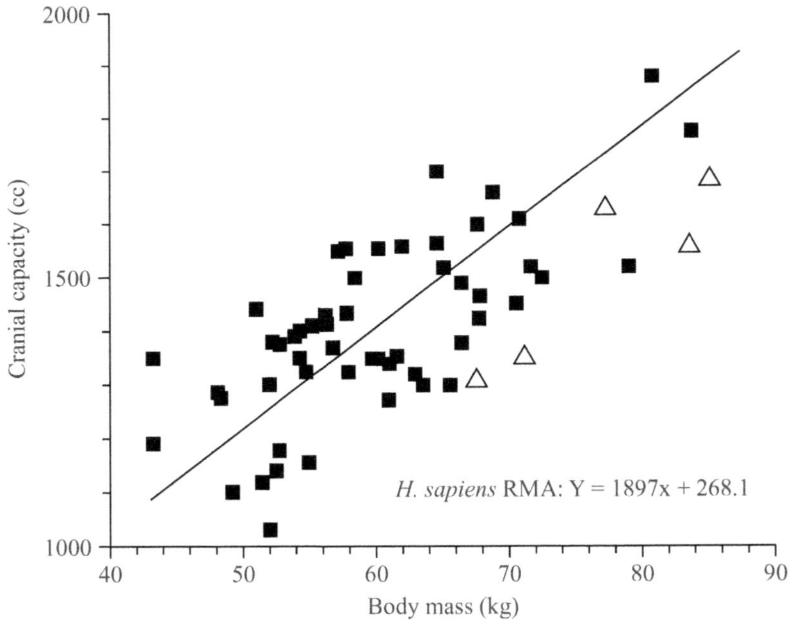

Figure 3. Scatter plot of endocranial capacity on estimated body mass, *Homo neanderthalensis* (triangles) and *H. sapiens* (squares). The reduced major axis regression line for *H. sapiens* is indicated.

level of gene flow (however clandestine) between human groups in contact *always* occurs. In this light, the likelihood that two symbol-bearing hominins would not at least sometimes recognize each other as potential mates stretches the bounds of credulity.

The third, and to my mind, most likely, possibility as to why there is so little paleontological evidence for genetic admixture between *H. sapiens* and *H. neanderthalensis* concerns the relative sizes of the Neanderthal versus modern human populations. In a series of papers, Relethford (1995, 1998; Relethford and Harpending, 1995) notes that during the Pleistocene, more humans lived in Africa than in the other Old World continents, due to the fact that vast areas of Eurasia were covered with glaciers, or were inhospitable frozen steppes. He has shown that even under a model of equal exchange of immigrants/emigrants across continents, humans outside of Africa are ultimately expected to appear more genetically (and presumably therefore, morphologically) African, simply as a consequence of the larger African population size. Such a disparity in relative population size between the Neanderthals and the expanding modern humans could therefore erase evidence of a Neanderthal contribution to subsequent populations.

Conclusions

An examination of interspecific hybridization among mammals reveals numerous cases of species that remain interfertile despite having diverged millions of years ago. In the mammalian species sampled, the most rapid onset of complete reproductive isolation is that of *Gazella thomsoni* and *G. rufifrons*, which molecular data indicate phylogenetically diverged ca. 1.4 million years ago (Brashares et al., 2000). Thus, while *Homo neanderthalensis* and *H. sapiens* are best considered separate evolutionary species, the fact that they only diverged ca. 700,000 to 250,000 years ago makes it extremely likely that members of the two species were capable of interbreeding to produce viable, fertile, offspring. Closely related, interbreeding species can be grouped into larger taxa known as syngameons; Neanderthals and modern humans appear to belong to such a taxon (other members might include *Homo erectus* and *H. floresiensis*). There are, however, relatively few data indicating a Neanderthal contribution to subsequent modern human populations. This dearth of data is most likely due, at least in part, to their small population size relative to that of the expanding modern human (i.e., African) population.

References

Adcock, G., Dennis, E., Easteal, S., Huttley, G., Jermin, L., Peacock, W., Thorne, A., 2001. Mitochondrial DNA sequences in ancient Australians: Implications for modern human origins. *Proc. Natl. Acad. Sci. U.S.A.* 98, 537–542.

Ballinger, S.W., Blankenship, L.H., Bickham, J.W., Carr, S.M., 1992. Allozyme and mitochondrial DNA analysis of a hybrid zone between white-tailed deer and mule deer (*Odocoileus*) in west Texas. *Biochem. Genet.* 30, 1–11.

Barton, N.H., 2001. The role of hybridization in evolution. *Mol. Ecol.* 10, 551–568.

Bermúdez de Castro, J.M., Arsuaga, J.L., Carbonell, E., Rosas, A., Martínez, I., Mosquera, M., 1997. A hominid from the Lower Pleistocene of Atapuerca, Spain: Possible ancestor to Neandertals and modern humans. *Science* 276, 1392–1395.

Bininda-Emonds, O.R.P., Gittleman, J.L., Purvis, A., 1999. Building large trees by combining phylogenetic information: a complete phylogeny of the extant Carnivora (Mammalia). *Biol. Rev.* 74, 143–175.

Brashares, J.S., Garland, T., Jr., Arcese, P., 2000. Phylogenetic analysis of coadaptation in behavior, diet, and body size in the African antelope. *Behav. Ecol.* 11, 452–463.

Bräuer, G., 1992. Africa's place in the evolution of *Homo sapiens*. In: Bräuer, G., Smith, F.H. (Eds.), *Continuity or Replacement: Controversies in Homo sapiens Evolution*. A. A. Balkema, Rotterdam, pp. 83–98.

Brownlow, C.A., 1996. Molecular taxonomy and the conservation of the red wolf and other endangered carnivores. *Conserv. Biol.* 10, 390–396.

Caramelli, D., Lalueza-fox, C., Vernesi, C., Lari, M., Casoli, A., Mallegni, F., Chiarelli, B., Dupanloup, I., Bertranpetit, J., Barbujani, G., Bertorelle, G. 2003. Evidence for a genetic discontinuity between Neandertals and 24,000-year-old anatomically modern Europeans. *Proc. Natl. Acad. Sci. U.S.A.* 100, 6593–6597.

Cathey, J.C., Bickham, J.W., Patton, J.C., 1998. Introgressive hybridization and nonconcordant evolutionary history of maternal and paternal lineages in North American deer. *Evolution* 52, 1224–1229.

Churchill, S.E., 1996. Particulate versus integrated evolution of the upper body in Late Pleistocene humans: A test of two models. *Am. J. Phys. Anthropol.* 100, 559–583.

Churchill, S.E., Smith, F.H., 2000. Makers of the Early Aurignacian of Europe. *Yrbk. Phys. Anthropol.* 43, 61–115.

Cracraft, J., 1983. Species concepts and speciation analysis. *Curr. Ornithol.* 1, 159–187.

Cracraft, J., 1989. Speciation and its ontology: the empirical consequences of alternative species concepts for understanding patterns and processes of differentiation. In: Otte, D., Endler, J.A. (Eds.), *Speciation and its Consequences*. Sinauer Associates, Sunderland, MA, pp. 28–59.

d'Errico, F., Soressi, M., 2002. Systematic use of manganese pigment by the Pech-de-l'Azé Neandertals: Implications for the origin of behavioral modernity. *J. Hum. Evol.* 42, A13.

d'Errico, F., Zilhão, J., Baffier, D., Julien, M., Pelegrin, J., 1998. Neandertal acculturation in Western Europe? A critical review of the evidence and its interpretation. *Curr. Anthropol.* 39, 1– 44.

De Queiroz, K., 1998. The general lineage concept of species, species criteria, and the process of speciation: A conceptual unification and terminological recommendations. In: Howard, D.J., Berlocher, S.H. (Eds.), *Endless Forms: Species and Speciation*. Oxford University Press, Oxford, pp. 57–75.

Duarte, C., Mauricio, J., Pettitt, P.B., Souto, P., Trinkaus, E., van der Plicht, H., Zilhão, J., 1999. The early Upper Paleolithic human skeleton from the Abrigo do Lagar Velho (Portugal) and modern human emergence in Iberia. *Proc. Natl. Acad. Sci. U.S.A.* 96, 7604–7609.

Franciscus, R.G., 1996. Neandertal nasal structures and upper respiratory tract "specialization." *Proc. Natl. Acad. Sci. U.S.A.* 96, 1805–1809.

Franciscus, R.G., Schoenebaum, C.R., 2000. Cross-sectional metric analysis of scapular axillary border morphology in recent humans and the Krapina Neandertals. *Am. J. Phys. Anthropol.* (Suppl. 30), 152.

Frayer, D.W., 1992. Evolution at the European edge: Neanderthal and Upper Paleolithic relationships. *Préhist. Europ.* 2, 9–69.

Frayer, D.W., Wolpoff, M.H., Thorne, A.G., Smith, F.H., Pope, G.G., 1993. Theories of modern human origins: The paleontological test. *Am. Anthropol.* 95, 14–50.

Gray, A., 1972. *Mammalian Hybrids. A Check-list with Bibliography*. 2nd Edition. Commonwealth Agricultural Bureaux, Farnham Royal, Buckinghamshire, UK.

Haldane, J.B.S., 1922. Sex ratio and unisexual sterility of hybrid animals. *J. Genet.* 12, 101–109.

Hall, R.L., 1978. Variability and speciation in canids and hominids. In: Hall, R.L., Sharp, H.S. (Eds.), *Wolf and Man: Evolution in Parallel*. Academic Press, New York, pp. 153–177.

Harvati, K., 2003. The Neanderthal taxonomic position: Models of intra- and inter-specific craniofacial variation. *J. Hum. Evol.* 44, 107–132.

Harvati, K., Frost, S.R., McNulty, K.P., 2004. Neanderthal taxonomy reconsidered: implications of 3D primate models of intra- and inter-specific differences. *Proc. Natl. Acad Sci. U.S.A.* 101, 1147–1152.

Hawks, J., Wolpoff, M.H., 2001. Paleoanthropology and the population genetics of ancient genes. *Am. J. Phys. Anthropol.* 114, 269–272.

Henshilwood, C.S., Marean, C.W., 2003. The origin of modern human behaviour: a review and critique of models and test implications. *Curr. Anthropol.* 44, 627–651.

Hilton, H., 1978. Systematics and ecology of the eastern coyote. In: Bekoff, M. (Ed.), *Coyotes: Biology, Behavior, and Management*. Academic Press, New York, pp. 209–228.

Holliday, T.W., 1997. Body proportions in Late Pleistocene Europe and modern human origins. *J. Hum. Evol.* 32, 423–448.

Holliday, T.W., 2000. Evolution at the crossroads: Modern human emergence in Western Asia. *Am. Anthropol.* 102, 54–68.

Holliday, T.W., 2003. Species concepts, reticulation, and human evolution. *Curr. Anthropol.* 44, 653–673.

Holloway, R.L., 1985. The poor brain of *Homo sapiens neanderthalensis*: See what you please. In:

Delson, E. (Ed.), *Ancestors: The Hard Evidence*. Alan R. Liss, New York, pp. 319–324.

Howell, F.C., 1994. A chronostratigraphic and taxonomic framework of the origins of modern humans. In: Nitecki, M.H., Nitecki D.V. (Eds.), *Origins of Anatomically Modern Humans*. Plenum Press, New York, pp. 253–319.

Hublin, J.-J., 1998. Climatic changes, paleogeography, and the evolution of the Neandertals. In Akazawa, T., Aoki, K., Bar-Yosef, O. (Eds.), *Neandertals and Modern Humans in Western Asia*. Plenum Press, New York, pp. 295–310.

Jablonski, N.G., 2004. The Evolution of human skin and skin color. *Ann. Rev. Anthropol.* 33, 585–623.

Jolly, C.J., 2001. A proper study for mankind: Analogies from papionin monkeys and their implications for human evolution. *Yrbk. Phys. Anthropol.* 44, 177–204.

Jolly, C.J., Woolley-Barker, T., Beyene, S., Disotell, T.R., Phillips-Conroy, J.E., 1997. Intergeneric hybrid baboons. *Int. J. Primatol.* 18, 597–627.

Kappelman, J., 1996. The evolution of body mass and relative brain size in fossil hominids. *J. Hum. Evol.* 30, 243–276.

King, W., 1864. The reputed fossil man of the Neanderthal. *Quart. Rev. Sci.*, 1, 88–97.

Klein, R.G., 2003. Whither the Neanderthals? *Science* 299, 1525–1527.

Krings, M., Stone, A., Schmitz, R.W., Krainitzki, H., Stoneking, M., Pääbo, S. 1997. Neandertal DNA sequences and the origin of modern humans. *Cell* 90, 19–30.

Krings, M., Capelli, C., Tschentscher, F., Geisert, H., von Haiseler Meyer, S., Grossschmidt, A., Possnert, G., Paunovic, M., Pääbo, S. 2000. A view of Neandertal genetic diversity. *Nat. Genet.* 26, 144–46.

Lahr, M.M., Foley, R., 1998. Towards a theory of modern human origins: Geography, demography, and diversity in recent human evolution. *Yrbk. Phys. Anthropol.* 41, 137–176.

Lehman, N., Eisenhawer, A., Hansen, K., Mech, L.D., Peterson, R.O., Gogan, P.J.P., Wayne, R.K., 1991. Introgression of coyote mitochondrial DNA into sympatric North American gray wolf populations. *Evolution* 45, 104–119.

Lotsy, J.P., 1925. Species or linneon. *Genetica* 7, 487–506.

MacFadden, B.J., 1992. *Fossil Horses: Systematics, Paleobiology, and Evolution of the Family Equidae*. Cambridge University Press, New York.

Mayden, R.L., 1997. The hierarchy of species concepts: The denouement in the saga of the species problem. In: Claridge, M.A., Dawah, H.A., Wilson, M.R. (Eds.), *Species: The Units of Biodiversity*. Chapman and Hall, New York, pp. 381–424.

Mayr, E., 1942. *Systematics and the Origin of Species from the Viewpoint of a Zoologist*. Columbia University Press, New York.

Mayr, E., 1963. *Animal Species and Evolution*. Belknap Press, Cambridge, MA.

Mayr, E., 2000. The biological species concept. In: Wheeler, Q.D., Meier, R. (Eds.), *Species Concepts and Phylogenetic Theory: A Debate*. Columbia University Press, New York, pp. 17–29.

Meijaard, E., Groves, C.P., 2004. Morphometrical relationships between south-east Asian deer (Cervidae, Tribe Cervini): evolutionary and biogeographic implications. *J. Zool., Lond.* 263, 179–196.

Mellars, P., 2005. The impossible coincidence. A single-species model for the origins of modern human behavior in Europe. *Evol. Anthrop.* 14, 12–27.

Mishler, B.D., Theriot, E.C., 2000. The phylogenetic species concept (*sensu* Mishler and Theriot): Monophyly, apomorphy, and phylogenetic species concepts. In: Wheeler, Q.D., Meier, R. (Eds.), *Species Concepts and Phylogenetic Theory: A Debate*. Columbia University Press, New York, pp. 44–54.

Nowak, R.M., 1978. Evolution and taxonomy of coyotes and related *Canis*. In: Bekoff, M. (Ed.), *Coyotes: Biology, Behavior, and Management*. Academic Press, New York, pp. 3–16.

Oakenfull, E.A., Clegg, J.B., 1998. Phylogenetic relationships within the genus *Equus* and the evolution of α and θ globin genes. *J. Mol. Evol.* 47, 772–783.

Oakenfull, E.A., Lim, H.N., Ryder, O.A., 2000. A survey of equid mitochondrial DNA: Implications for the evolution, genetic diversity and conservation of *Equus*. *Conserv. Genet.* 1, 341–355.

Ovchinnikov, I.V., Götherström, A, Romanova, G.P., Kharitonov, V.M., Lindén, K., Goodwin, W., 2000. Molecular analysis of Neanderthal DNA from the northern Caucasus. *Nature* 404, 490–493.

Pitra, C., Fickel, J., Meijaard, E., Groves, P.C., 2004. Evolution and phylogeny of Old World deer. *Mol. Phylogen. Evol.* 33, 880–895.

Purvis, A., Bromham, L., 1997. Estimating the transition/transversion ratio from independent pairwise comparisons with an assumed phylogeny. *J. Mol. Evol.* 44, 112–119.

Rak, Y., 1986. The Neanderthal: A new look at an old face. *J. Hum. Evol.* 15, 151–164.

Rak, Y., 1998. Does any Mousterian cave present evidence of two hominid species? In Akazawa, T., Aoki, K., Bar-Yosef, O. (Eds.), *Neandertals and Modern Humans in Western Asia*. Plenum Press, New York, pp. 353–366.

Relethford, J.H., 1995. Genetics and modern human origins. *Evol. Anthropol.* 4, 53–63.

Relethford, J.H., 1998. Genetics of modern human origins and diversity. *Ann. Rev. Anthropol.* 27, 1–23.

Relethford, J.H., Harpending, H.C., 1995. Ancient differences in population size can mimic a recent African origin of modern humans. *Curr. Anthropol.* 36, 667–674.

Ritz, L.R., Glowatzki-Mullis, M.-L., MacHugh, D.E., Gaillard, C., 2000. Phylogenetic analysis of the Tribe Bovini using microsatellites. *Animal Genet.* 31, 178–185.

Roy, M.S., Girman, D.J., Taylor, A.C., Wayne, R.K., 1994. The use of museum specimens to reconstruct genetic variability and relationships of extinct populations. *Experientia* 50, 551–557.

Ruff, C.B., Trinkaus, E., Holliday, T.W., 1997. Body mass and encephalization in Pleistocene *Homo. Nature* 387, 173–176.

Schillaci, M.A., Froehlich, J.W., 2001. Nonhuman primate hybridization and the taxonomic status of Neanderthals. *Am. J. Phys. Anthropol.* 115, 157–166.

Schwartz, M.K., Pilgrim, K.L., McKelvey, K.S., Lindquist, E.L., Claar, J.J., Loch, S., Ruggiero, L.F., 2004. Hybridization between Canada lynx and bobcats: Genetic results and management implications. *Conserv. Genet.* 5, 349–355.

Serre, D., Langaney, A., Chech, M., Teschler-Nicola, M., Paunovic, M., Mennecier, P., Hofreiter, M., Possnert, G., Paabo, S., 2004. No evidence of Neandertal mtDNA contribution to early modern humans. *PLoS Biol.* 2, 313–317.

Shea, J.J., 2003. Neandertals, competition, and the origin of modern human behavior in the Levant. *Evol. Anthropol.* 12, 173–187.

Shreeve, J., 1995. *The Neandertal Enigma: Solving the Mystery of Modern Human Origins*. William Morrow, New York.

Simpson, G.G., 1961. *Principles of Animal Taxonomy*. Columbia University Press, New York.

Skinner, A.R., Blackwell, B.A.B., Martin, S., Ortega, A., Blickstein, J.I.B., Golovanova, L.V., Doronichev, V.B. 2005. ESR dating at Mezmaiskaya Cave, *Russia. Appl. Radiat. Isot.* 62, 219–224.

Smith, F.H., 1984. Fossil hominids from the Upper Pleistocene of Central Europe and the origin of modern Europeans. In: Smith, F.H., Spencer, F. (Eds.), *The Origins of Modern Humans: A World Survey of the Fossil Evidence*. Alan R. Liss, New York, pp. 137–209.

Smith, F.H., 1985. Continuity and change in the origin of modern *Homo sapiens. Z. Morph. Anthrop.* 75, 197–222.

Smith, F.H., 1992. The role of continuity in modern human origins. In: Bräuer, G., Smith, F.H. (Eds.), *Continuity or Replacement: Controversies in Homo sapiens Evolution*. Balkema, Rotterdam, pp. 145–156.

Stringer, C., 2002. Modern human origins: Progress and prospects. *Phil. Trans. R. Soc. Lond.* B 357, 563–579.

Stringer, C.B., Andrews, P., 1988. Genetic and fossil evidence for the origin of modern humans. *Science* 239, 1263–1268.

Tattersall, I., 1992. Species concepts and species identification in human evolution. *J. Hum. Evol.* 22, 341–349.

Tattersall, I., 1999. *The Last Neanderthal: The Rise, Success, and Mysterious Extinction of our Closest Human Relatives*. Revised Edition. Westview Press, Boulder, CO.

Templeton, A.R., 1989. The meaning of species and speciation: A genetic perspective. In: Otte, D., Endler, J.A. (Eds.), *Speciation and its Consequences*. Sinauer Associates, Sunderland, MA, pp. 3–27.

Trinkaus, E., 1987. The Neandertal face: Evolutionary and functional perspectives on a recent hominid face. *J. Hum. Evol.* 16, 429–443.

Trinkaus, E., 1995. Near Eastern late archaic humans. *Paléorient* 21, 9–23.

Trinkaus, E., Zilhão, J., 2002. Phylogenetic implications of Lagar Velho 1. In: Zilhão, J., Trinkaus, E. (Eds.), *Portrait of the Artist as a Child: The Gravettian Human Skeleton from the Abrigo do Lagar Velho and Its Archeological Context*. Instituto Português de Arqueologia, Lisbon, Trabalhos de Arqueologia 22, pp. 497–518.

Vilà, C., Leonard, J.A., Götherström, A., Marklund, S., Sandberg, K., Lidén, K., Wayne, R.K., Ellegren, H., 2001. Widespread origins of domestic horse lineages. *Science* 291, 474–477.

Wayne, R.K., Jenks, S.M., 1991. Mitochondrial DNA analysis implying extensive hybridization of the endangered red wolf *Canis rufus. Nature* 351, 565–568.

Wayne, R.K., Van Valkenburgh, B., O'Brien S.J., 1991. Molecular distance and divergence time in carnivores and primates. *Mol. Biol. Evol.* 8, 297–319.

Wayne, R.K., Lehman, N., Allard, M.W., Honeycutt, R.L., 1992. Mitochondrial DNA variability of the gray wolf: Genetic consequences of population decline and habitat fragmentation. *Conserv. Biol.* 6, 559–569.

White, M.J.D., 1973. *Animal Cytology and Evolution*. 3rd Edition. Cambridge University Press, Cambridge.

Wiley, E.O., 1981. *Phylogenetics: The Theory and Practice of Phylogenetic Systematics*. Wiley, New York.

Wiley, E.O., Mayden, R.L., 2000. The evolutionary species concept. In: Wheeler, Q.D., Meier, R. (Eds.), *Species Concepts and Phylogenetic Theory: A Debate*. Columbia University Press, New York, pp. 70–89.

Wood, B., Collard, M., 1999. The human genus. *Science* 284, 65–71.

Wood, B., Richmond, B.G., 2000. Human evolution: Taxonomy and paleobiology. *J. Anat.* 196, 19–60.

Wolpoff, M.H., 1999. *Paleoanthropology*. 2nd Edition. McGraw-Hill, Boston.

Wolpoff, M.H., 2003. Comment to Holliday. *Curr. Anthropol.* 44, 666–667.

Xu, X. Gullberg, A., Arnason, U., 1996. The complete mitochondrial DNA (mtDNA) of the donkey and mtDNA comparisons among four closely related mammalian species-pairs. *J. Mol. Evol.* 43, 438–444.

Zeh, D.W., Zeh, J.A., 2000. Reproductive mode and speciation: the viviparity-driven conflict hypothesis. *BioEssays* 22, 938–946.

17. Speciation by distance and temporal overlap: a new approach to understanding Neanderthal evolution

J.-L. VOISIN
Institut de Paléontologie Humaine
1 rue René Panhard
75013 Paris, France
jeanlucv@mnhn.fr

Keywords: Neanderthal, modern human, speciation by distance, post-zygotic isolation, Europe, Near East, extinction, morphological cline

Abstract

Neanderthals are the best-known fossil hominid group, but at the same time many aspects of their evolution are still poorly understood. The variation of numerous characters in Neanderthal populations shows a geographical gradient. From west to east, characters become less and less Neanderthal-like and more and more modern human-like. Moreover, in Central Europe and the Near East, post-Neanderthal populations still exhibit some Neanderthal features, which is not the case in Western Europe. The spread of the first humans into Europe involved differentiation of this species by distance, whereas consecutive populations were linked by gene flow. Hence, from Western Europe to the Near East, there was a succession of human populations that developed, over time, Neanderthal characters that were more and more marked from east to west. Then, modern humans spread rapidly into Europe at about 40,000 years ago, but at least in the western part of the continent, no convincing evidence of hybridization with Neanderthals has been found. By contrast, interbreeding was still possible in the eastern part of Europe and in the Near East, but became less and less so towards the west. This hypothesis implies that the ancestors of Neanderthals arrived and evolved in Europe at a time when gene flow between Western Europe and Near Eastern populations was very limited. Hence, Near East Neanderthals cannot be interpreted as the result of a migration of a European population toward the east, but as a continuum in space and time of European inhabitants. Thus, as they moved westwards, modern humans integrated local populations in the Near East and Central Europe and replaced populations in Western Europe.

Introduction

Although Neanderthals are among the best-known fossil hominids, many aspects of their evolutionary history, especially their extinction and taxonomic position relative to modern humans, are still poorly understood. There are two main schools of thought on this last

K. Harvati and T. Harrison (eds.), Neanderthals Revisited: New Approaches and Perspectives, 299–314.

topic: (1) Neanderthals and modern humans are two distinct species (e.g., Rak, 1993; Hublin et al., 1996; Stringer and McKie, 1996; Bermúdez de Castro et al., 1997; Krings et al., 1997; Stringer, 1998, 2002; Bocquet-Appel and Demars, 2000; Hublin, 2000; Arsuaga et al., 2001; Bräuer, 2001; Schillaci and Froehlich, 2001; Rak et al., 2002; Harvati, 2003; Harvati et al., 2004); and (2) Neanderthals and modern humans are a single species, with or without two subspecies (Thoma, 1965; Trinkaus, 1983, 1991; Smith et al., 1989a, 2005; Smith, 1991; Smith and Trinkaus, 1991; Frayer, 1992; Wolpoff and Caspari, 1996; Duarte et al., 1999; Wolpoff et al., 2000; Relethford, 2001, 2003; Ahern et al., 2002, 2004; Curnoe and Thorne, 2003). Explanations of Neanderthal extinction depend in great part on how scholars consider their taxonomic status. If Neanderthals and modern humans belong to the same species, then Neanderthal morphology disappeared because Neanderthals were genetically absorbed into modern human populations. On the other hand, if Neanderthals and modern humans were two distinct species, the disappearance of the former is likely the result of competition with modern humans when they arrived in Europe. However, some scholars consider that Neanderthals and modern humans may never have met (d'Errico et al., 1998; Zilhão and d'Errico, 2000; d'Errico and Sánchez Goni, 2003), and that Neanderthal extinction was not related to their taxonomic status. Whatever the case, in the following analysis and discussion I will simply treat these two human groups as distinct populations without taking a position on their taxonomy.

In this study, I propose a new interpretation of the relationships between these two human groups. First, I briefly present an overview of Neanderthal characters and their variation in western and eastern populations. Then, I explain this variation in the context of "speciation by distance" and the migration of modern humans into Europe.

A West to East Morphological Cline

At the transition between the Middle and Upper Paleolithic in Western Europe, all fossil humans clearly belong to one of two groups: Neanderthals or modern humans. Everyone agrees that there were two distinct populations in this region of the world (whether they belong to two different species or not). Since 1999, the Lagar Velho child (Duarte et al., 1999) has been at the center of discussion, being the earliest human fossil in Western Europe about which no consensus exists. As we will see below, the Lagar Velho fossil fits well with my hypothesis. However, in Eastern Europe, and more evidently in the Near East, the two populations are less clear-cut. As noted by Smith et al. (1989a: 50) "there is little evidence of evolutionary trends in the modern human direction among the west European Neanderthals. ... However, in central Europe, there are possible indications of diachronic trends within the Neanderthals, in the direction of the modern European condition." The debate begins with early remains, such as the Zuttiyeh fossil, which are alternatively considered pre-*sapiens* (Vandermeersch, 1989a), related to Asian Paleolithic populations (Sohn and Wolpoff, 1993), or pre-Neanderthal (Smith et al., 1989a; Simmons et al., 1991). In the same way, more recent remains are considered to belong only to archaic *Homo sapiens* (Arensburg, 1989; Smith, 1991; Wolpoff, 1999; Arensburg and Belfer-Cohen, 1998; Kramer et al., 2001), with no Neanderthals existing in the Near East. Alternatively, others consider that Neanderthals do exist in the Near East (e.g., Condemi, 1991; Rak, 1998; Stringer, 1998; 2002; Trinkaus, 1983, 1991). This disagreement is primarily due to the variation of Neanderthal morphology from east to west. In Western Europe, Neanderthal morphology is well marked and easily distinguishable from that of modern humans, while differences are less pronounced in the Near East. In other words, "Neanderthal features are not uniformly spread across the Neanderthal range, with sharp boundaries with other contemporary populations. Instead, they vary clinally, diminishing in frequency to the south east and east. In the Levant, it has been seriously questioned whether the specimens should be called 'Neanderthal' at all because they share few diagnostic features with the Europeans [Neanderthal]" (Wolpoff et al., 2004: 529). Other authors have also noted this west to east morphological cline (Vandermeersch, 1989b; Smith, 1991; Rak, 1993; Nara, 1994; Arensburg and Belfer-Cohen, 1998; Moncel and Voisin, 2006).

Osteological Evidence

Table 1 lists characters of the cranium and mandible, postcranial skeleton, and overall stature that show a clinal variation from Western Europe to the Near East. For this study, I used characteristics and data taken from the literature. This is not an exhaustive list, but it does offer examples to illustrate that a morphological cline exists. A more extensive study is currently in progress.

CRANIUM AND MANDIBLE

In their overall proportions, skulls of Near Eastern Neanderthals look more "modern" than those of their Western European counterparts (Table 1). The sharply pointed mastoid process is a Neanderthal autapomorphy that is found in all western individuals. On the contrary, in the Near East this morphology is found only in Shanidar 1 and Tabun 1, but is absent in Shanidar 2 and 5 and in Amud 1 (Vandermeersch, 1981; Trinkaus, 1988; Condemi, 2005). Moreover, in Amud 1 and Shanidar the mastoid process looks more modern than in any other Neanderthal population (Suzuki, 1970). In the same way, Western European Neanderthals have a less rounded occipital region, with a pronounced bun, compared with Central European and Near Eastern Neanderthals (Vandermeersch, 1981; Piveteau, 1983; Trinkaus, 1983; Smith, 1991; Habgood, 2003). The frontal region is larger in Near Eastern than in Western Neanderthals. The difference is not great, but there is no overlap between the two populations (Vandermeersch, 1989b). The cranial vault is higher in Near Eastern Neanderthals than in the Western group (Vandermeersch, 1981; Condemi, 1992), and Amud 1 falls well within the Upper Paleolithic range of variation for this feature (Suzuki, 1970). All hominids, except *Homo sapiens sapiens*, lack a chin on the mandible, but, according to Suzuki (1970) and Bar-Yosef and Vandermeersch (1991), some Near Eastern Neanderthals, like Amud 1, display an incipient development of it. Mid-facial prognathism is less pronounced in Near Eastern than in western Neanderthals (Piveteau, 1983; Habgood, 2003). Habgood (2003) shows that it is possible to separate classic Neanderthals from those from Central Europe on the basis of the overall morphology of the skull and the mandible by multivariate analysis. However, most characters, such as mid-facial prognathism, or sharp mastoid processes, are similar in Central and Western European Neanderthals (Habgood, 2003).

POSTCRANIAL SKELETON

Near Eastern Neanderthals more closely resemble modern humans postcranially than do Western Neanderthals (Table 1). Clavicular morphology is quite different in modern humans and Neanderthals in posterior view (Voisin 2000, 2001, 2004), although the Kebara and Krapina 143 clavicles display a modern morphology (Voisin, 2004). The scapula, which is considered the best postcranial bone for characterizing Neanderthals (Vandermeersch, 1981; Heim, 1982b), displays a morphological cline from west to east, especially in the configuration of its axillary

Table 1. Skull, postcranial, and body proportion differences between western and eastern Neanderthals

	West Europe	*Near East*	*Modern human*	*Authors*
Mastoid process	Sharp pointed	Sharp pointed morphology is not present on all fossils	Never sharp pointed	Vandermeersch, 1981; Trinkaus, 1988
Frontal width	Average = 107.4 mm Min = 106 mm Max = 109 mm	Average = 112.5 mm Min = 110 mm Max = 115 mm	Average = 109 mm Min = 98 mm Max = 113 mm	Vandermeersch, 1981, 1989b
Occipital region	Less rounded with a pronounced torus (i.e., less modern)	More rounded with a torus less pronounced or absent (i.e., more modern)	Rounded without any torus	Trinkaus, 1983; Vandermeersch, 1981, 1989b
Height of the cranial vault (porion-bregma)	Average = 112.5 mm Min = 111 mm Max = 114 mm	Average = 118.5 mm Min = 116 mm Max = 121 mm	Average = 117.4 mm Min = 98 mm Max = 122.5 mm	Vandermeersch, 1981; Condemi, 1992
Position of the auditory meatus	Far from the modern position in regard to the zygomatic arch	Near the modern position in regard to the zygomatic arch	Low in regard to the zygomatic arch	Suzuki, 1970; Vandermeersch, 1989b
Chin	Absent	Incipient	Present	Suzuki, 1970; Bar-Yosef and Vandermeersch, 1991
Clavicle morphology	Two curvatures in dorsal view	Some clavicle show only one curvature in dorsal view, like Modern humans	One curvature in dorsal view (the inferior one)	Voisin, 2000, 2001, 2004
Axillary sulcus of scapula	Dorsal	Bisulcate or ventral	Ventral, sometimes bisulcate	Frayer, 1992; Nara, 1994; Voisin 2000 Hambücken, 1997;
Radius shaft	High curvature	Slight curvature	Slight curvature	Arensburg and Belfer-Cohen, 1998
Pubic length relative to body size	Very long (outside modern range of variation)	Short (within modern range of variation)	Short	Rosenberg (1998)
Stature (of male)	Average = 165.4 cm Min = 162 cm Max = 172	Average = 171.2 cm Min = 163.9 cm Max = 176.5 cm	(Qafzeh and Skhul) Average = 185.2 cm Min = 183.5 cm Average = 187 cm	Vandermeersch, 1981, 1989b
Thorax width	Very large	Smaller (but slightly larger than modern human)	Little bit smaller than the Near East Neanderthal	Endo and Kimura, 1970; Trinkaus, 1983
Limbs	Shorter	Longer	Longer	Trinkaus, 1981
Cold adapted body proportion	More Specialized	Less specialized	Less specialized	Churchill, 1998

border. In Western Europe, all scapulae share a dorsal *axillary sulcus*, but in Central Europe and in the Near East this *sulcus* can be dorsal or bisulcate (Frayer, 1992; Nara, 1994). This feature is important because it relates to arm movements (Boule, 1912; Fraipont, 1927; Stewart, 1962; Trinkaus, 1977; Voisin, 2000) and post-natal growth (Heim, 1982a; Madre-Dupouy, 1991).

Other parts of the postcranial skeleton show differences between western and eastern Neanderthals that make the latter appear closer to *Homo sapiens* than classical Neanderthals. For example, the radius and ulna shafts are straighter in Near Eastern Neanderthals and close to those of Skhul IV and VII or Předmostí males (Endo and Kimura, 1970; Arensburg and Belfer-Cohen,

1998). Also, the Neanderthal upper limb in the Near East is gracile rather than robust (Hambücken, 1995). On the basis of the morphology of the distal extremity of the humerus, Hambücken (1997) separates Neanderthal humeri into two groups: a "classic" group (including La Chapelle-aux-Saints, Combe-Grenal, La Ferrassie, Régourdou, Saint-Césaire, Neanderthal and Spy) and a Mediterranean one (including Hortus, Krapina and Lezetxiki). Both of these groups are more robust in their overall morphology than the humeri of Near Eastern Neanderthals. The length of the pubis, relative to body size, is greatest in Western Europe (with La Ferrassie 1 being outside the range of modern human) and the shortest in the Near East (with Tabun C1 falling within the variation of modern humans). Neanderthals from Central Europe are between these two extremes (with Krapina 208 falling in the upper part of modern human range of variation) (Rosenberg, 1998).

STATURE

Eastern Neanderthals are taller than Western Neanderthals, and the former are closer in estimated stature to individuals from Skhul and Qafzeh (Vandermeersch, 1981, 1989b) (Table 1). This is correlated with body proportions that are adapted to warmer climates, with the Near Eastern Neanderthal populations having longer limbs and a smaller thorax (Endo and Kimura, 1970; Trinkaus, 1981, 1983; Churchill, 1998).

The morphology of the Krapina fossils is typically Neanderthal, but most of the metric values are at the lower extreme of the range of variation in western Neanderthals (Smith and Trinkaus, 1991). Hence, they may not look identical to classic Neanderthals. Although the Vindija remains are fragmentary, and the stature of the individuals cannot be precisely estimated, their overall morphology is less robust than that of classic Neanderthals (Smith and Trinkaus, 1991; Trinkaus and Smith, 1995).

Stature and postcranial morphology show the same west to east cline as skull characters. In other words, the further west that Neanderthals originate, the more they display classical Neanderthal traits. As Smith and Trinkaus (1991: 255) wrote "En Europe centrale, il existe des données importantes qui font penser que la reconnaissance d'une différence morphologique qualitative est moins évidente qu'en Europe occidentale" (In central Europe, there are important data leading to the conclusion that morphological qualitative differentiation is less marked than in Western Europe [my translation]).

Most skull differences are found between European (including Western and Central Europe) and Near East Neanderthal populations, while postcranial characters display a more gradual clinal change from west to east.

NEANDERTHAL CHARACTERS IN POST-NEANDERTHAL POPULATIONS

According to a number of authors, such as Smith et al. (1989b, 2005), Frayer (1992, 1997), Wolpoff et al. (2001, 2004), Trinkaus et al. (2003b), Janković et al. (2006), Ahern (this volume) and Hawks (this volume), some morphological characters in early modern Europeans reflect a Neanderthal influence (Tables 2 and 3). These traits exhibit a higher frequency in early modern Europeans than in later Europeans and non-European Pleistocene samples. This pattern, used to infer a Neanderthal contribution to early modern Europeans, is found only in post-Neanderthal populations of Eastern Europe (Smith, 1991; Smith and Trinkaus, 1991; Frayer, 1992; Smith et al., 2005), and no worker has demonstrated such a contribution to Western European populations (Gambier, 1989; Smith et al., 1989b; Smith, 1991; Smith and Trinkaus, 1991; Frayer, 1992; Hublin et al., 1996; Trinkaus, 2001; Trinkaus et al., 2003). The most striking

Table 2. Frequency of features present in different human populations (from Frayer, 1992; Smith et al., 2005)

	Suprainiac fossa	*Occipital bun*	*H-O*
European Neanderthals	100 (24/24)	80 (8/10)	52.6 (10/19)
Skhul / Qafzeh	14.3 (1/7)	0 (0/5)	/
Early Upper Palaeolithic	38.5 (10/26)	68.4 (13/19)	44.4 (4/9)
Late Upper Palaeolithic	23.7 (9/38)	/	5.3 (2/38)
Mesolithic	19.3 (31/161)	/	1.9 (3/161)
Medieval Hungarians	5.9 (14/237)	/	1.4 (3/208)

Values represent the % of specimens in which the features is present. Number in ( ) indicates the number of individuals exhibiting the feature followed by the sample size. H-O is the occurrence of the horizontal-oval mandibular foramen.

Table 3. Frequencies (%) of axillary scapular border types in Neanderthal, Early Upper Palaeolithic, Late Upper Palaeolithic, Mesolithic and Modern European samples (from Frayer, 1992)

	Dorsal	*Bisulcate*	*Ventral*
Neanderthal	64.1	23.9	12
Early Upper Palaeolithic	12.3	62.4	25.3
Late Upper Palaeolithic	16.8	27.7	55.5
Mesolithic	7.4	18.2	74.4
Modern European	1.2	14.4	84.4

According to Frayer, the high frequency of bisulcate axillary border demonstrates the Neanderthal contribution to the gene pool, because this feature is intermediate between the two other morphologies.

example is the supraorbital torus, which shows a continuous reduction through time in Central Europe without any clear difference between the latest Neanderthals and the first modern human populations. On the contrary, there are sharp boundaries between the last Neanderthals and the first modern humans in Western Europe (Smith et al., 1989b). Although the Lagar Velho child has been interpreted as a hybrid between Neanderthals and modern humans (Duarte et al., 1999), it does not provide convincing evidence for this because: (1) all Neanderthal characters may not be present or may not reach their classical morphology in a four-year-old child, so it becomes difficult to determine if some features are the result of hybridization (Tattersall and Schwartz, 1999); and (2) we do not know whether we are dealing with an F1 generation or not (Tattersall and Schwartz, 1999). Moreover, a hybrid that died before reaching reproductive age could also be interpreted as having a low fitness (see below).

What about DNA?

Since the work of Krings et al. (1997) on Neanderthal mtDNA, other studies of ancient mtDNA have followed (Krings et al., 1999; Ovchinnikov et al., 2000; Scholz et al., 2000; Caramelli et al., 2003; Serre et al., 2004; Serre and Pääbo, this volume). According to these authors, the differences observed between Neanderthal and modern human mtDNA support the interpretation that these two human groups are distinct species, although they do not entirely rule out the possibility of gene flow between them. However, the differences may be due to numerous factors and may not only reflect the replacement of one population by another (Hawks and Wolpoff, 2001). They could show the replacement of an original mtDNA by a new one within the same population through introgression (Mounolou, 1989; Hawks and Wolpoff, 2001). Moreover, mtDNA differences between Neanderthals and modern humans are less profound than the ones observed between two of the three subspecies of *Pan troglodytes* (Relethford, 2001; Barriel and Tillier, 2002).

Phylogenetic trees generated from mtDNA data may be incongruent with those using nuclear DNA (Sota and Vogler, 2001), most notably because selection pressures on these two genomes are not identical. At least in humans, mtDNA is under a high selective pressure, and this invalidates the hypothesis of neutral selection with a constant rate of substitution (Curnoe and Thorne, 2003; Hawks, this volume). In other words, phylogenetic trees obtained from mtDNA may not correctly reflect the evolutionary relationships of Neanderthals and modern humans. In addition, there are numerous difficulties in extracting ancient DNA, especially due to its incompleteness and short length (Cooper and Wayne, 1998), but also because it is fragile. Ancient DNA amplification creates damage that produces mutation artefacts that may artificially enhance differences between Neanderthal and modern human mtDNA (Hansen et al., 2001; Hofreiter et al., 2001; Gutiérrez et al., 2002). Therefore, ancient DNA does not settle the debate about the systematic status of Neanderthals, and more work is needed, especially on the post-mortem diagenesis of DNA (Hofreiter et al., 2001; Geilg, 2002; Smith et al., 2003; Mitchell et al., 2005; Salamon et al., 2005).

An Extreme Example of Speciation by Distance: The Ring Species

In order to explain the east-west morphological cline in Neanderthal populations, as well as their relationships to modern humans, it is useful to look at a peculiar type of speciation: speciation by circular overlap or "ring species". "Ring species provide dramatic evidence that normal genetic divergence within one species can build up to a sufficient level to generate two species" (Ridley, 2004: 388). What is a ring species? Among vertebrates, good examples of ring species are few. These include the Californian salamander *Ensatina eschscholtzii* (Ridley, 2004), the herring gull *Larus argentatus* and lesser black-backed gull *Larus fuscus* (Mayr, 1974). Perhaps the best example is the greenish warbler *Phylloscopus trochiloides* (Wake, 2001) that lives in forests across much of northern and central Asia (Figure 1). In central Siberia, two distinct forms, *P. trochiloides viridanus* and *P. trochiloides plumbeitarsus* are sympatric without interbreeding, and therefore may be considered two species. These two forms are nevertheless connected by a chain of interbreeding populations encircling the Tibetan plateau to the south (*P. trochiloides ludlowi, P.t. trochiloides, P.t. obscuratus*), and traits change gradually in consecutive populations (Irwin et al., 2001a). There is no obvious species boundary along this chain, and the two terminal "species" *viridanus* and *plumbeitarsus* are connected by gene flow (Irwin et al., 2001b). The "species" *trochiloides* has expanded northward following two pathways, one on the east, the second on the west of the Tibetan plateau (Figure 1), evolving several differences as they moved north (Irwin et al., 2001a, b). These include: (1) morphological differences (most notably in their wing bars); (2) song differences (males sing both for attracting females and defending their territories; females of *viridanus* and *plumbeitarsis* do not recognize the song of males of the other form); and (3) genetic differences.

This example shows how differences between two extreme populations (here *P. trochiloides viridanus* and *P. trochiloides plumbeitarsus*) can be important and affect the phenotype as well as the genotype. Between the two forms living in central Siberia morphological traits change gradually in consecutive populations encircling the Tibetan plateau, in the same manner as those of western to eastern Neanderthals. Thus, just before the spread of modern humans into Europe about 40,000 years ago, there was a chain of

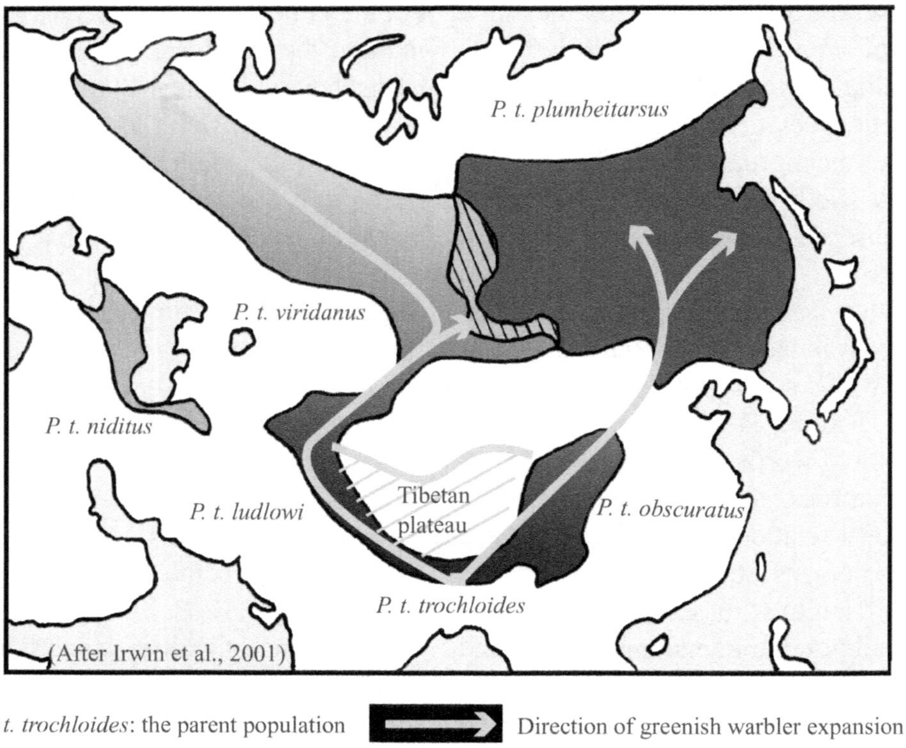

Figure 1. The greenish Warblers (*Phylloscopus trochiloides*) ring species. The break in the population in west China is inferred to be recent and caused by deforestation (after Irwin et al., 2001).

human populations throughout Europe and the Near East, more or less connected by gene flow (the gene flow rate would have varied as the ice sheets expanded and receded). When modern humans migrated westwards into Europe, they met more and more Neanderthal-like human populations. In Western Europe they encountered classic Neanderthals and were no longer able to interbreed with them, except in rare instances (see below).

In other words, the meeting of the two populations took place in two steps (Figure 2): (1) The spread of the first human populations into Europe, involved a clinal differentiation of this species, where each consecutive population was linked by gene flow. Hence, from Western Europe to the Near East, there was a succession of human populations that developed, over time, Neanderthal characters that became more and more marked from east to west; (2) In Western European Neanderthal populations, differentiation reached a level that did not allow interbreeding with modern humans. In Central Europe, gene flow was still possible, as shown by the persistence of Neanderthal features in post-Neanderthal populations (Smith, 1991; Smith and Trinkaus, 1991; Stringer, 1992; Frayer, 1992, 1997; Wolpoff and Caspari, 1996; Ahern et al., 2004; Wolpoff et al. 2004). This scenario is analogous to the sympatric populations of Greenish Warbler (*P. t. viridanus* and *P. t. plumbeitarsus*) to the north of the Tibetan plateau in Siberia. The two human populations in Western Europe were morphologically too different to allow admixture between them.

What Was the Level of Genetic Separation Between Western Neanderthals and Modern Humans?

The very low frequency or lack of admixture in Western Europe could have arisen in

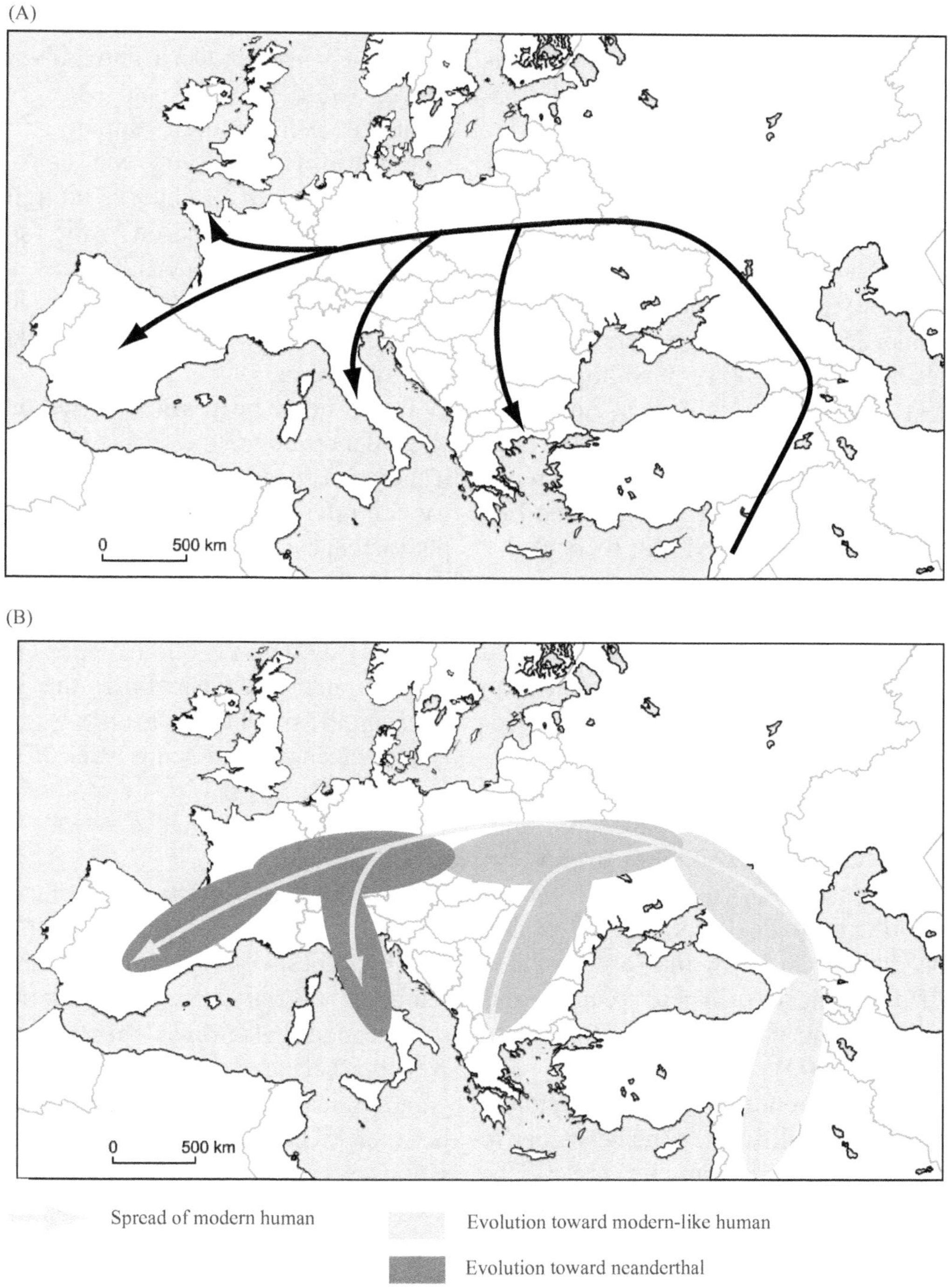

Figure 2. (A) the spread of the first human population into Europe. (B) differentiation of human populations, producing less Neanderthal-like people in the east (light grey), increasingly more Neanderthal-like in Eastern Europe, and classical Neanderthals in Western Europe (dark grey). Bright grey arrows symbolise the rapid spread of the first wave of modern humans into Europe.

numerous ways. The best-known mechanism is pre-zygotic isolation (genetic incompatibility and/or no mating between individuals of the two populations). However, other biological forces can create reproductive isolation, whereby hybrids are sterile or have a low fitness (i.e., the Lagar Velho child, see below for a discussion). In this latter case, hybrid individuals would not participate in gene flow between the two populations. Post-zygotic isolation between modern humans and western Neanderthals could have taken several forms (Ridley, 2004): (1) vanishing or low fitness of the male (Haldane's rule); (2) increased level of isolation between two populations by natural selection (reinforcement); and (3) interactions among several gene loci by epistasis (Dobzhansky-Muller theory).

Post-zygotic isolation would allow some degree of admixture between the two human groups, but it would have resulted in limited gene flow. Moreover, in humans, culture can also effectively contribute to reproductive isolation, with groups rejecting people with different behaviors. It is likely that cultural and biological factors worked together to limit gene flow between the two human groups.

Pre-zygotic isolation fits well with a classic view of the biological species concept, but often the distinction between closely related species is not so clear-cut. There are numerous ways of reaching genetic incompatibility (Schilthuizen, 2001; Ridley, 2004), and species level differences could exist prior to genetic incompatability. The time necessary to attain incompatibility is variable and can be very long; up to 4 myrs in some primates (see Holliday, this volume). Thus, in Western Europe, differences between the two human groups could have reached the species level without also reaching pre-zygotic isolation. This hypothesis excludes any large genetic contribution by Neanderthals to early modern human in Europe, which fits well with the DNA evidence (Caramelli et al., 2003; Serre et al., 2004; Serre and Pääbo, this volume).

In this way, is it possible to resolve the debate about the Lagar Velho child and its peculiar characters? Duarte et al. (1999) contend that it is a modern human-Neanderthal hybrid, while Tattersall and Schwartz (1999) regard it as a modern human without any admixture. However, if one considers the Lagar Velho child as a hybrid, it might be possible to infer that it had a lower fitness than Neanderthals and modern humans, as it died before reaching reproductive age. More fossils from this time period are needed to test this hypothesis.

It will never be possible, based on the fossil evidence alone, to establish beyond a doubt if there was post- or pre-zygotic isolation, but we can infer what is most probable. The morphological evidence implies that Neanderthals and modern humans in Western Europe may have behaved as two distinct species, most probably by post-zygotic isolation. In Eastern Europe and the Near East, the separation between the two human groups was apparently less clear-cut and some level of admixture was possible.

Neanderthal Evolution and Migration

The hypothesis of speciation by distance and temporal overlap between modern humans and Neanderthals implies that the ancestors of Neanderthals arrived and evolved in Europe, a geographical dead end, and that gene flow between Western and Eastern European populations was limited. Moreover, the effect of gene flow would have been more important in the eastern than in the western part of Europe because of the low density of Neanderthal populations (Mellars, 1998) and the asymmetric distribution of hunter-gatherer populations (Demars, 1996). Hence, Near Eastern Neanderthals should not be interpreted as the result of migrations of Neanderthal populations toward the East, but as a continuum in space and time. This interpretation allows us to explain why Near Eastern Neanderthals display such marked differences from Western European Neanderthals. It is also more consistent with the archaeological evidence (i.e., a lack of evidence of European cultural intrusion) than the notion that Neanderthals migrated into the Near East (Ahern, personal communication).

Conclusion

The evidence presented here indicates that there was a morphological cline (in skull form, postcranial skeleton, and stature) from west to east in Neanderthal populations. The farther those populations lived to the west, the more they displayed pronounced Neanderthal characters. Moreover, Neanderthal features seem to persist in Central Europe and Near East post-Neanderthal populations.

Ancient DNA studies do not settle the debate about the relationship of Neanderthals and modern humans because several alternative explanations may account for the observed differences; not only replacement of one population by the other. These explanations could be: (1) that mtDNA and nuclear DNA trees are not always congruent because of differences in selection pressures; (2) introgression; and (3) problems with ancient DNA conservation and extraction that introduce artificial differences between the two human populations.

In order to explain the peculiar distribution of characters in Neanderthals, as well as in post-Neanderthal populations, a two-phase model is proposed. First, an initial spread of human populations into Europe, followed by a clinal differentiation. This led to a succession of populations distributed from the Near East to Western Europe in which, over time, Neanderthal characters became increasingly

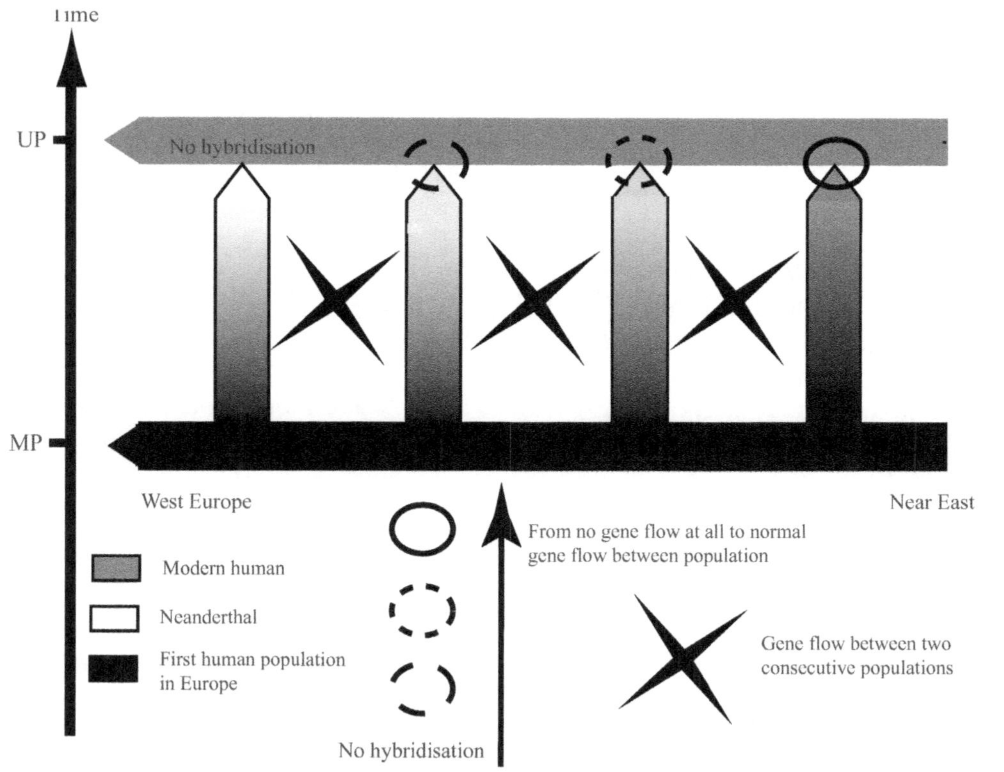

Figure 3. Gene flow between Neanderthal and modern human populations. UP: Upper Palaeolithic and MP: Middle Palaeolithic.

more marked from east to west. In other words, a speciation by distance could have occurred within the first European inhabitants. Second, when modern human spread into Europe at around 40,000 years ago, they met populations with more and more pronounced Neanderthal characters as they moved westward. Admixture was probably still possible in the Near East and in Central Europe, but in Western Europe differentiation between the two human groups reached such a level that admixture was no longer possible as a result of pre- or post-zygotic isolation (Figure 3).

Isolation between modern humans and Western Neanderthals would probably not have been achieved at this time. In other words, only post-zygotic isolation would have existed between the two human populations, and occasional admixture may still have occurred, although hybrids may have had a low fitness. The main basis for this assumption is the long time that is usually needed to attain pre-zygotic isolation in primates.

Acknowledgments

I would first like to thank Terry Harrison and Katerina Harvati for their invitation to the meeting and their helpful comments. I also thank Susan Antón and Jim Ahern for their helpful comments and suggestions, as well as everyone who supported my work, and my wife for her unconditional support.

References

Ahern, J.C.M., Lee, S.H., Hawks, J.D., 2002. The late Neanderthal supraorbital fossils from Vindija cave, Croatian: a biased sample? *J. Hum. Evol.* 43, 419–432.

Ahern, J.C.M., Karavanic, I., Paunovic, M., Jankovic, I., Smith, F.H., 2004. New discoveries and interpretations of hominid fossils and artifacts from Vindija Cave, Croatia. *J. Hum. Evol.* 46, 25–65.

Arensburg, B., 1989. New skeletal evidence concerning the anatomy of middle Palaeolithic populations in the middle east: the Kebara skeleton. In: Mellars, P., Stringer, B. (Eds.), *The Human Revolution*. Edinburgh University Press, Edinburgh, pp. 165–171.

Arensburg, B., Belfer-Cohen, A., 1998. Sapiens and Neanderthals: Rethinking the Levantine middle paleolithic hominids. In: Akazawa, T., Aoki, K., Bar-Yosef, O. (Eds.), *Neanderthals and Modern Humans in Western Asia*. Plenum Press, New York, pp. 311–322.

Arsuaga, J.L., Martínez, I., Gracia, A., 2001. Analyse phylogénétique des hominidés de la Sierra de Atapuerca (Sima de los Huesos et Gran Dolina TD-6): l'évidence crânienne. *L'Anthropologie (Paris)*. 105, 161–178.

Barriel, V., Tillier, A.M., 2002. L'enfant de Mezmaiskaya (Caucase) examiné dans une double perspective paléogénétique et paléoanthropologique. *Bull. Mém. Soc. Anthrop.* 14, 163–191.

Bar-Yosef, O., Vandermeersch, B., 1991, Premiers hommes modernes et néandertaliens au Proche-Orient: chronologie et culture. In: Hublin, J.J., Tillier, A.M. (Eds.), *Aux Origines d'Homo sapiens*. P.U.F., Paris, pp. 217–250.

Bermúdez de Castro, J.M., Arsuaga, J.L., Carbonell, E., Rosas, A., Martínez, I., Mosquera, M., 1997. A hominid from Lower Pleistocene of Atapuerca, Spain: possible ancestor to Neanderthals and modern humans. *Science* 276, 1392–1395.

Bocquet-Appel, J.P., Demars, P.Y., 2000, Neanderthal contraction and modern human colonization of Europe. *Antiquity* 74, 544–552.

Boule, M., 1912. L'homme fossile de La Chapelle-aux-Saints. *Ann. Paléontol.* 7, 21–192.

Bräuer, G., 2001. The "Out-of-Africa" model and the question of regional continuity. In: Tobias, P.V., Raath, M.A., Moggi-Checchi, J., Doyle, G.A. (Eds.), *Humanity from African Naissance to Coming Millennia*. Firenze University Press, Florence, pp. 183–189.

Caramelli, D., Lalueza-Fox, C., Vernesi, C., Lari, M., Casoli, A., Mallegni, F., Chiarelli, B., Dupanloup, I., Bertranpetit, J., Barbujani, G., Bertorelle, G., 2003. Evidence for a genetic discontinuity between Neanderthals and 24,000-year-old anatomically modern Europeans. *Proc. Natl. Acad. Sci. U.S.A.* 100, 6593–6597.

Churchill, S.E., 1998, Cold adaptation, heterochrony, and Neanderthals. *Evol. Anthropol.* 46–61.

Condemi, S., 1991. Some considerations concerning Neanderthal features and the presence of

Neanderthals in the Near East. *Riv. Antropol.* 69, 27–38.
Condemi, S., 1992. *Les Hommes Fossiles de Saccopastore et leur Relation Phylogénétiques.* CNRS, Paris.
Condemi, S., 2005. The Neanderthal from Le Moustier and European Neanderthal variability. In: H. Ullrich (Ed.), *The Neanderthal Adolescent Le Moustier* 1, *New Aspects, New Results*. Berliner Beiträge zur Vor und Frühgeschichte, Neue Folge Band 12, Staatliche Museen zu Berlin Preussischer Kulturbesitz, Berlin, pp. 321–327.
Cooper, A., Wayne, R., 1998. New uses for old DNA. *Curr. Opin. Biotechnol.* 9, 49–53.
Curnoe, D., Thorne, A., 2003. Number of ancestral human species: a molecular perspective. *Homo* 53, 201–224.
Demars, P.Y., 1996. Démographie et occupation de l'e-space au Paléolithique Supérieur et au Mésolithique en France. *Prehist. Eur.* 8, 3–26.
d'Errico, F., Sánchez Goni, M.F., 2003. Neanderthal extinction and the millennial scale climatic variability of OIS 3. *Quat. Sci. Rev.* 22, 769–788.
d'Errico, F., Zilhão, J., Julien, M., Baffier, D., Pelegrin, J., 1998, Neanderthal acculturation in Western Europe? *Curr. Anthropol.* 39, S1-S43.
Duarte, C., Mauricio, J., Pettitt, P.B., Souto, P., Trinkaus, E., Plicht (van der), H., Zilhão, J., 1999. The early Upper Paleolithic human skeleton from the Abrigo do Lagar Velho (Portugal) and modern human emergence in Iberia. *Proc. Natl. Acad. Sci. U.S.A.* 96, 7604–7609.
Endo, B., Kimura, T., 1970. Postcranial skeleton of the Amud man. In: Suzuki, H., Takai, F., (Eds.), *The Amud Man and his Cave Site*. The University of Tokyo Press, Tokyo, pp. 231–406.
Fraipont, C., 1927. Sur l'omoplate et le sacrum de l'homme de Spy. *Rev. Anthropol.* (Paris) 37, 189–195.
Frayer, D.W., 1992. The persistence of Neanderthal features in post-Neanderthal Europeans. In: Bräuer, G., Smith, F.H., (Eds.), *Continuity or replacement; Controversies in Homo sapiens Evolution*. A.A. Balkema, Rotterdam, pp. 179–188.
Frayer, D.W., 1997. Perspectives on Neanderthals as ancestors. In: Clark, G.A., Willermet, C.M. (Eds.); *Conceptual Issues in Modern Human Origins Research*. Aldine de Gruyter, New York, pp. 220–234.
Gambier, D., 1989. Fossil hominids from the early Upper Palaeolithic (Aurignacian) of France. In: Mellars, P., Stringer, C.B. (Eds.), *The Human Revolution*. Edinburgh University Press, Edinburgh, pp. 194–211.
Geigl, E.M., 2002. On the circumstances surrounding the preservation and analysis of very old DNA. *Archaeometry* 44, 337–342.
Gutiérrez, G., Sánchez, D., Marín, A., 2002. A reanalysis of the ancient mitochondrial DNA sequences recovered from Neanderthal bones. *Mol. Biol. Evol.* 19, 1359–1366.
Habgood, P.J., 2003. *A morphometric investigation into the origin(s) of anatomically modern humans*. BAR International Series 1176, Oxford.
Hambücken, A., 1995. Etude du degré de robustesse des os longs du membre supérieur des Néandertalien. *Bull. Mém. Soc. Anthrop.* 7, 37–47.
Hambücken, A., 1997. La variabilité géographique des néanderthaliens: apport de l'étude du membre supérieur. *Anthropol. Préhist.* 108, 109–120.
Hansen, A.J., Willerslev, E., Wiuf, C., Mourier, T., Arctander, P., 2001. Statistical evidence for miscoding lesions in ancient DNA templates. *Mol. Biol. Evol.* 18, 262–265.
Harvati, K., 2003. The neanderthal taxonomic position: models of intra- and inter-specific craniofacial variation. *J. Hum. Evol.* 44, 107–132.
Harvati, K., Frost, S.R., McNulty, P., 2004. Neanderthal taxonomy reconsidered: implication of 3D primate models of intra- and interspecific differences. *Proc. Natl. Acad. Sci. U.S.A.* 101, 1147–1152.
Hawks, J., Wolpoff, M.H., 2001. Paleoanthropology and the population genetics of ancient genes. *Am. J. Phys. Anthropol.* 114, 269–272.
Heim, J.L., 1982a. *Les Enfants Néandertaliens de la Ferrassie. Etude Anthropologique et Analyse Ontogénique des Hommes de Neanderthal.* Masson, Paris.
Heim, J.L., 1982b. Les hommes fossiles de la Ferrassie II. *Arch. Inst. Paléont. Hum.* 38, 1–272.
Hofreiter, M., Jaenicke, V., Serre, D., von Haeseler, A., Pääbo, S., 2001. DNA sequences from multiple amplifications reveal artifacts induced by cytosine deamination in ancient DNA. *Nucleic Acids Res.* 29, 4793–4799.
Hublin, J.J., 2000. Modern-nonmodern hominid interactions: a Mediterranean perspective. In: Bar-Yosef, O., Pilbeam, D. (Eds.), *The Geography of Neanderthals and Modern Humans in Europe and the Greater Mediterranean*. Peabody Museum of Archaeology and Ethnology, Bulletin 8, Harvard, pp. 157–182.
Hublin, J.J., Spoor, F., Braun, M., Zonneveld, F., Condemi, S., 1996. A late Neanderthal associated with Upper Palaeolithic artefacts. *Nature* 381, 224–226.

Irwin, D.E., Bensch, S., Price, T.D., 2001a. Speciation in a ring. *Nature* 409, 333–337.

Irwin, D.E., Irwin, J.H., Price, T.D., 2001b. Ring species as bridges between microevolution and speciation. *Genetica* 112–113, 223–243.

Janković, I., Karavanić, I., Ahern, J.C.M., Brajković, D., Lenardić, J.M., Smith, F.H., 2006. Vindija cave and the modern human peopling of Europe. *Coll. Antropol.* 30, 315–319.

Kramer, A., Crummett, T.L., Wolpoff, M.H., 2001. Out of Africa and into the Levant: replacement or admixture in Western Asia? *Quat. Int.* 75, 51–63.

Krings, M., Stone, A., Schmitz, R.W., Krainitzki, H., Stoneking, M., Pääbo, S., 1997. Neanderthal DNA sequences and the origin of modern humans. *Cell* 90, 19–30.

Krings, M., Geisert, H., Schmitz, R.W., Krainitzki, H., Pääbo, S., 1999. DNA sequence of the mitochondrial hypervariable region II from the Neanderthal type specimen. *Proc. Natl. Acad. Sci. U.S.A.* 96, 5581–5585.

Madre-Dupouy, M., 1991. Principaux caractères de l'enfant néandertalien du Roc de Marsal, Dordogne (France). *L'Anthropologie (Paris)*. 95, 523–534.

Mayr, E., 1974. *Populations, Espèces et Evolution.* Hermann, Paris.

Mellars, P., 1998. The impact of climatic changes on the demography of late Neanderthal and early anatomically modern populations in Europe. In: Akazawa, T., Aoki, K., Bar-Yosef, O., (Eds.), *Neanderthals and Modern Humans in Western Asia*. Plenum Press, New York, pp. 493–507.

Mitchell, D., Willerslev, E., Hansen, A., 2005, Damage and repair of ancient DNA. *Mutat. Res.* 571, 265–276.

Moncel, M.H., Voisin, J.L., (2006). Les "industries de transition" et le mode de spéciation des groupes néandertaliens en Europe entre 40–30 ka. *C. R. Palevol* 5,182–192.

Monoulou, J.C., 1989. Evolution et biologie des populations. Perspectives offertes par l'étude de l'ADN mitochondrial. In: Bons, J., Delsol, M. (Eds.), *Evolution Biologique quelques Données Actuelles*. Boubée, Paris, pp. 9–19.

Nara, T., 1994. Etude de la variabilité de certains caractères métriques et morphologiques des néandertaliens. Ph.D. Dissertation, Université de Bordeaux I, Talence.

Ovchinnikov, I.V., Götherström, A., Romanova, G.P., Kharitonov, V., Lidén, K., Goodwin, W., 2000. Molecular analysis of Neanderthal DNA from the northern Caucasus. *Nature* 404, 490–493.

Piveteau, J., 1983. *Origine et Destinée de l'Homme*. Masson, Paris.

Rak, Y., 1993. Morphological variation in *Homo neanderthalensis* and *Homo sapiens* in the Levant; a biogeographic model. In: Kimbel, W.H., Martin, L.B., (Eds.), *Species, Species Concepts, and Primate Evolution*. Plenum Press, New York, pp. 523–536.

Rak, Y., 1998. Does any Mousterian cave present evidence of the two hominid species? In: Akazawa, T., Aoki, K., Bar-Yosef, O. (Eds.), *Neanderthals and Modern Humans in Western Asia*. Plenum Press, New York, pp. 353–366.

Rak, Y., Ginzburg, A., Geffen, E., 2002. Does *Homo neanderthalensis* play a role in modern human ancestry? The mandibular evidence. *Am. J. Phys. Anthropol.* 119, 199–204.

Relethford, J.H., 2001. Absence of regional affinities of Neanderthal DNA with living humans does not reject multiregional evolution. *Am. J. Phys. Anthropol.* 115, 95–98.

Relethford, J.H., 2003. *Reflections of our Past*. Westview, Boulder, CO.

Ridley, M., 2004. *Evolution*, 3rd Edition. Blackwell, Oxford.

Rosenberg, K.R., 1998. Morphological variation in west Asian postcrania. In: Akazawa, T., Aoki, K., Bar-Yosef, O. (Eds.), *Neanderthals and Modern Humans in Western Asia*. Plenum Press, New York, pp. 367–379.

Salamon, M., Tuross, N., Arensburg, B., Weiner, S., 2005. Relatively well preserved DNA is present in the crystal aggregates of fossil bones. *Proc. Natl. Acad. Sci. U.S.A.* 102, 13783–13788.

Schillaci, M.A., Froehlich, J.W., 2001. Nonhuman primate hybridization and the taxonomic status of Neanderthals. *Am. J. Phys. Anthropol.* 115, 157–166.

Schilthuizen, M., 2001. *Frogs, Flies & Dandelions, the Making of Species*. Oxford University Press, New York.

Scholz, M., Bachmann, L., Nicholson, G.J., Bachmann, J., Giddings, I., Rüschoff-Thale, B., Czarnetzki, A., Pusch, C.M., 2000. Genomic differentiation of Neanderthals and anatomically modern man allows a fossil–DNA-based classification of morphologically indistinguishable hominid bones. *Am. J. Hum. Genet.* 66, 1927–1932.

Serre, D., Langaney, A., Chech, M., Teschler-Nicola, M., Paunovic, M., Mennecier, P., Hofreiter, M., Possnert, G., Pääbo, S., 2004. No evidence of Neanderthal mtDNA contribution to early modern human. *PLoS Biol.* 2, 313–317.

Simmons, T., Falsetti, A.B., Smith, F.H., 1991. Frontal bone morphometrics of southwest Asian Pleistocene hominids. *J. Hum. Evol.* 20, 249–270.

Smith, C.I., Chamberlain, A.T., Riley, M.S., Stringer, C.B., Collins, M.J., 2003. The thermal history of human fossils and the likelihood of successful DNA amplification. *J. Hum. Evol.* 45, 203–217.

Smith, F.H., 1991. The Neanderthals: evolutionary dead ends or ancestors of modern people? *J. Anthropol. Res.* 47, 219–238.

Smith, F.H., Trinkaus, E., 1991. Les origines de l'homme moderne en Europe Centrale: un cas de continuité. In: Hublin, J.J., Tillier, A.M., (Eds.), *Aux Origines d'Homo sapiens*. P.U.F., Paris, pp. 251–290.

Smith, F.H., Falsetti, A.B., Donnelly, S.M., 1989a. Modern human origins. *Yrbk. Phys. Anthropol.* 32, 35–68.

Smith, F.H., Simek, J.F., Harrill, M.S., 1989b. Geographic variation in supraorbital torus reduction during the later Pleistocene (c. 80 000–15 000 BP). In: Mellars, P., Stringer, B. (Eds.), *The Human Revolution*. Edinburgh University Press, Edinburgh, pp. 172–193.

Smith, F.H., Jankovic, I., Karavanic, I., 2005. The assimilation model, modern human origins in Europe, and the extinction of Neanderthals. *Quatern. Int.* 137, 7–19.

Sohn, S., Wolpoff, M.H., 1993. Zuttiyeh face: a view from the east. *Am. J. Phys. Anthropol.* 91, 325–347.

Sota, T., Vogler, A.P., 2001. Incongruence of mitochondrial and nuclear gene trees in the carabid beetles *Ophomopterus*. *Syst. Biol.* 50, 39–59.

Stewart, T.D., 1962. Neanderthal scapulae with special attention to the Shanidar Neanderthals from Iraq. *Anthropos* 57, 779–800.

Stringer, C.B., 1992. Replacement, continuity and the origin of *Homo sapiens*. In: Bräuer, G., Smith, F.S. (Eds.), *Continuity or Replacement; Controversies in Homo sapiens Evolution*. A.A. Balkema, Rotterdam, pp. 9–24.

Stringer, C.B., 1998. Chronological and biogeographic perspectives on later human evolution. In: Akazawa, T., Aoki, K., Bar-Yosef, O. (Eds.), *Neanderthals and modern humans in Western Asia*. Plenum Press, New York, pp. 29–37.

Stringer, C.B., 2002. Modern human origins: progress and prospects. *Phil. Trans. R. Soc. Lond.* B. 357, 563–579.

Stringer, C.B., McKie, R., 1996. *African Exodus – The Origins of Modern Humanity*. Pimlico, London.

Suzuki, H., 1970. The skull of the Amud man. In: Suzuki, H., Takai, F. (Eds.), *The Amud Man and his Cave Site*. The University of Tokyo Press, Tokyo, pp. 123–206.

Tattersall, I., Schwartz, J.H., 1999. Hominids and hybrids: the place of Neanderthals in human evolution. *Proc. Natl. Acad. Sci. U.S.A.* 96, 7117–7119.

Thoma, A., 1965. La définition des néanderthaliens et la position des hommes fossiles de Palestine. *L'Anthropologie (Paris)* 69, 519–534.

Trinkaus, E., 1977. A functional interpretation of the axillary border of the Neanderthal scapula. *J. Hum. Evol.* 6, 231–234.

Trinkaus, E., 1981. Neanderthal limb proportions. In: Stringer, C.B. (Ed.): *Aspect of Human Evolution*. Taylor and Francis, London, pp. 187–224.

Trinkaus, E., 1983. *The Shanidar Neanderthals*. Academic Press, New York.

Trinkaus, E., 1988. The evolutionary origins of the Neanderthals or, why were there Neanderthals? In: Trinkaus, E., (Ed.): *L'homme de Néandertal*. Actes du colloque international de Liège (4–7 Décembre 1986). ERAUL n°30, Liège, pp. 11–29.

Trinkaus, E., 1991. Les hommes fossiles de la grotte de Shanidar, Irak: évolution et continuité parmi les hommes archaïques tardifs du Proche-Orient. *L'Anthropologie (Paris)* 95, 535–572.

Trinkaus, E., 2001. Paleobiological perspectives on the early Upper Paleolthic human transition in the northwestern Old World. *Bull. Mém. Soc. Anthrop.* 13, 311–322.

Trinkaus, E., Smith, F.H., 1995. Body size of the Vindija Neanderthals. *J. Hum. Evol.* 28, 201–208.

Trinkaus, E., Marks, A.E., Brugal, J.P., Bailey, S.E., Rink, W.J., Richter, D., 2003a. Later Middle Pleistocene human remains from the Almonda karstic system, Torres Novas, Portugal. *J. Hum. Evol.* 45, 219–226.

Trinkaus, E., Milota, S., Rodrigo, R., Mircea, G., Moldovan, O., 2003b. Early modern human cranial remains from the Peştera cu Oase, Romania. *J. Hum. Evol.* 45, 245–253.

Vandermeersch, B., 1981. *Les Hommes Fossiles de Qafzeh (Israël)*. Cahiers de Paléoanthropologie CNRS, Paris.

Vandermeersch, B., 1989a. The evolution of modern humans: recent evidence from southwest Asia. In: Mellars, P., Stringer, C.B. (Eds.), *The Human Revolution*. Princeton University Press, Princeton, pp. 155–164.

Vandermeersch, B., 1989b. Homogénéité ou hétérogénéité des néandertaliens. In: Giacobini G. (Ed.), *Hominidae: Acte du 2ème Congrès International de Paléontologie Humaine, 1987.* Jaca Book, Turin, pp. 311–317.

Voisin, J.L., 2000. L'épaule des hominidés. Aspects architecturaux et fonctionnels, références particulières à la clavicule. Ph.D. Dissertation, Museum National d'Histoire Naturelle, Paris.

Voisin, J.L., 2001. Evolution de la morphologie claviculaire au sein du genre *Homo*, conséquence architecturale et fonctionnelle sur la ceinture scapulaire. *L'Anthropologie* (Paris) 105, 449–468.

Voisin, J.L., 2004. Clavicule: approche architecturale de l'épaule et réflexions sur le statut systématique des néandertaliens. *C. R. Palevol.* 3, 133–142.

Wake, D.B., 2001. Speciation in the round. *Nature* 409, 299–300.

Wolpoff, M.H., 1999. *Paleoanthropology*, 2nd Edition. McGraw-Hill, New York.

Wolpoff, M.H., Caspari, R., 1996. Why aren't Neanderthals modern humans? In: Bar-Yosef, O., Cavalli-Sforza, L.L., March, R.J., Piperno, M. (Eds.), XIIIème Congrés de l'U.I.S.P.P., 5, The Lower and Middle Palaeolithic. ABACO, Forli, pp. 133–156.

Wolpoff, M.H., Hawks, J., Caspari, M., 2000. Multiregional, not multiple origins. *Am. J. Phys. Anthropol.* 112, 129–136.

Wolpoff, M., Hawks, J., Frayer, D., Hunley, K., 2001. Modern human ancestry at the peripheries: a test of the replacement theory. *Science* 291, 293–297.

Wolpoff, M., Mannheim, B., Mann, A., Hawks, J., Caspari, R., Rosenberg, K.R., Frayer, D.W., Gill, G.W., Clark, G., 2004. Why not the Neanderthals? *World Archaeol.* 36, 527–546.

Zilhão, J., d'Errico, F., 2000. La nouvelle "bataille aurignacienne". Une révision critique de la chronologie du Châtelperronien et de l'Aurignacien ancien. *L'Anthropologie (Paris)* 104, 17–50.

18. The Neanderthal-*H. sapiens* interface in Eurasia

C.B. STRINGER
Dept of Palaeontology,
The Natural History Museum,
London SW7 5BD, UK
c.stringer@nhm.ac.uk

Keywords: Neanderthal, *Homo sapiens*, Aurignacian, climate, Paleolithic, species extinction

Abstract

The general framework and the factors behind the demise of the Neanderthals are still fiercely debated, and there remain many uncertainties in the data. While accelerator dating has purged the record of spurious fossils and confirmed the ages of others, it is likely that many of our current "dates" for the last Neanderthals and the earliest moderns in Europe are minimum ages, from the perspectives of both calibration and contamination by more recent radiocarbon. While the Aurignacian probably does reflect a dispersal of modern humans, it may not represent the oldest such dispersal into Europe. And while much new morphological data support a specific distinction for *H. neanderthalensis*, nevertheless the modern and Neanderthal lineages may be better characterized as allotaxa. Regarding the factors behind Neanderthal extinction, these are likely to have been many and varied, but almost certainly included the unstable climatic context of the period between 25–40,000 years ago. Finally, taking a wider context on the Neanderthal – *sapiens* relationship, we should remember that these events in western Europe were only the endpoints of hundreds of thousands of years of possible competition and interaction between these evolving lineages.

Introduction

The fate of the Neanderthals remains one of the most fascinating topics in paleoanthropology, and the last ten years have seen the appearance of a wealth of new data relevant to the enduring conundrum of their demise. Yet the general framework and the factors behind their disappearance are still fiercely debated, and there remain many uncertainties in the data. In this review I will examine what we have (and have not) learnt about events in Europe from some of

K. Harvati and T. Harrison (eds.), Neanderthals Revisited: New Approaches and Perspectives, 315–324.

these new findings, and then change focus to look at the wider picture of the relationship of Neanderthals and modern humans.

CHRONOLOGY

According to current thinking, the Neanderthals disappeared about 30,000 radiocarbon years ago, after a short period of overlap and possible interaction with the *H. sapiens* of the Upper Paleolithic (in Europe = Cro-Magnons). However, one of the most serious problems is the lack of fine chronological control over the relevant time frame, and it is highly likely that many of the dates commonly quoted for last appearances (Neanderthals) and first appearances (Cro-Magnons) are underestimates of the real ages when both analytical limitations and calibration are taken into account – for example, there is growing evidence for a FAD for Cro-Magnons of at least 40,000 years (calibrated). However, the application of direct accelerator dating to the fossil record has purged it of a number of supposed Cro-Magnon fossils from sites such as Engis (Hedges et al., 1996), Vogelherd (Conard et al., 2004), Velika Pećina (Smith et al., 1999) and Hahnöfersand (Terberger et al., 2001), but has confirmed the ages of some others.

The earliest well-dated early moderns in Europe are now those from Mladeč in the Czech Republic (~32,000 radiocarbon years old: Wild et al., 2005) and Oase in Romania (~36,000 radiocarbon years: Trinkaus et al., 2003). The former fossils can be associated with Aurignacian artefacts, whereas the Oase specimens have no archaeological context, but show a suite of unusual features compared with later Europeans. These include the morphology of the inferior nasal margin, the large posterior dentition with complex cusp morphologies, the wide ascending ramus of the mandible, and the unilateral occurrence of a horizontal-oval mandibular foramen (Trinkaus et al., 2003). The latter has been interpreted as a sign of possible gene flow from Neanderthals, but overall some of the morphologies are more reminiscent of Pleistocene (even Middle Pleistocene) African samples. Another potential early modern fossil is the Kent's Cavern 4 maxilla from England, which has a direct accelerator date of about 31,000 radiocarbon years (Stringer, 1990). However, this was one of the first fossil hominin dates produced by the Oxford Accelerator, and recent redatings of other fossils using improved pretreatments to remove contamination have often yielded older dates (Bronk Ramsey et al., 2004). These have included mammal fauna from the same levels as Kent's Cavern 4, suggesting that its real age is likely to exceed 35,000 radiocarbon years, and placing it in the time range of British leaf point industries rather than the Aurignacian. Unfortunately, it is too incomplete to allow further direct dating, but the specimen is now undergoing reanalysis in an attempt to establish its modern or Neanderthal affinities.

Now that we have finally escaped the straitjackets of Middle Paleolithic = Neanderthal and Upper Paleolithic = Cro-Magnon, we should also be open to the possibility that the Aurignacian does not mark the earliest dispersal of *H. sapiens* into Europe, not because it was not made by Cro-Magnons, as some have claimed, but because there could have been (perhaps unsuccessful) pre-Aurignacian dispersals, at least in eastern Europe. A possible example of this is the Bohunician, which appears to predate the Aurignacian and have technological links with Levantine and Nile Valley "transitional" industries, but which so far has no reliably associated fossil material (Tostevin, 2000).

The Neanderthals and their extinction

In the last few years we have seen an accumulation of data which, in my view, strongly supports the recognition of a distinct Neanderthal lineage with its own evolutionary history, one which warrants a species-level distinction based on morphology. This work includes comparative studies of Neanderthal ontogeny (e.g., Ponce de León and Zollikofer, 2001), cranial morphology (e.g., Harvati et al., 2004) and dental morphology (e.g., Bailey, 2004). Molecular data (presently only mitochondrial) also supports a distinctive population history for Neanderthals (e.g., Serre et al., 2004), with estimated divergence dates from the modern human lineage consistent with those derived from paleontology (Middle Pleistocene, ~ 500 ka). However, both the molecular divergence and the appearance of the first apomorphies probably provide a maximum age for any species separation, and it may be safer to consider the Neanderthal-*sapiens* relationship at the level of allotaxa (*sensu* Grubb, 1999), as already argued by Jolly (2001).

Regarding the demise of the Neanderthals, some workers argue that incoming early modern populations genetically absorbed them, while others suggest that they became extinct, perhaps accompanied by only a trivial amount of gene flow. Explanations for their extinction have ranged widely from suggestions of disease or warfare through to economic competition from early modern humans. While some workers have questioned the extent to which the two populations ever overlapped in Europe (e.g., Zilhão and d'Errico, 1999), there is disputed evidence of interstratification of Châtelperronian and Aurignacian archaeology in France (Gravina et al., 2005), which I accept as proxies for the presence of Neanderthals and modern humans. Up to now, the view of climate change in Europe at this time has been rather simplistic, leading to climatic factors being ignored in relation to Neanderthal

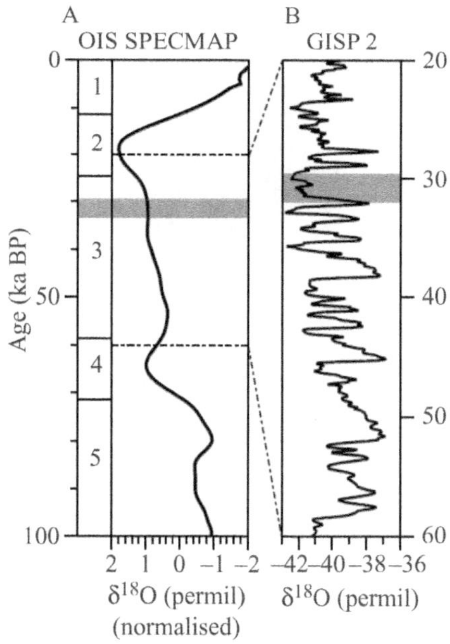

Figure 1. Oxygen isotope history of the last glacial-interglacial cycle. (a): SPECMAP marine record. B: Millennial scale fluctuations recorded by the GISP2 Greenland ice-core for the interval 20–60 ka BP. Note that 1A indicates that Neanderthal extinction (~30,000 radiocarbon years, marked in grey) preceded the last glacial maximum by several millennia, and apparently occurred during a time of relative climatic stability, whereas the more detailed record of 1B shows much greater complexity around this time. Reproduced with permission from van Andel, T., Davies W. (Eds.) 2003.

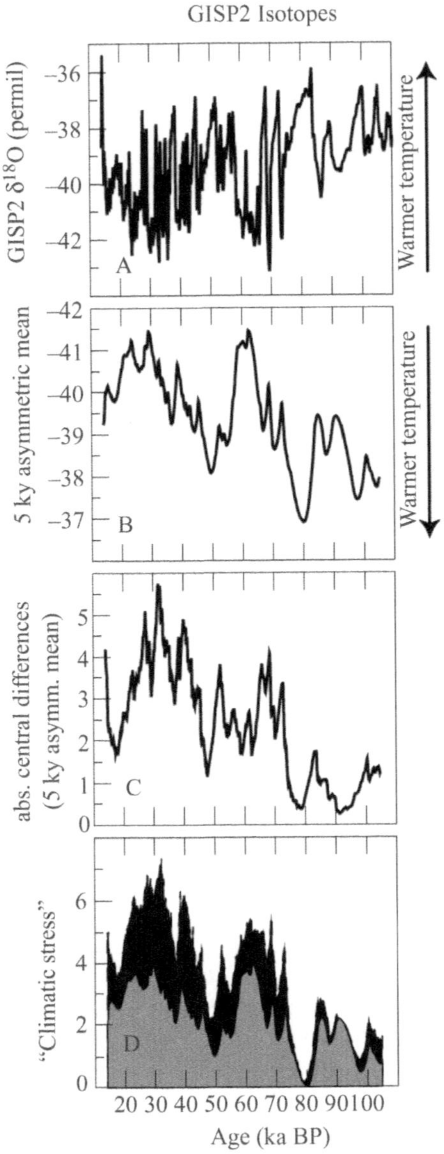

Figure 2. Panel (A) shows the raw GISP2 oxygen isotope data, plotted with warm equivalent temperatures up. See The Greenland Icesheet Project 2 summary web site, http://www.gisp2.sr.unh.edu/GISP2/. Reproduced with permission from van Andel, T., Davies W. (Eds.) 2003. Panel (B) shows a smoothed version of (A), which was obtained by integrating asymmetrically over a 5 ky long window, where the largest weight is set at a given age, with linearly decreasing weights down to zero weight 5 ky before each point. Panel (B) is inverted with respect to panel (A). Panel (C) shows the absolute values of the central differences of panel (A), smoothed as in panel (B). Panel (D) shows the sum of panels (B) and (C), after the curves were normalized by setting the highest absolute value and the lowest central difference value to zero, and dividing by their respective standard deviations. Different shades of black and grey indicate the relative contribution from panels (B) and (C) to create the final stress curve in (D). Reproduced with permission from van Andel, T., Davies W. (Eds.) 2003.

extinction: the Neanderthals became extinct before the last glacial maximum, had weathered such conditions before, and were physically equipped to cope with the cold (Figure 1A). Thus many explanations have instead focused on the direct impact of modern humans on the Neanderthals, and their inherent superiority. However, with the availability of rich paleoclimatic records from terrestrial and marine sediments, it has been possible to examine the period of their extinction in much greater detail. The data show a startling complexity in climatic change at this time (Figure 1B), and this has led to new hypotheses

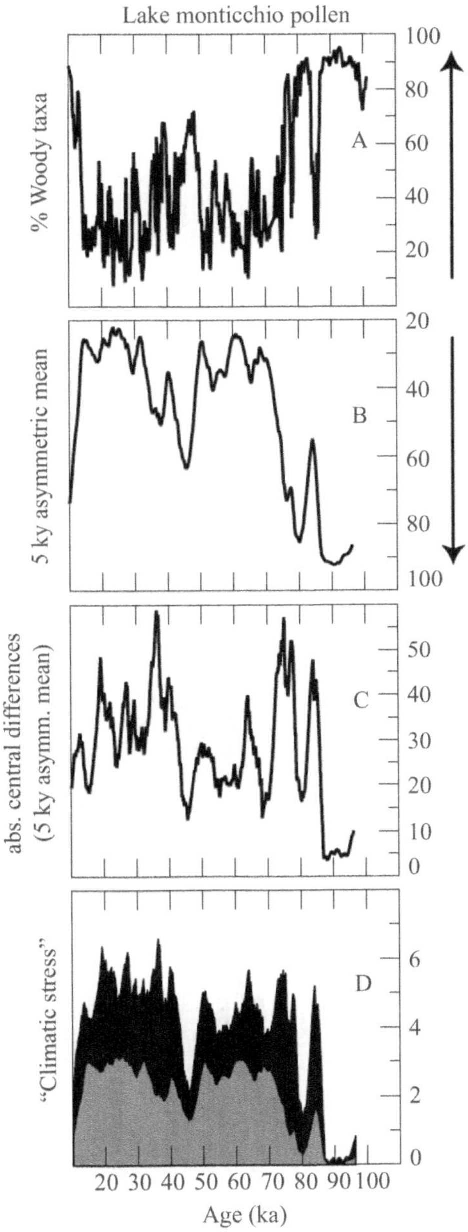

Figure 3. Equivalent to Figure 2, but with pollen data from Lago Grande di Monticchio, Italy (see Allen et al., 2000). Reproduced with permission from van Andel, T., Davies W. (Eds.) 2003.

about Neanderthal extinction involving paleoclimatic or paleoecological factors (see e.g., d'Errico and Sánchez Goñi, 2003; Finlayson, 2004).

In my own case, in collaborative work from the Stage 3 Project based at Cambridge (van Andel and Davies, 2003; Stringer et al., 2003), we produced modeled data for the potential effect of millennial-scale climatic oscillations on the Neanderthals. We first tested that oxygen isotope fluctuations in Greenland Ice Core (GISP2) records could reasonably serve both as a proxy for North Atlantic temperatures, and for those of continental Europe. We then used the GISP2 data as a basis for further analyses (Figure 2A). As an additional control, we also analyzed penecontemporaneous pollen records from Lake Monticchio in Italy (Figure 3A) as a further proxy for European continental paleotemperatures. Of course many other factors such as precipitation, wind chill and snow cover would have had important impacts on the human populations of Europe, but as these are extremely difficult to combine in a model, we chose a deliberately simple approach that looked only at temperature change (assuming low temperatures were more stressful) and the rate of temperature change (assuming a high rate of change in either direction was stressful). We derived smoothed curves of temperature change (Figures 2B, 3B) and rate of temperature change (Figures 2C, 3C) through both records, and the two components were then summed. The resultant two independently derived "stress curves" (Figures 2D, 3D) could then be directly compared (Figure 4A). Although the earliest parts of the derived stress curves do not agree well, perhaps because of dating or correlation problems, the

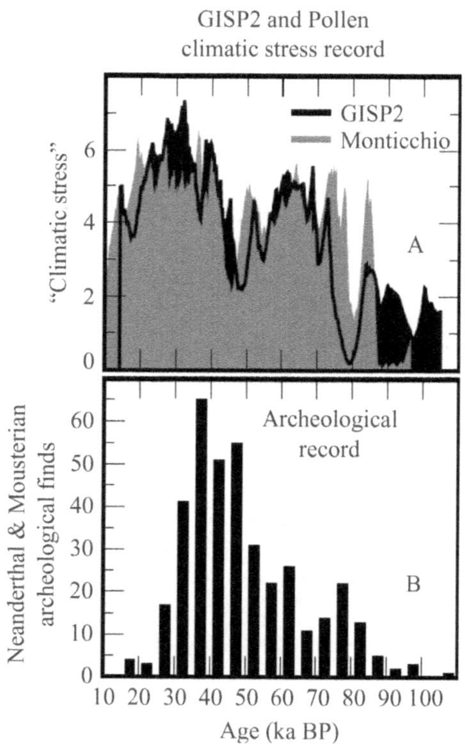

Figure 4. A. Superimposed stress curves from GISP2 oxygen isotope data and Lago Grande di Monticchio pollen data. B. Histogram of radiometric dates for Middle Paleolithic sites and fossils from the Stage 3 database. Reproduced with permission from van Andel, T., Davies W. (Eds.) 2003.

later parts both show two significant stress peaks, with the most severe and prolonged at around 30 ka. The period of reduced stress indicated at around 45 ka could perhaps have encouraged Cro-Magnon dispersals into Europe and Neanderthal population expansions, as is hinted at even from the rather poor record of reliable dates from this time (Figure 4B), but the subsequent peak of cumulative climatic stress could indeed have played a significant part in the extinction of the Neanderthals. Such stressful conditions would have affected both Neanderthal and Cro-Magnon populations, heightening resource competition where their ranges overlapped, in environments with reduced carrying capacity. The evidence suggests that only *H. sapiens* came through these crises. Perhaps the Gravettian was a successful adaptive response to the challenges of those stressful conditions, whereas Neanderthal and Aurignacian populations succumbed?

The Wider Picture

Regarding geographical factors, we need to remember that the events in western Europe that attract most academic and popular attention were only the endpoints of hundreds of thousands of years of possible competition and interaction between the evolving Neanderthal lineage in western Eurasia and the *sapiens* lineage in Africa, and that the Levant, in particular, was potentially a region of regular contact. Looked at through time, the repeated lack of paleontological or archaeological visibility of both lineages suggests that they regularly underwent local extinctions, and these were probably caused by many different factors. Equally, the interactions of these populations over that time may well have run the whole gamut of possible scenarios from conflict, competition or avoidance, through coexistence but little contact, to peaceful interactions and even hybridization. As also suggested by Hublin (1998) and Stringer (2002) common behavioral features of the Neanderthal and modern human lineages, such as Mode 3 technology and intentional burial, could have spread via areas of overlap in western Asia.

Prepared Core (Mode 3) technology seems to have appeared in both Europe and Africa by about 300 ka, and depending on the dating of the relevant levels of Tabun Cave, may well have been in the Levant by this time as well. It is currently unclear whether it originated in one region and spread, or whether it developed independently in more than one location. The Mode 3 Hypothesis of Lahr and Foley (2001) argued for a relatively late divergence between Neanderthals and *H. sapiens*, at around 250 ka, linking this with the development of Mode 3 (Prepared Core or Levallois) technology. In their view, this archaeological innovation first appeared in the African species "*H. helmei*" (the type fossil of which is the Florisbad cranium). In Lahr and Foley's view, "*H. helmei*" evolved from *H. heidelbergensis* in Africa and then dispersed to give rise to modern humans in Africa and the Neanderthals in Eurasia. Thus "*H. helmei*" would have carried the newly derived Mode 3 technology with it during its late Middle Pleistocene dispersal. However, the hypothesis requires the development of Neanderthal apomorphies only *after* the appearance of Mode 3 artefacts. There is enough evidence of Neanderthal features in European populations associated with pre-Mode 3 handaxes (e.g., at Swanscombe: Stringer and Hublin, 1999; and Atapuerca Sima de los Huesos: Bischoff et al., 2003) to falsify this hypothesis.

In the case of presumed intentional burials, the oldest examples known occur in the Levant in Neanderthal (Tabun C1) and early modern (Skhul and Qafzeh) contexts, but current dating only has the resolution to place them as approximately contemporaneous (Grün and Stringer, 2000; Grün et al., 2005), and thus it is not possible to say which populations may have originated the practice.

However, only the early modern human sites appear to have any symbolic associations.

Concluding Remarks

The considerable challenge now facing us is how best to use the growing body of climatic, environmental, chronological, behavioral, morphological and genetic data to test, and where appropriate finally exclude, models of Neanderthal and early modern population histories. The promise of better quality radiocarbon accelerator dates (Bronk Ramsey et al., 2004) and accurate calibration of dates (Hughen et al., 2004), coupled with the use of marker tephras (Fedele et al., 2004), will strengthen our abilities to reconstruct significant events in the critical time range of Neanderthal – Cro-Magnon overlap in Europe. Discoveries such as Oase hold the promise of completely new windows on the first modern humans in Europe. And remarkable discoveries from much further afield (Morwood et al., 2005) should remind us to always check our consideration of Neanderthal-*sapiens* problems against wider perspectives, since we ultimately need to integrate European data into the bigger picture. We have some exciting years ahead of us.

References

Allen, J.R.M., Watts, W.A., Huntley, B., 2000. Weichselian palynostratigraphy, paleovegetation and paleoenvironment; the record from Lago Grande di Monticchio, southern Italy. *Quat.Int.* 73/74, 91–110.

Bailey, S., 2004. A morphometric analysis of maxillary molar crowns of Middle-Late Pleistocene hominins. *J. Hum. Evol.* 47, 183–198.

Bischoff, J.L., Shamp, D.D., Aramburu, A., Arsuaga, J.L., Carbonell, E., Bermúdez de Castro, J.M., 2003. The Sima de los Huesos hominids date to beyond U/Th equilibrium (>350 Kyr) and perhaps to 400–500 Kyr: new radiometric dates. *J. Archaeol. Sci.* 30, 275–280.

Bronk Ramsey, C., Higham, T., Bowles, A., Hedges, R., 2004. Sample processing – improvements to the pretreatment of bone at Oxford. *Radiocarbon* 46, 155–164.

Conard, N., Grootes, P., Smith, F., 2004. Unexpectedly recent dates for human remains from Vogelherd. *Nature* 430, 198–201.

d'Errico, F., Sánchez Goñi, M.F., 2003. Neandertal extinction and the millennial scale climatic variability of OIS 3. *Quat. Sci. Rev.* 22, 769–788.

Fedele, F.G., Giaccio, B., Isaia, R., Orsi, G., 2004. The Campanian Ignimbrite eruption, Heinrich event 4, and the Palaeolithic in Europe: a high-resolution investigation. In: Robock, A., Oppenheimer, C. (Eds.), *Volcanism and the Earth's Atmosphere*. American Geophysical Union. Geophys. Monograph 139, 301–325.

Finlayson, C., 2004. *Neanderthals and Modern Humans*. Cambridge University Press, Cambridge.

Gravina, B., Mellars, P., Bronk Ramsey, C., 2005. Radiocarbon dating of interstratified Neanderthal and early modern human occupations at the Châtelperronian type-site. *Nature* 438, 51–56.

Grubb, P., 1999. Evolution processes implicit in distribution patterns of modern African mammals. In: Bromage, T., Schrenk, F. (Eds.), *African Biogeography, Climate Change, and Human Evolution*. Oxford University Press, New York, pp. 150–164.

Grün, R., Stringer, C., 2000. Tabun revisited: revised ESR chronology and new ESR and U-series analyses of dental material from Tabun C1. *J. Hum. Evol.* 39, 601–612.

Grün, R., Stringer, C., McDermott, F., Nathan, R., Porat, N., Robertson, S., Taylor, L., Mortimer, G., Eggins, S., McCulloch, M., 2005. U series and ESR analyses of bones and teeth relating to the human burials from Skhul. *J. Hum. Evol.* 49, 316–334.

Harvati, K., Frost, S.R., McNulty, K.P., 2004. Neanderthal taxonomy reconsidered: implications of 3D primate models of intra- and inter specific differences. *Proc. Natl. Acad. Sci. U.S.A.* 101, 1147–1152.

Hedges, R., Pettitt, P., Bronk-Ramsey, C., van Klinken, G.,1996. Radiocarbon dates from the Oxford AMS system: Archaeometry datelist 21. *Archaeometry* 38, 181–207.

Hublin, J.-J., 1998. Climatic changes, paleogeography, and the evolution of the Neandertals. In: T. Akazawa, Aoki, K., Bar-Yosef, O., (Eds), *Neandertals and Modern Humans in Western Asia*. Plenum, New York, pp. 295–310.

Hughen, K., Lehman, S., Southon, J., Overpeck, J., Marchal, O., Herring, C., Turnbull, J., 2004. ^{14}C activity and global carbon cycle changes over the past 50,000 years. *Science* 303, 202–207.

Jolly, C.J., 2001. A proper study for mankind: analogies from the papionin monkeys and their implications for human evolution. *Yrbk Phys. Anthropol.* 44, 177–204.

Lahr, M., Foley, R., 2001. Mode 3, *Homo helmei*, and the pattern of human evolution in the Middle Pleistocene. In: Barham, L., Robson Brown, K. (Eds.), *Human Roots: Africa and Asia in the Middle Pleistocene*. Western Academic & Specialist Press, Bristol, pp. 23–40.

Morwood, M., Brown, P., Jatmiko, Sutikna, T., Saptomo, E., Westaway, K., Due, R., Roberts, R., Maeda. T., Wasisto, S., Djubiantono, T., 2005. Further evidence for small-bodied hominins from the Late Pleistocene of Flores, Indonesia. *Nature* 437, 1012–1017.

Ponce de León, M.S., Zollikofer C.P.E., 2001. Neanderthal cranial ontogeny and its implications for late hominid diversity. *Nature* 412, 534–538.

Serre, D., Langaney, A., Chech, M., Teschler-Nicola, M., Paunovic, M., Mennecier, P., Hofreiter, M., Possnert, G, Pääbo, S., 2004. No evidence of Neandertal mtDNA contribution to early modern humans. *PLoS Biol.* 2, 313–317.

Smith, F., Trinkaus, E., Pettitt, P., Karavanic, I., Paunovic, M., 1999. Direct radiocarbon dates for Vindija G1 and Velika Pećina Late PleistocenehHominid remains. *Proc. Natl. Acad. Sci. U.S.A.* 96, 12281–12286.

Stringer, C.,1990. Hominid remains – an up-date. Volume 3: British Isles and East Germany. Université Libre de Bruxelles, pp. 1–40.

Stringer, C., 2002. Modern human origins – progress and prospects. *Phil. Trans. R. Soc., Lond.* (B) 357, 563–579.

Stringer, C., Hublin J.-J., 1999. New age estimates for the Swanscombe hominid, and their significance for human evolution. *J. Hum. Evol.* 37, 873–877.

Stringer, C., Pälike, H., van Andel, T., Huntley, B., Valdes, P., Allen, J., 2003. Climatic stress and the extinction of the Neanderthals. In van Andel, T., Davies, W. (Eds.), *Neanderthals and Modern Humans in the European Landscape during the Last Glaciation*. McDonald Institute Monographs, Cambridge, pp. 233–240.

Terberger, T., Street, M., Bräuer, G., 2001. Der menschliche Schädelrest aus der Elbe bei Hahnöfersand und seine Bedeutung für die Steinzeit Norddeutschlands. *Archäol. Korrespond.* 31, 521–526.

Tostevin, G., 2000. The Middle to Upper Paleolithic transition from the Levant to Central Europe: *in situ* development or diffusion? In: Weniger, G.-C., Orschiedt, J. (Eds.), *Neanderthals and Modern Humans: Discussing the Transition. Central and Eastern Europe from 50,000–30,000 BP*. Neanderthal Museum, Düsseldorf, pp. 90–109.

Trinkaus, E., Milota, S., Rodrigo, R., Mircea, G., Moldovan, O., 2003. Early modern human cranial remains from the Peştera cu Oase, Romania. *J. Hum. Evol.* 45, 245–253.

van Andel, T., Davies, W. (Eds.), 2003. *Neanderthals and modern humans in the European landscape during the last glaciation*. McDonald Institute Monographs, Cambridge.

Wild, E., Teschler-Nicola, M., Kutschera, W., Steier, P., Trinkaus, E., Wanek, W., 2005. Direct dating of Early Upper Paleolithic human remains from Mladeč. *Nature* 435, 332–335.

Zilhão, J., d'Errico, F., 1999. The chronology and taphonomy of the earliest Aurignacian and its implications for the understanding of Neandertal extinction. *J. World Prehist.* 13, 1–68.

Index

VERTEBRATE PALEOBIOLOGY AND PALEOANTHROPOLOGY

PUBLISHED AND FORTHCOMING TITLES:

Neanderthals Revisited: New approaches and perspectives
Edited by K. Harvati and T. Harrison
ISBN: 978-1-4020-5120-3, 2006
A Volume in the Max-Planck Institute Sub-series in Human Evolution

The Evolution and Diversity of Humans in South Asia: Interdisciplinary studies in archaeology, biological anthropology, linguistics and genetics
Edited by M. Petraglia and B. Allchin
ISBN: 978-1-4020-5561-4, 2007

Dental Perspectives on Human Evolution: State of the art research in dental paleoanthropology
Edited by S.E. Bailey and J-J. Hublin
ISBN: 978-1-4020-5844-8, 2007
A Volume in the Max-Planck Institute Sub-series in Human Evolution

Hominin Environment in the East African Pliocene: An Assessment of the Faunal Evidence
Edited by R. Bobe, Z. Alemseged and A.K. Behrensmeyer
ISBN: 978-1-4020-3097-0, 2007

Deconstructing Olduvai: A Taphonomic Study of the Bed I Sites
By M. Domínguez-Rodrigo, C.P. Egeland and R. Barba Egido
ISBN: 978-1-4020-6150-9, 2007

GPSR Compliance

The European Union's (EU) General Product Safety Regulation (GPSR) is a set of rules that requires consumer products to be safe and our obligations to ensure this.

If you have any concerns about our products, you can contact us on ProductSafety@springernature.com

In case Publisher is established outside the EU, the EU authorized representative is:

Springer Nature Customer Service Center GmbH
Europaplatz 3
69115 Heidelberg, Germany

Batch number: 10824522

Printed by Printforce, the Netherlands